THE PARTICLE IMAGE VELOCIMETRY – CHARACTERISTICS, LIMITS AND POSSIBLE APPLICATIONS

Edited by **Giovanna Cavazzini**

The Particle Image Velocimetry – Characteristics, Limits and Possible Applications
Edited by Giovanna Cavazzini

Published by InTech
Janeza Trdine 9, 51000 Rijeka, Croatia

Publishing Process Manager Marija Radja

Technical Editor Teodora Smiljanic

Cover Designer InTech Design Team

Additional hard copies can be obtained from orders@intechopen.com

The Particle Image Velocimetry – Characteristics, Limits and Possible Applications,
Edited by Giovanna Cavazzini
p. cm.
ISBN 978-953-51-0625-8

For EU product safety concerns:
IN TECH d.o.o., Prolaz Marije Krucifikse Kozulić 3, 51000 Rijeka, Croatia, info@intechopen.com or visit our website at intechopen.com.

Meet the editor

Dr Giovanna Cavazzini received her degree in Mechanical Engineering in 2003 from the University of Padova, and her doctorate with European Label in Energetics in 2007 from the University of Padova. She started working with PIV in 2005 at the Ecole Nationale Superieure Arts et Métiers ParisTech de Lille with emphasis to the application of the PIV in the turbo-machinery field with focus to the study of unsteady turbulent phenomena. In 2009 she recevied the qualification of Maitre de Conference from the Ministère de l'Enseignement supérieur et de la Recherche in France and from 2011 she is aggregate professor of Turbomachinery and Energetic Systems at the University of Padova.

Contents

Preface

The Particle Image Velocimetry is undoubtedly one of the most important technique in Fluid-dynamics since it has a spatial character that allows to obtain a direct and instantaneous visualization of the flow field in a non-intrusive way. It represents the point of arrival of an intense research on the laser techniques and its origin dates back at the beginning of 80's when Meynart, a doctoral student of the Von Karman Institute, while explorying the Laser Speckle Velocimetry technique, guessed the importance of capturing the movements of individual particles instead of speckles. The rising PIV technique immediately captured the scientific world interest but its spread was initially limited by technical problems related to the limits of the available technologies.

A first significant boost to the spreading of the PIV technique was provided by the migration from photographic to video-graphic recording. This allowed not only to increase the maximum number of PIV images that could be acquired in a single experimental test but also to improve the interrogation process. However, the applications were still limited to analyses of flow fields having two-dimensional characteristics since the technique allowed to determine only the two velocity components belonging to the laser sheet plane.

In the midi-90's the introduction of the stereographic PIV gave a further great impulse to the use and popularity of the PIV technique since it enabled the possibility of determining by stereographic imaging of the particles the third velocity component, normal to the planar laser sheet. This further improvement, combined with the rising and continuous development of the available computational capacity, greatly favored the spread of the technique in a wide number of research fields, from aerodynamics to medicine, from biology to turbulence researches, from aerodynamics to combustion processes.

The book is aimed at presenting the PIV technique and its wide range of possible applications so as to provide a reference for researchers who intended to exploit this innovative technique in their research fields. Several aspects and possible problems in the analysis of large- and micro-scale turbulent phenomena, two-phase flows and polymer melts, combustion processes and turbo-machinery flow fields, internal waves and river/ocean flows were considered.

The book is managed as follows. The first part is focused on the characteristics, advantages and limits of the PIV technique, with reference both to 2D and 3D analyses. Moreover, the latest improvements in the post-processing analysis methods are presented with particular reference to the turbo-machinery field.

The other sections present an excursus of the possible applications of the PIV technique. In particular, the second section focuses on the studies of large-scale flows considering combustion processes, fluid-dynamic phenomena and flows in polymer melts. In the third section two examples of micro-piv applications are presented, whereas in the forth section the combination of the PIV technique with the Particle Tracking Velocimetry for the definition of velocities and trajectories in internal waves, stratified ocean and river flows is considered.

Eng. Cavazzini Giovanna, Ph.D.
Department of Industrial Engineering
University of Padova
Italy

Section 1

The PIV Technique: Characteristics, Limits and Post-Processing Methods

1

Stereoscopic PIV and Its Applications on Reconstruction Three-Dimensional Flow Field

L. Gan
Department of Engineering, University of Cambridge
United Kingdom

1. Introduction of stereoscopic PIV

Various concepts involving in the stereoscopic PIV are very briefly summarised in this section. For more details, readers are recommended to read Lavision (2007); Prasad (2000); Raffel et al. (2007). Stereoscopic PIV adopts two digital cameras viewing at the same laser illuminated plane[1] from two different angles to resolve the three velocity components on the plane; see figure 1. Sometimes it is also called 2D3C (two-dimension three-component) PIV. The basic principle of stereoscopic PIV is similar to a pair of human eyes simultaneously observing an object to capture its movement in a plane as well as in the third direction. One major difference to the two-dimensional PIV is that the illuminated plane cannot be too thin, because the third component needs to be resolved. It should allow most of the particles to remain in the illuminated volume after the PIV Δt, to give valid cross-correlation signals for calculating the third component.

1.1 Principle

In this arrangement, the two cameras simultaneously accept a pair of laser exposures to do normal two-dimensional PIV independently. Because the common field of view (FOV) of the

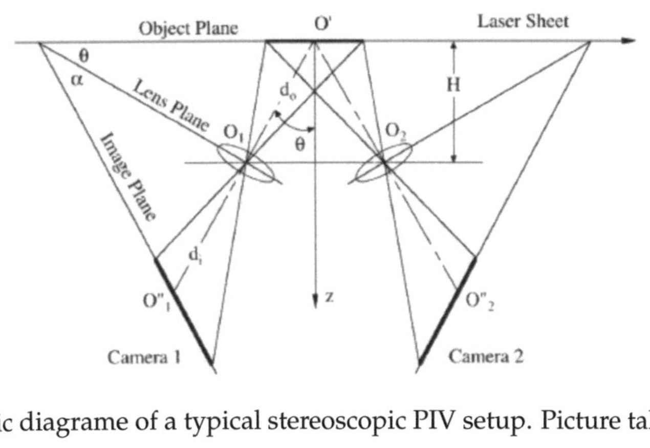

Fig. 1. A schematic diagrame of a typical stereoscopic PIV setup. Picture taken from Prasad (2000).

[1] Strictly speaking, it is not a plane, but a very thin volume with a typical thickness of $2 - 5mm$.

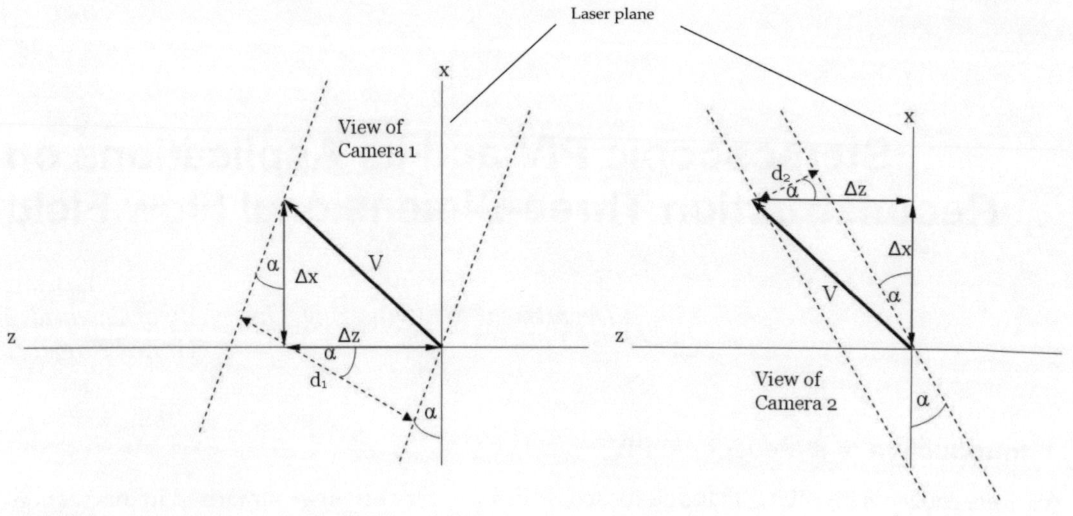

Fig. 2. Reconstructions of displacements seen from the two cameras.

two cameras are maximized[2], the resultant velocity vector fields from the two cameras are combined and from which the 2D3C field is reconstructed, provided a successful stereoscopic calibration having been applied.

The relationships between the real particle displacements (Δx and Δz) and the pixel displacements seen in each of the cameras (d_1 and d_2) can be worked out simply by considering the geometries, as shown in figure 2. For camera 1,

$$d_1 = \Delta z \cos\alpha + \Delta x \sin\alpha; \tag{1}$$

for camera 2,

$$d_2 = \Delta z \cos\alpha - \Delta x \sin\alpha; \tag{2}$$

therefore the true displacements can be re-written in terms of pixel displacements, once the viewing angle is known, as

$$\Delta x = \frac{d_1 - d_2}{2\cos\alpha}$$
$$\Delta z = \frac{d_1 + d_2}{2\sin\alpha}. \tag{3}$$

The displacements in the physical space can be obtained by a simple calibration of d_1 and d_2 to their corresponding displacements in physical space.

However, in most situations, the viewing angles are difficult to measure and moreover, if the experiments are carried out in water, as the viewing angle deviates from 90°, the image distortion becomes severer. This requires a robuster calibration process.

[2] The two cameras can be located either on the same side (FB [forward-backward scattering] setup, like it shown in figure 1) or on different sides (FF [forward-forward scattering] or BB [backward-backward scattering] setup) of the plate. In the first case, the two FOVs are not possible to be the same; while the second type of setup it is possible to make the FOV almost the same.

Fig. 3. Calibration plate images viewed from the two cameras. The viewing angle of the two cameras in this case is about 30° each.

1.2 Calibration

To calibrate the cameras for stereoscopic PIV, one typically needs to take an image of the calibration plate (a plate with a two-dimensional array of circular or cross-shaped markers, which is necessary for correcting distortions), then shift the plate in the plane-normal direction for a small amount and take a second image of it. The amount of shift should be comparable to the laser sheet thickness, as mentioned above, typically around $2-5mm$, depending on and limiting by various situations. Because the amount of shift needs to be very accurate, it is recommended to use a two-level calibration plate as it shown in figure 3, without any need to shift.

To take into account the possible distortion, a third-ordered polynomial fitting function is applied for mapping the global coordinates of the markers on the plate to the associated pixel locations of them. One can also adopts higher ordered polynomial functions, but usually third-order is sufficient. If a third-ordered polynomial function is used, the mapping function typically looks like equation 4, as used in Lavision (2007).

$$\begin{bmatrix} X \\ Y \end{bmatrix} = \begin{bmatrix} a_o + a_1 s + a_2 s^2 + a_3 s^3 + a_4 t + a_5 t^2 + a_6 t^3 + a_7 st + a_8 s^2 t + a_9 st^2 \\ b_o + b_1 s + b_2 s^2 + b_3 s^3 + b_4 t + b_5 t^2 + b_6 t^3 + b_7 st + b_8 s^2 t + b_9 st^2 \end{bmatrix}, \tag{4}$$

where X and Y are physical coordination array of the markers on the calibration plate, with the image size $nx \times ny$, the origin location (x_o, y_o) and the normalized pixel location:

$$s = \frac{2(x - x_o)}{nx}$$
$$t = \frac{2(y - y_o)}{ny}. \tag{5}$$

To solve this matrix for the unknown coefficients, one needs sufficient number of markers; if there are more markers than necessary, a least-squares fit will be applied. This fitting process is carried out for each image in the two different normal (z) locations for both cameras (two sets of coefficients for each camera). After the calibration is done, the angles of the line-of-sight

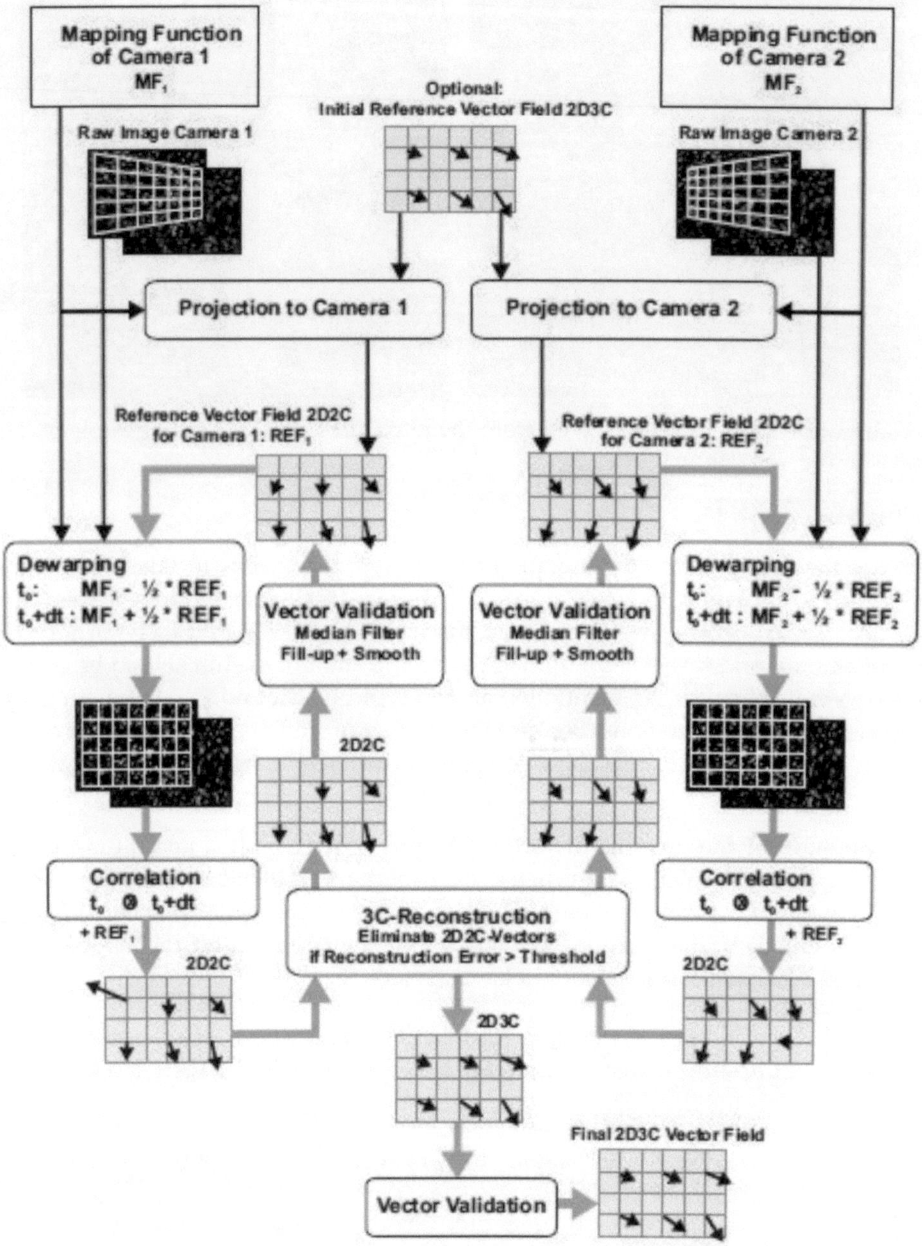

Fig. 4. The flowchart of the computation of 2D3C vector fields adopted in Lavision (2007).

(α in equation 3) of each camera should be known, or one can use this mapping function to reconstruct the 2D3C velocity field directly based on the two-dimensional PIV vectors calculated in each camera, e.g. figure 4 illustrates how Lavision (2007) does this computation.

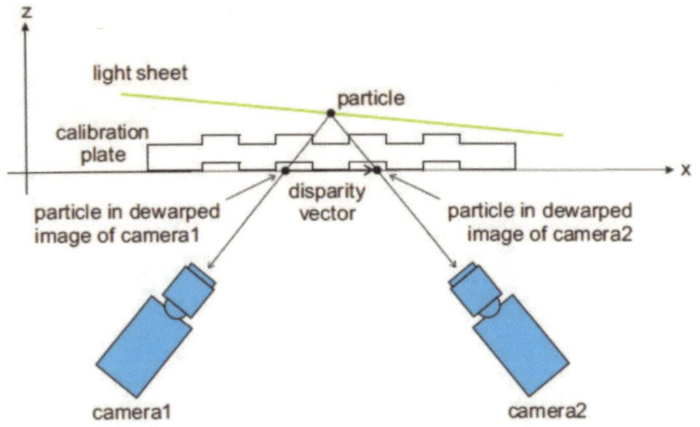

Fig. 5. The concept of disparity vectors, due to the mis-alignment of the laser sheet and the calibration plate surface. Figure taken from Lavision (2007).

1.3 Self-calibration

Stereoscopic self-calibration is useful when there is an imperfection in the calibration process discussed in section 1.2: the laser sheet may not align perfectly with the calibration plate surface. In this case, disparity vectors will appear when the images of camera 1 and camera 2 are dewarped[3] and combined by the original calibration; see figure 5. The original calibration coefficients (in the third-ordered polynomial function) will be modified and updated after self-calibration iterations such that the length of the disparity vectors will be minimized (ideally zero disparity vectors everywhere).

The prerequisite for self-calibration is that the calibration process should be done fairly well and one has about 20 or more good quality[4] particle field images from the stereoscopic recordings for the purpose of statistical convergence. A single particle image is typically divided into a grid of 7×7 (or more) sub-windows. In each of these windows, a two-dimensional disparity vector is calculated by seeking the highest disparity signal in the cluster (with respect to the centre of the window). A typical disparity map of an image is shown in figure 6. Based on the disparity map generated by such figures, the calibration coefficients are updated. Then this updated calibration coefficients are used to compute an updated disparity map, and after which calibration will be updated again. Usually a good self-calibration process requires two to three such iterations. To search the possible matching particles, a method similar to particle tracking technique is used (Wieneke, 2005; 2008).

1.4 Focusing problem

Because the two cameras view the FOV at an angle, usually Scheimpflug adapters are needed. When the object, lens and image planes intersect, the FOV in the object plane will be in focus. Depending on the limits of different types of the Scheimpflug adapters, compensation between the camera aperture size and the laser light intensity needs to be considered. The

[3] The concept of dewarping is not introduced in this chapter, readers are referred to Lavision (2007); Prasad (2000); Raffel et al. (2007).

[4] Good quality particle images should be well focused, with proper particle size and density and low background noise level, etc. Image preprocessing is also recommended before applying self-calibration.

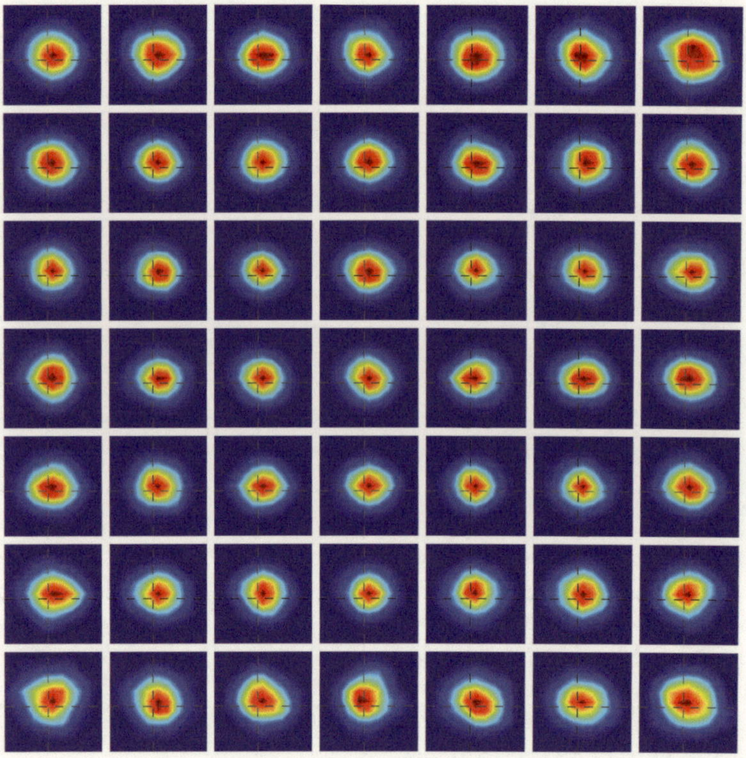

Fig. 6. A typical disparity map. The image window is divided into 7×7 sub-windows. From the peak intensity of each sub-window, a two-dimensional disparity vector can be defined, with respect to the centre of the sub-window. Each sub-window shown in this figure is typically 11×11 $pixel^2$.

viewing angle between a single camera and the laser sheet normal direction usually should not be larger than 60°, above which the distortion can be too strong and the focusing is difficult even a Scheimpflug adapter is used. In some circumstances, prisms (Prasad, 2000) can also be adopted to help reduce the distortion.

2. An application of stereoscopic PIV: three-dimensional velocity field reconstruction

This section presents an application of stereoscopic PIV to reconstruct a three-dimensional turbulent vortex ring. It is a critical assessment of such an application because a turbulent vortex ring is fully three-dimensional with no preferential direction in the azimuthal plane. Moreover, there is an important property- the translational velocity of a ring is unambiguous.

A vortex ring in laboratory studies is usually generated by an impulsive ejection of fluid through a nozzle or an orifice into a quiescent environment. Circulation based Reynolds number is defined as:

$$Re = \frac{\Gamma_{slug}}{\nu} = \frac{U_p L}{2\nu}. \quad (6)$$

where U_p is the piston velocity and L is the length of the discharged slug, Γ_{slug} is the slug circulation and ν is the kinematic viscosity of the fluid.

Vortex rings which are well-formed at the nozzle or orifice exit, under certain conditions, undergo an instability along the core circumference in the form of azimuthal waves. This type of instability is an important feature of the transition from laminar to turbulent rings. Although previous investigations of this type of waviness have been made in laminar rings or rings during the transition stage from laminar to turbulent Maxworthy (1977); Saffman (1978); Shariff et al. (1994); Widnall & Tsai (1977), the results show that azimuthal waves are also observed in fully turbulent vortex rings and this technique can successfully capture this feature.

2.1 Theoretical background

2.1.1 Taylor's hypothesis

Taylor's hypothesis (Taylor, 1938; Townsend, 1976) states that "if the velocity of the airstream which carries the eddies is very much greater than the turbulent velocity, one may assume that the sequence of changes in U at the fixed point are simply due to the passage of an unchanging pattern of turbulent motion over the point". In other words, if the relative turbulence intensity u' is assumed to be small enough compared to the mean advection speed U:

$$\frac{u'}{U} \ll 1, \tag{7}$$

the time-history of the flow signal from a stationary probe can be regarded as that due to advection of a frozen spatial pattern of turbulence past the probe with the mean advection speed U, i.e.

$$u\left(x,t\right) = u\left(x - U\Delta t, t + \Delta t\right), \tag{8}$$

where Δt is the time delay and should not be a too large value. Taylor's hypothesis is effectively a method to transfer the time dependent measurement results to a spatial domain. The selection of the correct velocity scales relevant to turbulent vortex rings are discussed in section 2.3.

2.1.2 Similarity model

All the figures presented in this chapter are plotted in the similarity coordinates derived from the similarity model (Glezer & Coles, 1990). The reason is because the original model was derived from Taylor's hypothesis. Any structure plotted in the similarity coordinates contains not only spatial but also temporal information: it can be transferred back to the physical coordinates at any point in a series of time history. It has been proved in Gan & Nickels (2010) that rings produced in this study are well predicted by this model.

If a proper pair of spatial and temporal origins z_o, t_o can be found, the length and velocity scales in a turbulent vortex ring in physical space can be shown to be self-similar, and they can be written in terms of the similarity variables, i.e.

$$\xi = (z - z_o)\left[\frac{\rho}{I\left(t - t_o\right)}\right]^{\frac{1}{4}} \qquad \eta = r\left[\frac{\rho}{I\left(t - t_o\right)}\right]^{\frac{1}{4}}; \tag{9}$$

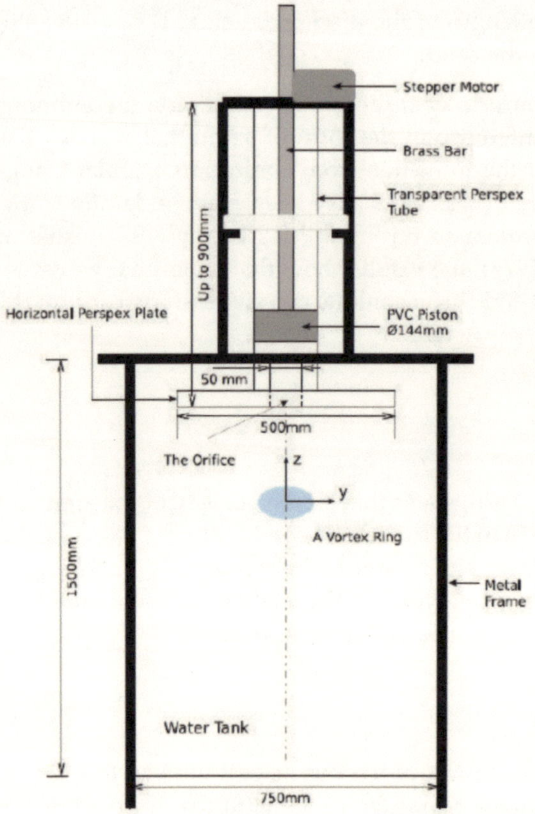

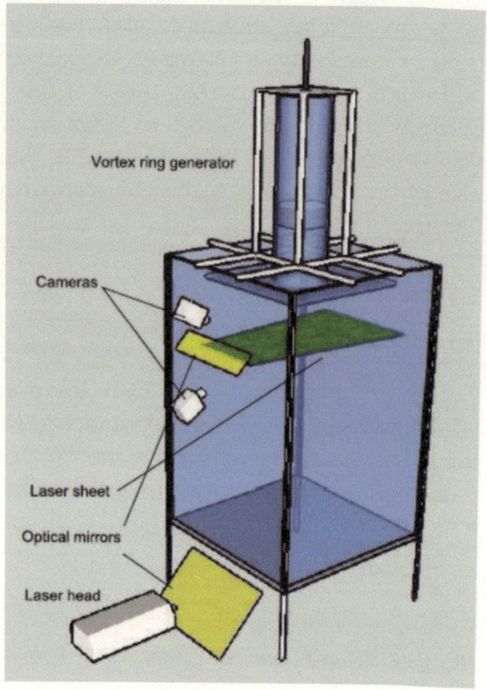

Fig. 7. Schematic diagram of the vortex ring generator. The diagram is not to scale. The coordinate system adopted in this experiment is also shown, where $z = 0$ is set at the orifice exit.

and from the streamfunction,

$$U = u\left(\frac{\rho}{I}\right)^{\frac{1}{4}}(t-t_o)^{\frac{3}{4}} \qquad V = v\left(\frac{\rho}{I}\right)^{\frac{1}{4}}(t-t_o)^{\frac{3}{4}}, \tag{10}$$

where I is the hydrodynamic impulse (assumed invariant); ρ is the density (a constant when incompressibility is assumed); u, v are the radial and axial velocity components respectively; U, V are the corresponding non-dimensional velocities. ξ and η are the non-dimensional similarity quantities for the length scales.

2.2 Experimental setup

A simple sketch of the ring generator is presented in figure 7.

The rectangular tank is made of $15mm$ thick perspex with a bottom cross-sectional area of $750mm \times 750mm$ and a height of $1500mm$. The top of the tank is uncovered. The tank is therefore transparent from all directions of views. Other important geometrical parameters of the vortex generator are labelled in figure 7 and is described in Gan et al. (2011).

The motion of the piston is driven by a stepper motor. The motor is able to drive the piston to move at a constant speed of up to $1000mms^{-1}$ with an acceleration and deceleration of about $1500mms^{-2}$.

The effective Re is set to 41280 and $L/D = 3.43$ in order to match the conditions in Glezer & Coles (1990) and Gan & Nickels (2010), because parts of the reconstruction validation will rely on the data in them. The piston velocity programme is approximately a top-hat shape: the piston acceleration/deceleration time is about 10% of the piston total movement duration. Rings produced at this combination of Re and L/D are well in the turbulent region on the transition map (see Glezer, 1988).

In this arrangement, the PIV (laser sheet) plane is located 6D (six orifice diameters) downstream of the orifice exit. The PIV system is provided by LaVision Ltd: a pair of high-speed Photron APX cameras are used as the image recording devices and the particle illumination is realised by a Pegasus PIV Laser which consists of a dual-cavity diode pumped Nd:YLF laser head and is capable of emitting a beam of $527nm$ wavelength and $10mJ$ energy. The laser beam is converted to a sheet by passing through a cylindrical diverging lens. The thickness of the sheet can be adjusted by changing the separation of a pair of telescope lenses housed before the cylindrical lens and is set to about $4 - 5mm$. The flow is seeded by $50\mu m$ diameter, silver-coated hollow glass spheres.

After the changes in refractive index are considered, the effective angle between the two cameras is approximately 120°. The two cameras run in single-frame single-exposure mode, and the operation frequency is set at $f = 600Hz$ giving a $\Delta t = 1.67\ ms$. The interrogation window size for time series cross-correlation process is set to $16 \times 16\ pixel^2$ with a 25% overlap to give a spatial resolution of $\Delta x = \Delta y \approx 1.70mm$ (based on vector spacing, $\approx$ 3.4% D) in the PIV plane. The FOV of the PIV plane is about $230 \times 164\ mm^2$.

2.3 Accuracy justification

An important prerequisite for the reconstruction of a turbulent vortex ring using Taylor's hypothesis is to identify appropriate scales of the fluctuating and convective velocities in equation 7.

The relevant convective velocity in this work is the ring advection speed, or celerity, and is obtained from the independent two-dimensional PIV results in Gan & Nickels (2010). The centroid-determined vortex ring radii r and celerities u_t for 50 realisations are reproduced in figure 8. From this figure the spatially averaged celerity (□ data) at various streamwise locations is obtained between $z = 5.5D$ to $z = 6.5D$ and the ensemble averaged ring celerity is obtained by a least-squares fit (blue line) giving: $u_t = 270.34\ mms^{-1} \approx 0.54\ U_p$.

The ring celerity also affects the spatial resolution in the z direction, Δz. The reconstruction of the planar velocity fields in the z direction is selected to give a similar spatial resolution in the $x - y$ plane.

The most straight-forward reconstruction process requires the entire cross-sectional area of the ring bubble to pass normally through the PIV plane with constant celerity. Figure 8 shows that

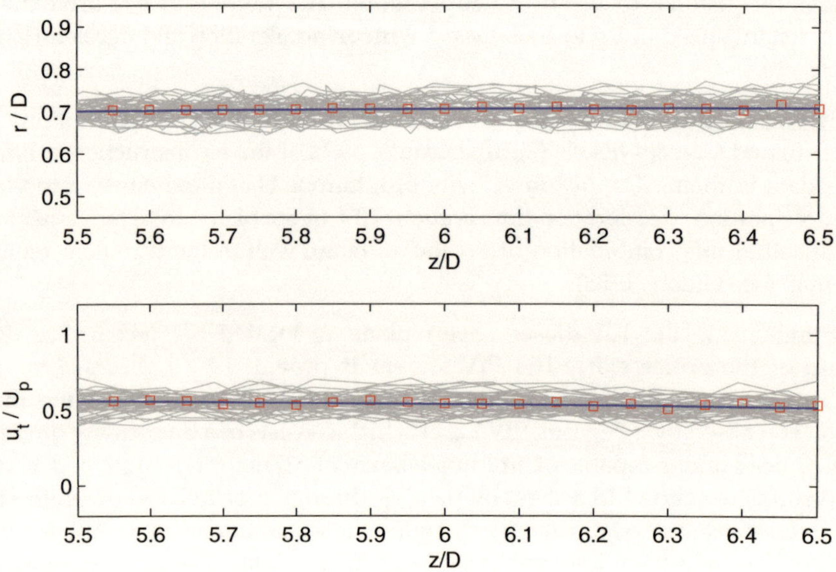

Fig. 8. Ring radius r and celerities u_t as functions of streamwise distances z for the ensemble of 50 realisations from two-dimensional PIV measurements. The grey lines are the traces of each of the 50 realisations; □: the ensemble averaged quantities; −: the first order least squares fit of □. Data are taken from Gan & Nickels (2010).

this condition can be approximated by the ensemble averaged values.[5] The celerities of single realisations are denoted by the grey traces and are always scattered as the ring centroid is changed by the turbulence. This results in a variation of $\pm 10\%$ of the □ value at each location.

If the averaged Reynolds stresses $-\overline{V'V'}$ of the ring bubble is used (from Gan & Nickels, 2010), taking the square root of these and the similarity value of the ring celerity, $U_t = 6.4$ (which can be scaled from u_t in physical space by equation 10), for rings with a $Re = 41280$ (see Table 1 in Gan & Nickels, 2010), equation 7 gives:

$$\frac{u'}{U} = \frac{V'}{U_t} < \frac{0.78}{6.4} \approx 0.12. \tag{11}$$

The maximum error can be estimated using the maximum fluctuation within the ring bubble V' which gives $V'/U_t \approx 0.33$. However, only small contour regions within the ring bubble have this value and these regions will have the largest uncertainty in the reconstruction. A rough estimation of the contribution from u_t fluctuation can also be given by figure 8, which is $u_t'/\overline{u_t} < \pm 10\%$, which according to equation 11 gives a value of 0.2.

A more rigorous method to assess the validity of Taylor's reconstruction is to check whether the material derivative of the velocity vector $D\mathbf{u}/Dt \approx 0$ which is satisfied if $u_t \gg u', v'$ neglecting pressure and viscous terms. The 2D PIV results in Gan & Nickels (2010) are used for

[5] Ideally the instantaneous convection velocity of the ring should be used. This requires simultaneous measurements normal to the measurement plane which were not available.

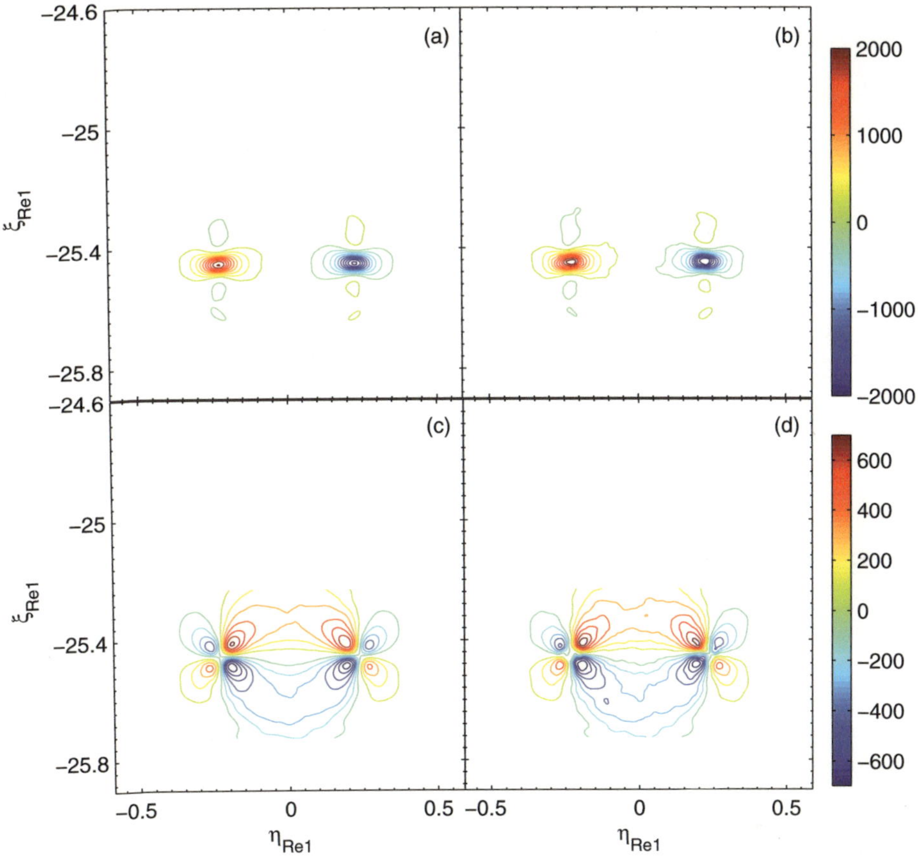

Fig. 9. Test of Taylor's hypothesis by comparing (a): $\left\langle u_t \frac{\partial u}{\partial z} \right\rangle$, (b): $\left\langle -\frac{\partial u}{\partial t} \right\rangle$, (c): $\left\langle u_t \frac{\partial v}{\partial z} \right\rangle$ and (d): $\left\langle -\frac{\partial v}{\partial t} \right\rangle$. Data obtained from two-dimensional PIV measurements. If the Taylor's reconstruction is perfect, contours of these two terms will be identical.

this purpose. The convection terms $\left\langle u_t \frac{\partial u}{\partial z} \right\rangle$ and $\left\langle u_t \frac{\partial v}{\partial z} \right\rangle$ are compared with the acceleration terms $\left\langle -\frac{\partial u}{\partial t} \right\rangle$ and $\left\langle -\frac{\partial v}{\partial t} \right\rangle$ in their non-dimensional forms in figure 9, where

$$\left\langle u_t \frac{\partial \mathbf{u}}{\partial z} \right\rangle = \left(u_t \frac{\partial \mathbf{u}}{\partial z} \right) \left(\frac{\rho}{I} \right)^{\frac{1}{4}} (t - t_o)^{\frac{7}{4}}$$
$$\left\langle -\frac{\partial \mathbf{u}}{\partial t} \right\rangle = \left(-\frac{\partial \mathbf{u}}{\partial t} \right) \left(\frac{\rho}{I} \right)^{\frac{1}{4}} (t - t_o)^{\frac{7}{4}} . \quad (12)$$

Figure 9 shows excellent agreement between the ensemble averaged acceleration and convection terms which lends confidence to the validity of the reconstruction. The good agreement obtained in the similarity coordinates means that Taylor's hypothesis is valid over the rings' time history where the similarity transformation holds.

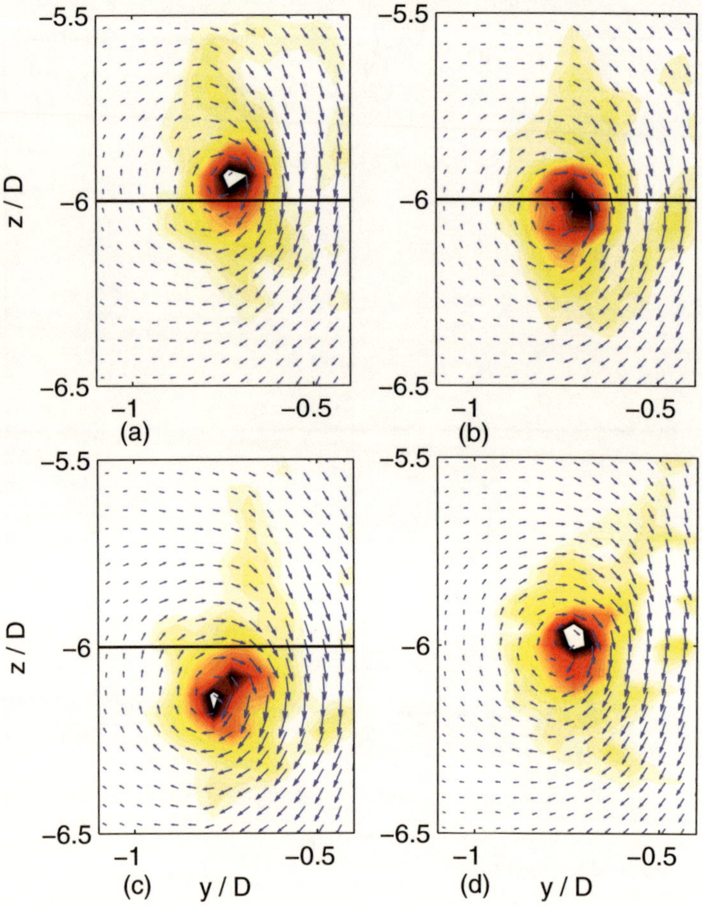

Fig. 10. An individual vortex ring recorded by two-dimensional PIV is used to test the Taylor's reconstruction process. (a), (b), (c): the three continuous snapshots showing the left core of a ring passing through the $6D$ station. The layers of the two-component velocity information at the $6D$ station are then stacked to reconstruct the core, in (d), in which the ensemble averaged ring celerity 270.34 mms^{-1}, instead of the one for this particular ring, is used. The vorticity contour levels: -240.0 (10.0) -20.0; $\rightarrow$ denotes the in-plane velocity vectors.

Figure 10 (a) to (c) plots an image sequence showing a single vortex ring core passing through the 6D station which are then used to reconstruct the vortex core in (d). The velocity vectors are overlaid with vorticity contours. Comparing (a) to (c) with (d) shows that the peak vorticity of the core is basically retained. The vorticity distribution of the inner core shows a slight change in figure 10 (c) but these are small changes and close to the maximum spatial resolution. The instantaneous vorticity is naturally very sensitive to the velocity fluctuation however the core radius remains similar, within $\pm 10\%$, see figure 11. The core is compact being about $0.2D$ in size, passes through the 6D station within about 37 ms. The vorticity surrounding the core is weaker and undergoing shedding-reattachment processes at a fairly fast pace (Gan, 2010). Any high speed fluctuations in the weaker regions are less frozen than the core but their features are well captured in the reconstructed image in figure 10 (d).

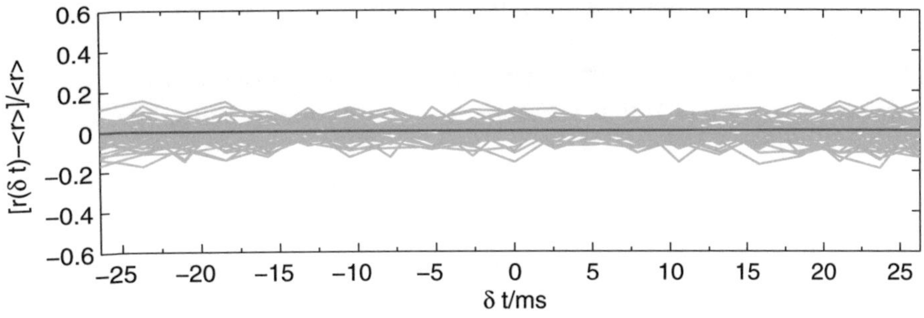

Fig. 11. The core radius variation as the ring passes through the testing station for all the 50 realisations from two-dimensional PIV measurements. The grey lines are the traces of each of the 50 realisations. $r(\delta t)$ denotes the radius of a single realisation as a function of the time t; $\langle r \rangle$ denotes the average radius of all the 50 rings. The threshold to determine the radius is set as $100s^{-1}$, which is about 50% of the peak vorticity intensity, see figure 18. Note that the core passes through the testing station within $\pm 19ms$.

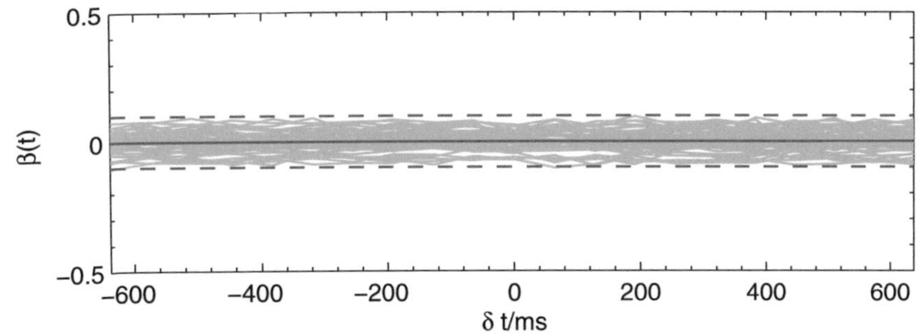

Fig. 12. The circulation variation as the ring passes through the testing station for all the 50 realisations from two-dimensional PIV measurements. The grey lines are the $\beta(t)$ traces for each of the 50 realisations, which is calculated by equation 13. The $\pm 10\%$ lines are also shown by the $--$ traces.

As the largest velocity fluctuations occur around the vortex core, the spatially averaged velocity of 50 realisations can be compared with the reconstructed value. To estimate the spatially averaged velocity error, a quantity $\beta(t)$ is introduced:

$$\beta(t) = \frac{\Gamma(t) - \Gamma_{re}}{\Gamma_{re}} = \frac{\oint_C \mathbf{u}(t) \cdot \mathbf{dl} - \oint_C \mathbf{u}_{\mathrm{re}} \cdot \mathbf{dl}}{\oint_C \mathbf{u}_{re} \cdot \mathbf{dl}}, \tag{13}$$

where Γ is the circulation obtained along a closed loop C enclosing an area of $2D$ and $1.5D$ in streamwise and spanwise directions respectively around each vortex core similar to that shown in figure 10. (t) is the time when the ring passes through the $6D$ station, similar to that shown in (a), (b) and (c) of figure 10. The subscript 're' indicates the reconstructed ring at the $6D$ station as shown in (d) of figure 10. The quantity $\beta(t)$ is determined for 50 realisations and shown in figure 12.

As the loops are of the same size, and the circulation is conserved, $\beta(t)$ compares the spatially-averaged variation between $\mathbf{u}(t)$ of each realisation with the reconstructed one. The resulting error is less than $\pm 10\%$, a similar value to the core dispersion in figure 8 and radius variation in figure 11. The results from these tests show that the expected error of the reconstruction is within $\pm 10\%$.

A more direct assessment is to investigate the divergence field of the reconstructed three-dimensional vortex ring. Figure 13 shows the p.d.f. (probability density function) of the divergence of the flow field, where in (b) and (d) the divergence is normalised by the norm of the local vector gradient tensor $(\nabla\mathbf{u} : \nabla\mathbf{u})^{1/2}$; in (c) κ is defined by:

$$\kappa = \frac{(\partial u/\partial x + \partial v/\partial y + \partial w/\partial z)^2}{(\partial u/\partial x)^2 + (\partial v/\partial y)^2 + (\partial w/\partial z)^2}. \tag{14}$$

It can be seen that most of the data points locate in regions very closed to divergence free, bearing in mind that no experiment is divergence free, due to the finite spatial resolution.

2.4 The reconstructed velocity field

A typical instantaneous stereoscopic PIV velocity - vorticity field is shown in figure 14. Only the in-plane velocity vectors are plotted from which the overlaid vorticity contours ω_k were determined. The level of turbulence is clearly captured by the asymmetry in the vector field and the fluctuations of vorticity.

In order to visualise the vortex ring structure, an appropriate scalar which might be used after Taylor's reconstruction of velocity field is the vorticity ($\nabla \times \vec{u}$) magnitude. The structure of the vortex ring bubble and the wake can then be observed in figure 15 with the strong level of turbulence illustrated by the degree of the wrinkling along the isosurfaces and streamlines. It must be pointed out that in order to visualise the ring bubble by streamlines, the ring needs to be put in a stationary frame of reference. This is achieved by simply subtracting the mean advection velocity of the ring at the PIV testing location[6] from the instantaneous velocity at every data point which has no effect on the vorticity field. In figure 15 the extent of irregularity in the streamline patterns can be distinguished upstream and downstream of the ring bubble with localised areas of high vorticity found in the wake. A wavy core is also observed, confirming the existence of azimuthal waves.

In order to visualise the relationship between the high and low vorticity fluctuations around the vortex ring core and the outer bubble respectively, two sections of the selected bubble vorticity isosurface are shown in figure 16. As expected the low intensity wraps around the core. Tube shaped vorticity isosurfaces of low intensity are shed into the wake showing some agreements with the numerical simulations of Bergdorf et al. (2007) and Archer et al. (2008) which show hairpin vortices being shed from the vortex bubble into the wake. Although, hairpin structures were not clearly observed in the current study. This could be due to the lower spatial resolution of the experiment or the different Reynolds numbers and initial conditions compared with the numerical simulations.

[6] It is assumed here that the ring advection velocity in the vicinity of the PIV testing location is constant. Figure 8 shows that it is a fairly reasonable assumption.

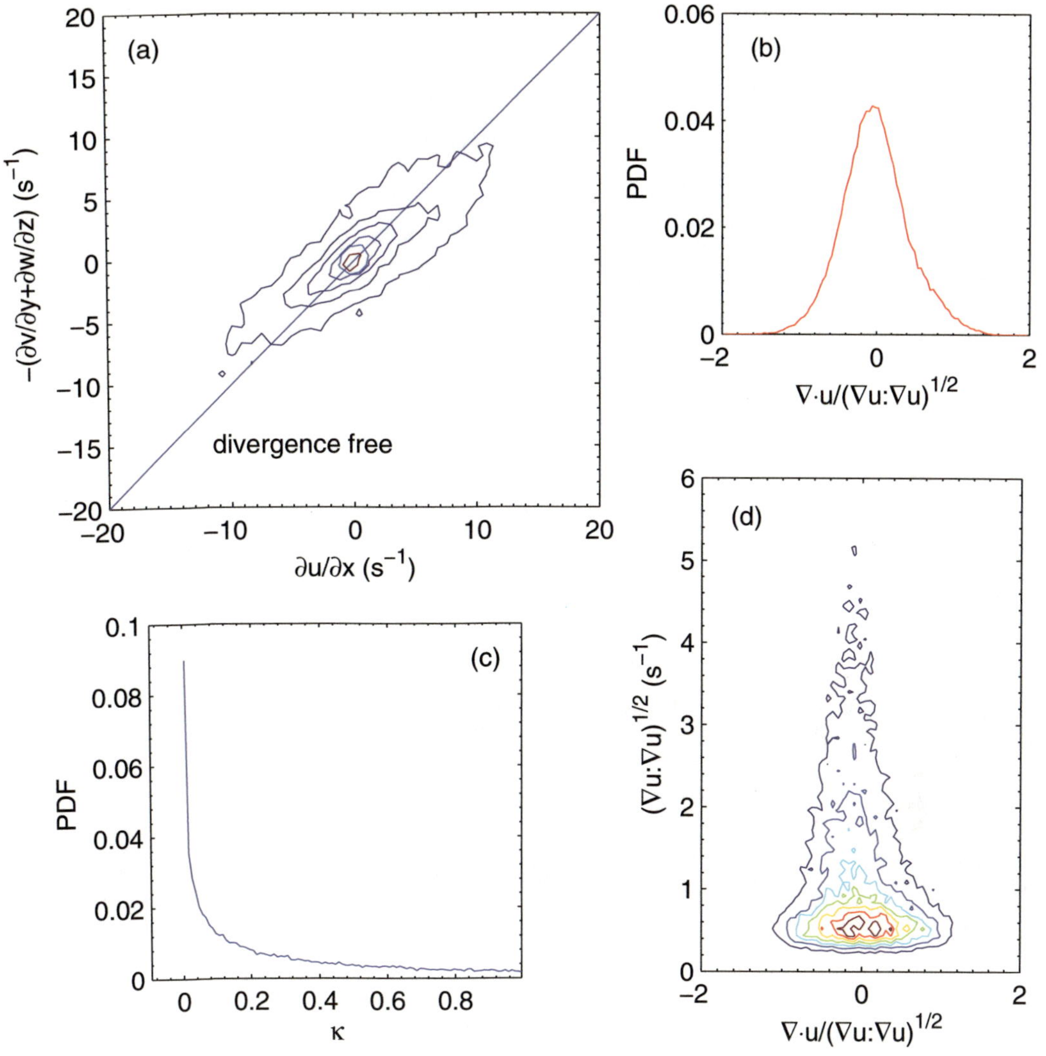

Fig. 13. The divergence of the reconstructed three-dimensional velocity field. (a) the joint p.d.f. between $\partial u/\partial x$ and $-(\partial v/\partial y + \partial w/\partial z)$, contour levels [0.0002 0.0007 0.002 0.007 0.02 0.1]; the straight line indicates divergence free. (b) the p.d.f. of $\nabla \cdot \mathbf{u}/(\nabla\mathbf{u} : \nabla\mathbf{u})^{-1/2}$. (c) the p.d.f. of κ. (d) the joint p.d.f. of $\nabla \cdot \mathbf{u}/\ (\nabla\mathbf{u} : \nabla\mathbf{u})^{1/2}$ and $(\nabla\mathbf{u} : \nabla\mathbf{u})^{1/2}$; contour levels [0.0003 : 0.0003 : 0.003].

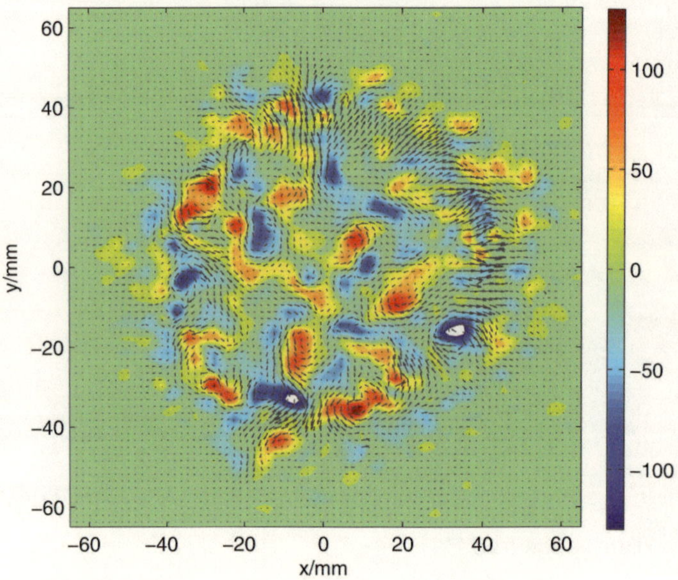

Fig. 14. A presentation of instantaneous velocity and vorticity in azimuthal plane in physical coordinates, when the ring centre is about to reach the PIV plane. Only the velocity vectors in the azimuthal plane are shown. The vorticity is in the streamwise direction, ω_k, zero level bypassed.

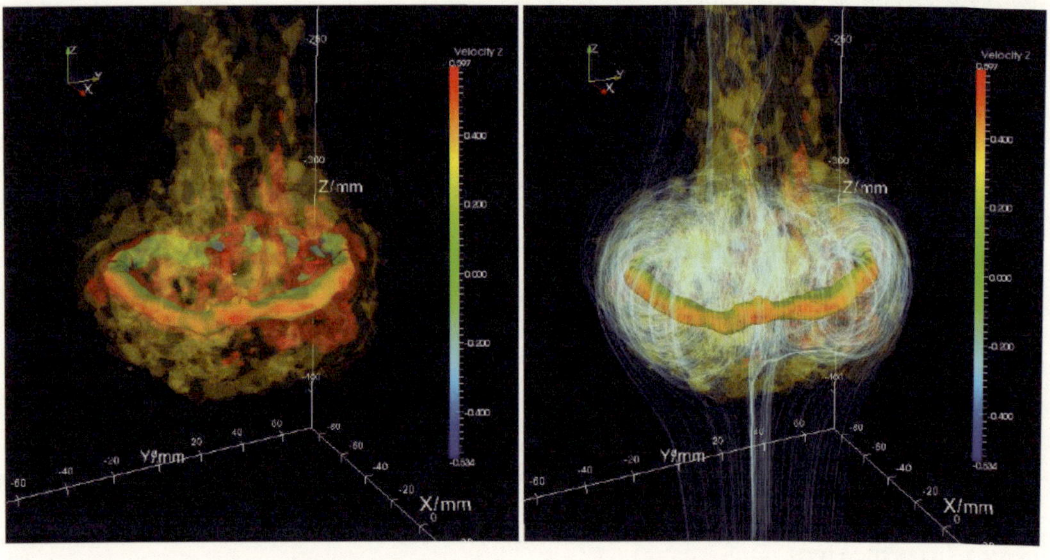

Fig. 15. A three-dimensional vorticity magnitude contour. Isosurface levels are $150s^{-1}$, $100s^{-1}$, $50s^{-1}$. Streamlines are shown in the second figure. The colour bar shows the streamwise velocity level on the core surface.

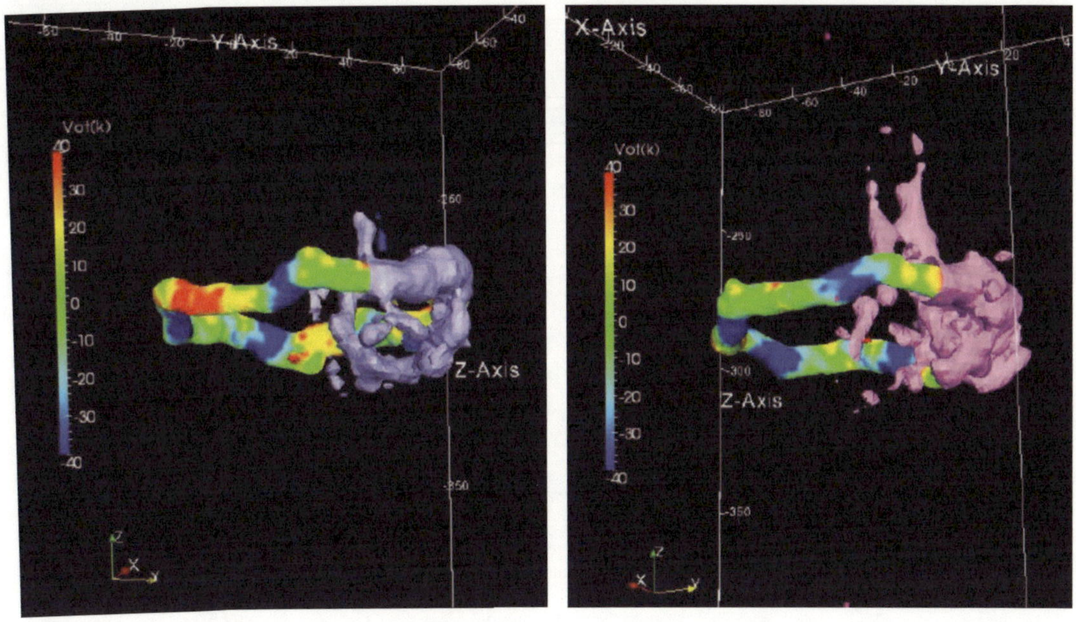

Fig. 16. Isosurfaces of the three-dimensional vorticity magnitude, showing the high intensity core and two portions of the low intensity vorticity blobs wrapping around the core and shed to the wake. Isosurface intensities: $150s^{-1}$ and $90s^{-1}$. Colour code on the core surface indicates the streamwise vorticity ω_k.

The three components of the vorticity, ω_i ω_j ω_k in the two central cross-sectional planes $y-z$ and $x-z$ are plotted in figure 17. The reconstructed azimuthal vorticity contours in (a) and (b) can be compared with the instantaneous vorticity contours obtained from the two-dimensional PIV results, shown in figure 18. Note that Figure 17 and 18 belong to different realisations. Because the rings are very turbulent, contour shapes are noisier especially for low level vorticity, and are expected to be different for different realisations due to the stochastic nature of the flow. However, the main features and the maximum intensities show good agreements with the main difference being that the reconstructed fields show less low level azimuthal vorticity. A long streamwise vorticity ω_k structure is observed in the three-dimensional view in figure 19, which agrees with the findings in Bergdorf et al. (2007). In the ring bubble region, the streamwise vorticity are found to wrap around the vortex core, in a manner such that the positive and the negative valued vortices are separated by each other. This can be seen by the colour code on the core isosurface in figure 16 and shows reasonable agreement with the results in Archer et al. (2008). However, the structure of the vorticity is difficult to identify in the three-dimensional view due to the effect of turbulence which breaks the isosurfaces thresholded at $\omega_k = 30s^{-1}$ into pieces. The colour code on the ω_k isosurface indicates the strength of the streamwise stretching term $S_k = (\vec{\omega} \cdot \nabla)\,\vec{u}_k$. The positive value of ω_k reflects the stretching of the streamwise vortex tubes. The negative valued streamwise vortex tubes are not shown but the structure and the stretching strength of these tubes was found to be similar.

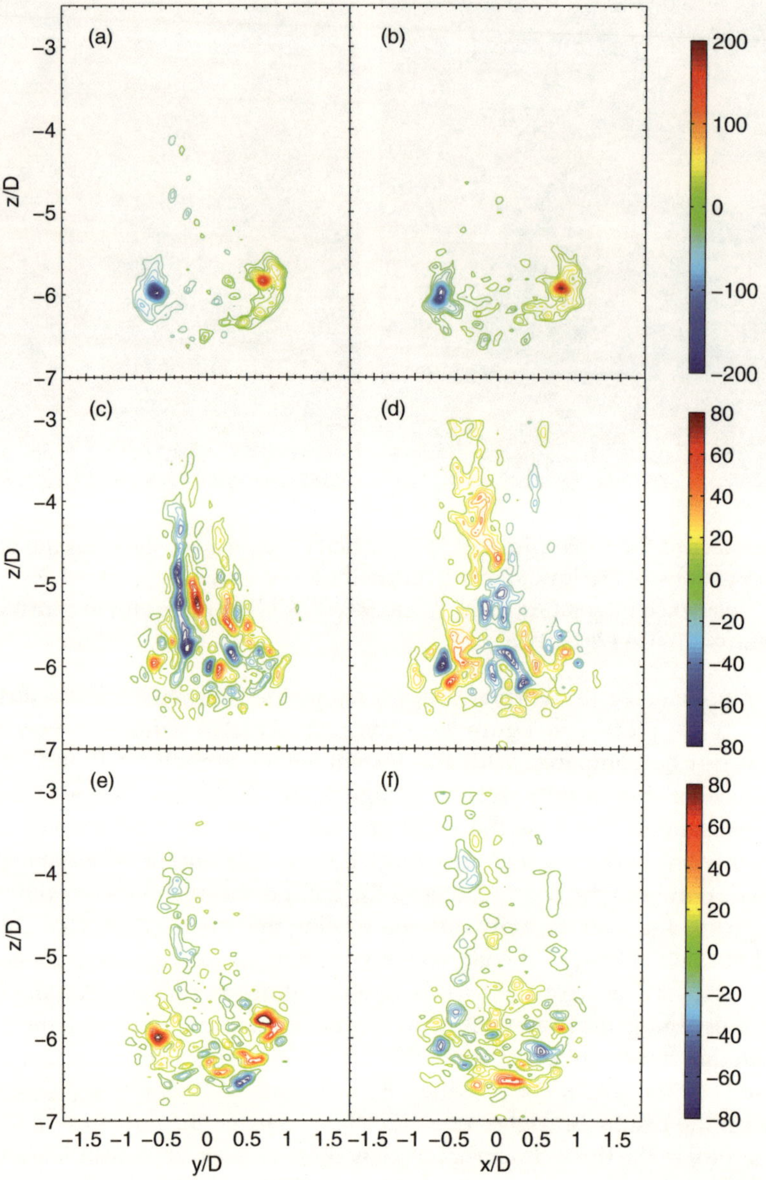

Fig. 17. The three components of vorticity in the two central cross-sectional planes $y - z$ and $x - z$. (a) and (b): azimuthal vorticity contours, ω_i, ω_j; (c) and (d): streamwise vorticity contours, ω_k; (e) and (f): spanwise vorticity contours, ω_j, ω_i.

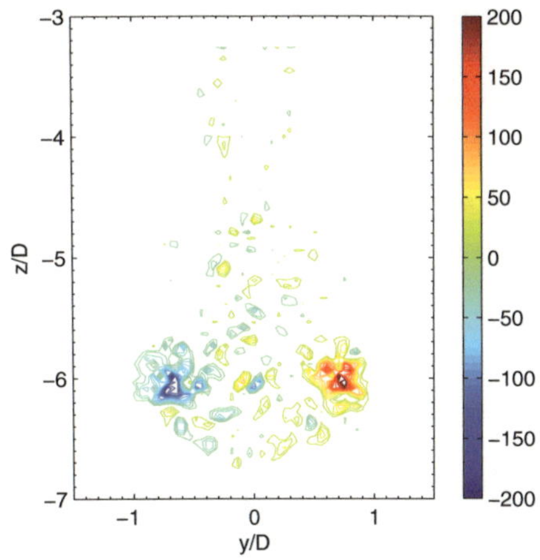

Fig. 18. The vorticity contour of an instantaneous realisation from the two-dimensional PIV experiments.

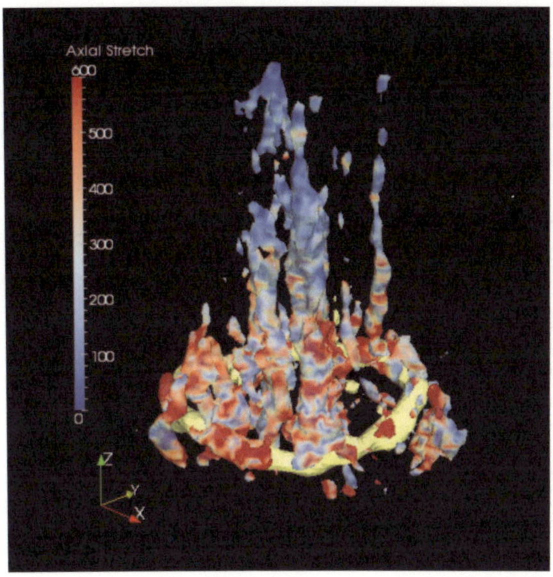

Fig. 19. The streamwise vorticity wraps around the core. The core position is shown by the yellow coloured isosurface of vorticity magnitude $\omega_m = 140s^{-1}$. The vorticity isosurfaces wrapping around the core are $\omega_k = 30s^{-1}$. Colour code on ω_k surface indicates the positive-valued streamwise (axial) vortex stretching term S_k.

2.5 The azimuthally averaged quantities

Statistically speaking, an ensemble-averaged vortex ring bubble tends to be axisymmetric[7]. It is interesting to compare the azimuthally averaged velocity components to those by two-dimensional ensemble averaged ones, which may also serve as a validation of the technique.

The azimuthally averaging process is applied in a cylindrical coordinate system in which the process begins at $\theta = 0$ and ends at $\theta = 2\pi$. The first step is to locate the 'best' centre point, or a proper axis of symmetry as the results from the azimuthally averaging process will depend heavily on the location of this axis. The axis is found by a best r.m.s. fit of the core area to a circle. After this, data format in Cartesian coordinates is converted to cylindrical coordinates. More details can be found in Gan et al. (2011).

The velocities are non-dimensionalised by equation 10 and presented in the similarity coordinates calculated from equation 9. Figure 20 shows the dimensionless velocities U_θ (radial), V_θ (axial), W_θ (azimuthal) and the dimensionless vorticity $\widehat{\omega}_\theta$, where:

$$\widehat{\omega}_\theta = \frac{\partial V_\theta}{\partial \eta} - \frac{\partial U_\theta}{\partial \xi}, \tag{15}$$

and θ here denotes an azimuthally averaging result, ξ, η denote the axial and radial direction in cylindrical coordinates respectively.

In figure 20, noise increases towards the centre due to the reduced number of data points for averaging[8]. However, the main area of interest is obviously the core region. The central region exhibits low azimuthal velocities in figure 20 c): a long region of positive mean velocity is observed in the wake while a weak negative mean velocity can be observed in the core centre region. The presence of a mean velocity in the core region is not surprising, a number of researchers have also observed such behaviour, nevertheless the magnitude is considerably smaller than both the convection velocity V_θ and the radial velocity U_θ components shown in figure 20 a) and b). The opposite sensed mean velocity in the inner region can be partly due to the noise and partly due to a possible mechanism of conservation of angular momentum of the vortex ring bubble. The streamline pattern in figure 15 also suggests the presence of a mean azimuthal velocity.

The dimensionless circulation Γ_θ can be computed by equation 16 from the contour plot in figure 20 d).

$$\Gamma_\theta = \int_S \widehat{\omega}_\theta \, d\xi d\eta, \tag{16}$$

where S denotes the entire area of the FOV in the similarity coordinates. A value of 7.08 is found, which is similar to the circulation from the two-dimensional PIV results, 6.87 (see Gan & Nickels, 2010).

[7] If one produces a large number of turbulent vortex rings from a circular orifice, and does an ensemble average, the resultant velocity field is approximately axisymmetric.

[8] Recall that the raw data is stored in the Cartesian coordinates; towards smaller radii, there are less data points.

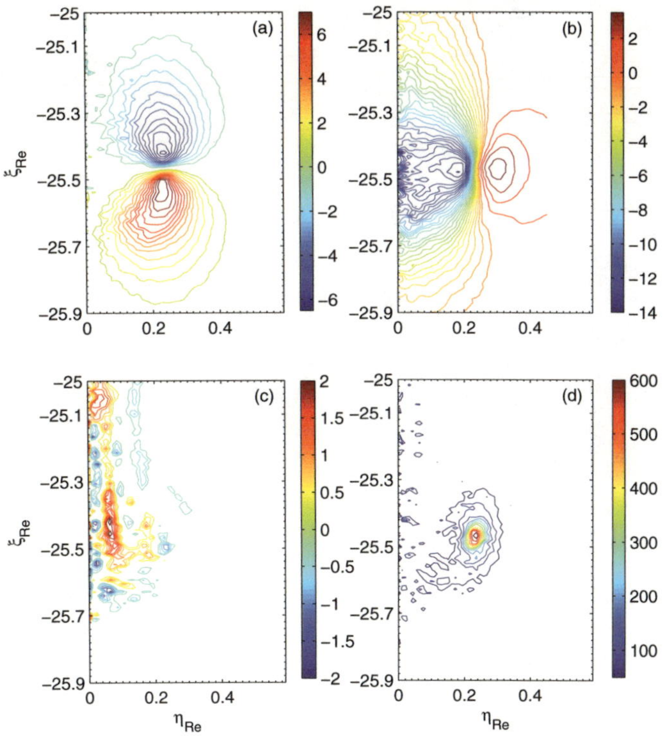

Fig. 20. The mean structures of velocity and vorticity calculated by the azimuthally averaging process: averaging along the θ direction in similarity coordinates. Contours shown are U_θ (radial), V_θ (axial), W_θ (azimuthal) and $\widehat{\omega}_\theta$ (azimuthal) in (a), (b), (c) and (d), respectively; the subscript θ denotes that the corresponding quantity is calculated by the azimuthally averaging process.

If the azimuthally averaged velocity components make sense it is also possible to compare the turbulence quantities, one can refer to Gan et al. (2011) for more details.

The turbulence production is believed to be closely related to the vortex stretching, this mechanism is expected to take place in the ring core and bubble windward regions (Gan & Nickels, 2010). The stretching effect can be assessed via the vorticity equation. In Cartesian coordinates, the stretching term $(\vec{\omega} \cdot \nabla)\,\vec{u}$ represents the vortex stretching in three principal directions, S_i, S_j, S_k:

$$\begin{pmatrix} S_i \\ S_j \\ S_k \end{pmatrix} = (\vec{\omega} \cdot \nabla) \begin{pmatrix} u_i \\ u_j \\ u_k \end{pmatrix}, \tag{17}$$

on ω_i, ω_j, ω_k respectively.

Because the flow field of a vortex ring is close to axisymmetric, the stretching of the vortex tubes in the windward bubble surface is expected to be orientated in the radial direction (due to the mean velocity direction). To better illustrate the stretching effect, the vorticity and the stretching vector in equation 17 are transferred from Cartesian coordinates to cylindrical coordinates, as for velocities: $\omega\,(i,j) \mapsto \omega\,(\theta,r)$, $S\,(i,j) \mapsto S\,(\theta,r)$, ω_k and S_k are unaffected.

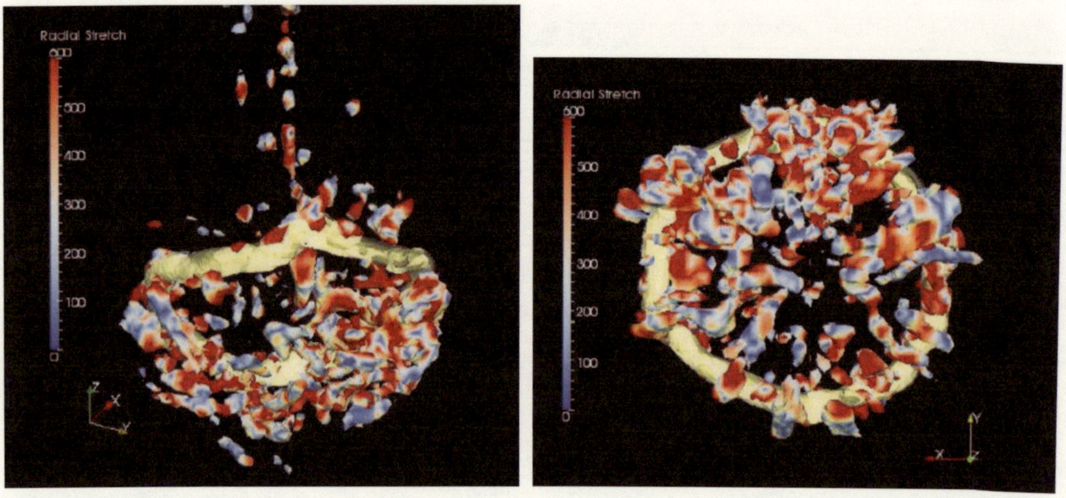

Fig. 21. Vortex ring core $\omega_\theta = -140s^{-1}$ shown by the yellow coloured isosurface; and the radial vorticity $\omega_r = \pm 30s^{-1}$. The colour code on the ω_r isosurface shows the magnitudes of the radial stretching term $|S_r|$, with the positive valued S_r dyed on the positive ω_r and negative valued S_r dyed on the negative ω_r.

The radial vorticity ω_r is plotted in figure 21, with the radial direction stretching term S_r coloured on the isosurface. Figure 21 supports the expectation that vortices are being stretched in the radial direction and this stretching mechanism prevails on the windward side of the bubble. The compression of vorticity, i.e. the correlation of the opposite sensed vorticity and the stretching value, is much weaker (less than 20%) compared with stretching and is not shown here. The vorticity in the windward bubble region can also be seen in figure 17, in Cartesian coordinates. The streamwise stretching, which has been mentioned in section 2.4, is believed to be responsible for the turbulence production in the wake region however this is very small overall.

3. Conclusions

The first part of this chapter introduces the 2D3C working principle of stereoscopic PIV and its calibration procedure and its self-calibration to increase the accuracy level. The second part of the chapter presents an application of stereoscopic PIV for reconstructing a fully three-dimensional vortex ring. It has shown that by fixing the laser sheet position and hence the two cameras' FOV, and reconstructing the fully three-dimensional velocity field by Taylor's hypothesis, the accuracy of the results is limited. In other words, due to the intrinsic nature of Taylor's hypothesis, the resultant velocity field is never really instantaneous. However, there are various aspects which can increase the potential accuracy of such reconstruction and thus deserve further discussion.

3.1 Possible accuracy improvements

In order to freeze the ring structure better, the results of equation 11 needs to be closer to zero. Nevertheless, it is expected that for fully turbulent vortex ring, the level of turbulence intensity u' scales with Reynolds number (because $u' \sim U_p$), hence scales with vortex ring circulation

Γ and the advection speed u_t. Simply increasing Reynolds number will not improve the situation. Certain treatments independent of the ring will be necessary. A possible way is to move the PIV measurement (cameras and the laser sheet together) against the ring advection direction - an 'active scanning' process. Equation 11 shows that by simply moving the PIV measurement plane at the same speed as the ring, it can bring down the ratio u'/u_t significantly (doubling the denominator). Moreover, to resolve the wake correctly, an active scanning seems to be compulsory.

The next question is that how fast the scan speed is optimal. Definitely scan with an infinite speed gives no error. However, first it is not allowed, even it is capable to; second, it may not be necessary in applications where Reynolds number is not too high. A proper quantity can be searched to judge what scan speed gives satisfactory accuracy level. For instance, a possible candidate can be the spatial velocity correlation tensor, or the structure function, which is expected to converge as the scan speed increases. Thus, a scan speed can be considered adequate and economic (in terms of hardware requirement and computational expense), above which the increase of the accuracy is less significant while the cost-benefit ratio is high. In different flow problems, the optimal scan speed is expected to vary. Similar to the accuracy assessment given in section 2.3, a crude estimation of the scan speed could be made from the eddy turn over time: if the mean flow is of the only interest, a moderate scan speed should be enough; if small eddies are to be resolved, higher speed is required.

It has been pointed out that there exists an upper limit of the active scanning speed, it is not allowed to go to infinity. In other words, it is never possible to capture a truly frozen flow field. It is because if the flow is really frozen, there will be no particle displacement, thus the PIV fails to work. In order to allow a particle displacement to compute the velocity, a minimum time duration of PIV Δt has to be given at one measurement station along the scanning path. In this aspect, this PIV Δt, despite its shortness - in order of millisecond, limits the highest scan speed.

Nevertheless there is a potential solution for this limitation: a second stereoscopic PIV system can be introduced, aligning with the first system in the scanning path, and with a spacing l to the first one; see figure 22.[9] Thus the value of l/U_{scan} effectively gives the PIV Δt. Theoretically, this dual-system arrangement allows the choice of the U_{scan} value to a much higher value: at a fixed U_{scan}, PIV Δt can be adjusted by setting the spacing l between the two systems (two laser sheets) very carefully. The highest scan speed in this case is then limited by the required spatial resolution in the scan direction, which is much more relaxed than the limit from the particle displacements. The high speed cameras available nowadays can work at several *kHz* frame rates, which allows very high scan speed while maintaining good spatial resolution in the scanning direction. By choosing a very high value of U_{scan}, one approaches a truly frozen and instantaneous flow structure.

3.2 Comparison to tomographic PIV

The state-of-the-art tomographic PIV is also a candidate to provide truly instantaneous three-dimensional velocity information of a flow field. Its working principle is developed

[9] Vortex ring study is definitely not the only customer of this method. It can be applied to many general flow problems. Figure 22 shows an application on a turbulent jet flow.

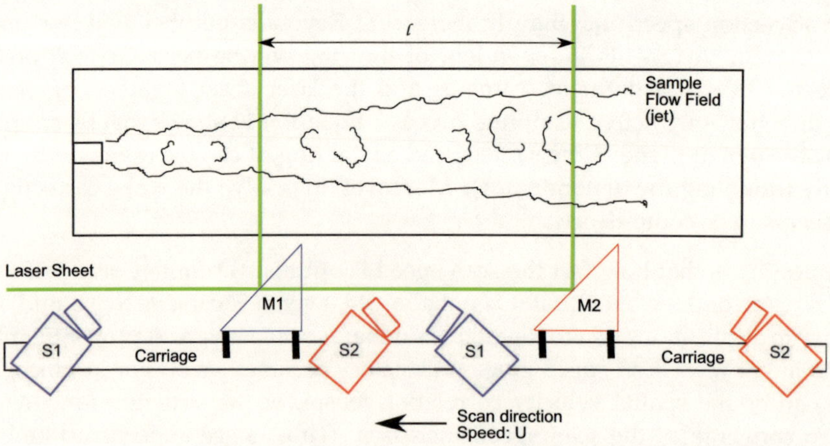

Fig. 22. A simple sketch of the proposed scan system including dual stereo recording PIV systems, denoted in the figure by S1 (M1) and S2 (M2). The two systems can be mounted on a rail-carriage system, so that they can move together, at a designed speed U_{scan}. M1 can be a beam splitter. l denotes the spacing of the two systems, U denotes the scan speed U_{scan}. This scan system can be applied to any flow problem. A sample flow field - turbulent jet flow - is shown.

from the medical tomographic applications. The illuminated particle volume is viewed by several cameras (the more cameras used, the more accurate the results will be) at the same time. Then the *most-likely* three-dimensional particle distribution in the volume is reconstructed by the tomographic algorithm, based on the intersections of lines of sight giving the estimated particle location; see Elsinga et al. (2006) for more details. Albeit its robustness of measuring the truly instantaneous three-dimensional velocity field, its major weakness lies in its relatively low signal-noise ratio compared to the two-dimensional and stereoscopic PIV. It is because tomographic reconstruction is an inverse problem, i.e. the three-dimensional information is reconstructed from its two-dimensional projections, regardless of how many cameras are used, the error or the noise (in particular ghost particles, which will be discussed later below) is inevitable.

In addition to this intrinsic problem, tomographic PIV would also encounter some difficulties when a large volume of flow is to be examined. To reach the same FOV as the current experiment, a minimum required FOV will be $100mm^3$. This means that first of all, this volume needs to be illuminated; the emission intensity of practical lasers would be very weak when it is diffused to such a large volume.

Second, it would be extremely difficult for the cameras to be focused on such a deep FOV (which means the aperture needs to be very small) while accepting enough light during very short laser emission time. Therefore, if tomographic PIV is to be used, one can only produce small scaled flows, but small scaled flow reduces the spatial resolution, probably to an undesired level. The spatial resolution is also limited by the interrogation volume (similar to interrogation window for planner PIV configuration) size. For the practical sparsity level of particle field, the commonly acceptable interrogation volume size is typically 48 $voxel^3$, while the planner PIV can easily reach 16 $pixel^2$. Thus the absolute spatial resolution limit of tomographic PIV is normally much lower than that of stereoscopic PIV with typical $voxel - pixel$ ratio of unit.

Moreover, although tomographic PIV can also provide temporal-resolved information, its working principle requires iteration and can only give *the most likely* (not true) particle distribution in the FOV (due to the intersection of lines of sight giving ghost particles, although the cross-correlation signals of which are weaker). This is a considerable error source of tomographic PIV, among others. In addition, compared with the active scanning method, with the same amount of information[10], to resolve temporal information, it must either be more computational expensive or less accurate.

4. References

Archer, P. J., Thomas, T. G. & Coleman, G. N. (2008). Direct numerical simulation of vortex ring evolution from the laminar to the early turbulent regime., *J. Fluid Mech.* 598: 201–226.

Bergdorf, M., Koumoutsakos, P. & Leonard, A. (2007). Direct numerical simulations of vortex rings at Re_Γ =7500., *J. Fluid Mech.* 581: 495–505.

Elsinga, G. E., Scarano, F., Wieneke, B. & van Oudheusden, B. W. (2006). Tomographic particle image velocimetry., *Exp. Fluids* 41: 933–947.

Gan, L. (2010). *PhD Dissertation: An experimental study of turbulent vortex rings using particle image velocimetry*, University of Cambridge.

Gan, L. & Nickels, T. B. (2010). An experimental study of turbulent vortex rings during their early development, *J. Fluid Mech.* 649: 467–496.

Gan, L., Nickels, T. B. & Dawson, J. R. (2011). An experimental study of turbulent vortex rings during: a three-dimensional representation, *Exp. Fluids* 51: 1493–1507.

Glezer, A. (1988). On the formation of vortex rings., *Phys. Fluids* 31: 3532–3542.

Glezer, A. & Coles, D. (1990). An experimental study of turbulent vortex ring., *J. Fluid Mech.* 211: 243–283.

Lavision (2007). *Product Manual-Davis 7.2 Software.*, Lavision GmbH, Gottingen.

Maxworthy, T. (1977). Some experimental studies of vortex rings., *J. Fluid Mech.* 81: 465–495.

Prasad, A. K. (2000). Stereoscopic particle image velocimetry., *Exp. Fluids* 29: 103–116.

Raffel, M., Willert, C., Wereley, S. & Kompenhans, J. (2007). *Particle Image Velocimetry-A Practical Guide, Second Edition*, Springer-Verlag, Berlin.

Saffman, P. G. (1978). The number of waves on unstable vortex rings., *J. Fluid Mech.* 84: 625–639.

Shariff, K., Verzicco, R. & Orlandi, P. (1994). A numerical study of three-dimensional vortex ring instabilities: viscous corrections and early nonlinear stage., *J. Fluid Mech.* 279: 351–375.

Taylor, G. I. (1938). The spectrum of turbulence., *Proc. R. Soc. London, Ser. A* 164: 476–490.

Townsend, A. A. (1976). *The structure of turbulent shear flow. Second Edition.*, Cambridge University Press, Cambridge.

Widnall, S. E. & Tsai, C. Y. (1977). The instability of the thin vortex ring of constant vorticity., *Philo Trans R Soc Lond A* 287: 273–305.

[10] If two scan systems are used, four cameras are needed, which is the same as tomo-PIV; while only two cameras are needed for one system scanning, which is only half of the information amount.

Wieneke, B. (2005). Stereo-piv using self-calibration on particle images., *Exp. Fluids* 39: 267–280.

Wieneke, B. (2008). Volume self-calibration for 3d particle image velocimetry., *Exp. Fluids* 45: 549–556.

2

Limits in Planar PIV Due to Individual Variations of Particle Image Intensities

Holger Nobach
Max Planck Institute for Dynamics and Self-Organization
Germany

1. Introduction

The basic algorithm of digital particle image velocimetry (PIV) processing (Keane & Adrian, 1992; Utami et al., 1991; Westerweel, 1993; Willert & Gharib, 1991) utilizes the cross-correlation of image sub-spaces (interrogation windows) for local displacement estimation from two consecutively acquired images of a tracer-particle-laden flow. A variety of image processing techniques using sub-pixel interpolations has been applied in the past to significantly improve both, the accuracy of the particle displacement measurement beyond the nominal resolution of the optical sensor and the spatial resolution beyond the nominal averaging size of image sub-spaces to be correlated. These include:

- sub-pixel interpolation of the correlation planes, e. g. the peak centroid (center-of-mass) method (Alexander & Ng, 1991; Morgan et al., 1989), the Gaussian interpolation (Willert & Gharib, 1991), a sinc interpolation (Lourenco & Krothapalli, 1995; Roesgen, 2003) or a polynomial interpolation (Chen & Katz, 2005), which reduce the "pixel locking" or "peak locking" effect (Christensen, 2004; Fincham & Spedding, 1997; Lourenco & Krothapalli, 1995; Prasad et al., 1992; Westerweel, 1998)
- windowing functions, vanishing at the interrogation window boundaries (Gui et al., 2000; Liao & Cowen, 2005), reducing the effect of particle image truncation at the edges of the interrogation windows to be correlated (Nogueira et al., 2001; Westerweel, 1997)
- direct correlation with a normalization, which so far has been realized in three ways: asymmetrically, with a small interrogation window from the first image correlated with a larger window in the second image (Fincham & Spedding, 1997; Huang et al., 1997; 1993a; Roháły et al., 2002), symmetrically, with two interrogation windows of the same size (Nobach et al., 2004; Nogueira et al., 1999) or bi-directional, combining an asymmetric direct correlation as above and a second direct correlation with a small interrogation window from the second image correlated with a larger window in the first image (Nogueira et al., 2001), originally introduced as a "symmetric" method, but nonetheless using image sub-spaces of different sizes
- iterative shift and deformation of the interrogation windows (Fincham & Delerce, 2000; Huang et al., 1993b; Lecordier, 1997; Scarano, 2002; Scarano & Riethmuller, 2000) with different image interpolation schemes as e. g. the widely used, bi-linear interpolation, or more advanced higher-order methods (Astarita, 2006; Astarita & Cardone, 2005; Chen & Katz, 2005; Fincham & Delerce, 2000; Lourenco & Krothapalli, 1995; Roesgen, 2003)

including the Whittaker interpolation (Scarano & Riethmuller, 2000; Whittaker, 1929), also known as sinc or cardinal interpolation and the bi-cubic splines, which have found wide acceptance

- image deformation techniques (Astarita, 2007; 2008; Jambunathan et al., 1995; Lecuona et al., 2002; Nogueira et al., 1999; Nogueira, Lecuona & Rodriguez, 2005; Nogueira, Lecuona, Rodriguez, Alfaro & Acosta, 2005; Scarano, 2004; Schrijer & Scarano, 2008; Tokumaru & Dimotakis, 1995), where the entire images are deformed accordingly to the assumed velocity field before the sub-division into interrogation windows to be correlated, also using different image interpolation techniques.

With iterative window shift and deformation or image deformation techniques, an accuracy of the order of 0.01 pixel or better has been reported (Astarita & Cardone, 2005; Lecordier, 1997; Nobach et al., 2005) based on synthetic test images. In contrast, the application to real images from experiments shows less optimistic results, where the limit usually observed is about 0.1 pixel. Only under special conditions, like in two-dimensional flows with carefully aligned light sheets, can better accuracy be achieved (Lecordier & Trinité, 2006).

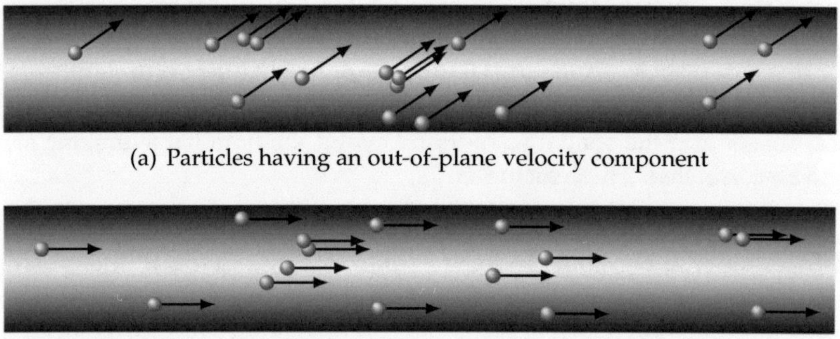

(a) Particles having an out-of-plane velocity component

(b) Two-dimensional flow aligned with the light sheet plane (only in-plane velocity components)

Fig. 1. Particles moving through a light sheet with an intensity profile

One reason for the different achievable accuracies in simulations and experiments may be the fact that in experiments, particles usually change their position within the light sheet (Fig. 1a). Therefore, the particles are illuminated differently in the two consecutive exposures. Additionally, the different illumination is individually different for each particle due to their different starting positions perpendicular to the light sheet plane. The result is an individual variation of particle intensities (further denoted as "intensity variations"), even in a homogeneous flow without any velocity gradient. Intensity variations can easily be seen in images from a variety of PIV applications, where some particles become brighter between the two exposures, whereas other particles, even if close by, become darker (Fig. 2). Simulations often assume that different particles can have different intensities, but not that the intensities can vary between subsequent exposures. This scenario can be realized in experiments only in two-dimensional flows with light sheets exactly aligned parallel to the flow field (Fig. 1b). Other sources of intensity variations could be an offset between the light sheets of the two illumination pulses or fluctuating scattering properties of the particles, e. g. non-spherical particles rotating in the flow.

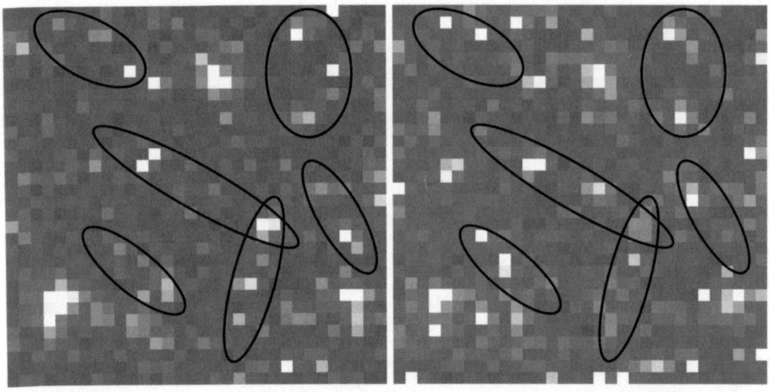

(a) First exposure (b) Second exposure

Fig. 2. Examples demonstrating individual particle intensity variations (marked regions, detail of public PIV images from the PIV challenge 2003, case A, axisymmetric turbulent jet in stagnant surrounding, images A001a and A001b)

Note that the effects of intensity variations are different from extern large scale illumination variations (Huang et al., 1997), the intensity variations only due to the different particle locations within the light sheet without relative changes between the exposures (Westerweel, 2000), or the loss-of-pairs and the degradation of the correlation peak due to out-of-plane motion (Keane & Adrian, 1990; 1992; Keane et al., 1995; Westerweel, 2000). While the loss-of-pairs and the degradation of the correlation peak increase the susceptibility to noise and the probability of outliers, the effect discussed here occurs additionally and directly affects the position of the correlation maximum and is a dominant limitation of the achievable accuracy in correlation-based image processing of planar PIV (Nobach, 2011; Nobach & Bodenschatz, 2009).

This study generally applies to Standard-PIV (two-dimensional, two-component, planar), independent of its application. This error principally applies also for Micro-PIV, where the particle images are large and may strongly overlap. However, the particle density and the intensity variations between consecutive images are small yielding a small effect of intensity variations. For Stereo-PIV, the errors are expeted to increase further compared to Standard-PIV due to the necessary coordinate transforms. Furthermore, the errors from the two perspectives are dependent due to the observation of identical particles within the same illumination sheet. In Tomo-PIV a reconstruction of three-dimensional particle locations preceds a three-dimentional correlation analysis. Since the overlap of particle images occurs in the projections only, the three-dimentional correlation should not be affected by this error. However, detailed studies about this error in Micro-PIV, Stereo-PIV and Tomo-PIV are still pending.

2. Effect of varying intensities

In PIV, the displacement of particle patterns between consecutive images is obtained from the peak position in the two-dimensional cross-correlation plane of the two images or image sub-spaces (interrogation windows). Assuming (i) a certain number of imaged particles in the interrogation window, each with different intensity, but with the same relative intensity in the two consecutive images and (ii) no truncation at the edges of the interrogation windows, the correlation peak is at the correct position, even if the particle images overlap and if the intensity of one entire image is scaled by a constant factor. Note the different meaning of "images", which are the entire images to be correlated, and "particle images", which are the spots at the particle positions. For demonstration, in Fig. 3a two images, each consisting of two well separated particle images (Airy discs), are correlated. The particles are at identical positions in the two images (no displacement between the images). The correct position of the correlation maximum at zero displacement can be seen clearly even for overlapping particle images and also with a constant scaling of one image (Fig. 3b).

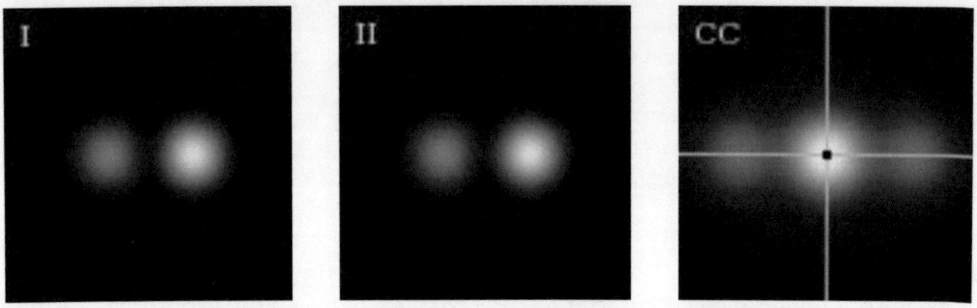

(a) Same intensity of the particle images in the two images with well separated particle images

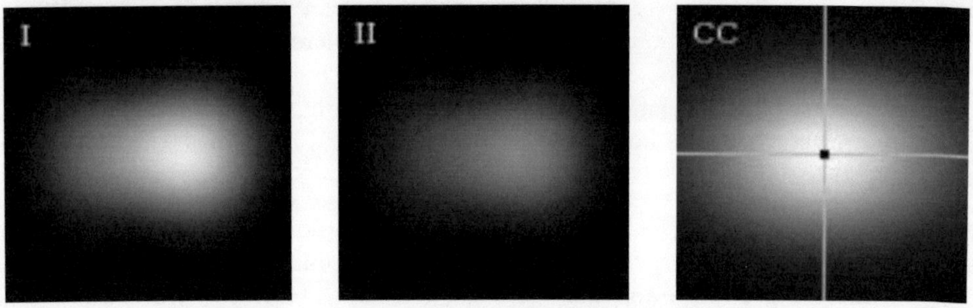

(b) One image intensity scaled and with overlapping particle images

Fig. 3. Intensity and cross-correlation function (CC with lines of zero displacement in x and in y direction respectively and with the correlation maximum marked with a black dot) of two images (I and II), each consisting of two particle images

This holds true also for the correlation of images with different relative amplitudes of the particle images, as long as the particle images do not overlap (Fig. 4a). With overlapping particle images and varying relative amplitudes (Fig. 4b), the maximum position of the correlation peak is shifted, yielding a biased displacement estimate, depending on the amplitudes of the particle images, widths, and overlap.

The consequence for PIV image processing is an additional error in displacement estimates, if the intensities of particle images vary between the consecutive PIV images, while the particle images overlap. This error is especially large for de-focussed particle images (where the particle images tend to overlap) and in the case of misaligned light sheets or flows with

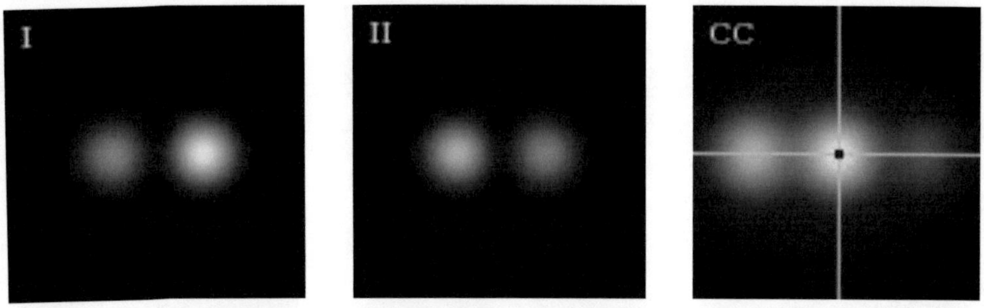

(a) Varying relative intensity of well separated particle images

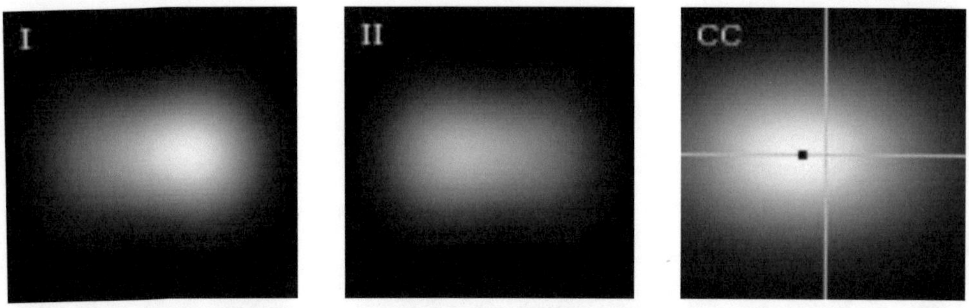

(b) Varying relative intensity of overlapping particle images yielding a correlation peak with a shifted maximum location

Fig. 4. Intensity and cross-correlation function (CC with lines of zero displacement in x and in y direction respectively and with the correlation maximum marked with a black dot) of two images (I and II), each consisting of two particle images

out-of-plane motion of the particles (where the illumination of individual particles changes between the two light pulses). This is almost independent of the particle number density as shown below.

While different intensities of particle images obviously occur if the particles move out-of-plane in e. g. a Gaussian illumination profile, this effect also occurs for a top-hat profile, if one of the two particle images is present in only one of the images (drop-off), as it occurs if one particle moves out of or enters the illumination plane. With a top-hat illumination profile, the amplitude of one of the particle images stays constant between the two exposures while the other particle image is absent in one of the two images. For well separated particle images (Fig. 5a) the correlation has its maximum at the correct position. As soon as the two particle images (in one of the two images) overlap, the correlation maximum is shifted (Fig. 5b).

3. Accuracy

To derive the dependence of the achievable accuracy on the intensity variations, computer simulated images have been used with varying parameters. The simulated particles are uniformly distributed within the light sheet and over the observation area. To consider the diffraction-limited imaging of small particles, the simulated particle images are represented by Airy functions (diameter given by the first zero value), integrated over the sensitive sensor

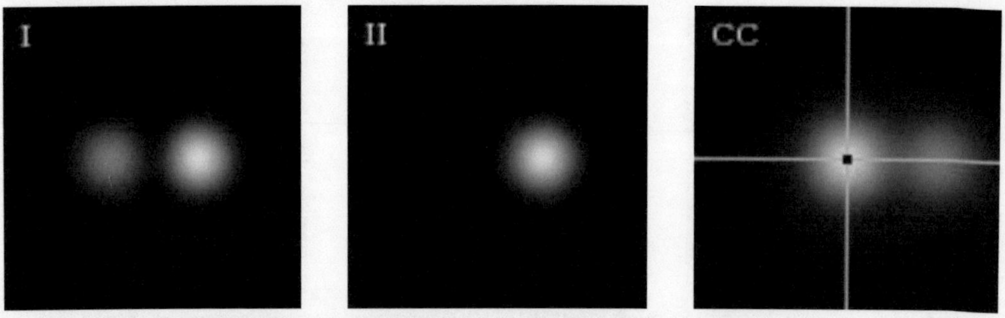

(a) Drop-off with well separated particle images

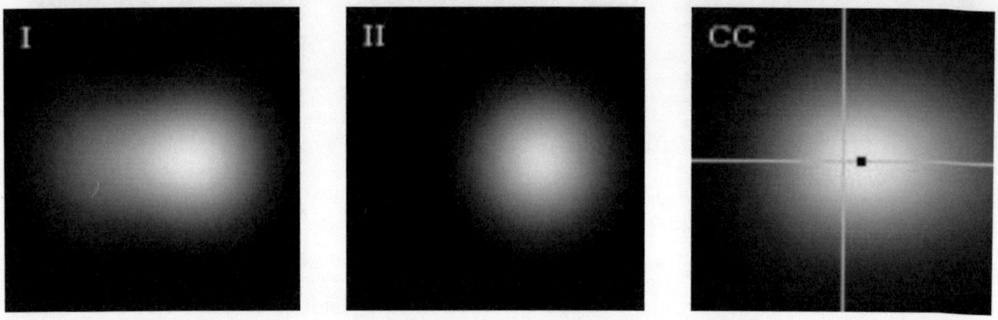

(b) Drop-off with overlapping particle images yielding a correlation peak with a shifted maximum location

Fig. 5. Intensity and cross-correlation function (CC with lines of zero displacement in x and in y direction respectively and with the correlation maximum marked with a black dot) of two images (I and II), one consisting of two particle images and one with only one particle image (particle image drop-off)

areas (pixels). The pixels are assumed to have a square shape with uniform sensitivity with a fill-factor of 1 (no gaps between the sensitive areas). All particle images get a random maximum intensity, equally distributed between zero and 1000 photo electrons (see comments about the noise below), corresponding to e. g. different sizes or reflectivity. The maximum intensity does not change between the exposures for only in-plane motion. With an out-of-plane motion, the particles change their position relative to the light sheet plane yielding different illumination of each individual particle in the two exposures. In this simulation, a top-hat profile of the light sheet illumination intensity is simulated, where the illumination changes only, if a particle enters or leaves the light sheet. The Airy functions of overlapping particle images are linearly superimposed. To investigate the error of the displacement estimation, a series of 1000 individual image pairs is generated for each of the following test cases. The displacement of the particles between the two exposures is randomly chosen between -1 and $+1$ pixel simulating a variety of sub-pixel displacements. Larger in-plane displacements can easily be eliminated by full-pixel shift of the interrogation windows (Scarano & Riethmuller, 1999; Westerweel, 1997; Westerweel et al., 1997). To isolate the effect of intensity variations from additional effects by e. g. velocity gradients, the simulated displacement is constant for all particles, imitating a homogeneous velocity field.

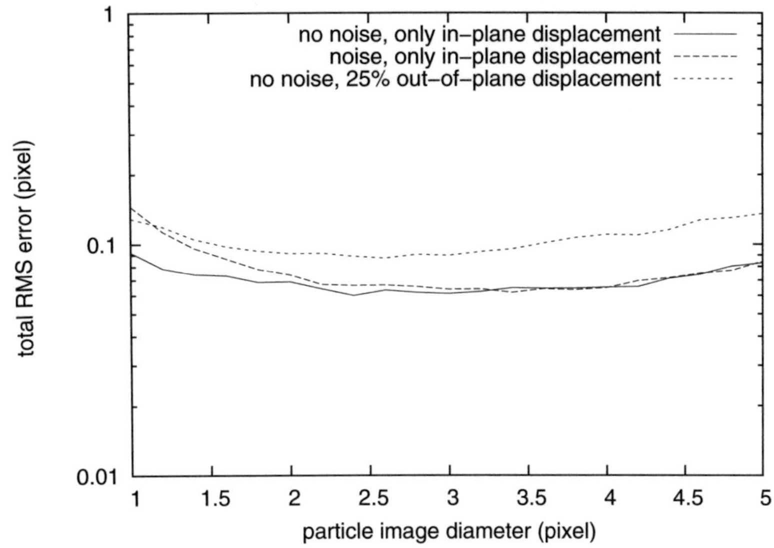

(a) A simple FFT estimation with full-pixel shift

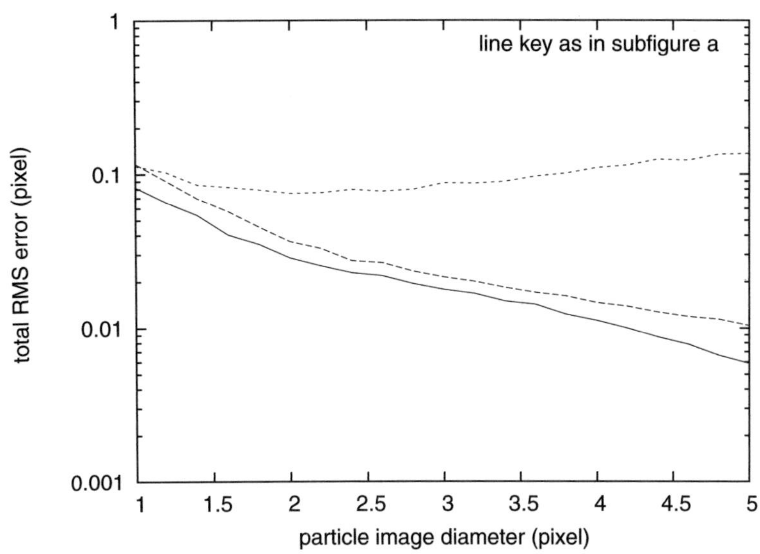

(b) Iterative sub-pixel interrogation window shift with bi-cubic splines image interpolation

Fig. 6. Total RMS error of the displacement estimate as a function of the particle image diameter (particle number density: 0.05 pixel^{-2}, interrogation window size: 16×16 pixels)

To demonstrate the dominating influence of the intensity variations on the accuracy of correlation-based PIV algorithms, in Fig. 6 the total RMS error over the particle image diameter is shown for three test cases: (i) for only in-plane motion (without noise), (ii) for only in-plane motion, but with strong photon noise (1000 photo electrons for the brightest particles giving about 32 electrons noise), read-out noise (RMS of 20 electrons) and quantization noise (10 electrons per count, yielding a mean gray value of 102 for the above mentioned 1000 photo electrons incl. read-out noise) and (iii) for an out-of-plane component of 25 % of the light sheet thickness. Fig. 6a shows the results for a simple displacement estimation utilizing the peak position of the cross-correlation of two interrogation windows with 16×16 pixels obtained by means of the fast Fourier transform (FFT). The sub-pixel location of the maximum is obtained by fitting a Gaussian function to the maximum of the correlation and its two direct neighbors in x and y direction separately. In Fig. 6b an iterative window shift method has been used alternatively. Starting with the displacement estimate obtained from the simple FFT-based method above, in the next and all following iteration steps, the two consecutive PIV images are re-sampled at positions shifted symmetrically by plus/minus half the pre-estimated displacement. For re-sampling the images at sub-pixel positions, bi-cubic splines are used for interpolation, widely accepted as one of the best methods so far (Raffel et al., 2007; Stanislas et al., 2008). The interpolation has been realized here with an 8×8 pixels kernel, requiring also the environment of the 16×16 pixels large interrogation window to be simulated. To keep the investigations simple and to isolate the influence of intensity variations, window deformation has not been implemented here to avoid other well known effects, such as limited spatial resolution or dynamic range issues, which may additionally influence the results. However, the conclusions are equally applicable to the case of velocity fields with gradients. In that case the other error sources sum.

The difference between the simulated displacement and that estimated by the above procedures gives individual estimation errors. >From the series of individual errors, an averaged RMS error is derived. In the interesting range of particle image diameters of 2 pixels and larger, for both algorithms, the influence of the out-of-plane displacements is significantly larger that the error due to the noise, making the intensity variations a dominating limitation of the achievable accuracy of planar PIV displacement estimation. The uncertainty of the estimated RMS values is about 21 % of the actual value. This value has been derived assuming independent estimates, yielding an estimation variance of the variance estimate of $2\sigma^4/N$ with N the number of estimates (1000 image pairs) and σ the true RMS value. The uncertainty of the shown graphs is then $\sqrt[4]{2/N}\,\sigma$.

The estimation accuracy can be improved in all three test cases by increasing the size of the interrogation windows, because the displacement errors average (Fig. 7a). The particle image diameter is set fix to 3 pixels. All other simulation and estimation parameters remain unchanged from the simulation above. The results are shown representatiovely for an iterative sub-pixel interrogation window shift with bi-cubic splines image interpolation only. For a constant particle number density, the RMS value decreases as the inverse of the linear dimension of the interrogation window. For large interrogation window sizes a transition towards a lower bound of the total RMS error is indicated. This lower bound is due to remaining interpolation errors in the correlation plane, which are independent of the size of the interrogation windows, and agrees with the findings in Fig. 6.

In contrast, varying the particle number density (Fig. 7b) has almost no effect on the RMS error in the case with an out-of-plane displacement. With only in-plane motion, the number of successfully correlated particle images increases linearly with the particle number density. For each particle, the correlation of the images has a small stochastic error, caused e. g. by image

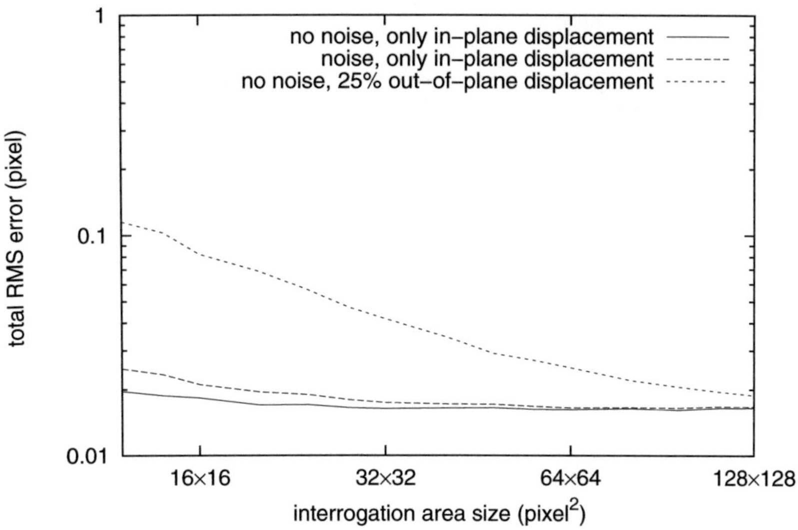

(a) Total RMS error as a function of the size of the interrogation windows

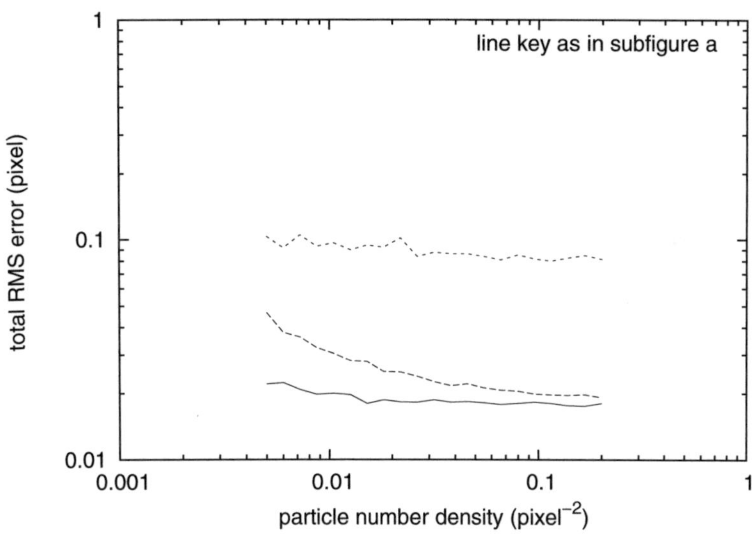

(b) Total RMS error as a function of the particle number density

Fig. 7. Total RMS error of the displacement estimates for an iterative sub-pixel interrogation window shift with bi-cubic splines image interpolation

noise, intensity interpolation over the pixel areas or by errors during image interpolation. The individual errors average over all particles in the interrogation window, yielding an RMS error decreasing with the square root of the particle number density. This complies with Westerweel (2000) (there for low particle densities). This error has a lower bound caused by interpolation errors in the correlation plane, which are independent of the particle number density. The image noise has been used to provoke large RMS errors in Fig. 7b, to make the range of RMS errors decreasing with the square root of the particle number density on top of the lower bound visible.

If there is a certain out-of-plane displacement, the previous errors are superimposed by the strong influence of intensity variations of overlapping particle images. In contrast to the number of successfully correlated particle images, the probability of overlapping particle images increases with the square of the particle number density. Each of these pairs of overlapping particle images contributes a stochastic error to the correlation. After averaging the individual errors of overlapping pairs of particle images over the number of particles in the interrogation window, these two contributions exactly compensate, and the observed error becomes independent of the particle number density.

A better view onto the influence of the out-of-plane displacement can be achieved by investigating the total RMS error as a function of the out-of-plane displacement (Fig. 8a). Here, a variety of commonly used algorithms has been simulated for comparison: a simple FFT-based estimation with full-pixel shift as above, the same algorithm but with a triangular window function applied to the interrogation windows, a symmetric direct correlation with normalization and iterative sub-pixel shift of the interrogation windows with either bi-linear, Whittaker or bi-cubic splines image interpolation. The different algorithms show the smallest errors for only in-plane motion, however, they have a large variation of achievable accuracy in this case. With increasing out-of-plane displacement, the total error increases approximately exponentially with decreasing difference (on the log scale) between the various algorithms. The large error of the method with the window function applied to the interrogation windows is originated in the smaller "effective" window size, which is for the triangular weighting function about half the nominal size of the window, amplifying the susceptibility to intensity variations. Also the iterative window shift with bi-linear interpolation shows large errors due to the pure quality of the bi-linear interpolation scheme.

For large displacements, also outliers occur. To separate the RMS error due to the limited accuracy and the dominating influence of outliers a simple outlier detection algorithm has been implemented. All displacement estimates outside a range of ± 1 pixel around the expected value are assumed to be outliers and are not taken into account for the calculation of the RMS error. From the number of outliers the probability of outliers is estimated. More reliable outlier detection algorithms based on statistical properties of the surrounding vector field as e. g. in Westerweel & Scarano (2005) could not be used in this simulation because only single displacement vectors are simulated. Starting at about 50 % out-of-plane displacement, the probability of outliers increases rapidly (Fig. 8b), limiting the useful range to a maximum out-of-plane displacement of about half the light sheet thickness for the given particle number density and interrogation window size. For the algorithm making use of the window function the onset of outliers is at smaller out-of-plane displacements due to the smaller effective size of the interrogation window. For the symmetric direct correlation with normalization the onset is shifted to larger displacements due to the better robustness of this procedure. The uncertainty of the estimated RMS values again is about 21 % of the actual value for small out-of-plane displacements ($\sqrt[4]{2/N_1}\,\sigma$, where N_1 is the number of validated estimates) and increases with larger out-of-plane displacements as N_1 the number of validated estimates decreases. The uncertainty of the outlier probability is $\sqrt{P(1-P)/N}$ with the true value P of the outlier probability and N the total number of estimates (1000 image pairs).

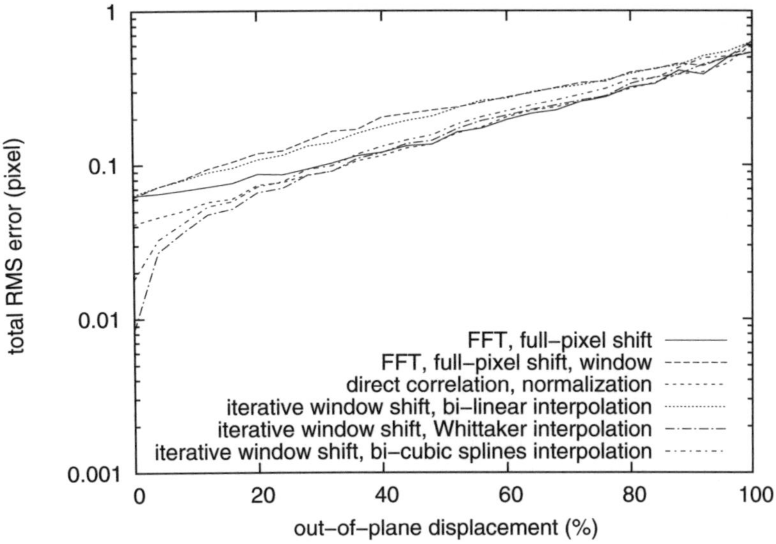

(a) Total RMS error

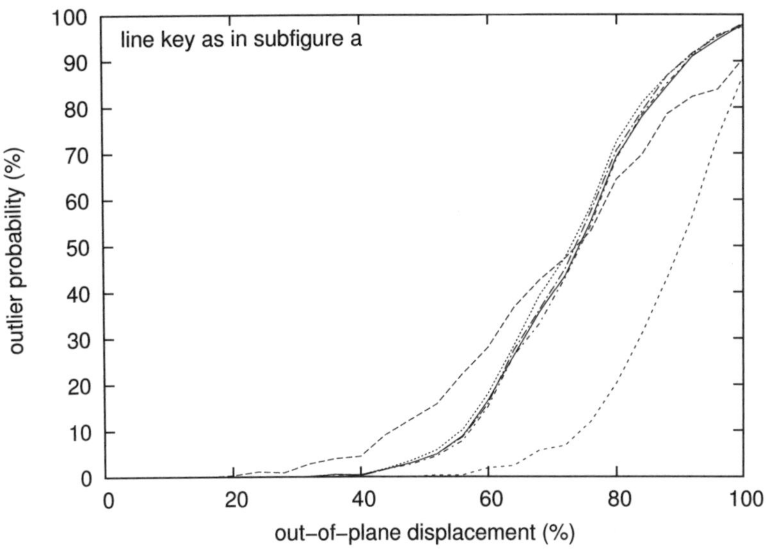

(b) Probability of outliers

Fig. 8. Properties of the displacement estimates as a function of the out-of-plane displacement (in percent of the light sheet thickness) for various PIV procedures (particle number density: 0.05 pixel^{-2}, interrogation window size: 16×16 pixels)

Experimental verification of the results given above requires a PIV setup with an adjustable beam shape (and width) and an adjustable out-of-plane component of the real velocity field. The first requirement can be realized with a video projector imaging different intensity profiles into the measurement volume using an additional collimation lens (Fig. 9). To achieve stable illumination, LCD technology is preferred. The projector with DLP technology used here realizes individual gray values by pulse width modulation, which causes illumination problems with PIV cameras at short exposure (integration) times. In the present study the exposure time has been set to 0.25 s, which corresponds to 30 illumination cycles of the DLP chip, since it works at a frame rate of 120 Hz. This long exposure time requires small velocities, which have been realized by moving a solid glass block on a 3D translation stage (Newport CMA12PP stepping motors and ESP300 controller). The glass block has a size of 5 cm × 5 cm × 8 cm and includes 54 000 randomly distributed dots in the inner 3 cm × 3 cm × 6 cm volume, corresponding to a particle density of 1 mm^{-3}.

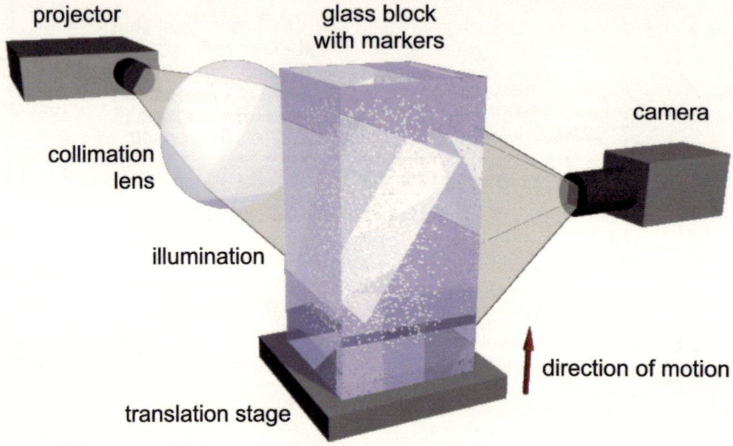

Fig. 9. Sketch of the experimental setup: A video projector is imaging different illumination profiles into the measurement volume, which is observed by a digital camera. A glass block with internal markers is translated vertically through the measurement volume.

Furthermore, an accurate synchronization of the in-plane and the out-of-plane translation through the light sheet is required. To avoid synchronization problems, the system has been inverted. The glass block moves along one axis of the translation stage, while the plane of illumination is tilted with respect to the axis of motion. During the translation of the glass block with a constant velocity of 0.1 mm/s through the observation area of the camera (Phantom V10), a series of 80 images of 480×480 pixels size has been taken at a frame rate of 0.8 Hz. By choosing the number of frames between the two images to be correlated, different out-of-plane components can be imitated. For details of the experiment see Nobach & Bodenschatz (2009). The images are available from the author. For a comparison to the previous simulation, the images taken with a 4mm wide top-hat illumination profile with a slope of 0.75 have been re-processed in this study.

Unfortunately, the precision of the translation stage and the motion of the glass block are not satisfactory. An *a priory* analysis discovered a frame-to-frame variation of the displacement. Additionally, a small perspective error has been found generating a velocity gradient in y direction. To compensate the displacement variations and the velocity gradient within the observation field, for each image pair, an *a priori* analysis with two large interrogation windows (352×192 pixels) with 50 % overlap in y direction has been taken as a reference

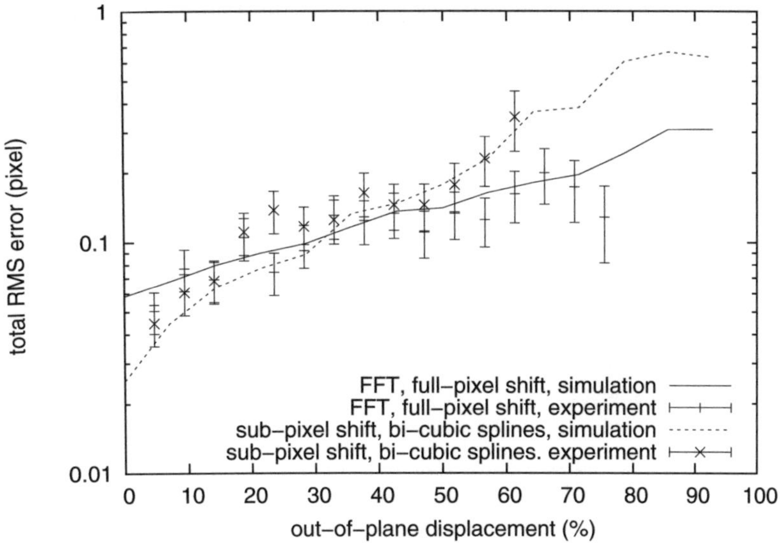

(a) Total RMS error

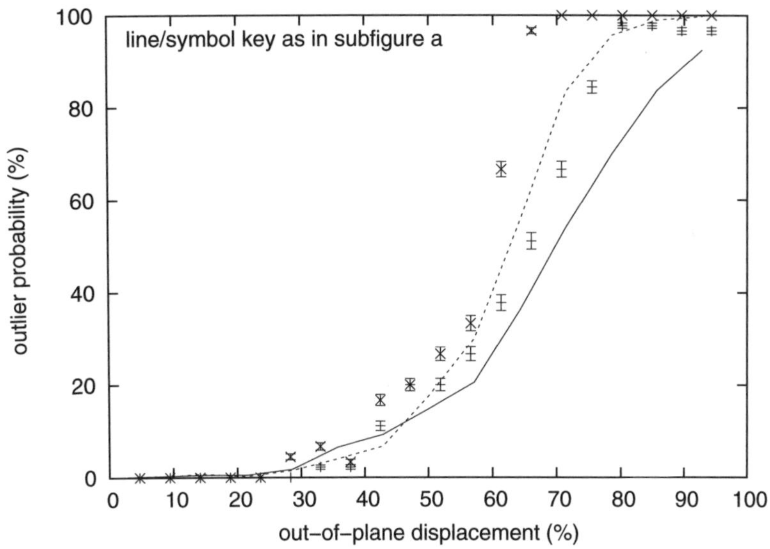

(b) Probability of outliers

Fig. 10. Comparison of the experiment and the simulation for an iterative sub-pixel interrogation window shift with bi-cubic splines image interpolation as a function of the out-of-plane displacement

to derive the mean displacement and the velocity gradients in y direction. The second PIV analysis is done with standard interrogation windows (32×32 pixels) in a 352×288 pixels large window, centered within the original observation area of 480×480 pixels. This area coincides with the area that is taken for the reference estimation. Based on the difference between the PIV analysis with standard interrogation windows and the reference estimation with large interrogation windows the RMS error is calculated. To suppress effects from the edges, the RMS analysis uses only valid vectors from a further reduced window (160×96 pixels), yielding 5×3 displacement estimates with interrogation windows of 32×32 pixels size. From these estimates the universal outlier detection (Westerweel & Scarano, 2005) can be used as a validation criterion with a threshold of 0.5 pixel plus 2 times the found median RMS derived from the neighboring vectors. For better statistics, all validated displacement vectors from all image pairs with the same number of frames between them, selected from the original series of 80 images, have been averaged.

For direct comparison of the experimental and simulated results in Fig. 10, the numerical simulation has been repeated with simulation parameters and processing and validation methods as for the experimental images (particle number density: 0.013 pixel^{-2}, interrogation window size: 32×32 pixels, iterative window shift and deformation, universal outlier detection). The uncertainty of the results of the simulation again are $\sqrt[4]{2/N_1}\,\sigma$ for the RMS values and $\sqrt{P(1-P)/N}$ for the outlier probability. The pendents for the measurements change with the distance between frames since the number of image pairs decreases with increasing distance between frames. Here error bars are given, showing the interval of plus/minus the RMS of expected uncertainty. Note that the expected uncertainty represents only random errors. Systematic errors or non-detected outliers are not included.

Except for a small shift of large probabilities of outliers towards smaller out-of-plane displacements, the results of simulated and experimentally obtained data agree, verifying both the effect of the intensity variations and the simulation procedure. Remaining deviations are possibly originated in cross-illumination of markers, interference and camera noise.

4. Resolution

To increase the spatial resolution of PIV processing beyond the size of the interrogation windows, overlapping the interrogation windows is an appropriate mean. Of course, this has limitations, since the image data of overlapping windows is not independent, however for moderate overlaps of about 50 % this works fine for all PIV algorithms, for PIV algorithms with windowing functions or iterative image deformation techniques the overlap can be even larger to obtain further increased spatial resolution. In the latter case, the deformation's degree of freedom is related to the grid of velocity estimates, independent of the interrogation window size. With a high overlap of neighboring interrogation windows the spatial resolution of iterative image deformation is governed by the grid spacing without loosing the robustness of the large interrogation windows. Therefore, this method is gained to improve the achievable spatial resolution of the PIV processing. Instabilities of this technique, occurring for high overlaps of interrogation windows due to negative responses in certain frequency ranges (Nogueira et al., 1999; Scarano, 2004) can be avoided either by applying appropriate spatial filters to the estimated velocity field or the application of appropriate windowing functions to the interrogation windows, which then have frequency responses with only non-negative values. Investigations of stability and spatial resolution of iterative image deformation applying either spatial filters or window functions can be found in Astarita (2007); Lecuona et al. (2002); Nogueira, Lecuona & Rodriguez (2005); Nogueira, Lecuona, Rodriguez, Alfaro & Acosta (2005); Scarano (2004); Schrijer & Scarano (2008).

To proof the gain of resolution by image deformation a series of 100 pairs of PIV images with 512×512 pixels each has been generated with a random in-plane displacement on a pixel-resolution (Gaussian distribution for each component and for each pixel with an RMS value of 0.5 pixel) and no out-of-plane motion. The particle images have random maximum intensities, equally distributed between zero and 1000 photo electrons, and Airy disk intensity profiles with 3 pixels diameter (defined by the first zero value of the Airy disk).

The images have been analyzed with an iterative window shift and first-order deformation technique (Scarano, 2002) with 32×32 and 16×16 pixels window size and an iterative image deformation with a triangular weighting applied to each PIV window of 32×32 pixels size. Except for the interrogation window size, the window function is identical to that in Nogueira et al. (1999), who apply the square of the triangular window to the product of the two PIV windows. To isolate the effect of decreasing the effective window size by weighting, the triangular weighting function has also been applied to the iterative window shift and deformation with a 32×32 pixels window. All methods use 10 iteration and a velocity estimation grid of 8×8 pixels corresponding to 75 % overlap for 32×32 pixels windows and 50 % overlap for the 16×16 pixels window.

>From the individual displacement estimates, which are interpolated with bi-cubic splines and re-sampled at all pixel positions, and the simulated displacement, which originally is given for all pixel positions, a two-dimensional coherent frequency response

$$C_{ij} = \frac{\left\langle U^*_{\mathrm{est},ij} U_{\mathrm{sim},ij} + V^*_{\mathrm{est},ij} V_{\mathrm{sim},ij} \right\rangle}{\left\langle U^*_{\mathrm{sim},ij} U_{\mathrm{sim},ij} + V^*_{\mathrm{sim},ij} V_{\mathrm{sim},ij} \right\rangle} \quad (1)$$

is calculated, where $U_{\mathrm{sim},ij}$ and $V_{\mathrm{sim},ij}$ are the two-dimensional Fourier transforms of the simulated u and the v displacement fields, $U_{\mathrm{est},ij}$ and $V_{\mathrm{est},ij}$ are the estimated counterparts, the * denotes the conjugate complex and $\langle\rangle$ denotes the ensemble average. The products and the coherent frequency function are calculated element-wise for the two-dimensional functions. From the two-dimensional coherent frequency response function a common (one-dimensional) one is derived by iteratively optimizing a one-dimensional function c_i so that the component-wise products $c_i c_j$ fit best the two-dimensional function C_{ij} with minimum L_2 norm.

Fig. 11a shows the frequency response function for only in-plane motion for the four investigated estimation procedures. With a rectangular weighting window, the frequency response clearly drops below zero at $1/32\,\mathrm{pixel}$ or $1/16\,\mathrm{pixel}$ corresponding to the interrogation window size of 32×32 or 16×16 pixels respectively. The triangular weighting window applied to a 32×32 pixels interrogation window leads to a frequency response function with only non-negative values, while the resolution increases beyond the nominal resolution of the interrogation window size, reaching almost an effective window size of half the nominal window size. The image deformation technique can further improve the spatial resolution, which then is limited by the velocity grid of 8×8 pixels. Fig. 12a shows the obtained bandwidth (−3 dB limit) as a function of the overlap of interrogation windows. Clearly, the image deformation technique gains most by increasing the density of the velocity estimation grid. Note, that the overlap of interrogation windows for a given grid of velocity estimates

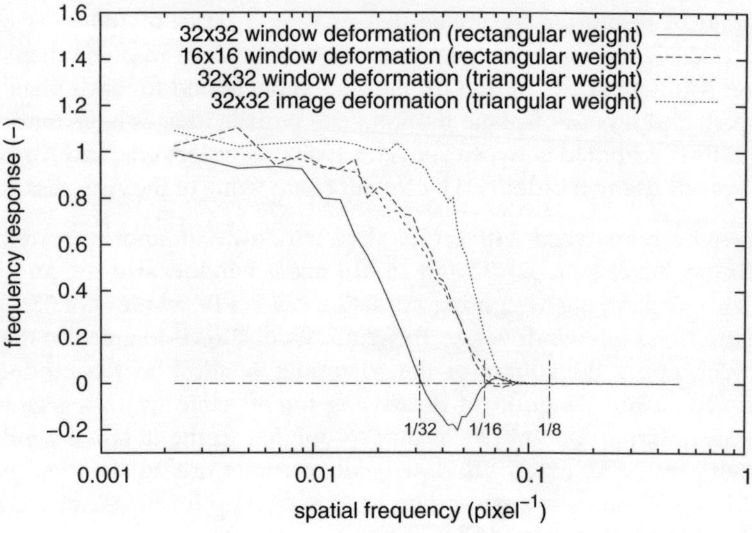

(a) Only in-plane motion

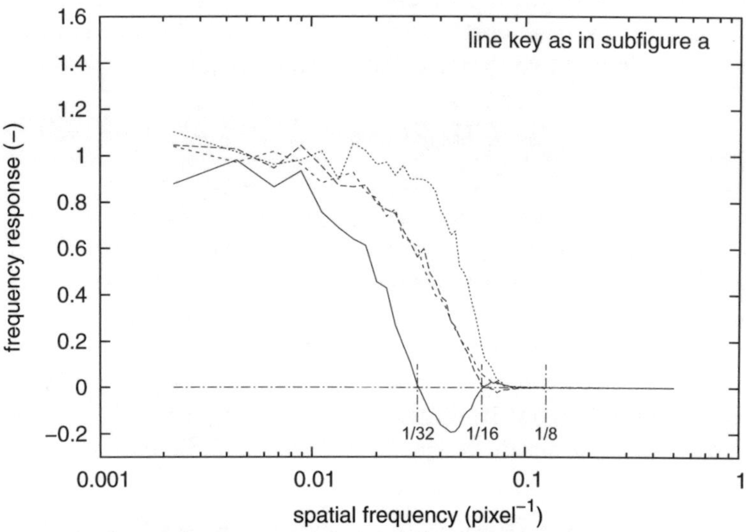

(b) Out-of-plane displacement of 25 % of the light sheet thickness

Fig. 11. Coherent frequency response for the different estimation procedures for a velocity estimation grid of 8×8 pixels corresponding to 75 % overlap for 32×32 pixels windows and 50 % overlap for the 16×16 pixels window (particle number density: 0.05 pixel^{-2})

changes with the size of the interrogation windows yielding a shifted overlap for the method with the 16×16 pixels interrogation window compared to the other methods.

Figs. 11b and 12b show the corresponding results for an out-of-plane displacement of 25 % of the light sheet thickness. There is no significant difference compared to Figs. 11a and 12a. Therefore, one can conclude that the intensity variations have no significant influence on the achievable resolution.

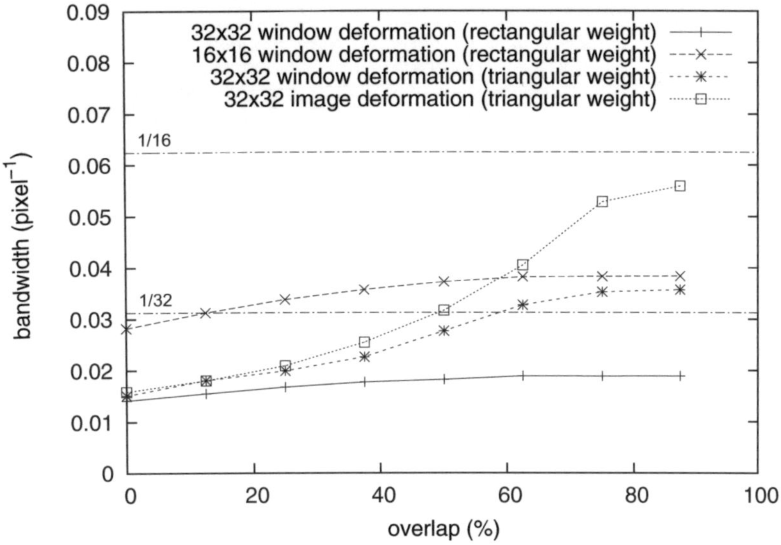

(a) Only in-plane motion

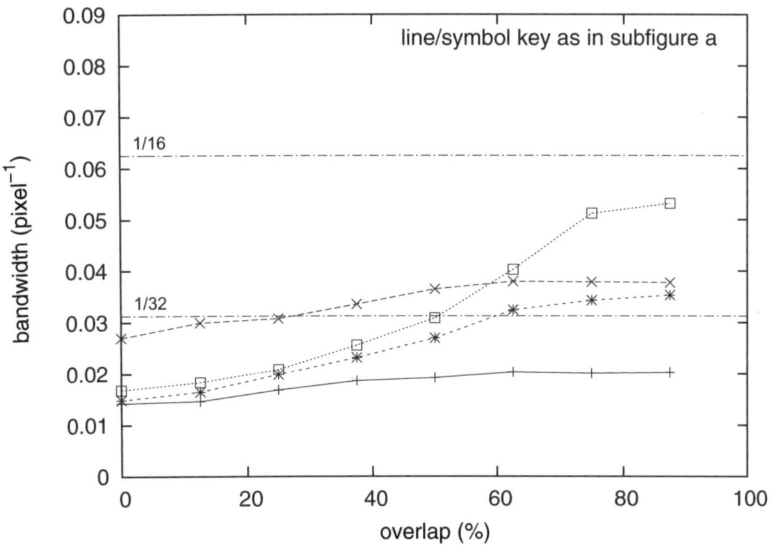

(b) Out-of-plane displacement of 25 % of the light sheet thickness

Fig. 12. Bandwidth as a function of the window overlap obtained from a series of simulated PIV images with a random in-plane displacement field (particle number density: 0.05 pixel^{-2})

5. RMS error versus resolution

However, taking into account the RMS errors, a significant influence of the intensity variations can be seen. Fig. 13a shows the obtained total RMS errors against the bandwidth with overlaps of interrogation windows varied between 0 and 87.5 %. The various methods cover different

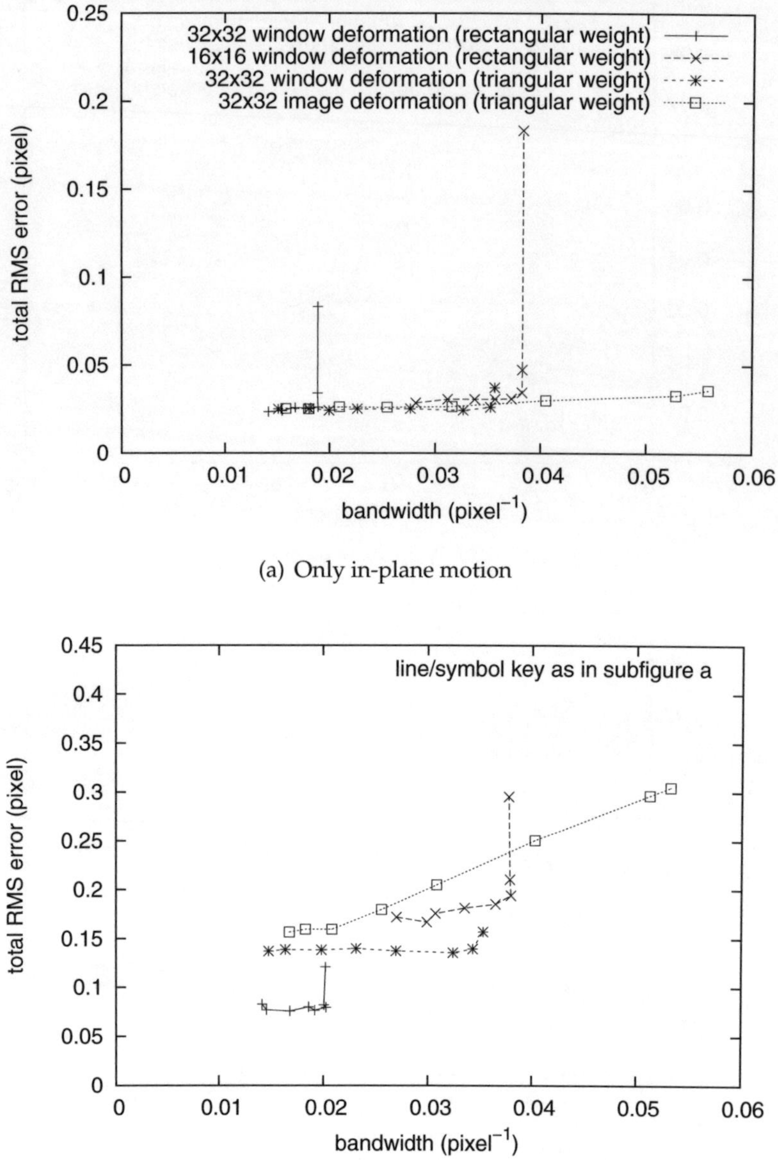

(a) Only in-plane motion

(b) Out-of-plane displacement of 25 % of the light sheet thickness

Fig. 13. Total RMS error against the bandwidth (particle number density: 0.05 pixel^{-2})

ranges of obtainable bandwidths and RMS errors yielding a lower bound of about 0.02 pixel, slightly increasing with the obtainable bandwidth.

For the window shift and deformation techniques with rectangular window function, the achievable bandwidth basically depends on the size of the interrogation windows. The bandwidth increases slightly with the overlap up to about 50 % overlap. For higher overlaps, the bandwidth stays constant and the RMS error rapidly increases. With the window shift and deformation technique with a triangular window function the bandwidth further increases up to about 75 % overlap, reaching almost the bandwidth of the window shift and deformation with a rectangular window function of half the size corresponding to an effective window size, which is half as large as the nominal size. Again, the RMS error increases rapidly for further increased overlaps. With the image deformation technique and strong overlap of interrogation windows the bandwidth can be increased further. The RMS error increases much less then with the other methods.

The picture changes completely in the presence of an out-of-plane component. Fig. 13b shows the results for an out-of-plane component of 25 % of the light sheet thickness. Again a rapid increase of the RMS error can be seen beyond 50 % overlap for iterative window shift and deformation techniques with rectangular window functions, respectively 75 % for a triangular window function. The image deformation reaches the highest bandwidth at strong overlaps. However, with the out-of-plane component the errors are much larger than with only in-plane motion and, additionally, the results for the various methods do not overlay any more. For the window shift and deformation techniques the achievable accuracy depends on the window function and the interrogation window size, as one has seen in Fig. 7. For the image deformation, the bandwidth continuously increases with the overlap, and the RMS error nearly linearly increases with the obtained bandwidth, but here for the prize of a larger RMS error in the entire range of overlaps and bandwidths compared to the other methods.

6. Conclusion

The effect of particle image intensities varying individually between the two consecutive images on the obtainable accuracy of a PIV system has been reviewed. Such intensity variations occur in experiments due to the motion of the particles in the intensity profile of the light sheet, misalignments of the two light pulses or changes of the particle's scattering properties between the two exposures. The error has been quantified for several commonly used PIV processing methods. This effect limits the obtainable accuracy of PIV measurements, even under otherwise ideal conditions and is much stronger than noise or in-plane loss of particle images. The commonly used best practice parameters for PIV experiments (particle image diameter around 3 pixels and out-of-plane components of not more than 25 % of the light sheet thickness) and the usually observed limit of about 0.1 pixel could be re-produced. This error is almost independent of the particle number density, but it strongly increases with increasing out-of-plane displacements, and decreases with increasing interrogation window size. In summary, besides under-sampling, the variations of the particle image intensities are an additional error, dominating the range of particle image diameters of larger than 2 pixels. This error leads to a basic limitation of the planar PIV technique and explains the accuracy limit of PIV of about 0.1 pixel usually seen in experiments. High-resolution image deformation techniques as in Nogueira et al. (1999) or Schrijer & Scarano (2008), with their small effective interrogation windows are especially affected in terms of the achievable accuracy, even if the achievable resolution does not change with intensity variations.

7. References

Alexander, B. F. & Ng, K. C. (1991). Elimination of systematic error in sub-pixel accuracy centroid estimation, *Opt. Eng.* 30: 1320–1331.

Astarita, T. (2006). Analysis of interpolation schemes for image deformation methods in PIV: effect of noise on the accuracy and spatial resolution, *Exp. in Fluids* 40: 977–987.

Astarita, T. (2007). Analysis of weighting window functions for image deformation, *Exp. in Fluids* 43: 859–872.

Astarita, T. (2008). Analysis of velocity interpolation schemes for image deformation methods in PIV, *Exp. in Fluids* 45: 257–266.

Astarita, T. & Cardone, G. (2005). Analysis of interpolation schemes for image deformation methods in PIV, *Exp. in Fluids* 38: 233–243.

Chen, J. & Katz, J. (2005). Elimination of peak-locking error in PIV analysis using the correlation mapping method, *Meas. Sci. Technol.* 16: 1605–1618.

Christensen, K. T. (2004). The influence of peak-locking errors on turbulence statistics computed from PIV ensembles, *Exp. in Fluids* 36: 484–497.

Fincham, A. & Delerce, G. (2000). Advanced optimization of correlation imaging velocimetry algorithms, *Exp. in Fluids* 29: S13–S22.

Fincham, A. M. & Spedding, G. R. (1997). Low cost, high resolution DPIV for measurement of turbulent fluid flow, *Exp. in Fluids* 23: 449–462.

Gui, L., Merzkirch, W. & Fei, R. (2000). A digital mask technique for reducing the bias error of the correlation-based PIV interrogation algorithm, *Exp. in Fluids* 29: 30–35.

Huang, H., Dabiri, D. & Gharib, M. (1997). On errors of digital particle image velocimetry, *Meas. Sci. Technol.* 8: 1427–1440.

Huang, H. T., Fiedler, H. E. & Wang, J. J. (1993a). Limitation and improvement of PIV; Part I: Limitation of conventional techniques due to deformation of particle image patterns, *Exp. in Fluids* 15: 168–174.

Huang, H. T., Fiedler, H. E. & Wang, J. J. (1993b). Limitation and improvement of PIV; Part II: Particle image distorsion, a novel technique, *Exp. in Fluids* 15: 263–273.

Jambunathan, K., Ju, X. Y., Dobbins, B. N. & Ashforth-Frost, S. (1995). An improved cross correlation technique for particle image velocimetry, *Meas. Sci. Technol.* 6: 507–514.

Keane, R. D. & Adrian, R. J. (1990). Optimization of particle image velocimeters. Part I: Double pulsed systems, *Meas. Sci. Technol.* 1: 1202–1215.

Keane, R. D. & Adrian, R. J. (1992). Theory of cross-correlation analysis of PIV images, *Applied Scientific Research* 49: 191–215.

Keane, R. D., Adrian, R. J. & Zhang, Y. (1995). Super-resolution particle imaging velocimetry, *Meas. Sci. Technol.* 6: 754–768.

Lecordier, B. (1997). *Etude de l'intéraction de la propagation d'une flamme prémélangée avec le champ aérodynamique, par association de la tomographie Laser et de la Vélocimétrie par Images de particules*, PhD thesis, l'Université de Rouen, France.

Lecordier, B. & Trinité, M. (2006). Accuracy assessment of image interpolation schemes for PIV from real images of particle, *Proc. 13th Int. Symp. on Appl. of Laser Techn. to Fluid Mechanics*, Lisbon, Portugal. paper 26.4.

Lecuona, A., Nogueira, J., Rodríguez, P. A. & Santana, D. (2002). Accuracy and time performance of different schemes of the local field correlation technique., *Exp. in Fluids* 33: 743–751.

Liao, Q. & Cowen, E. A. (2005). An efficient anti-aliasing spectral continuous window shifting technique for PIV, *Exp. in Fluids* 38: 197–208.

Lourenco, L. & Krothapalli, A. (1995). On the accuracy of velocity and vorticity measurements with PIV, *Exp. in Fluids* 18: 421–428.

Morgan, J. S., Slater, D. C., Timothy, J. G. & Jenkins, E. B. (1989). Centroid position measurements and subpixel sensitivity variations with the MAMA detector, *Applied Optics* 28(6): 1178–1192.

Nobach, H. (2011). Influence of individual variations of particle image intensities on high-resolution PIV, *Exp. in Fluids* 50: 919–927.

Nobach, H. & Bodenschatz, E. (2009). Limitations of accuracy in PIV due to individual variations of particle image intensities, *Exp. in Fluids* 47: 27–38.

Nobach, H., Damaschke, N. & Tropea, C. (2004). High-precision sub-pixel interpolation in PIV/PTV image processing, *Proc. 12th Int. Symp. on Appl. of Laser Techn. to Fluid Mechanics*, Lisbon, Portugal. paper 24.1.

Nobach, H., Damaschke, N. & Tropea, C. (2005). High-precision sub-pixel interpolation in particle image velocimetry image processing, *Experiments in Fluids* 39: 299–304.

Nogueira, J., Lecuona, A. & Rodríguez, P. A. (1999). Local field correction PIV: on the increase of accuracy of digital PIV systems, *Exp. in Fluids* 27: 107–116.

Nogueira, J., Lecuona, A. & Rodríguez, P. A. (2001). Identification of a new source of peak locking, analysis and its removal in conventional and super-resolution PIV techniques, *Exp. in Fluids* 30: 309–316.

Nogueira, J., Lecuona, A. & Rodriguez, P. A. (2005). Limits on the resolution of correlation PIV iterative methods. Fundamentals, *Exp. in Fluids* 39: 305–313.

Nogueira, J., Lecuona, A., Rodriguez, P. A., Alfaro, J. A. & Acosta, A. (2005). Limits on the resolution of correlation PIV iterative methods. Practical implementation and design of weighting functions, *Exp. in Fluids* 39: 314–321.

Prasad, A. K., Adrian, R. J., Landreth, C. C. & Offutt, P. W. (1992). Effect of resolution on the speed and accuracy of particle image velocimetry interrogation, *Exp. in Fluids* 13: 105–116.

Raffel, M., Willert, C., Wereley, S. & Kompenhans, J. (2007). *Particle Image Velocimetry — a practical guide*, Springer.

Roesgen, T. (2003). Optimal subpixel interpolation in particle image velocimetry, *Exp. in Fluids* 35: 252–256.

Roháły, J., Frigerio, F. & Hart, D. P. (2002). Reverse hierarchical PIV processing, *Meas. Sci. Technol.* 13: 984–996.

Scarano, F. (2002). Iterative image deformation methods in PIV, *Meas. Sci. Technol.* 13: R1–R19.

Scarano, F. (2004). On the stability of iterative PIV image interrogation methods, *Proc. 12th Int. Symp. on Appl. of Laser Techn. to Fluid Mechanics*, Lisbon, Portugal. paper 27.2.

Scarano, F. & Riethmuller, M. L. (1999). Iterative multigrid approach in PIV image processing with discrete window offset, *Exp. in Fluids* 26: 513–523.

Scarano, F. & Riethmuller, M. L. (2000). Advances in iterative multigrid PIV image processing, *Exp. in Fluids* 29: S51–S60.

Schrijer, F. F. J. & Scarano, F. (2008). Effect of predictor-corrector filtering on the stability and spatial resolution of iterative PIV interrogation, *Exp. in Fluids* 45: 927–941.

Stanislas, M., Okamoto, K., Kähler, C. J., Westerweel, J. & Scarano, F. (2008). Main results of the third international PIV challenge, *Exp. in Fluids* 45: 27–71.

Tokumaru, P. T. & Dimotakis, P. E. (1995). Image correlation velocimetry, *Exp. in Fluids* 19: 1–15.

Utami, T., Blackwelder, R. F. & Ueno, T. (1991). A cross-correlation technique for velocity field extraction from particulate visualization, *Exp. in Fluids* 10: 213–223.

Westerweel, J. (1993). *Digital Particle Image Velocimetry: Theory and Application*, Delft University Press, Delft, The Netherlands.

Westerweel, J. (1997). Fundamentals of digital particle image velocimetry, *Meas. Sci. Technol.* 8: 1379–1392.

Westerweel, J. (1998). Effect of sensor geometry on the performance of PIV interrogation, *Proc. 9th Int. Symp. on Appl. of Laser Techn. to Fluid Mechanics*, Lisbon, Portugal. paper 1.2.

Westerweel, J. (2000). Theoretical analysis of the measurement precision in particle image velocimetry, *Exp. in Fluids* 29: S3–S12.

Westerweel, J. & Scarano, F. (2005). Universal outlier detection for PIV data, *Exp. in Fluids* 39: 1096–1100.

Westerweel, J., Dabiri, D. & Gharib, M. (1997). The effect of discrete window offset on the accuracy of cross-correlation analysis of digital PIV recordings, *Exp. in Fluids* 23: 20–28.

Whittaker, J. M. (1929). The Fourier theory of the cardinal functions, *Proc. - R. Soc. Edinburgh Sect. A Math.* 1: 169–176.

Willert, C. E. & Gharib, M. (1991). Digital particle image velocimetry, *Exp. in Fluids* 10: 181–193.

3

PIV Measurements Applied to Hydraulic Machinery: Cavitating and Cavitation-Free Flows

Gabriel Dan Ciocan and Monica Sanda Iliescu
Université Laval, Laboratoire de Machines Hydrauliques
Canada

1. Introduction

Hydraulic machinery is an ideal field of application for Particle Image Velocimetry in terms of scientific interest, due to the complexity of the flow behaviour and to the need of detailed unsteady experimental data simultaneously recorded over large sections of the flow field. Within the same machine, a whole range of phenomena are encountered in the different components: wake patterns, separation, rotating vortex structures, vortex breakdown, etc. The unsteady interactions between the stationary and rotating frames, both upstream and downstream the runner, contribute to the efficiency loss. The generated quasi-periodic fluctuations overlay onto the average flow field, which may be symmetrical or not, issuing an unpredictable dynamic behaviour with respect to the operating regime. Whilst the intrinsic parameters of the local phenomena (e.g. sheared flow mixing length) vary, the flow topology may be modified drastically for conditions situated relatively close to one another in terms of head, flow rate and efficiency. The mapping of the unsteady velocity fields and corresponding turbulence levels is thus an essential tool in the analysis of these complex phenomena. The PIV technique opens large perspectives in the analysis of internal flows in hydraulic machinery, providing valuable insight towards an extensive understanding of the underlying physical mechanisms. Nevertheless, the use of PIV systems in this context is one of the most challenging applications, due to the structural constrains related to the optical access to the measurement areas, to the spatial and temporal scales of the phenomena that are to be investigated, two phase flow structure in cavitating regime and also due to the industrial aspects of the application.

In this book chapter it is proposed for presentation a development for hydraulic machines, based on our extensive experience and on the current bibliography in the field. The focus will be on two directions: rotor-stator interactions and two-phase flows in hydraulic turbines taking into account flow periodicity, turbulence and cavitation.

The research on hydraulic machinery is mainly performed on reduced scale models. The internal geometry needs to be respected because it is essential for the step-up of flow phenomena taking place on the model to actual prototype conditions. Furthermore, the model operation conditions must comply with the IEC 60193 Standard, which rules the methodology for model acceptance testing of hydraulic turbines, storage pumps and pump-turbines, and provides guidelines for the model-to-prototype transposition.

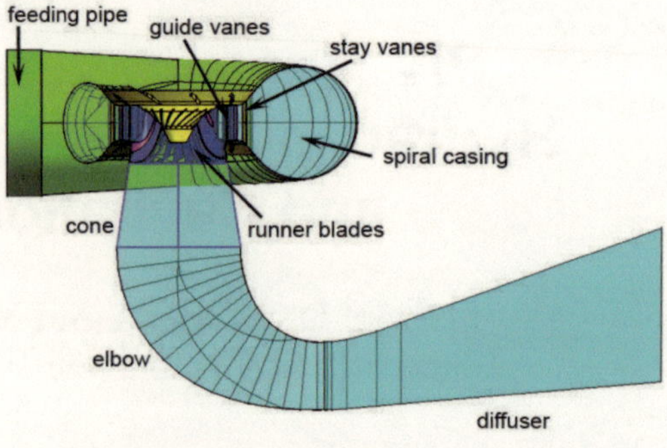

Fig. 1. Cross-section of a Francis turbine scale model

Fig. 2. Development of a cavitating helical vortex downstream the runner of a Francis turbine at partial load

An example of hydraulic geometry of a medium-head radial turbine is presented in Fig. 1. For this kind of turbine, the runner has constant-pitch fixed blades and the guide vanes are adjustable. The environment around the machine is also complex because it concerns the operating elements of the machine, supports of the test bench, model instrumentation to operate the machine. The working fluid is water; also thus optical interfaces are required for measurements based on imaging techniques. Their design must take into account the internal geometry of the model, while minimizing distortion.

Another major topic to be considered is the flow topology and diversity, with various unsteady phenomena taking place: wake propagation, vortex breakdown and machine-circuit resonance, rotor-stator interaction, runner flow behaviour. The 2D PIV and 3D PIV, in cavitation-free or two phase flow are the ideal tool to characterize these complex flows - see Fig. 2.

The proposed chapter covers all aspects of the development of a stereoscopic PIV experiment for the investigation of unsteady flows in hydraulic machinery, illustrated by the main applications of the authors' experience (see bibliography). The following topics are discussed in details:

- Interface design criteria and optimisation of the optical access taking into account local optical distortions and perspective effects;
- Calibration devices and practical procedures for accuracy assessment;
- Experiment setup for measurements in the static and rotating frames: synchronisation of the acquisition with the predominant physical phenomena; adjustment of the acquisition parameters with respect to the local flow conditions and to the global operating range of the machine;
- Requirements for image quality optimisation in 2D/3D-PIV experiments in single and two-phase flows: uniformization of the laser illumination in the measuring section, solutions to prevent localised reflections in the active viewing area, specificities related to measurements in two phase flows such as background illumination;
- Image processing in two-phase flows: filtering methodology and morphological operations to extract the relevant flow features;
- Data processing tools to determine the 3D velocity fields, with emphasis on particle detection on textured backgrounds and related masking techniques;
- Accuracy study and validation of measurement results, in terms of velocity fields and geometrical features extracted in two-phase flow conditions;
- Physical analysis of the flow (wake propagation, vortex core detection techniques, reconstruction of vapours volume, etc).

To conclude the chapter, best practice guidelines are provided for the setup of PIV experiments in hydraulic turbo-machinery, comparatively in single-phase flows and cavitation conditions.

All this works are done in the frame of PhD works of the authors: (Ciocan 1998) performed at Institute National Politechnique de Grenoble, France and (Iliescu 2007) performed at Ecole Polytechnique Federale de Lausanne, Switzerland. Other ulterior developments were performed (Tridon et al. 2008), (Tridon et al. 2010), (Gagnon et al. 2008), (Beaulieu et al. 2009) and (Houde et al. 2011) in collaboration or under the coordination of the authors.

2. General setup for hydraulic machines applications

2.1 General experimental set-up

A Dantec MT 3D-PIV system is used for measuring the three-dimensional unsteady velocity fields. The equipment presented herein corresponds to a classical medium-frequency PIV system. The system components belong to four categories of devices: illumination, particles introduced in the flow, image recording and data processing unit - see Fig. 3.

2.2 Illumination system

Two laser units with individual power supplies and cooling loops deliver a high-energy laser beam, which is transformed into a plane for locally illuminating the flow seeded with tracer particles.

The laser has an Yttrium Aluminum Garnet crystal, doped with triply ionized Neodymium, as lasing medium (Nd:YAG). While stimulated with a flash lamp, it emits photons of 1064 nm wavelength. The infrared output light is converted to the visible spectrum by frequency doubling with a birefringent crystal, up to the green radiation wavelength, 532 nm. The laser delivers pulses of 250 µs at a frequency of 8Hz. A Q-switch mechanism releases only 8 ns

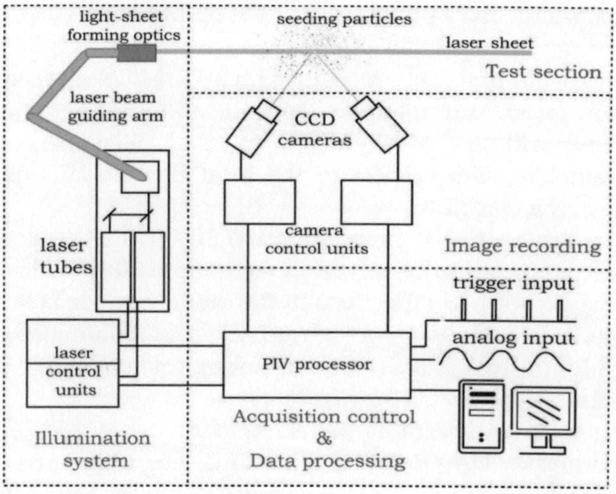

Fig. 3. Typical layout of a 3D-PIV system

out of the total pulse duration, thus producing a stroboscopic effect on the particles in the test area. For obtaining the desired delay between two pulses, two laser cavities with the same characteristics are used. Thus, the time interval between two successive impulses can be easily adjusted within 1 μs ÷ 100 ms range, depending on the local flow characteristics or the phenomenon, which is to be captured. The characteristics of the illumination system used for the current experiments are summarized in Table 1.

Type	NewWave Gemini	Peak energy	60 mJ	Pulse duration	8 ns
Lasing medium	Nd:YAG	Power	25 MW	Pulse repetition rate	8 to 20 Hz
Output wavelength	532nm	Beam diameter	6mm	Pulse separation rate	300 ns to 300 ms

Table 1. Laser unit specifications

For facilitating the optical access to the test section, the laser beam is transmitted through an articulated light guide with high transfer ratio and resistant to the high laser energy employed in water experiments. The characteristics of the beam guide are given in Table 2.

Type	5 flexible joints mirrors
Optical transmission	>90%
Maximum input pulse energy	500 mJ for 10 ns pulse and 6-12 mm beam
Maximum input beam diameter	12 mm

Table 2. Beam-guiding arm specifications

A beam expander mounted at the end of the arm transforms the input laser beam into a light sheet. A series of lenses allow adjusting the thickness and divergence angle of the laser sheet. The characteristics of the optics assembly are given in Table 3.

Narrow angle lens opening	20°
Wide angle lens opening	40°
Thickness adjuster compression factor for 2D PIV	0.67
Thickness adjuster expansion factor for 3D PIV	1.5
Focus lens	focus adjuster module

Table 3. Laser sheet optics assembly specifications

2.3 Image recording

The enlighten field is visualized by two double-frame progressive scan interline CCD cameras, with an active matrix of 1280x1024 with 8 bit depth, see Table 4. The active area of light-sensitive cells is doubled by a second array of storage cells, for increasing the data transfer rate. The acquisition frequency in double-frame mode is 4.5 Hz. The CCD chip is cooled, which gives a higher sensitivity and signal-to-noise ratio enhancement in low lighting conditions. Two Nikon objectives with focal length of 24mm and 60mm are used depending on the geometrical characteristics of the camera setup. measurement area dimensions and optical path.

CCD chip	cooling system, anti-blooming protection, pixels binning	Active area	1280 x 1024	Resolution	8-12 bit
		Pixel width and pitch	3.4µm x 6.7µm	Frame rate	4.5 Hz in double-frame mode

Table 4. CCD camera specifications

2.4 Acquisition control and data processing

The control and synchronization of the laser, cameras and external trigger input, as well as the raw vector field processing, are realized with a specific processor, Dantec MT's FlowMap2200. The main advantage is the integrated correlator unit, which performs real-time raw vector maps processing by a cross-correlation technique applied on the double-frame images. In this way a qualitative vector field validation can be rapidly performed during acquisition.

Dantec MT's FlowManager software version 4.5, along with 3D-PIV software module, have been used for data acquisition, validation and processing. Data post-processing and visualization tools are realized in Matlab.

2.5 Seeding

Spherical particles in borosilicate glass, are used as flow tracers; a silver coating improves their scattering characteristic. The relative density of 1.4 against the water one and the average size of 10µm allow these particles to accurately follow the flow. Their refractive index is 1.52. The melting point is high, 740°C, which makes them suitable for a broad range of applications.

2.6 Calibration

For the correct evaluation of the 3D displacement of the particles, a mapping of the measurement volume onto the two cameras' images and the definition of the overlapping zone of the two fields of view are necessary.

The camera calibration consists in defining the coefficients of a mathematical model that relates the real spatial locations in the measurement plane to the corresponding positions in the recording plane. This model includes the geometrical and optical characteristics of the cameras set-up, taking into account the optical distortions due to perspective imaging, lens aberrations, interposed media with different refractive indices. The images of a plane target with equally spaced markers, moved in five transversal positions (corresponding to the laser sheet thickness) are stored to have volume information. The corresponding positions of the points in all the image plane allows to determine the optical transfer function by a least squares fitting algorithm.

2.6.1 Optical distortion correction

For the optical access of the cameras and of the laser, the test model is equipped with polymethyl methacrylate (PMMA) windows, with a refractive index of 1.44. The inner face of the windows follows the hydraulic profile of the test model, while their external face is flat, for minimizing the optical distortions. The optical interfaces used for the PIV measurements in the cone will be detailed in the next paragraphs.

The Scheimpflung correction is applied for perspective imaging rectification. It consists in rotating the CCD plane relatively to the lens plane for reducing the perspective effect by balancing the optical path difference between points near and far away from the camera axis. The focus plane, lens plane and CCD sensor plane are made coincident using a mechanism which allows individual rotation of these components.

The angle at which the CCD chip plane needs to be tilted about the lens plane can be calculated with the camera's focal length and its geometrical position, i.e. the camera tilt angle and the distance from the lens to the measurement plane. In our case, this value can only be used as a rough approximation, because the optical path is distorted while passing through air, PMMA and water. The final adjustment is then realized by compensating the blur on the lateral edges of the image and bringing the entire view into focus.

2.6.2 Optical transfer functions

The optical path from the measurement zone to the camera crosses 3 media with different refractive indices: water 1.33. PMMA 1.44 and air 1. In our particular case, the optical windows are not parallel to the cameras' plane. To correct the optical distortion between the measurement zone and the corresponding image, two analytical functions have been tested:

- direct linear transform :

$$\begin{bmatrix} k_x \\ k_y \\ k_0 \end{bmatrix} = \begin{bmatrix} A_{11} & A_{12} & A_{13} & A_{14} \\ A_{21} & A_{22} & A_{23} & A_{24} \\ A_{31} & A_{32} & A_{33} & A_{34} \end{bmatrix} \cdot \begin{bmatrix} X \\ Y \\ Z \\ 1 \end{bmatrix} \tag{1}$$

- 3rd order polynomial for the XY directions in the plane of the target and parabolic in the out-of-plane direction Z:

$$\begin{bmatrix} x \\ y \end{bmatrix} = \overrightarrow{A_{000}} + \left(\overrightarrow{A_{100}} \cdot X + \overrightarrow{A_{010}} \cdot Y + \overrightarrow{A_{001}} \cdot Z\right) + \left(\overrightarrow{A_{110}} \cdot XY + \overrightarrow{A_{101}} \cdot XZ + \overrightarrow{A_{011}} \cdot YZ\right) +$$
$$+\left(\overrightarrow{A_{200}} \cdot X^2 + \overrightarrow{A_{020}} \cdot Y^2 + \overrightarrow{A_{002}} \cdot Z^2\right) + \left(\overrightarrow{A_{300}} \cdot X^3 + \overrightarrow{A_{210}} \cdot X^2Y + \overrightarrow{A_{201}} \cdot X^2Z\right) + \quad (2)$$
$$+\left(\overrightarrow{A_{030}} \cdot Y^3 + \overrightarrow{A_{120}} \cdot XY^2 + \overrightarrow{A_{021}} \cdot Y^2Z\right) + \left(\overrightarrow{A_{102}} \cdot XZ^2 + \overrightarrow{A_{012}} \cdot YZ^2 + \overrightarrow{A_{111}} \cdot XYZ\right)$$

The accuracy of the model in transposing the space coordinates in the image plane and the accuracy in detecting the 3rd velocity component have been analysed, taking into account the effect of the geometry and refractive index of the optical interface. In addition, a comparison between the two types of calibration has been made. The 3rd order polynomial model in XY, parabolic in Z has been chosen. The relative errors between the three methods are in the range 2 to 10 %, with higher errors near the measurement zone boundary.

The calibration process includes the following steps:

- The target is placed in the test section and its spatial position is checked to fit the accuracy limits.
- The test section is filled with water, the cameras are positioned in stereoscopic arrangement The position and angle of the cameras, relatively to the target surface are adjusted and the lenses are chosen such that the field of view recovers the most part of the target surface and their overlapping area is maximized. The perspective correction is realized by rotating the CCD plane relatively to the lens plane until all the camera's field of view is in focus.
- Images of the target are acquired for different transversal positions of the target. The target surface is evenly illuminated with high-power projectors. The image accuracy is checked by applying an image-processing algorithm for markers detection. The coefficients of the transfer function, for mapping the real-world coordinates onto the camera image, are calculated with the position information.
- The laser sheet is aligned with the target surface and its opening angle and thickness are adjusted taking into account the uniform energy distribution over the entire measurement zone.
- The calibration target is removed from the test section, the hydraulic circuit is refilled, and seeding particles are introduced in the flow.

2.6.3 Calibration target definition

For the calibration, two targets were tested – see Table 5:

- a two-dimensional target with black dots on a white plane support, which is displaced in five vertical positions at 1 mm +/-0.01 distance;
- a volumetric target with white dots on dark background and placed on two separate levels at an offset of 5 mm.

An error analysis is made for a same measurement position at the outlet of the draft tube, using the two calibration targets successively. For the same operating conditions, the comparison between mean 3D velocity fields obtained with the two calibrations is shown in Fig. 4 for the linear transform and for the polynomial transfer functions.

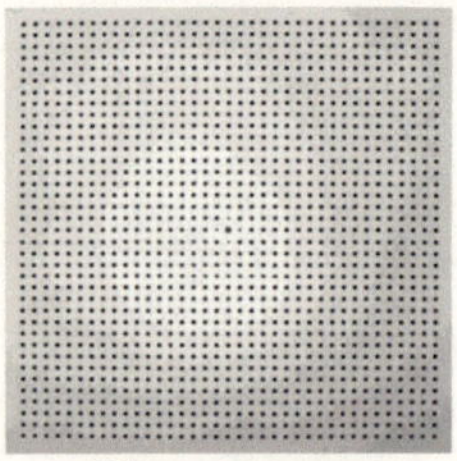

Plane target

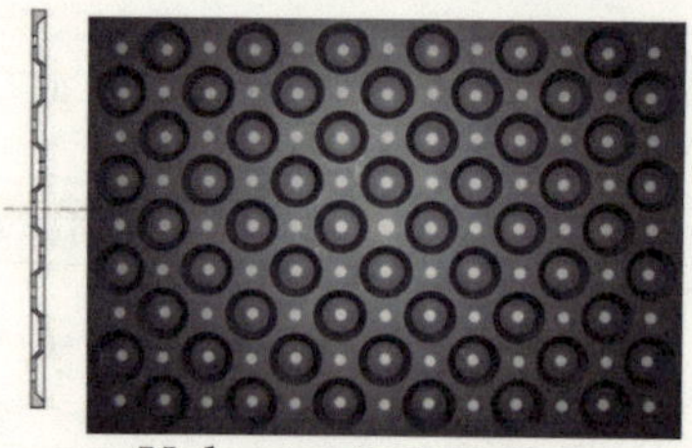

Volumetric target

active area	200x200 mm	270x190 mm
dot spacing	5 mm	20 mm
dot diameter	2 mm	5 mm
reference marker diameter	2.7 mm	7 mm
axis marker diameter	1.3 mm	5 mm
Level spacing	-	4 mm

Table 5. 2D and volumetric calibration targets dimensions

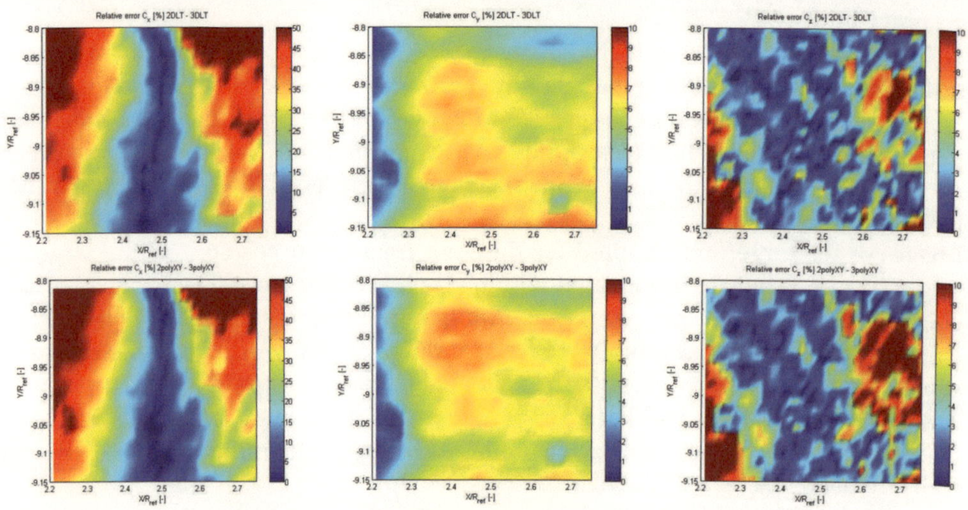

Fig. 4. Relative error for calibration with the 2D and the 3D targets, in the case of linear (top row) and polynomial (bottom row) transfer functions

$$\varepsilon = \frac{C_{2\text{Dtarget}} - C_{3\text{Dtarget}}}{C_{2\text{Dtarget}}} \cdot 100\,[\%] \tag{3}$$

The high error values obtained for the comparison between the measurements obtained with the two targets shows that the 3D target cannot be used in our configuration. The reasons, which explain this difference of accuracy, are the following:

- for the volumetric target, the dots positions cover only 2 traverse planes on the vertical direction (Z), thus only a linear correction can be performed along Z, and the larger spacing of the dots results in a lack of information for the in-plane positions;
- the position of the cameras is not symmetrically placed relatively to the target plane, by accessibility reasons;

- the optical access window is not parallel to the image plane;
- the tangential velocity component (perpendicular to the main flow direction) is small compared with the axial flow velocity. only 10%, and thus the error is amplified.

Taking into account these considerations, the final choice was to use the 2D target, with a spatial resolution of 5mm for the in-plane positions and 1mm in the out-of-plane direction.

2.6.4 Target positioning system

A slight displacement of the target during calibration has been noticed. In this context, a theoretical and an experimental study have been performed on the stability of the target in the fluid medium, both in rotation and in translation, and on its influence on the measurement accuracy.

Target rotation

Denoting C the velocity in the reference frame OXYZ and C' the velocity in a frame rotated by an angle δ around one of the axes, the relative error for each component is:

$$\varepsilon = \left|\frac{C - C'}{C}\right| \cdot 100\ [\%] \qquad \varepsilon = \left|1 - \frac{C'}{C}\right| \cdot 100\ [\%] \tag{4}$$

- rotation around Ox:

$$\begin{bmatrix} C_y' \\ C_z' \end{bmatrix} = \begin{bmatrix} C_y & -C_z \\ C_z & C_y \end{bmatrix} \cdot \begin{bmatrix} \cos\delta \\ \sin\delta \end{bmatrix}; \quad \begin{bmatrix} \varepsilon_y \\ \varepsilon_z \end{bmatrix} = \begin{bmatrix} 1 & -C_z/C_y \\ 1 & C_y/C_z \end{bmatrix} \cdot \begin{bmatrix} 1-\cos\delta \\ \sin\delta \end{bmatrix} \tag{5}$$

- rotation around Oy:

$$\begin{bmatrix} C_x' \\ C_z' \end{bmatrix} = \begin{bmatrix} C_x & -C_z \\ C_z & C_x \end{bmatrix} \cdot \begin{bmatrix} \cos\delta \\ \sin\delta \end{bmatrix}; \quad \begin{bmatrix} \varepsilon_x \\ \varepsilon_z \end{bmatrix} = \begin{bmatrix} 1 & -C_z/C_x \\ 1 & C_x/C_z \end{bmatrix} \cdot \begin{bmatrix} 1-\cos\delta \\ \sin\delta \end{bmatrix} \tag{6}$$

- rotation around Oz:

$$\begin{bmatrix} C_x' \\ C_y' \end{bmatrix} = \begin{bmatrix} C_x & -C_y \\ C_y & C_x \end{bmatrix} \cdot \begin{bmatrix} \cos\delta \\ \sin\delta \end{bmatrix}; \quad \begin{bmatrix} \varepsilon_x \\ \varepsilon_y \end{bmatrix} = \begin{bmatrix} 1 & -C_y/C_x \\ 1 & C_x/C_y \end{bmatrix} \cdot \begin{bmatrix} 1-\cos\delta \\ \sin\delta \end{bmatrix} \tag{7}$$

Thus the error depends only on the rotation angle and velocity components ratio. These ratios vary in the range [0 1] for Cz/Cy, [1 20] for Cz/Cx, and [1 30] for the in-plane components Cy/Cx. The relative errors for rotation around each axis are given in Fig. 5, for angles ranging between 0 and 2°. The black line denotes 3% error limit.

For rotation around the OX axis:

- Cy is sensitive if Cz<Cy, but the admissible angle drops under 1° if Cz>Cy;
- Cz is very sensitive to rotation if Cz<<Cy, but for Cz~=Cy the tolerance approaches 1.7°;
- Cz is the most sensitive component, which gives a tolerance of 0.1-0.2°.

For rotation around the OY axis:

- Cx is the less sensitive if Cz<Cx (tolerance >2°), but the limit decreases under 0.2° if Cz>>Cx, slowly until Cz=3.5Cx (1.7° to 0.5°) and steeply as the ratio Cz/Cx increases;
- Cz is very sensitive to rotation if Cz<Cx (limit 0.1°), but for Cz>>Cx the tolerance exceeds 1.7°;
- Cx is the most sensitive component, being the smallest, which gives a tolerance of 0.2°.
- For rotation around the OZ axis:
- Cx is more sensitive as the ratio Cy/Cx increases;
- Cy is almost insensitive if Cy>Cx;
- Cx is the most sensitive component, tolerance 0.1°.

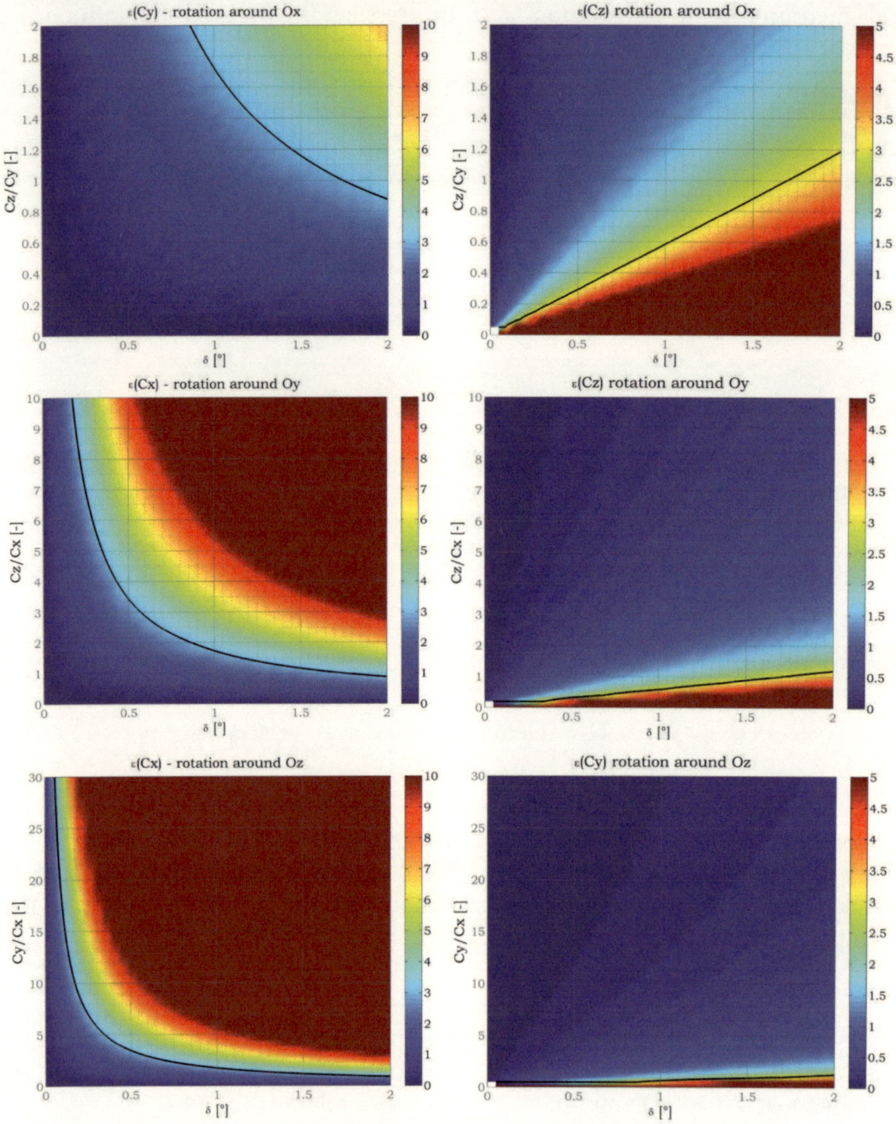

Fig. 5. Relative error for an accidental target rotation during calibration

The transversal and out-of-plane velocity components (Cx and Cz) are the most sensitive to accidental rotation of the target during calibration, imposing a limit of 0.1° in rotation about the three axes.

Target transversal displacement

The necessary resolution of the target displacement in the out-of-plane direction has been evaluated by accuracy analysis for calibrations with 1 mm and 0.5 mm spacing between successive positions of the target along the Z axis.

$$\varepsilon = \frac{C_{1\text{mm}} - C_{0.5\text{mm}}}{C_{1\text{mm}}} \cdot 100[\%] \tag{8}$$

The error distributions show that the uncertainty does not exceed 3% for the velocity components in the cross-section plane, Cx and Cz, and is smaller than 0.5 % for the component in the main flow direction, Cy. On the other hand, the polynomial transfer function linear in Z and parabolic in XY is very sensitive to the target positions spacing in the out-of-plane direction, reaching 5% for the small velocity components Cx and Cz. The calibration with 1 mm target displacement in the out-of-plane direction has been chosen for the PIV measurements in most of the turbine sections. The offset was reduced to 0.5mm in configurations where drastic tolerances were required for the 3rd velocity component.

Target traversing system

A specific system for placing the target inside the draft tube has been developed. For an uncertainity of 3% on the velocity field measurement, the tolerances are 0.1° in rotation and 0.01 mm for the relative target displacement in the out-of-plane direction.

2.7 Acquisition parameters

A typical PIV experiment procedure consists in setting several acquisition parameters, choosing and testing the trigger signal, acquiring the raw image and/or velocity data, validating the vector field, calculating the 3rd velocity component and finally data output.

The parameters that need to be adjusted for each measurement setup are:

- image quality – laser energy level and camera aperture settings;
- seeding particles density in the measurement area;
- spatial resolution – dimensions of the analysis window;
- time interval between successive frames;
- number of unsteady acquisitions.

2.7.1 Image quality

For the CCD camera used in this experiment, the first frame's exposure takes up to 132 µs, while the second frame is exposed during the entire read-out sequence of the first frame, 111 ms. It means that, for short time delays between laser pulses, the background gray level on the second frame will increase sensibly. In order to broaden the dynamic range, the ambient light level is set to minimum during the data acquisition. As the time delay varies according to the local flow field characteristics, the laser's energy level is balanced for each experiment, in order to reach uniform brightness on both images. Moreover, the cameras are equipped with high-pass filters for the emission wavelength of the laser: 532 +/- 5 nm.

Another important topic is the uneven energy distribution along the laser sheet, which can have several causes:

- non-uniform Gaussian energy distribution over the beam cross-section;
- misalignment of the internal mirrors of the light-guide;
- slight rotation of the sheet-forming optics against the thickness adjuster;
- lens aberrations;
- bad focus of the laser sheet, which leads to thickness variation over the test section;
- interface geometry, which can lead to important distortions depending also on the refractive index of the material.

All these parameters are taken into account and adjusted accordingly for each measurement setup, prior to data acquisition.

2.7.2 Spatial resolution

The measurement zone is divided in small analysis areas, for which a local velocity is calculated with the mean displacement of seeding particles between two successive frames, and the vector's application point is chosen at the center of the area. This grid defines the spatial resolution of the PIV measurement in the 2D case.

In a 2D configuration, the spatial resolution would be given by the distance between two successive vector anchor points in pixels, divided by the scale factor pixels/mm. If a linear dependence exists between the CCD chip dimensions and the measurement zone dimensions, then the scale factor depends on the field of view, on which the camera is focused, i.e. the aperture setup and the optical characteristics of the encountered media.

In a stereoscopic setup, the correspondence image - field of view is not linear anymore, because of the geometrical and perspective distortions due to camera setup. The velocities are calculated on a grid defined on the common zone of the two fields of view, by applying the calibration relationship on the neighboring vectors in both 2D fields, thus solving a linear or nonlinear system of four equations with three unknowns.

The spatial resolution is chosen such as the interpolation bias has a minimum influence on the vector value uncertainty - 2.4x2.5x3mm.

2.7.3 Seeding density

For a measurement to be successful, a minimum of two matching particles should be present in the analysis areas on both frames. According to the Nyquist criterion, the average number of correlated particles in the analysis window should not exceed five. This can be verified statistically checking the overall distribution of the correlation peaks width, which corresponds to the number of matching particle pairs in the correspondent analysis areas. It is checked that the average value fits within 3 – 5 for each measurement setup.

For the instantaneous velocity fields it is very difficult to insure a homogeneous seeding distribution of particles in all interrogation windows, particularly in measurements zones with strong velocity gradients or vortices. To improve the particles traceability, an overlapping of 25% of the analysis areas has been considered. The lack of particles or bad correlations is accounted for during vector field post-processing by correlation peak height and vector size criteria.

2.7.4 Time interval

The time delay between two laser pulses must be adapted to the local characteristics of the flow. The prior knowledge of the local velocity range and flow field structure is useful for setting an initial guess for the time delay setting.

The time interval must be chosen as small as possible to insure a good sensitivity for small particle displacements, to fit into the limits of the sub-pixel interpolation resolution, and as large as possible that the particle remains in the laser sheet width and within the interrogation area limits during the two camera exposures. The time delay optimization is particularly challenging in zones with strong velocity gradients, such as backflow zones or vortices. Typical values in our experimental conditions are between 50 - 300 µs.

2.7.5 Synchronization

In a turbine, the runner rotation forces the periodic behavior of the flow. One way to reconstruct the shape of a signal acquired by an unsteady measurement technique is the phase-locking, i.e. triggering the acquisition with a reference signal at different time delays within the event's period. In our specific case, the reference signal is the runner frequency, delivered by an optical encoder mounted on the shaft. The optical encoder delivering a signal per rotation has been used for acquisition synchronization for the operating points near the best efficiency conditions.

The PIV laser can be operated in window triggering mode, which means that the laser burst can be advanced if the trigger signal comes within a fixed time interval before the expected bursting moment. Although, the acquisition frequency is limited by the camera frequency in double-frame mode: 4.5 Hz. This slow-down only influences the experiment duration for achieving convergence, since the acquired signal is sampled at the same phase.

The synchronization of the laser and cameras with the external event is insured by the processor unit. The timing diagram is presented in Fig. 6.

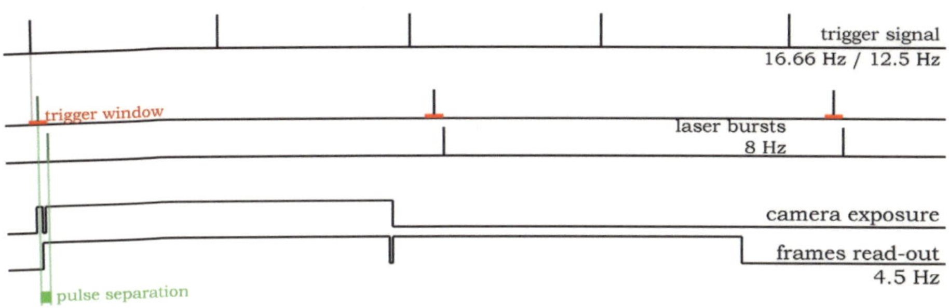

Fig. 6. Timing diagram for the PIV data acquisition and synchronization with an external frequency

2.8 Data validation and post-processing

2.8.1 Data validation criteria

The raw 2D vector fields resulting from cross-correlation can contain velocity values that have not been correctly detected, which bias the subsequent statistical analysis. A series of validation methods have been used for removing the outliers:

- Area masking - a digital mask is applied on the zones covered by obstacles in the field of view of one or both cameras and the corresponding vectors are eliminated;
- Correlation quality validation - the first highest peak in the correlation plane is considered as the signal peak, while the second one comes from the image noise. The signal-to-noise ratio should be about 20%. The first peak's height depends on the seeding density in the interrogation area, as well as on the image quality, i.e. uniform illumination and good spot-background contrast;
- Range validation - depending on the local characteristics of the flow field, the vector's length or its components may be limited;
- Moving average - through an iterative process, a vector is replaced with the average of its m×n neighbors,

$$\bar{C}(x,y)=\frac{1}{mn}\sum_{i=x-\frac{m-1}{2}}^{x+\frac{m-1}{2}}\ \sum_{j=y-\frac{n-1}{2}}^{y+\frac{n-1}{2}} C(i,j) \tag{9}$$

if the difference between them exceeds a percentage of the maximum difference in the vector field,

$$\left\|C(x,y)-\bar{C}(x,y)\right\|>\alpha\max_{x,y}\left\|C(x,y)-\bar{C}(x,y)\right\| \tag{10}$$

This filter has a smoothing effect, thus it is used with caution when strong velocity gradients are present.

2.8.2 Validation with LDV data

Systematic errors, coming from calibration accuracy, image quality, cross-correlation, vector validation, interpolation for 3rd component calculation, repetitiveness and reference frame transform, have been addressed in the previous chapters and are evaluated to 3% of the mean velocity value.

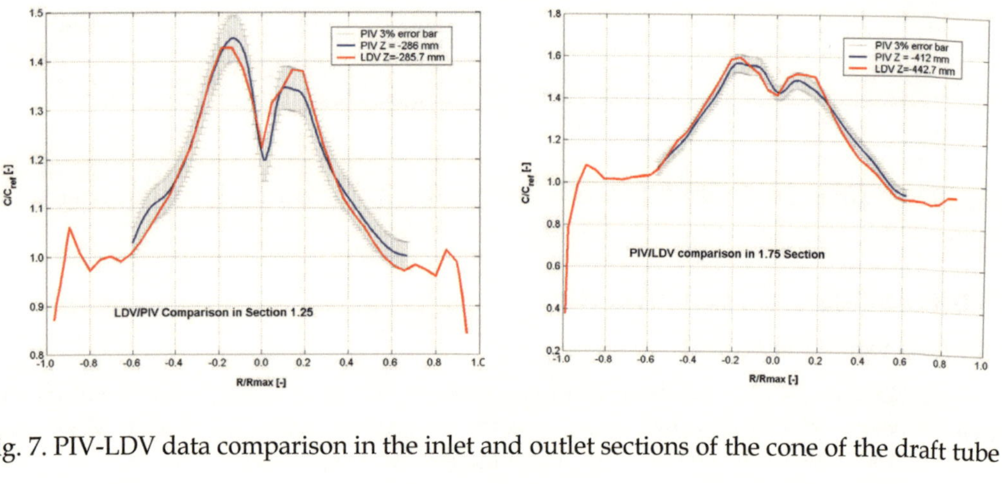

Fig. 7. PIV-LDV data comparison in the inlet and outlet sections of the cone of the draft tube

However, even if all of the previous considerations are fulfilled, the uncertainty of 3% holds only if the optical deformations of the image, due to the optical interfaces, are sufficiently small to be corrected by the polynomial function. To check this condition, a comparison was made between the PIV measurements and LDV measurements see Fig. 7. In this way, the global accuracy of the PIV measurement in this configuration is assessed.

The similar spatial resolution for the two measurement systems does not induce a supplementary error source: the measurement volume for the PIV measurement is a parallelepiped of 2.4 x 2.4 x 3 mm3 and for the LDV measurement is an ellipsoid of 2.4 x 0.2 x 0.2 mm^3. The result shows a good agreement, within 3%, for the entire PIV measurement field.

3. PIV measurements for rotor-stator interaction investigations

3.1 Phenomenology in pumps and turbines

The rotor-stator interaction is a complex set of phenomena stemming from the nature of the machine itself - such as periodic rotation instabilities and blade interactions - (Ciocan et al. 1996). The unsteady phenomena related to the interactions between the static and mobile parts often affect machine efficiency. Under the influence of periodic constraints, pressure and velocity fluctuations generate stress fluctuations and induce vibrations that contribute to material fatigue of the components and may also issue hydraulic noise. Blade interactions differentiate from other phenomena encountered in turbomachinery by their independence of the operation conditions. Whichever the turbomachine type and its operating regime, these interactions occur under the form of an unsteady secondary flow superposed onto the average stationary flow.

In hydraulic machinery (pumps, turbines and pump-turbines), the following phenomena, briefly described hereafter, are to be considered: potential interactions, wake interactions, von Kármán vortex interactions, three-dimensional viscous interactions and instabilities of the efficiency curve:

- Potential interactions are characterized by a non-uniform distribution of the unsteady pressure field in the gap between the static and rotating parts (e.g. guide-vanes and runner). This type of interaction has a non-convective character and its influence extends towards upstream as well as downstream the gap - see (Mesquita et al. 1999).
- Wake interactions are determined by the non-uniformity of the velocity profiles downstream blade casades (e.g. at the runner outlet in a centrifugal pump) - see (Iliescu et al 2004). The main source of this non-uniformity is the shear layer created at the trailing edge by the combination of the boundary layers developed on the blade's pressure and suction sides. The velocity defect decays due to viscous effects, and its mixing length depends on the blade shape, local flow velocity and turbulence level. This phenomenon is of purely convective nature and it has no influence upstream its location.
- Von Kármán vortex street interactions take place downstream blade cascades with blunt trailing edge, in low turbulence conditions. Their shedding frequency is generally expressed in terms of Strouhal number, which depends on the characteristic blade dimensions and on the local flow velocity. If the vortex shedding frequency reaches resonance with the natural frequencies of the system, it may lead to structural vibrations.

- Viscous three-dimensional interactions are secondary flows in the inter-blade channels, such as passage vortex, corner vortex, horseshoe vortex and tip gap vortices. They are closely related to the shape of the different hydraulic passages, which force variations in the flow direction and local modifications of the boundary layer topology. It is likely that these phenomena are the main contribution to local efficiency loss in specific areas of the operating range. Certain machines (pumps and pump-turbines) often exhibit an unstable zone of the efficiency characteristic, with discontinuities of up to 3%, accompanied or not by hysteresis - see (Ciocan et al. 2001).

The accurate prediction of these phenomena is essential at the design stage, and PIV measurements are the ideal tool to investigate them. The first experiment set up to analyze these interactions is described in (Ciocan et al. 2006).

3.2 Experimental set-up

2D-PIV measurements have been performed in the guide-vanes channel of a pump-turbine model of specific speed nq=66. Several operating conditions were investigated in both pump and turbine regimes, covering a large portion of the operating range.

To obtain the velocity field evolution in the guide-vane channels, the measurement section has been chosen such as to cover the space between two wicket gates. This area is determined by the 'visibility' zone, i.e. the area which is accessible by a laser sheet through windows embedded in the spiral casing walls - see Fig. 8. Two stay vanes that obstructed the visual access to the measurement section have been removed. Views through two separate windows were necessary in order to cover the full extent of the measurement domain.

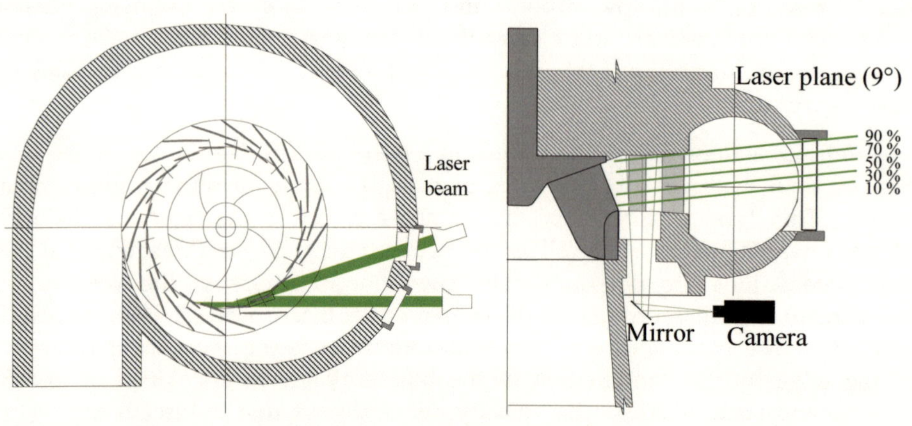

Fig. 8. Top and side views of the measurement sections and laser sheet positions for the PIV experiment in the guide-vanes channel of a pump-turbine

Five sections have been analyzed along the height of the channel. Considering the important optical constraints imposed by previous LDV investigations in the same section, the imaging window was tilted by 9° with respect to the machine's axis. To enable the most adequate conditions for the PIV experiment, the laser planes were also tilted of 9° in the radial direction to match the central window level - see Fig. 9.

Fig. 9. Overview of the PIV setup. Detail on the right side shows the image reflected by the inclined mirror, as seen from the CCD camera standpoint

To obtain a sharp image with a CCD camera, the focusing plane must be in the plane of the laser sheet. Another requirement for a 2D-PIV experiment is to set the camera imaging axis normal to the measurement plane. To ensure minimum optical distortion, the different interfaces traversed by the scattered light must be smooth and parallel to the laser sheet as well. In the present case, the camera is installed on a mobile support with two degrees of freedom - translation and rotation. Considering the camera's focal length and visual access difficulties, a plane mirror was placed in front of the window to redirect the optical path horizontally towards the camera, as illustrated in Fig. 9.

The mirror must meet high quality standards (flatness of the surface and homogeneity of the reflective coating), to avoid additional errors in the measurement process. The solution adopted for this experiment was a price-quality compromise: Pyrex wafer with ALMGF2 coating, $\lambda/2$ flatness. The mirror was mounted on the same support as the camera, aligned at 45° with respect to the lens. The CCD-mirror assembly was oriented parallel to the laser sheet and the support axis was secured horizontally. The final adjustment was made by optimizing the sharpness of the acquired images, with an estimated maximum uncertainty of 0.6°.

Regarding the laser sheet positioning, in a first approach the optical path of the laser sheet was calculated and it was planned to adjust its position in-situ using a device attached to the light-guiding arm. The precision was not satisfactory, thus it has been decided to mark directly the position of the plane directly onto the wicket gates, for each vertical position. The laser sheet contour was traced on the opposite faces of two guide vanes that form a channel. First tests revealed scattering problems on the shroud, runner blades and guide vanes. The issue was corrected by painting these bright metallic surfaces with a matte black dye. Several tests were made to determine the appropriate dye for the intended objectives - reflection attenuation and durability under repeated laser pulse operation. A black permanent marker was chosen.

3.3 Main results

The measurements have been acquired synchronously with the runner rotation, for multiple phase angles corresponding to portions of a single runner inter-blade channel. Ten angular positions were investigated, evenly distributed at 7.2°. For each spatial position, downstream the runner and between the guide vanes, the two dominant velocity components have been measured, corresponding to the radial and tangential directions. The average velocities are calculated with 250 PIV frame pairs by phase.

According to the ergodicity assumption for a stationary flow, the mean velocity was estimated by the temporal mean of the ensemble (eq. 11), and the standard deviation (eq. 12) for each component of the mean velocity is calculated as follows:

$$c = \Sigma c(i) / N \tag{11}$$

$$\sigma^2 = \left[\Sigma\left(c(i) - c\right)^2\right] / (N-1) \tag{12}$$

c(i) – measured instantaneous velocity
c – statistical average velocity
N – number of samples in each spatial position
σ - standard deviation (rms) of the mean value

Using the rms values, the turbulent kinetic energy can be estimated. Having only two components, it is assumed that isotropy conditions are met, i.e. the rms of the missing component is of the same order of magnitude as the two measured components. Furthermore, the periodic velocity fluctuations (deterministic) are negligible with respect to the turbulent fluctuations (random), thus the specific energy of the turbulent structures in the flow may be calculated using the individual standard deviation of the velocity components. In these conditions, the turbulent kinetic energy can be calculated as follows:

$$k = \tfrac{3}{4}\left(\sigma_r^2 + \sigma_t^2\right) \tag{13}$$

k - turbulent kinetic energy for the synchronous velocity component
σ_r - rms of the mean radial velocity
σ_t - rms of the mean tangential velocity

A typical result for the steady velocity field in turbine mode is presented in Fig. 10. For the measured range of operating points, ± 20% around the nominal flow rate, the guide vanes direct correctly the mean flow towards the runner inlet and back flows or detachment zones have not been observed.

The mean velocity profiles in pump mode, in a guide vanes channel ad mid height, are presented in Fig. 11 for a stable operating condition, while Fig. 12 shows the velocity profiles for an unstable portion of the operating range.

A phase averaging technique is used to analyze the periodic fluctuations synchronuous with the runner. An optical encoder provides the time basis, as a reference runner position. According to the Reynolds decomposition scheme (eq. 14), an unsteady velocity field can split in three parts: the temporal mean of the ensemble $\overline{c}$, a periodic fluctuation with respect to the mean $(\tilde{c} - \overline{c})$ and a random turbulent fluctuation c' . The process is illustrated in Fig. 13. A comparison between PIV and LDV phase average technique is presented in Fig. 14

$$c(i) = \overline{c} + (\tilde{c} - \overline{c}) + c' \tag{14}$$

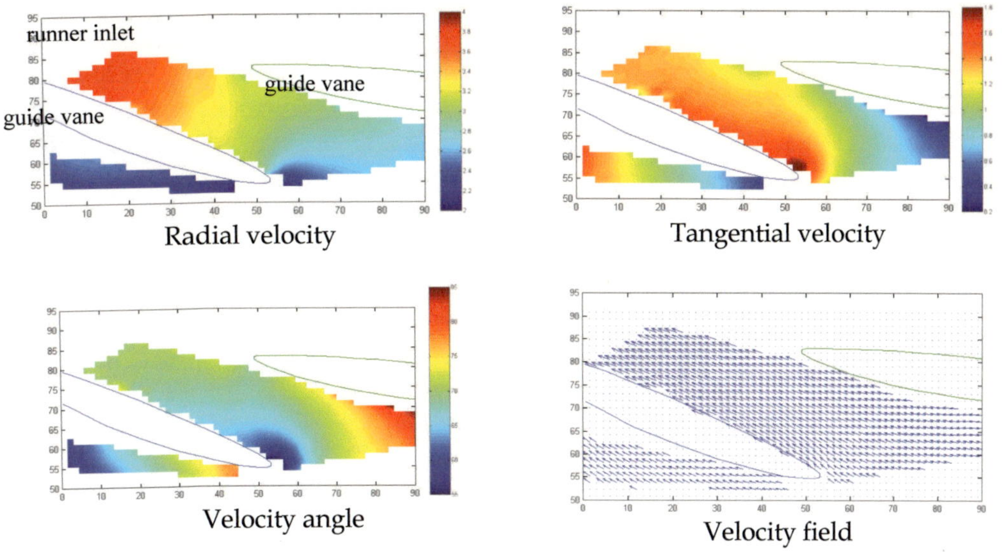

Fig. 10. Mean velocity field in the guide vanes channel for optimum flow rate Qn in turbine mode; mid-channel height; flow direction towards the runner

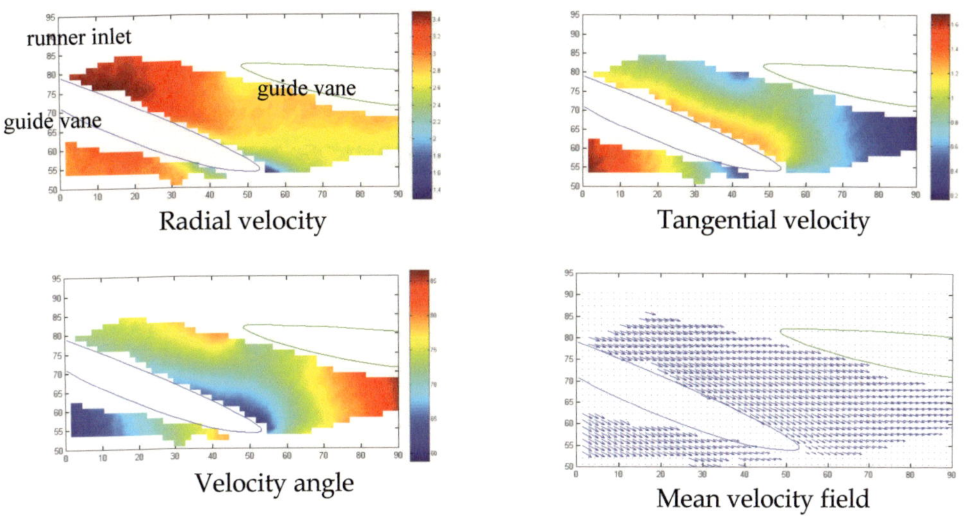

Fig. 11. Mean velocity field in the guide vanes channel for optimum flow rate Qn in pump mode; mid-channel height; flow direction exiting the runner

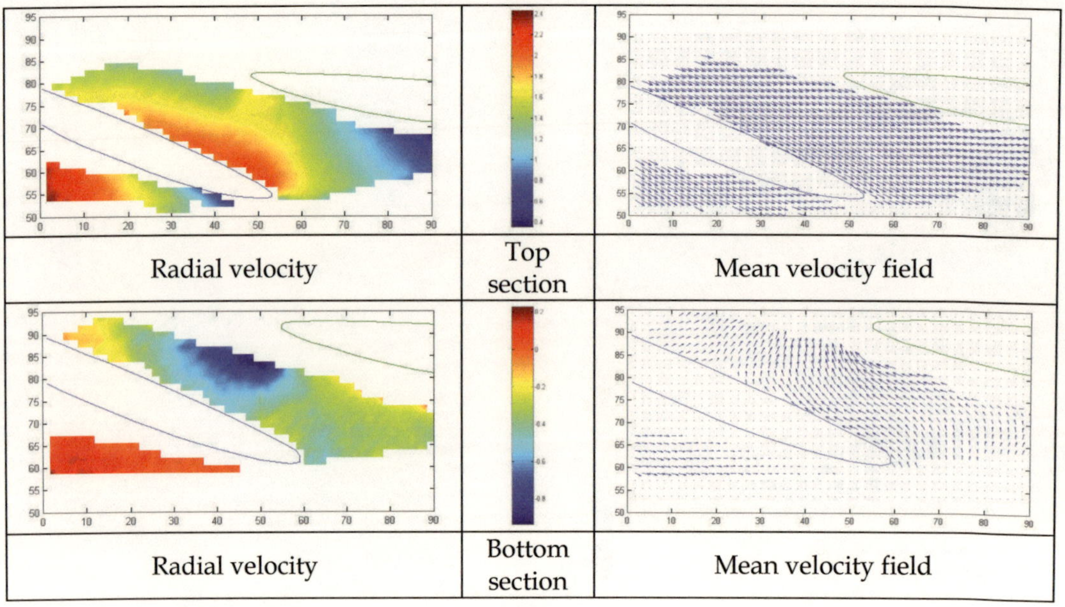

Fig. 12. Mean velocity field in the guide vanes channel for partial load 80%Qn, in pump mode corresponding to an unstable zone on the efficiency curve; planes at 90% and 10% of the guide-vanes channel's height respectively

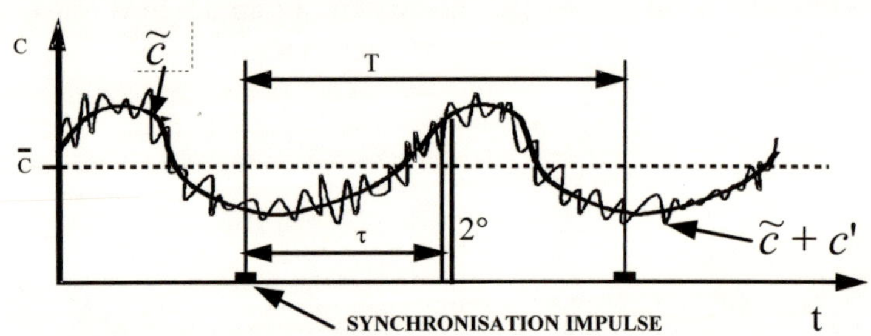

Fig. 13. Reynolds decomposition of a velocity signal

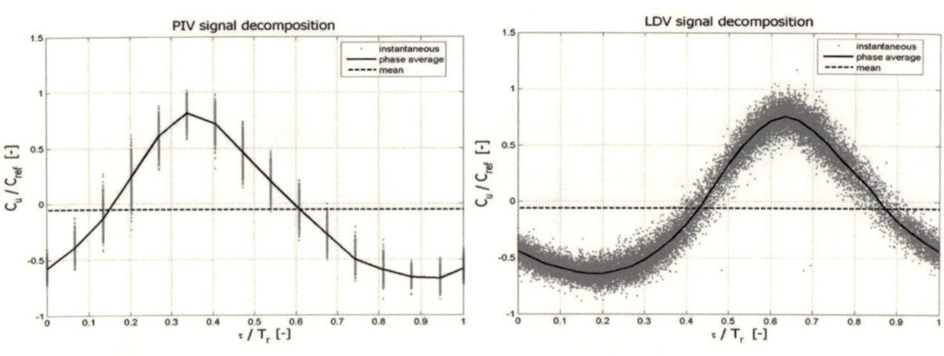

Fig. 14. Comparison of phase averaging technique for a PIV and a LDV experiment; data collected in a same spatial location and same operating conditions

All the measurements can be superposed in a representative channel due to the good periodicity of the blade-to-blade channels; this periodicity is checked for each measurement point.

The specific turbulent kinetic energy is the summation of the normal stress components. Direct measurement provides two orthogonal components of the velocity field. By assuming turbulence isotropy, the 3 normal stress components have the same order of magnitude, the specific turbulent kinetic energy becomes:

$$\bar{k}^* = \frac{3}{4}(\bar{c}_x'^2 + \bar{c}_z'^2) \tag{15}$$

$$\tilde{k}^* = \frac{3}{4}(\tilde{c}_x'^2 + \tilde{c}_z'^2) \tag{16}$$

Distinction is made between the total kinetic energy (eq. 15) and the turbulent kinetic energy content of the periodic fluctuations (eq. 16). This calculation is a better approach of the specific turbulent kinetic energy because the velocity fluctuation synchronous with the runner position is a deterministic phenomenon and this fluctuation does not concern the specific turbulent energy.

At the runner inlet the difference between the $\bar{k}^*$ and $\tilde{k}^*$ calculations is close to 15%, less important for the design operating point and more important for the off-design operating points.

Fig. 15 and Fig. 16 show the unsteady flow propagation in the guide-vane channel for 5 runner positions. This unsteady velocity field distribution shows that the fluctuation of the velocity components at the runner inlet is the same in all the guide vanes channel.

The blade passage perturbs the flow and therefore leads to a pressure fluctuation at the runner inlet. This pressure fluctuation induces a velocity fluctuation synchronous with the runner position. The direction of propagation of this perturbation is not the same for the two components: in the radial direction for the radial velocity and in tangential direction for the tangential velocity in the same direction with the runner rotation. The suction side of the blade produces a suction effect and an increase of the radial velocity. The pressure side of the blade induces a decrease of the radial velocity and a deviation of the flow direction. The fluctuating component $\tilde{c} - \bar{c}$ is the same on the two velocity components and corresponds to 30% of the mean velocity.

For the operating conditions corresponding to the unstable region of the efficiency curve, the same wake phenomenon is observed in the bottom half of the channel, but presents higher fluctuations than the optimum operation condition – see Fig.17.

In the top half of the channel, the flow is uneven. Analyzing a sequence of synchronous velocity fields, von Karman vortex structures are detected, propagating in the guide-vanes channel, as well as flow tendency to re-enter the runner. The von Karman vortex frequency does not seem to relate to the runner rotation frequency, and the runner wake isn't present either. A comparison of the flow topology in different operation conditions is presented in Fig. 18.

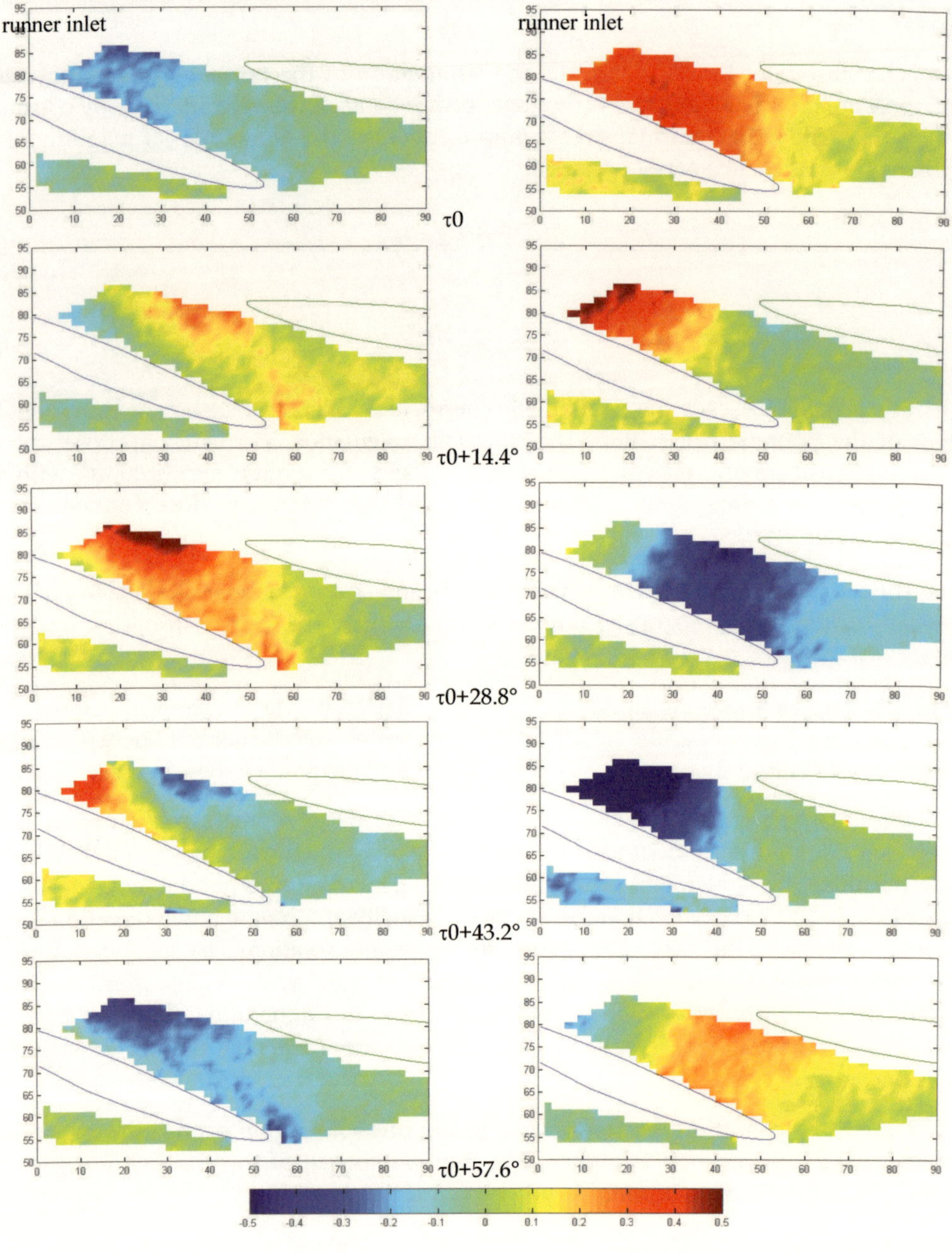

Fig. 15. Velocity field synchronous with the runner rotation $\tilde{c}(\tau)$ for different runner positions - turbine operation Q_n ; tangential velocity on left column and radial velocity on right column

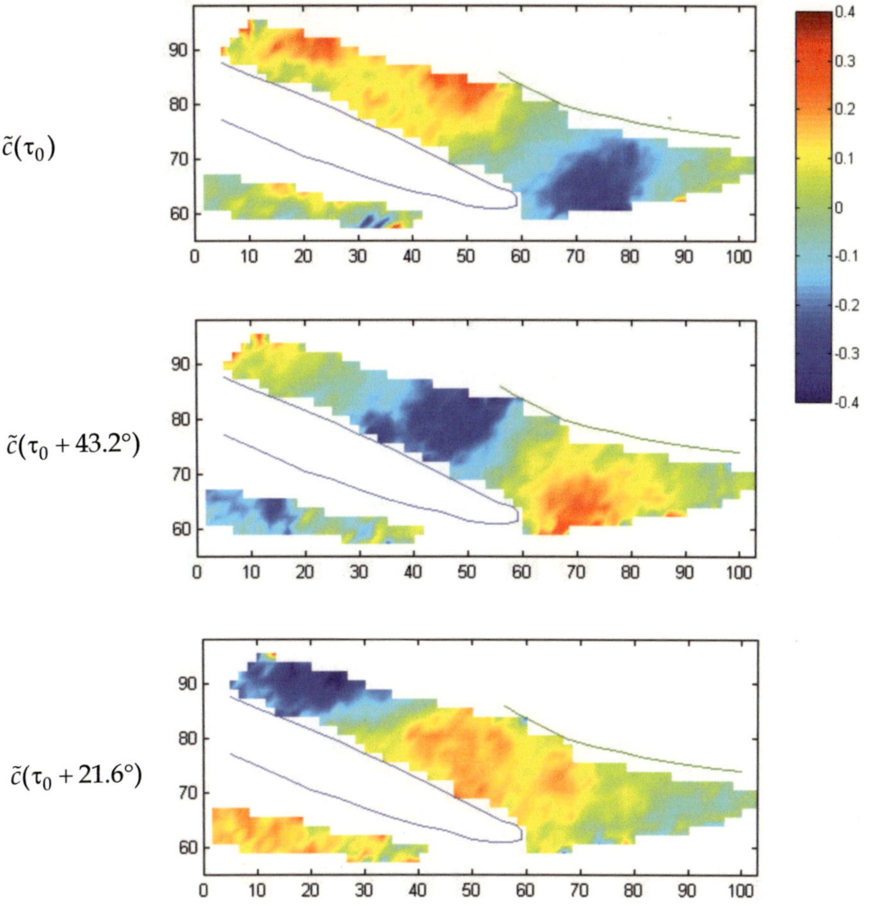

Fig. 16. Velocity field synchronous with the runner rotation $\tilde{c}(\tau)$ for nominal flow rate Q_n

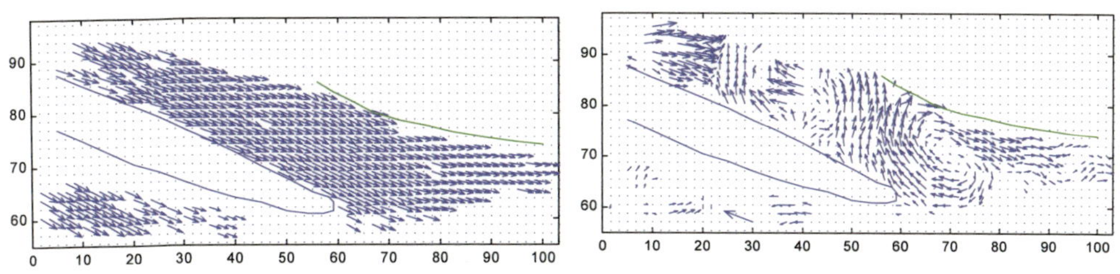

Fig. 17. Instantaneous velocity fields at 80% of the channel height: optimum flow rate Qn (left) and partial load 0.8Qn (right)

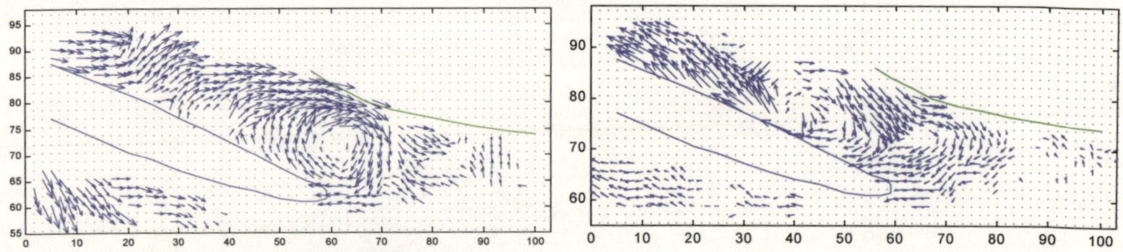

Fig. 18. Instantaneous velocity fields at 80% of the channel height: optimum flow rate Qn (left) and partial load 0.8Qn (right)

4. PIV measurements in two phase flow

4.1 Two-dimensional PIV in two-phase flow

4.1.1 Principle

PIV has been applied successfully to two-phase flows in the case of bubble flows or mixing experiments. Starting from these examples, a new two-phase PIV application was developed for simultaneous measurement of the flow velocity field and the volume of a compact unsteady vapors cavity - the rope that forms downstream the runner of a hydraulic turbine in low-pressure conditions - see Fig. 19. Specific image acquisition and filtering procedures are implemented for the investigation of cavity volume and its evolution related to the under-pressure level in the draft tube to quantify the development of the rope in the diffuser cone of a Francis turbine scale model.

The new technique is an adaptation of the PIV method for two-phase flows. Using fluorescent particles and corresponding cut-off filters on the two cameras, the two wavelengths can be separated. Fluorescence is an optical phenomenon that occurs when a molecule or atom goes to a higher energy level under the influence of incident radiation and emits radiation as the system relaxes, at a higher wavelength. The emission can be in the visible or ultraviolet spectrum if an electronic transition is involved or in the infrared range if it is a vibrational transition.

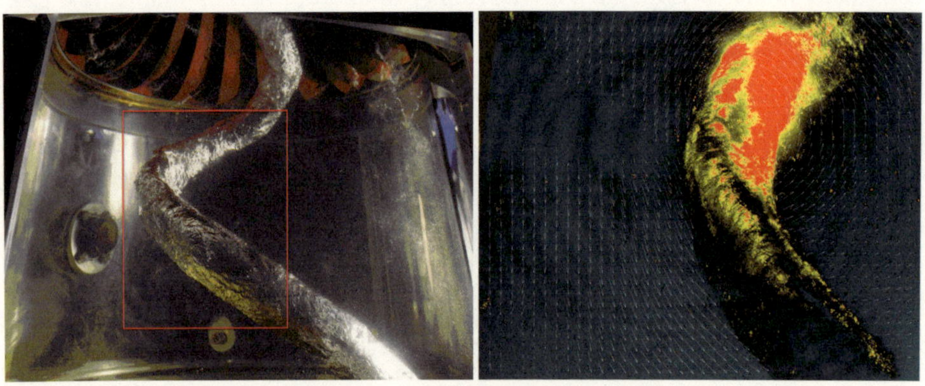

Fig. 19. Vapors-core vortex development in the cone at part load – example of measured unsteady velocity field overlaid on the corresponding rope image

For these measurements, rhodamine (Tetraethylrhodamine hydrochloride - RhB) particles of 1-10μm diameter, receiving 532nm and emitting at 594nm wavelength are introduced in the flow. Using fluorescent particles, which are excited by the laser wavelength, but emit at a higher wavelength, along with a longpass filter, prevents reflections in the laser wavelength to reach the cameras. In this way, the strong reflections from the rope interface, from residual bubbles in the flow or due to the optical interfaces are strongly attenuated and the vector field can be determined by cross-correlation from the second camera. The first camera has an antireflection-coated filter, focused on the laser wavelength 532 ±5 nm, to record the reflections in the laser light from the rope-water interface, and the second camera has a cut-off filter, 580nm, on the emission wavelength of fluorescent particles. The vortex core boundary can then be extracted by image processing of the first camera image and the velocity field is extracted of the second camera image.

4.1.2 Measurement set-up

The laser, cameras and processor used for the 2D-PIV experiment are adapted to the present measurement set-up. Optical filters have been mounted on the cameras. The two cameras' fields of view have been made coincident through a mirrors system, see Fig. 20.

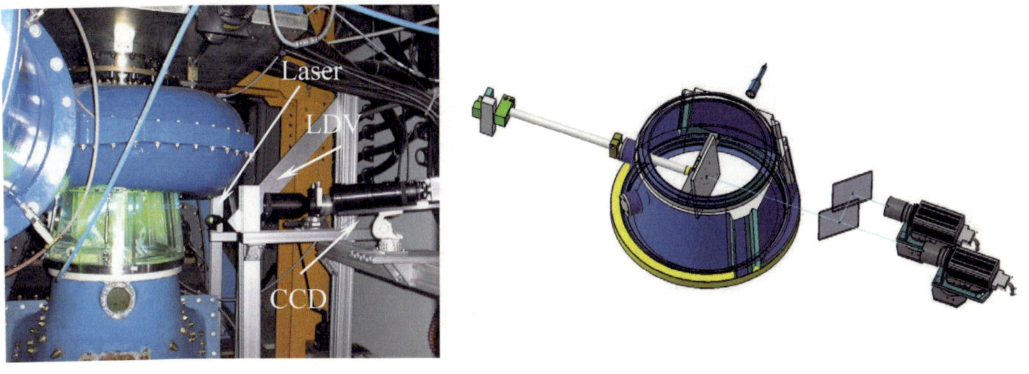

Fig. 20. Experimental setup for 2D-PIV measurements in two-phase flow

4.1.3 Calibration

Since the PIV system is set up in 2D configuration, the calibration would only be necessary for establishing the scale factor. Nevertheless, in this case, the distortions due to the cone's wall shape give an uneven distribution of the scale factor over the image. The distortions are diminished by a flat external surface, but the internal geometry had to be preserved for the hydraulic path, see Fig. 21. This conical surface induces distortion near the image edges. For this reason, a polynomial model is found to be suitable for these measurements.

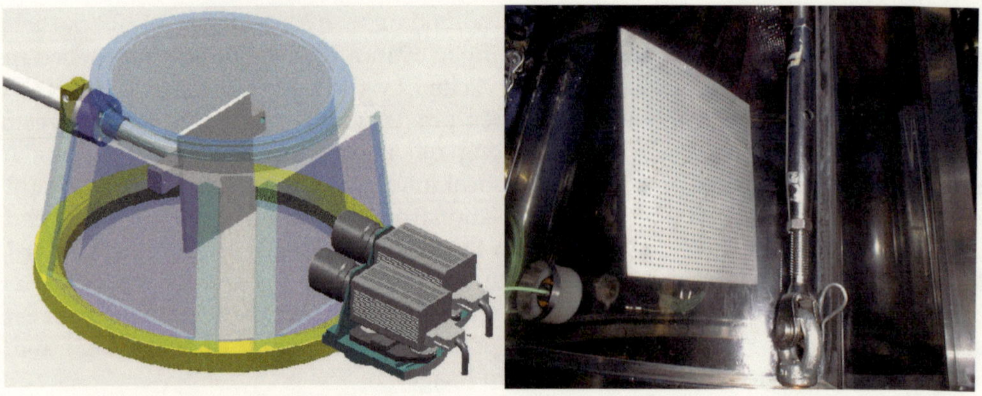

a) calibration setup b) calibration target position

Fig. 21. Image calibration setup for 2D-PIV measurements in two-phase flow

4.1.4 Acquisition parameters

The laser energy distribution, camera opening and exposure, seeding density, time interval between pulses and number of acquisitions are adjusted as described in the paragraph 2.

Synchronization

For determining the helical shape of the rope, it is necessary to synchronize the image acquisition with the rope position. The frequency of rope precession is influenced by the σ cavitation level and vapours content in its core. In certain operating conditions it can change over one revolution. Therefore, the triggering system cannot be based on the runner rotation but should detect the rotation frequency of the rope.

The technique to detect the rope precession is based on the measurement of the pressure pulsation generated at the cone wall by the precession of the rope. It has the advantage to work even for cavitation-free conditions when the rope is no longer visible, and, for this reason, it has been selected as trigger for the PIV data acquisition. To characterise the cavitation level, the σ - cavitation number is defined:

$$\sigma = \frac{P_a + P_d - P_v}{\frac{1}{2}\rho C^2} \tag{17}$$

Pa = atmospheric pressure, Pd = water pressure, Pv = water vapour pressure, C = flow velocity and ρ = water density

The pressure signals power spectra corresponding to the σ values are represented in the waterfall diagram in Fig. 22. For this operating point, the frequency of the rope precession is decreasing with the σ value.

The correspondence between the wall pressure signal breakdown and the rope spatial position has been validated through the optical detection of the rope passage using a LDV probe. By reducing the gain, the photomultiplier of the LDV system delivers a signal each time the rope boundary or bubbles intersect the LDV measuring volume. Small bubbles follow the rope interface for a brief period, for less than 10% of the rope period but the minimum interval for two successive PIV measurements is limited by the maximum CCD cameras frequency, i.e. 4.5 Hz, such as to avoid new acquisitions triggered by the bubble passage just after the rope passage in front of the pressure sensor. The PIV acquisition is performed at constant phase delay value with respect to the vortex trigger signal – see Fig. 23. The influence of the vortex period variation for this kind of phase average calculation is checked, and fits within the same uncertainty range as the measurement method: 3%.

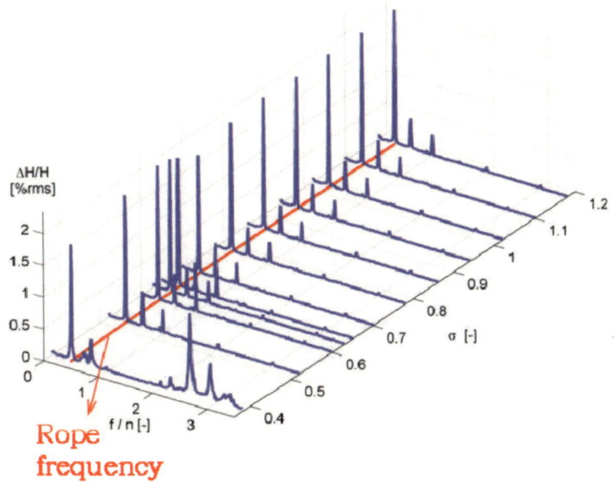

Fig. 22. Waterfall diagram of the power spectra of the wall pressure fluctuations

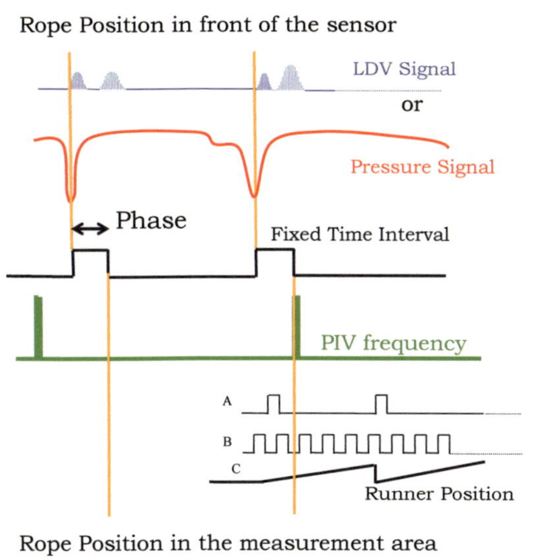

Fig. 23. Trigger signals diagram

4.1.5 Image processing

Image processing is applied to the first camera image, to obtain the rope shape. The distortion correction is performed with the calibration transform prior to the image processing step. Using a polynomial optical transfer function, the deformed raw image is mapped onto the real measurement cross-section, cf. Fig. 24. Each pixel on the image has now a linear dependency to its real-worls coordinates through a constant scale factor.

Fig. 25 presents the steps that are considered for filtering the image noise and for calibrating the contour relatively to the image brightness and rope boundary reflection:

- the zone where the rope appears on the image is separated from the rest by high-pass spatial filtering;
- the contrast is enhanced by histogram equalization;
- the noise removal is performed by non-linear adaptive band-pass filtering of the image on sliding neighborhoods of 8x8 pixels;
- the brightness amplification of the rope image is performed by histogram shifting towards higher values, weighted logarithmically;
- the rope background separation is obtained by threshold the gray intensity values;
- the local minima on the binary image are filled to smooth the rope contour;
- the spurious bright spots on the image are detected and filtered in order to keep only the largest foreground area, representative for the rope and eliminate the bubbles outside the rope;
- the rope boundary line is extracted. In the near vicinity of the intersection of the laser sheet and the rope, the boundary line is smoothed by a spline fit and the rope center point and the rope boundaries are determined. Then for each image the rope center position and the rope diameter are available;
- the aberrant images are filtered on criteria of minimum/maximum dimension of the rope area and rope diameter.

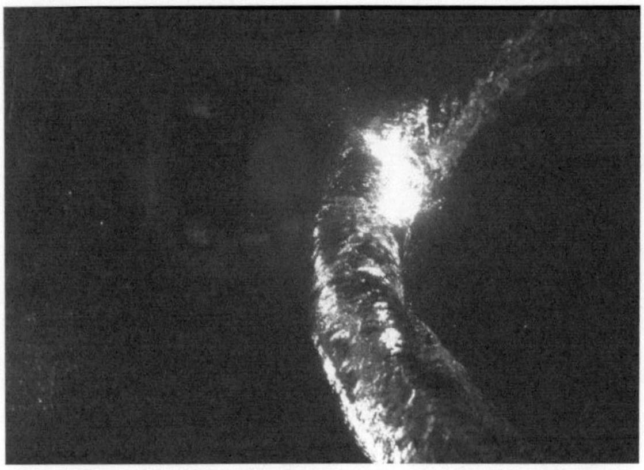

Fig. 24. Raw image of the rope, after distorsion compensation with a polynomial transfer function

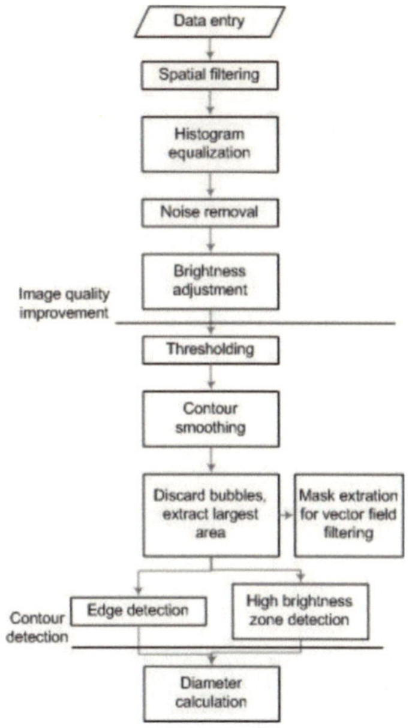

Fig. 25. Image processing flowchart

Spatial filtering

By spatial filtering, the zone of the image where the rope appears is extracted from the input image, see Fig. 24. The image-processing algorithm will only be applied on this zone, with the benefit of reducing the computing time, due to smaller amount of data. Since almost 50% of the input image is rejected, the contribution to the image histogram of dark background pixels and of the noise coming from seeding particles in the discarded area is strongly reduced, thus improving the grey-level distribution, see Fig. 26.

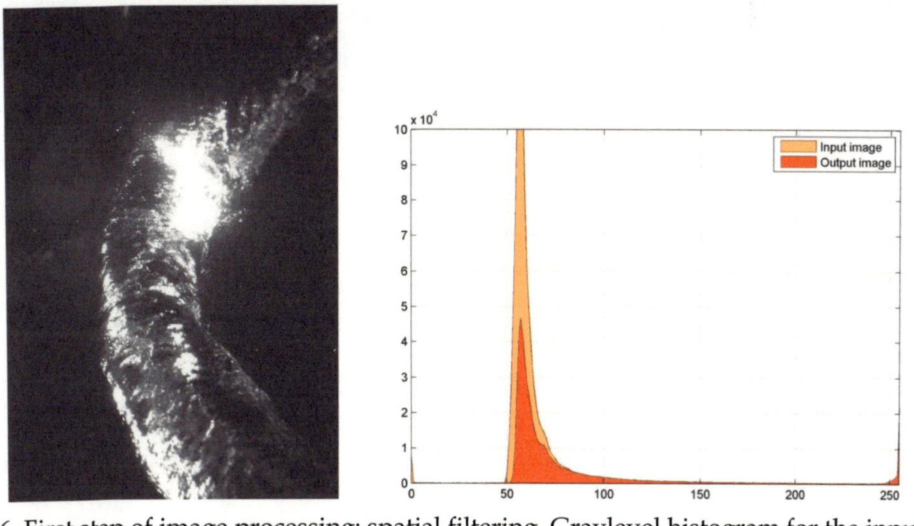

Fig. 26. First step of image processing: spatial filtering. Graylevel histogram for the input image and after the spatial filter is applied

Noise reduction

Since the camera records the reflection in the laser wavelength, the reflections from the seeding particles will be apparent on the image as well. In order to reduce the noise, a median filter is applied on sliding neighbourhoods adapted to the particle reflections size on the image. This filter replaces the central pixel with the median value of the surrounding ones. It has the advantage of preserving the shape and location of edges, but removes the "salt and pepper" like noise induced by seeding particles by reducing and spreading the grey intensity values. The dynamic range of the input image intensities will not be extended, as the median filter does not generate new grey values, Fig. 27 left.

Brightness adjustment

The nonlinear filtering is followed by an intensity adjustment through a linear transfer function for the grey values of the input and output images, Fig. 27 centre. The result is presented in Fig. 27 right.

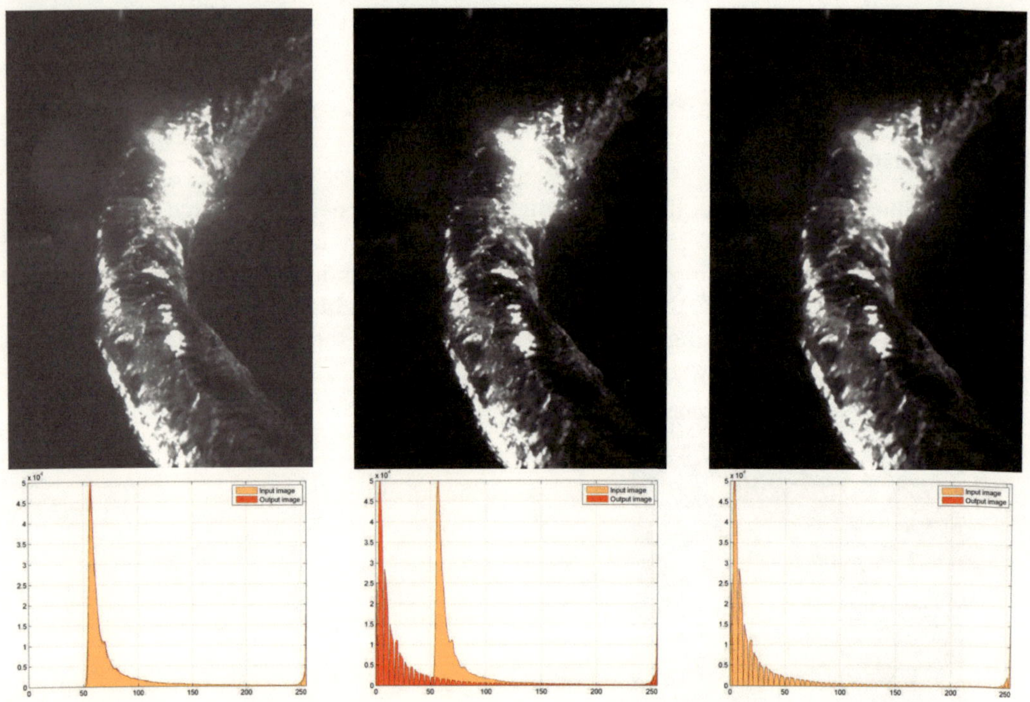

Fig. 27. Image processing steps: noise removal (left), intensity adjustment (centre) and nonlinear filtering (right). State of the graylevel histogram before&after each step

Rope diameter detection

Due to pressure and velocity fields' unsteady variations in the cone in low-charge operating conditions, as well as vapors compressibility, the rope surface exhibits irregularities. The laser sheet, encountering this uneven surface of the vapor-water interface, is reflected diffusely and gives a strongly illuminated area on the image. An antireflection-coated optical filter on the camera prevents image blurring, by reflecting part of the incident light scattered by the vapors-core boundary, as well as by residual bubbles and tracer particles in the flow. Nevertheless, the zone where the laser sheet reaches the rope boundary can be identified as the area with concentrated high intensity values. Thus, the best approximation of the real rope diameter, at the intersection with the inclined laser plane, would be the distance between edges along a line inclined at ~30° about the horizontal, see Fig. 28.

The influence of the reflection zone where the laser sheet reaches the rope boundary is overcame by cutting the gray intensity profile along the diameter line found in the first approximation at 94% of the maximum value, and keeping the value on the right as the new starting point for the diameter line, see Fig. 29. The centre of the rope in the measurement plane is considered to be the centre of mass of the final diameter line, see Fig. 29.

At last, a statistical analysis on the position and diameter of the rope in the measurement plane is performed for each phase - see (Iliescu et al. 2003). The coherence of the result is based on dimensions criteria check - Fig. 30. The rate of validated images is 95%.

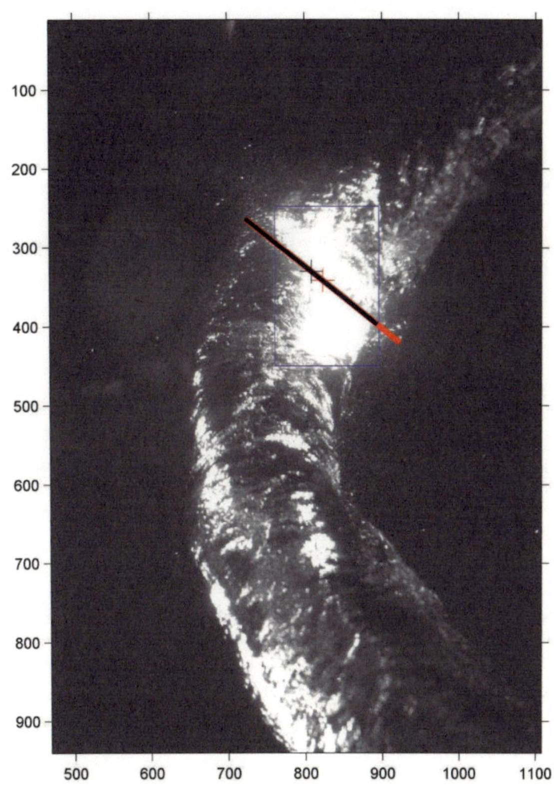

Fig. 28. Rope diameter detection and validation

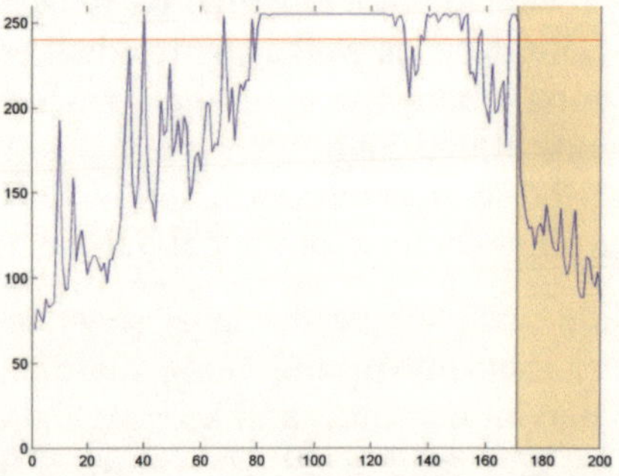

Fig. 29. Intensity profile along the approximate diameter

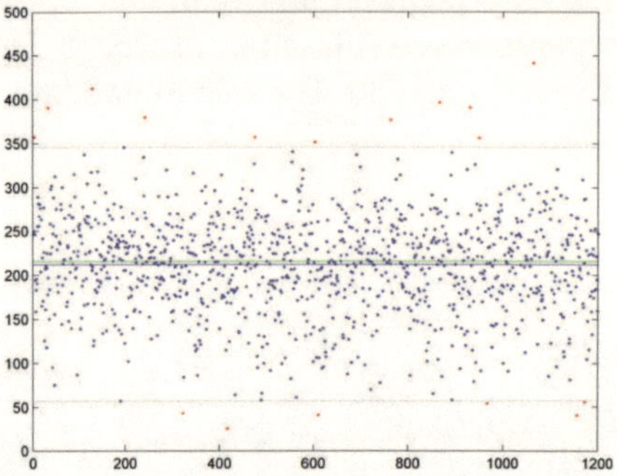

Fig. 30. Lower and upper thresholds for discarding aberrant diameter values

For each image, the intersection of the laser plane with the rope is detected and the values of the rope centre and diameter are transformed into real coordinates, for each validated image. Considering the mean values for each phase, the rope volume is reconstructed by the phase averaging technique.

Binary mask extraction

The binary image also provides a mask for the velocity fields. The pixel values on areas corresponding to each interrogation area for the velocity vector evaluation are averaged and then converted back to binary values. This digital mask thus obtained will be multiplied with the vector values and the zone covered by the rope is eliminated.

Due to the uneven intensity distribution on the zone occupied by the rope, particular image processing steps should be considered for separating the bright rope area from the dark background pixels.

First, the Canny edge-detection method - see (Canny 1986) is applied on the gray level image resulting from the previous steps. This method is based on the computation of the intensity gradient in both horizontal and vertical directions. Applying two threshold values on it gives a binary map of the contours between contrasting portions of the image, Fig. 31a. The boundaries are than thickened by extrapolation, Fig. 31b. Vertical and horizontal nonlinear filters are applied for linking the edges Fig. 31c. After holes filling, Fig. 31d is obtained. Spurious bubbles in the binary image are eliminated and only the largest foreground area is retained, Fig. 31e. Fig. 32f presents the rope contour overlaid on the image.

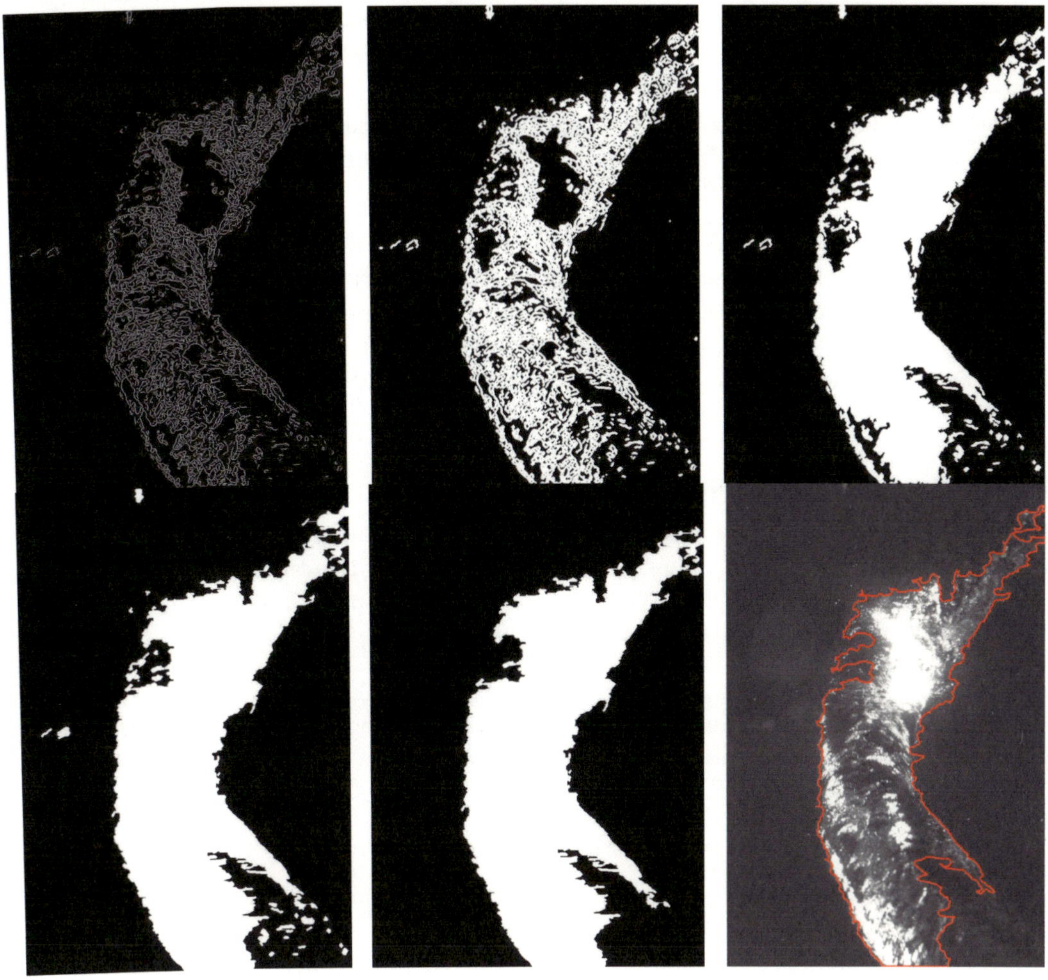

Fig. 31. Image processing steps: edge detection (a), edge dilation (b), edge linking (c), filling connected edges (d), bubble removal from the background (e) and extracted rope contour overlaid on the input graylevel image (f)

4.1.6 Vector field processing

For the velocity field calculation the same processing and validation methods like in paragraph 4.1.6. are applied, after the extraction of the rope mask from the image - see Fig. 34 and Fig. 35.

4.2 Tree-dimensional PIV in two-phase flow

4.2.1 Principle

The new method is an adaptation for two-phase flows of the stereoscopic PIV technique. It allows obtaining simultaneously the unsteady 3D velocity field and the rope shape. Using fluorescent particles, which are excited by the laser wavelength, but emit at a higher wavelength, along with high-pass filters mounted on the cameras, prevents reflections in the laser wavelength to reach the CCD chip and allows recording the light scattered by particles. In this way, the strong reflections from the rope boundary, from residual bubbles in the flow or due to the optical interfaces are eliminated.

Furthermore, using backward illumination in a third wavelength renders a darker rope shape on a brighter background with a good contrast. It makes the cavity profile sharp enough for an accurate detection of the rope edges using fewer processing steps for image enhancement. The constraint consists in adjusting the luminosity level of the background, such that the bright reflections from seeding particles to be displayed with a good contrast as well.

4.2.2 Measurement equipment

The laser, cameras and processor used for the 3D-PIV experiment are adapted to the present measurement set-up. Optical filters have been mounted on the cameras' lenses. For these measurements, rhodamine (RhB) particles of 1-10μm diameter, receiving 532nm and emitting at 594nm wavelength are used as flow-field tracers.

Nominal wavelength	Spectral bandwidth	Luminous intensity	Viewing angle	Operating voltage	Operating current	Power consumption
587 nm	15 nm	280 mcd	120°	10.5 VDC	300 mA	3.2 W

Table 6. LED-array characteristics

Two panels of LEDs, Osram OS-LM01A-Y –Table 6–, placed in front of each camera behind the cone, insure the backside illumination. Arrays of 22x24 LEDs wired in parallel are mounted on two plates, and connected to the same power supply system and synchronization board. A diffusing screen is placed in front of the LED panels.

4.2.3 Calibration & acquisition parameters

The distortions due to the cone wall shape, giving an uneven distribution of the scale factor over the image, are integrated in a calibration model. The distortions are diminished by a flat external surface, but the internal geometry of the hydraulic path had to be preserved, see Fig. 32. This conical surface induces distortion near the image edges. For this reason, a polynomial model, parabolic in Z, is found to be suitable for these measurements.

The laser energy distribution, camera opening and exposure, seeding density, time interval between pulses and number of acquisition are carefully adjusted. The synchronization process is the same as for 2D-PIV in two-phase flow. The image acquisition is triggered by the rope passage in front of a pressure sensor mounted on the cone's wall - see paragraph 4.1.4.

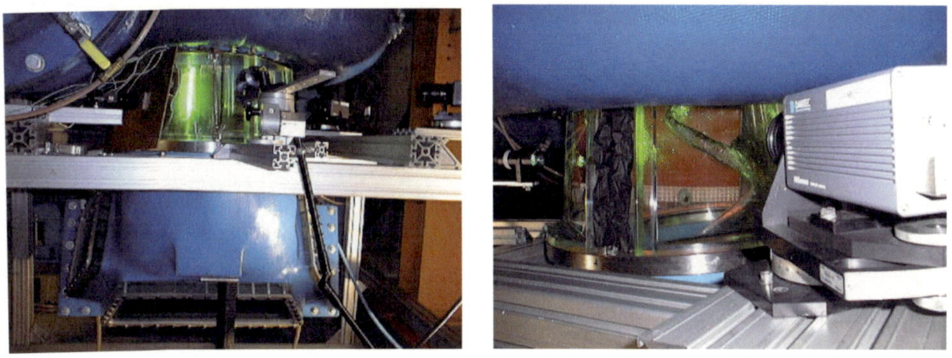

Fig. 32. Experimental setup for 3D-PIV measurements in two-phase flow

4.2.4 Image processing

The gray level distribution on the raw image histogram, Fig. 33, shows three separate peaks corresponding to the rope shadow, background light and particles.

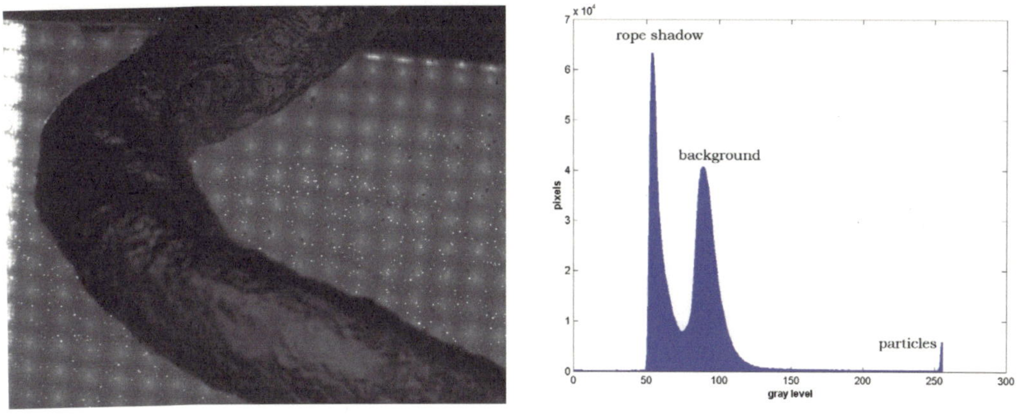

Fig. 33. Raw image of the rope and corresponding histogram

A series of image processing steps, summarized in Fig. 34, are applied on both cameras' raw images, to obtain the rope contour.

The first objective is to extract the rope shape from the recorded images, and transform it into a binary mask used for outliers' removal from the vector field. The second one is to detect the rope diameter at its intersection with the laser plane within the measurement area limits.

The distortion correction is performed with the calibration transform prior to the image processing sequence. Using a polynomial optical transfer function, the deformed raw image is mapped onto the real measurement position - Fig. 35a. Each pixel on the image depends now linearly on the real coordinates through a constant scale factor.

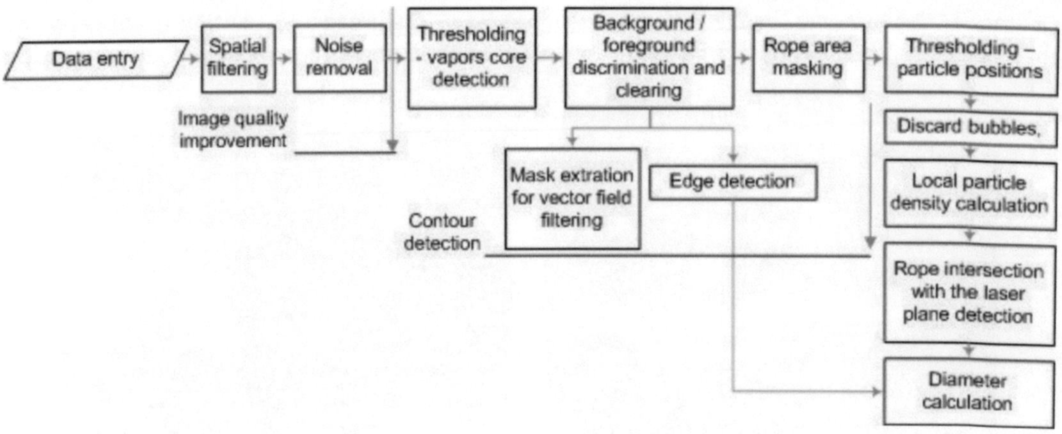

Fig. 34. Image processing flowchart

Image enhancement

The zone where the rope appears is extracted from the input image by spatial filtering – Fig. 35b. Almost 40% of the input image is rejected and the image processing algorithm will only be applied on this zone. A median filter is then applied on sliding neighborhoods of 8x8 pixels of the image, in order to remove the particles from the background, Fig. 35c, and smooth the corresponding peak at high intensity levels on the histogram - Fig. 35d.

Binary mask extraction

An adaptive threshold – denoted with a vertical line in Fig. 35d – is applied on the gray-level distribution for separating the rope area from the background pixels. The resulting binary image is presented in Fig. 35e. Lower gray values corresponding to the contour of the rope are converted to black, while brighter intensities are converted to white values.

In the next step, spurious spots are removed from the white background and bright zones corresponding to reflections from the rope boundary facing the camera are removed from the rope foreground, giving a smooth rope area – Fig. 35f. This binary image will be used as mask on the corresponding velocity field obtained by cross-correlation from the same raw image and its pair.

Rope diameter detection

In the current case, the passage of the rope through the laser plane does not provide reflections on the image like in the case of 2D two-phase PIV, thus another solution for detecting this zone had to be used. Particles are present on the image around the rope, except at the rear of the zone where the compact vapors volume obstructs the laser light passage. Therefore, calculating the particles density on the image leads to the expected solution.

The zone occupied by the rope is subtracted from the image by applying the mask in Fig. 35f to the raw image in Fig. 35c. An adaptive threshold applied on the resulting image, Fig. 35g, gives the positions of seeding particles surrounding the rope profile - Fig. 35h. Furthermore, larger objects are eliminated from the image based on area selection criteria –

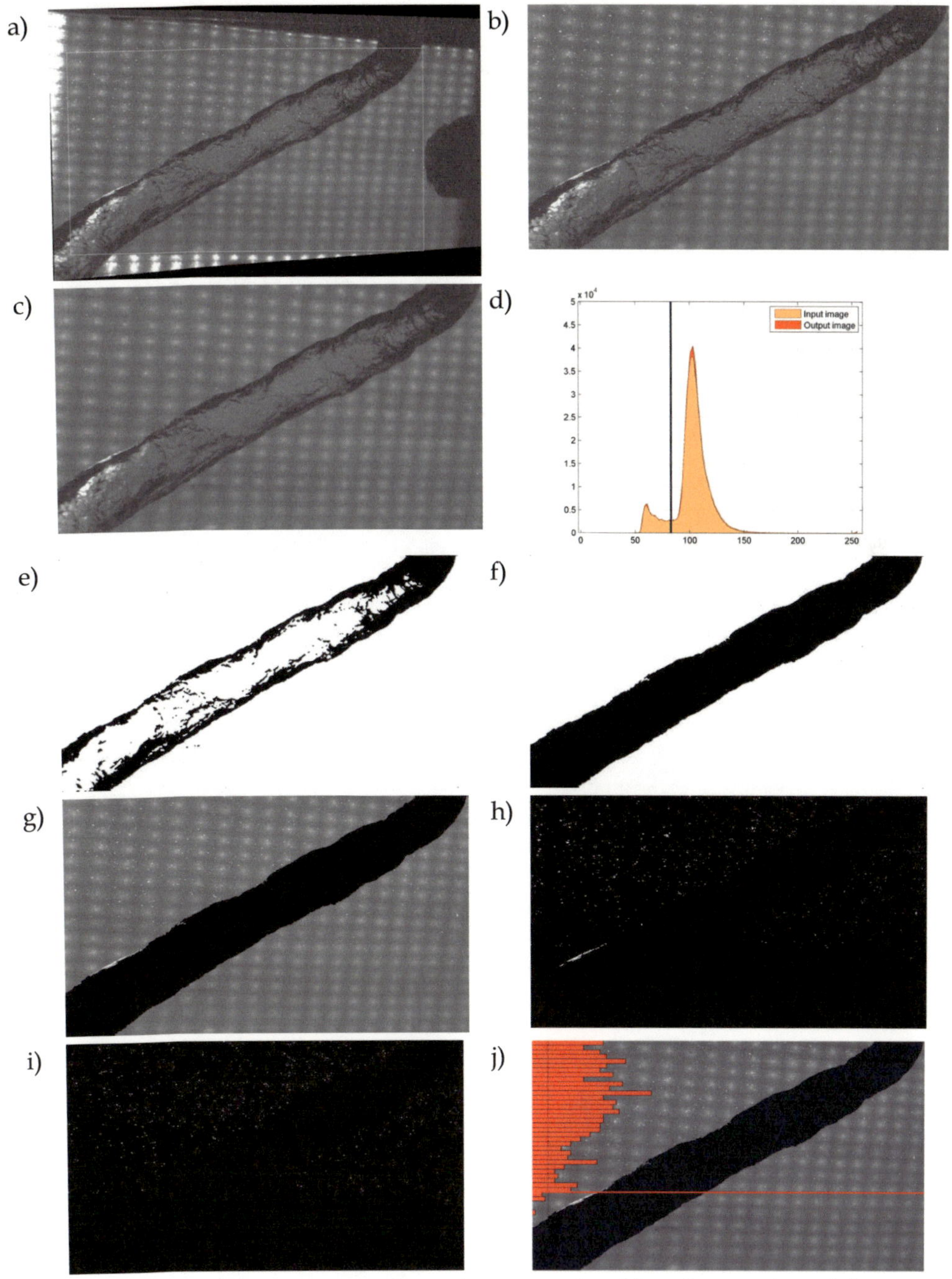

Fig. 35. Raw image of the rope, after distorsion correction (a); only a section of the active area is used for subsequent image proccessing (b); Image processing stages: noise removal (c) and respective histogram of the filtered image (d); binary image obtained by adaptive thresholding (e); binary mask extraction (f); masking of the rope area (g); detection of particle locations (h); particles selection based on maximum diameter criteria (i); calculation of particle density on horizontal slices (j)

Fig. 35i. The particle density is then calculated on horizontal slices; their distribution is shown in Fig. 35j. An adaptive threshold based on local minima of particle density gives the intersection limit of the rope with the laser sheet - illustrated by a horizontal red line in Fig. 35j.

The boundary of the rope is recovered from black-white transitions in the binary mask in Fig. 35f. The contour of the rope projection onto the camera imaging plane is outlined in red on the raw image in Fig. 36. A linear fit is applied on the median line (blue) in the proximity of the horizontal limit determined previously. The rope diameter at the intersection with the laser sheet is calculated within the rope boundary limits in the direction perpendicular on the linear fit - yellow line in Fig. 36 - and the center is considered at their intersection. The geometrical parameters of the intersection of the rope with the laser plane (diameter and centre position) are detected for each image, and the values are transformed into real coordinates through the scale factor. Considering the mean values for each phase, the rope shape is reconstructed spatially by the phase averaging technique, see (Iliescu et al 2008).

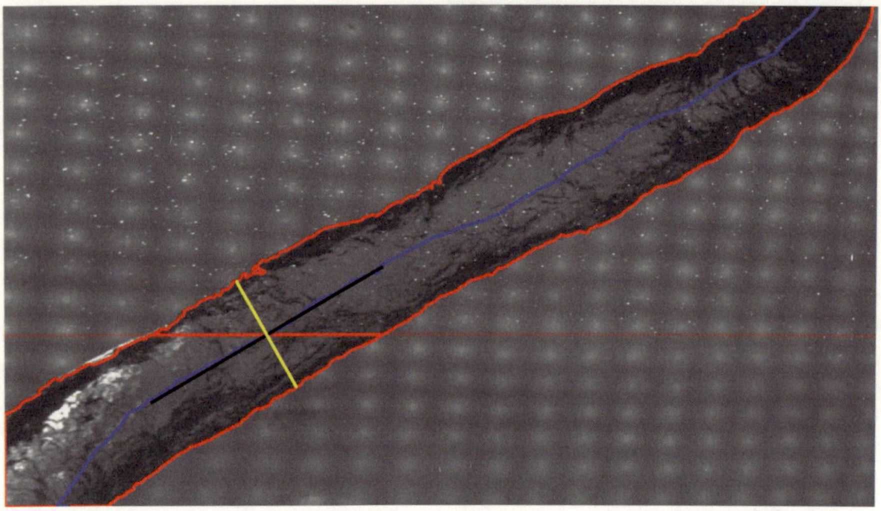

Fig. 36. Rope diameter detection (yellow), normal to the centerline of the vapours core (blue/black), at the intersection with the laser sheet (red)

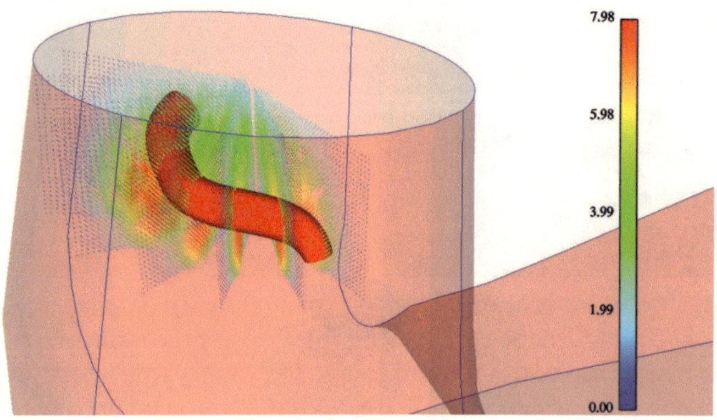

Fig. 37. Phase averaged vectors field and measured vapors core rope for extreme cavitation factor at partial load

4.2.5 Vectors field processing

For the velocity field calculation, the raw vectors maps are processed by cross-correlation of the two frames from each camera. The raw values are filtered based on range and peak validation criteria. The distortions of position coordinates and particles displacements are corrected with the calibration transform. In order to eliminate the aberrant vectors in the region of the rope or due to the shadow produced by bubbles, the binary mask obtained previously by image processing is applied on the vector maps by multiplication. The statistical convergence is achieved at 1200 velocity fields for 3% uncertainty.

4.3 Main results

The 2D- and 3D-PIV measurements in two-phase flow described herein were the first experiments of this kind in hydraulic turbomachinery operating under cavitation conditions - see Fig. 37, see (Iliescu et al 2003). They provide data on the rope volume and the surrounding velocity field, acquired simultaneously and in unsteady regime. The spatial position of the rope was also obtained, both in cavitation and cavitation-free conditions. This set of measured geometrical characteristics served to derive an analytical expression (eq. 18-eq. 20) for the pseudo-temporal vortex rope pattern - Fig. 38.

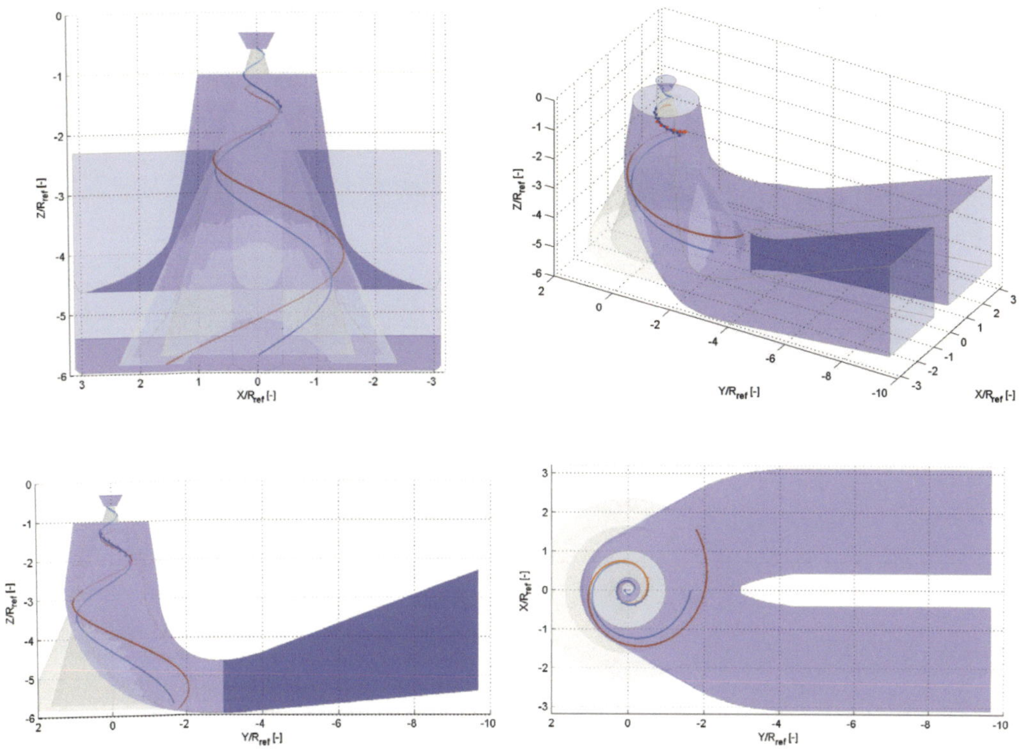

Fig. 38. Reconstructed vortex centreline for a Francis turbine operating at partial load: single phase flow (blue) and in cavitation conditions (red)

$$\vec{r} = \vec{r}(\theta) = x(\theta)\vec{i} + y(\theta)\vec{j} + z(\theta)\vec{k} \tag{17}$$

$$r = r_o b^{(\theta-\theta_o)/2\pi} \tag{18}$$

$$\begin{cases} x = r_o b^{\theta/2\pi} \cos\theta \\ y = r_o b^{\theta/2\pi} \sin\theta \\ z = z_O - r_o(b^{\theta/2\pi} - 1) / \tan\beta \end{cases} \tag{19}$$

The main parameters are given in Table 7, for the investigated operation conditions: cavitation-free flow and maximum vapours core volume.

		Cavitation-free vortex	Vapors-core vortex
Initial radius	ro	0.09	0.15
Initial depth	zo	-0.615	-1.2
Rate of radial growth	b	3.2	4
Cone angle	β	17°	25.5°

Table 7. Conical helical vortex model parameters

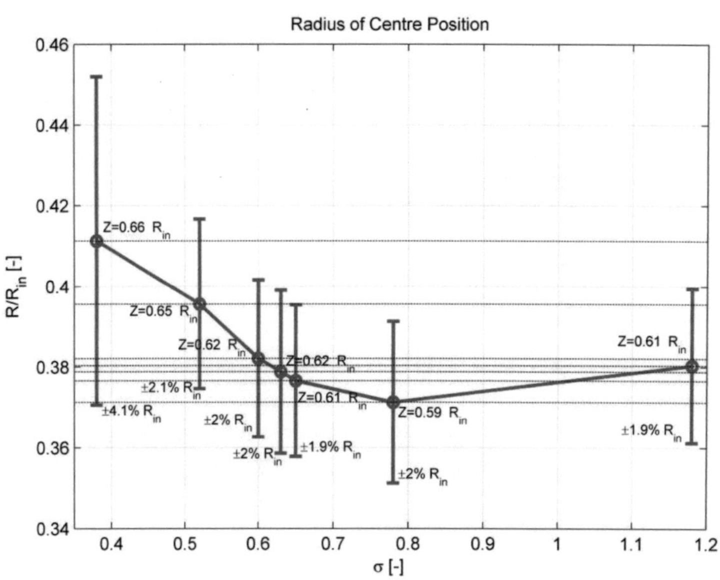

Fig. 39. Rope centre variations versus the cavitation number σ

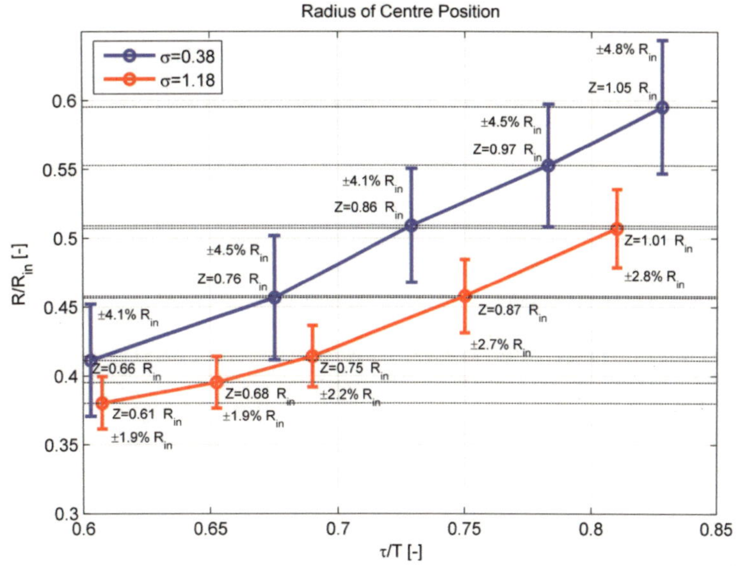

Fig. 40. Vortex position variation for two σ values versus the vortex phase τ

The unsteady analysis of the results the standard deviation of the vortex centre position representative for the flow stability. The decrease of the standard deviation of the rope centre with the σ value, Fig. 39, shows that the flow becomes unstable when the volume of the rope cavity increases. For σ = 0.380, the location of the rope center has an unsteady spatial variation of ~8% of the local radius of the cone - see Fig. 40. The resonance at this σ value induces a lost of stability of the rope shape and spatial position are illustrated by the increase of standard deviation of the vortex center position - twice, from 2% to 4.1% of the radius.

Fitting a linear equation to the vapors-core vortex radius in the measured positions, added to the vortex filament model, allows reconstructing the rope volume in the draft tube - see Fig. 41. In this way, starting with the experimental results, the rope shape and position is completely parameterized - see (Ciocan et al 2009).

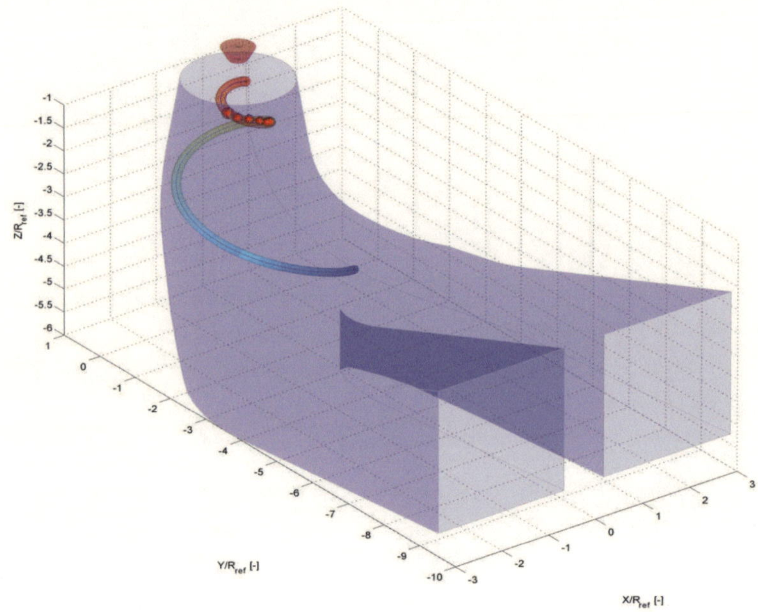

Fig. 41. Analytical representation of the vortex rope

In the cone cross-section, the rope center radius is increasing following the phase evolution: the rope goes closer to the wall with the cone depth - see Fig. 42. This evolution is confirmed by the increasing of the wall pressure fluctuation synchronous with the rope rotation, with the cone depth. The standard deviation corresponding to the rope position is quasi-constant for all phases (depth) in the measurement zone.

The procedure to establish the phase average of the rope diameter, already described, is applied for 7 σ values. The rope diameter is represented versus the σ value, as well as the rope center position in the measurement zone. Associated with these values, the standard deviation of the rope diameter and vortex center position is calculated.

The rope diameter is decreasing with the σ value - Fig. 43. The standard deviation of the rope diameter is related to the rope diameter fluctuations and represents the rope volume variations. The physical significance of this calculation is related to the axial pressure waves that produce a volume variation of the rope, due to the local changing of the pressure distribution. The standard deviations are quasi-constant for all σ values at 2.5% of R, except for the value 0.380, where it increases at 4.3% of R. For σ = 0.380 the rope area reported to the local cone section area has a variation between 0.5% and 1.2%.

In fact, for the 0.380 σ value it was pointed out, by hydro-acoustic simulation, that a pressure source located in the inner part of the draft tube elbow induces a forced excitation. This excitation represents the synchronous part of the vortex rope excitation. An eigen frequency of the hydraulic system is also excited at 2.5 of the runner rotation frequency. The plane waves generated by the pressure source, propagating in all the hydraulic circuit, induce consecutively a decreasing and an expansion of the vapors volume of the rope, which explains the increasing of the rope diameter standard deviation for this σ value. Reported to the phase evolution, the rope volume and its standard deviation remain quasi-constant.

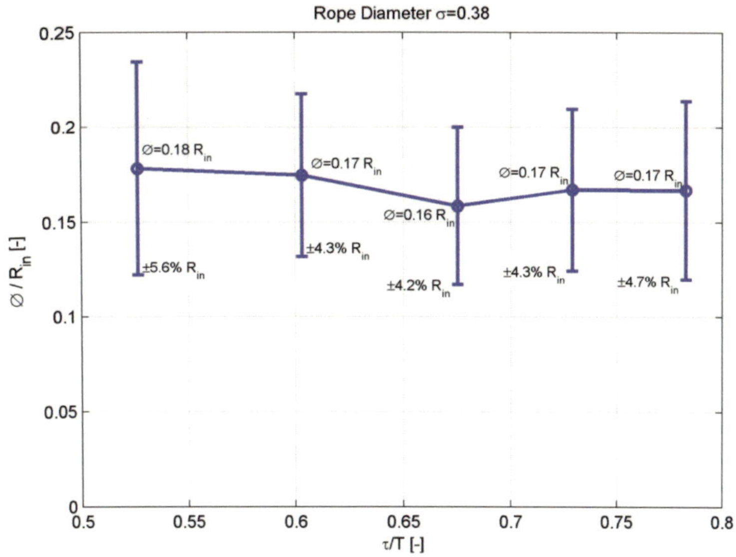

Fig. 42. Rope diameter variations versus the vortex phase τ

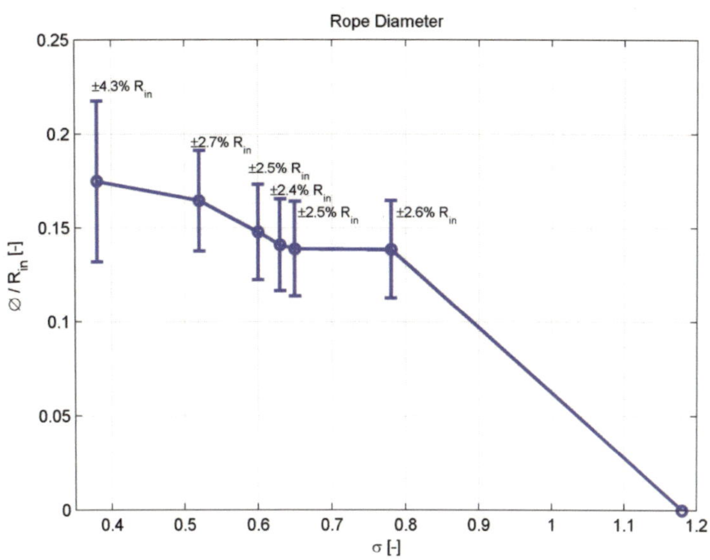

Fig. 43. Rope diameter variations versus the σ value

5. Conclusions

The PIV measurement system is a valuable asset for flow dynamics investigations in turbomachines. The PIV measurements, either 2D or 3D, give access to steady values of velocity and turbulent kinetic energy, as well as to their periodic fluctuations. With the help of specific data analysis tools, a wide range of phenomena can be analysed, such as vortex detection and pattern tracking, wake propagation and dissipation, rotor-stator interactions.

However, results with the required accuracy are only possible by rigorous developments. The main parameters influencing the measurement accuracy have been identified and their impact on the final result analyzed. A careful choice of the most favorable optical configuration, flawless calibration setup and adequate adjustment of the acquisition parameters are the ingredients of a successful PIV experiment. Furthermore, for investigations in hydraulic turbomachinery the original geometry of the hydraulic profile should be preserved. This requirement adds a degree of complexity due to strong optical deformations and possible local discontinuities in the optical interface, which are difficult to compensate.

Through a specific implementation of the standard PIV technique in two-phase flows, it was possible to evaluate the unsteady flowfield generated by the partial load vortex rope which develops downstream a turbine runner in cavitation conditions, together with the vortex core topology. Specific developments were necessary for the PIV application in cavitation conditions, both in the 2D and the 3D configurations. New image processing tools have been created, allowing simultaneous measurements of the velocity field and related vapours core characteristics. Based on the experimental results, an analytical description of the rope has been determined. The impact of resonance on the rope behaviour was analysed as well.

These measurements represent a valuable database used for theoretical developments as well - see (Susan-Resiga et al. 2006), (Ciocan et al. 2008) the numerical simulations as boundary conditions or validation data - see (Guedes et al. 2002) and (Ciocan et al. 2007).

6. References

Beaulieu S., Deschenes C., Iliescu M., Ciocan G.D., (2009) "Study of the Flow Field Through the Runner of a Propeller Turbine using Stereoscopic PIV" 3rd IAHR International Meeting of the Workgroup on Cavitation and Dynamic Problems in Hydraulic Machinery and Systems, Brno, Czech Republic, October 14-16

Canny J., (1986) "A Computational Approach to Edge Detection," IEEE Transactions on Pattern Analysis and Machine Intelligence, Vol. PAMI-8, No. 6, pp. 679-698.

Ciocan G.D., (1998) "Contribution à l'analyse des écoulements 3D complexes en turbomachines", PhD Thesis, Institut National Polytechnique de Grenoble

Ciocan G.D., Avellan F., (2004) "Flow Investigations in a Francis Draft Tube: Advanced Experimental Methods" Proceedings of the 3rd Conference of Romanian Hydropower Engineers, Bucarest, Roumanie, Mai 28-29

Ciocan G.D., Iliescu M.S., (2009) "3D PIV Measurements in two phase flow", 4th International Conference on Energy and Environment, The 1st International Symposium on Green Energy, May, 14-16 – invitated paper

Ciocan G.D., Iliescu M.S. (2008) "3D PIV Measurements in Two Phase Flow and Rope Parametrical Modeling", 24th IAHR Symposium on Hydraulic Machinery and Systems, Foz do Iguassu, Brazil, October 27-31

Ciocan G.D., Iliescu M.S., Vu T., Nnennemann B., Avellan F., (2007) "Experimental Study and Numerical Simulation of the Flindt Draft Tube Rotating Vortex" - Journal of Fluid Engineering, ISSN: 0098-2202, vol. 129, p. 146-158

Ciocan G.D., Kueny J-L., (2006) "Experimental Analysis of the Rotor-Stator Interaction in a Pump-Turbine", 23rd IAHR Symposium on Hydraulic Machinery and Systems, Yokohama, Japan, October 17 – 21

Ciocan G.D., Kueny J-L, Mesquita A.A., (1996) "Steady and unsteady flow pattern between stay and guide vanes in a pump-turbine" - Proceedings of the XVIII International Symposium on Hydraulic Machinery and Cavitation, IAHR, vol. 1, p. 381-390, Valencia, Spain, 16-19 September

Ciocan G.D., Mauri S., Arpe J., Kueny J-L., (2001) "Etude du champ instationnaire de vitesse en sortie de roue de turbine - Etude expérimentale et numérique" - La Houille Blanche no. 2, p. 46-59

Gagnon J.M., Ciocan G.D., Deschênes C., Iliescu M.S., (2008) "Experimental investigation of runner outlet flows in an axial turbine using LDV and stereoscopic PIV", 24th IAHR Symposium on Hydraulic Machinery and Systems, Foz do Iguassu, Brazil, October 27-31

Guedes A., Kueny J-L., Ciocan G.D., Avellan F., (2002) "Unsteady rotor-stator analysis of a hydraulic pump-turbine – CFD and experimental approach" – Proceedings of the XXI International Symposium on Hydraulic Machinery and Cavitation, IAHR, p.767-780, Lausanne, Suisse, 9-12 September

Houde S., Iliescu M.S., Fraser R., Deschênes C., Lemay S., Ciocan G.D., (2011) "Experimental and Numerical Analysis of the Cavitating Part Load Vortex Dynamics of Low-Head Hydraulic Turbines" Proceedings of ASME-JSME-KSME Joint Fluids Engineering Conference, paper AJK2011-FED, Hamamatsu, Shizuoka, Japan, July 24-29

Iliescu M.S., Ciocan G.D., Avellan F., (2008) "Two Phase PIV Measurements of a Partial Flow Rate Vortex Rope in a Francis Turbine" – Journal of Fluids Engineering, ISSN: 0098-2202, Volume 130, Issue 2, pp. 146-157

Iliescu M.S. (2007) "Large scale hydrodynamic phenomena analysis in turbine draft tubes", EPFL Thesis n° 3775

Iliescu M.S., Ciocan G.D., Avellan F. (2004): "Experimental Study of the Runner Blade-to-Blade Shear Flow Turbulent Mixing in the Cone of Francis Turbine Scale Model", 22nd IAHR Symposium on Hydraulic Machinery and Systems, Stockholm, Sweden, June 29 – July 2

Iliescu M., Ciocan G.D., Avellan F., (2003) "2 Phase PIV Measurements at the Runner Outlet in a Francis Turbine" The 2003 Joint US ASME-European Fluids Engineering Summer Conference, Honolulu, Hawaii, USA, July 6-10, – Award in the Student Paper Contest

Mesquita A.A., Ciocan G.D., Kueny J-L., (1999) "Experimental Analysis of the Flow between Stay and Guide Vanes of a Pump-Turbine in Pumping Mode" – Journal of the Brazilian Society of Mechanical Sciences and Engineering, ISSN: 1678-5878, vol. 21, no. 4, p. 580-588

Susan-Resiga R., Ciocan G.D., Anton I., Avellan F., (2006) "Analysis of the Swirling Flow Downstream a Francis Turbine Runner" – Journal of Fluids Engineering, vol. 128, p. 177-189

Tridon S., Barre S., Ciocan G.D., Leroy P., Ségoufin C. (2010), "Experimental Investigation of Draft Tube Flow Instability", Institute of Physics (IoP) Conf. Series: Earth and Environmental Science 12

Tridon S., Barre S., Ciocan G.D., Tomas L., (2010) "Experimental Analysis of the Swirling Flow in a Francis Turbine Draft Tube: Focus on Radial Velocity Component

Determination" - European Journal of Mechanics – B/Fluids, ISSN: 0997-7546, vol. 25, Issue 4, pp. 321-335

Tridon S., Ciocan G.D., Barre S., Tomas L., (2008) "3D Time-resolved PIV Measurements in a Francis Draft Tube Cone", 24th IAHR Symposium on Hydraulic Machinery and Systems, Foz do Iguassu, Brazil, October 27-31

4

Post-Processing Methods of PIV Instantaneous Flow Fields for Unsteady Flows in Turbomachines

G. Cavazzini[1], A. Dazin[2], G. Pavesi[1], P. Dupont[2] and G. Bois[2]
[1]*Department of Mechanical Engineering, University of Padova, Padova,*
[2]*Laboratoire de Mécanique de LILLE (UMR CNRS 8107),*
Arts et Métiers ParisTech, École Centrale de Lille,
[1]*Italy*
[2]*France*

1. Introduction

Among the experimental techniques, the particle image velocimetry (PIV) is undoubtedly one of the most attractive modern methods to investigate the fluid flow in a non-intrusive way and allows to obtain instantaneous fluid flow fields by correlating at least two sequential exposures. This technique was successfully applied in several fields in order to study high complex three-dimensional flow velocity fields and to provide a significant experimental data base for the validation of combined numerical analysis models.

However, on one side, experimental limits and possible perturbing phenomena could negatively affect the PIV experimental accuracy, altering the real physics of the studied fluid flow field. On the other side, the huge amount of data obtainable by means of the PIV technique requires properly post-processing tools to be exploited in an in-depth study of the fluid-dynamical phenomena.

To avoid misinterpretation of the phenomena, complex cleaning techniques were developed and applied at the different steps of the PIV processing, starting from the acquired images (background subtraction, mask application, etc.) so as to increase the signal to noise ratio, and finishing to the instantaneous flow fields by means of statistical methods applied in order to identify residual spurious vectors [Raffel et al., 2002]. Even though all these methods allows to obtain a good filtering of the instantaneous flow fields, however they are not able to completely eliminate all the outliers in the results since the removal criteria are always dependent on the choice of a threshold value [Heinz et al., 2004; Westerweeel, 1994; Westerweel and Scarano, 2005].

To overcome this problem, the most common approach is to average the instantaneous PIV flow fields so as to improve the quality of the resulting flow field reconstruction and to more easily identify the flow field characteristics in the investigated area.

Several averaging methods were proposed and applied in literature. However their effectiveness in reducing the spurious vector number is strictly connected with the flow field characteristics, the experimental set-up and the acquisition characteristics of the PIV instrumentation.

The most simple but less accurate averaging procedure is undoubtedly the classical time average of a suitable number of instantaneous velocity fields, whose effectiveness is greatly affected by the quality of the starting velocity fields. To overcome the limits of this classical method still maintaining a similar approach, Meinhart et al. (2000) proposed to determine the time average of the instantaneous correlation functions so as to determine with greater precision the correlation peak and hence the average velocity. Even though this method allows to increase the quality of the resulting averaged flow field, however it is not able to overcome the essential limit of the time-averaging methods, that is their inapplicability to unsteady flow fields and in particular to fluid-dynamical structures having a formation rate different from the framing rate of the camera. In addition to this, the method loses all the information about the evolution in time of the flow field, allowing to obtain only the averaged one.

To overcome these limits of the time-averaging methods and in particular their dependence from the framing rate of the camera, several phase-averaging methods were developed [Geveci et al. 2003; Perrin et al. 2007; Raffel et al. 1995, 1996; Schram and Riethmuller, 2001-2002; Ullum et al. 1997; Vogt et al. 1996; Yao and Pashal 1994]. These methods reorder and average the instantaneous flow fields on the basis of a proper phase, characterizing the development of the investigated phenomena so as to obtain a phase-averaged time series. These approaches, even though partially overcome the limits of the time-averaging methods, do not represent an universal solution to the problem of the data validation, since they require the characteristic frequencies of the phenomena to be known beforehand or to be determinable by combination with further experimental measurements (for example, pressure signals post-processed by spectral analysis). Moreover, they fail in case of not-periodical or frequency-combined structures, developing in the flow field.

In the first part of the chapter, a validation method of PIV results was proposed to critically analyse the quality and the meaningfulness of the experimental results in a PIV analysis on unsteady turbulent flow fields, commonly developing in turbomachines. The procedure was tested on the results of a classical phase-averaging method and was subdivided into three main steps: a convergence analysis to verify the fairness of the number of acquired images; an analysis of the probability density distribution to verify the repeatability of the velocity data; an evaluation of the maxima errors associated with the velocity averages to quantitatively analyse their trustworthiness. The procedure allowed to statistically verify the meaningfulness of the average flow field in unsteady flow conditions and to identify possible zones characterized by a low accuracy of the averaging method results.

In the second part of the chapter, a particular averaging method of PIV velocity fields was proposed to experimentally capture and visualize the unsteady flow field associated with an instability developing in a turbomachine with a known movement velocity. According to this method, the PIV flow fields was properly spatially moved according to its development velocity and was averaged on the basis of their new location. This procedure allowed to combine and average the flow fields in a frame moving with the instability so as to obtain a global visualization of the instability characteristics.

2. Validation method of PIV results

The experimental results, on which the validation procedure was tested, were obtained in a 2D/2C PIV measurement campaign carried out on one diffuser blade passage of a centrifugal pump (fig. 1).

Fig. 1. Schematic representation of the centrifugal pump. The dotted line indicates the investigated diffuser blade passage.

All the details about the test rig and the measurement devices, being outside the interest of this work, are not here reported, but can be found in previous studies [Wuibaut et al., 2001-2002].

As regards the images acquisition and processing, two single exposure frames were taken each two complete revolutions of the impeller and 400 instantaneous flow fields were determined for various operating conditions at different heights (Fig. 2). A home-made software was used to treat and process the images so as to increase the signal to noise ratio (background subtraction, mask application, etc.) and a detailed cleaning procedure was applied to the instantaneous flow field to remove possible spurious vectors.

Since the turbulent phenomena under investigation were expected to be periodically associated with the impeller passage frequency, a phase-averaging technique based on this frequency was applied the instantaneous flow fields.

2.1 Convergence history

The first parameter to be considered to verify the meaningfulness of an averaged flow field is undoubtedly the number of acquired images, whose choice is generally affected by two conflicting aims. On one side, the meaningfulness of the averaged flow field that is favoured by a great number of acquired images; on the other side, the reduction of the acquisition time and of the required data storage capacity, increasing with the images number.

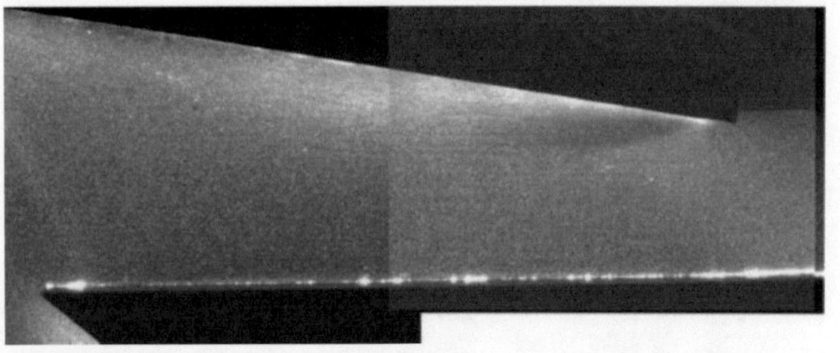

Fig. 2. Seeding of the blade passage as seen by PIV cameras with an overlapping (black parts are the walls of the diffuser passage).

So, to determine a suitable number of images to be acquired, a convergence analysis, similar to that suggested by Wernert and Favier (1999), has to be applied.

This analysis studies the evolution in time of the average $\overline{C_N}(x,y)$ and of standard deviation $\varepsilon_N(x,y)$ of the absolute velocity C(x,y) over an increasing number of flow fields:

$$\overline{C_N}(x,y) = \frac{1}{N}\sum_{i=0}^{N} C(x,y,t_0 + i\Delta t) \tag{1}$$

$$\varepsilon_N(x,y) = \sqrt{\frac{1}{N-1}\sum_{i=1}^{N}\left(\overline{C_i}(x,y) - \overline{C_{N\max}}(x,y)\right)^2} \tag{2}$$

where N is the progressive number of flow fields (N=1,...N_{max}), N_{max} is the total number of determined flow fields, Δt is the sampling period, t_0 is the initial instant, $C(x,y,t_0+i\Delta t)$ is the absolute velocity at the coordinates (x, y) of the flow field (i+1), $\overline{C_i}(x,y)$ is the average of the absolute velocity determined over 'i' flow fields at the coordinates (x, y) and $\overline{C_{N\max}}(x,y)$ is the average of the absolute velocity over the total number of acquired flow fields at the coordinates (x, y).

The analysis of the evolution in time of the average velocity and of its standard deviation allows to verify the existence of a minimum number of flow fields to be averaged so as to obtain a meaningful averaged flow field. For example, the convergence history of fig. 3 is characterized by an asymptotic behaviour of the average and standard deviation with a asymptotic value reached after about 300 flow fields. This number represents the minimum number of flow fields to be determined in order to obtain a meaningful result. A greater number would not change the resulting average velocity and would not increase its meaningfulness.

A different behaviour characterized the convergence history of fig. 4, where both the average velocity and its standard deviation are not clearly stabilized after 400 flow fields. The velocity tends to zero and the standard deviation is of the order of the average velocity,

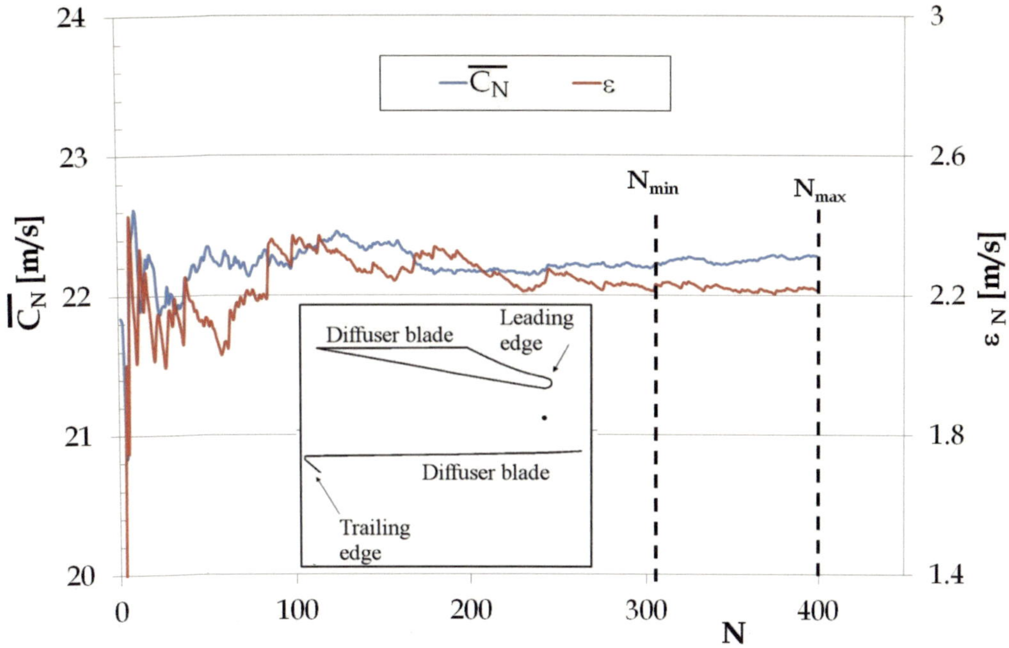

Fig. 3. Convergence history in a point located at the entrance of the diffuser passage at mid-span.

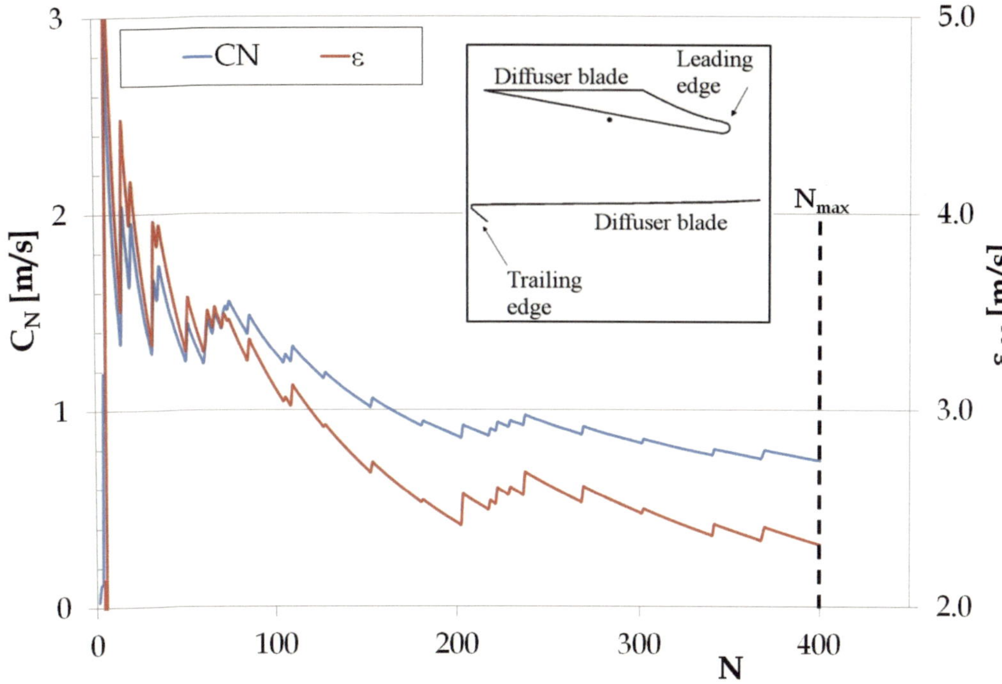

Fig. 4. Convergence history in a point located in the middle of the diffuser passage at mid-span near the blade pressure side.

highlighting a great perturbation of the instantaneous velocity values around the average one.

Before indistinctly increasing the number of images to be acquired, a critical analysis of the convergence history is necessary so as to consider the reasons of the non-convergent trend. The development of turbulence flows or unsteady structures in the zone of the reference point and/or experimental problems such as laser reflections or seeding problems should be considered. In fig. 4, as the reference point is located near the pressure side of the pump diffuser blade, laser reflection problems as well as spurious vectors owing to the boundary-layer development could not be excluded. The trend to zero of the progressive averaged velocity and the great values of the standard deviation support this hypothesis. In this case, the acquisition of a higher number of images would not probably guarantee the achievement of a meaningful averaged velocity value.

Hence, the convergence analysis appears to provide useful information about the proper choice of the number of images to be acquired, but allows only preliminary hypotheses on the quality of the results.

2.2 Probability density distribution

A phase-averaging method is based on the hypothesis that the experimental values to be averaged are repeated measures of the same experimental quantity. However, the repeatability of the experimental measurements in an investigated area could be invalidated by the possible development of non-periodical fluid-dynamical phenomena and by possible experimental problems. The lack of this repeatability, negatively affecting the accuracy of the phase-averaging method, is highlighted by a non-Gaussian probability density distribution of the experimental values. Therefore, the second step of the validation procedure is the analysis of the probability density distribution of the determined velocity values.

Since the aim of the analysis was to verify the Gaussianity of this distribution, no hypothesis on its form can be done. Hence, the probability density function has to be estimated using non-parametric kernel smoothing methods, with no hypothesis on the original distribution of the data [Bowman and Azzalini, 1997].

Figure 5 reports three examples of possible probability density distributions of velocity values, translated to have zero mean value. In fig. 5a, the classical symmetric bell-shape of the Gaussian distribution testifies the repeatability of the corresponding experimental measures. Moreover, the great values of the probability density function demonstrates the meaningfulness of the determined average velocity. In contrast, fig. 5b shows an asymmetric distribution of the data with multiple maxima and a wide dispersion of the values. In this case, the velocity average, corresponding to the abscissa $\mu = 0$, cannot be considered as meaningful, even if a higher number of images will be acquired, as the repeatability of the measures is not guaranteed.

The analysis of the probability density distribution could also allow to identify possible experimental problems. Indented bell-shaped distribution, as that reported in fig. 5c, clearly indicates the presence of peak-locking problems in the images acquisition. The peak-locking does not affect the mean velocity flow field but only its fluctuating part [Christensen, 2004]

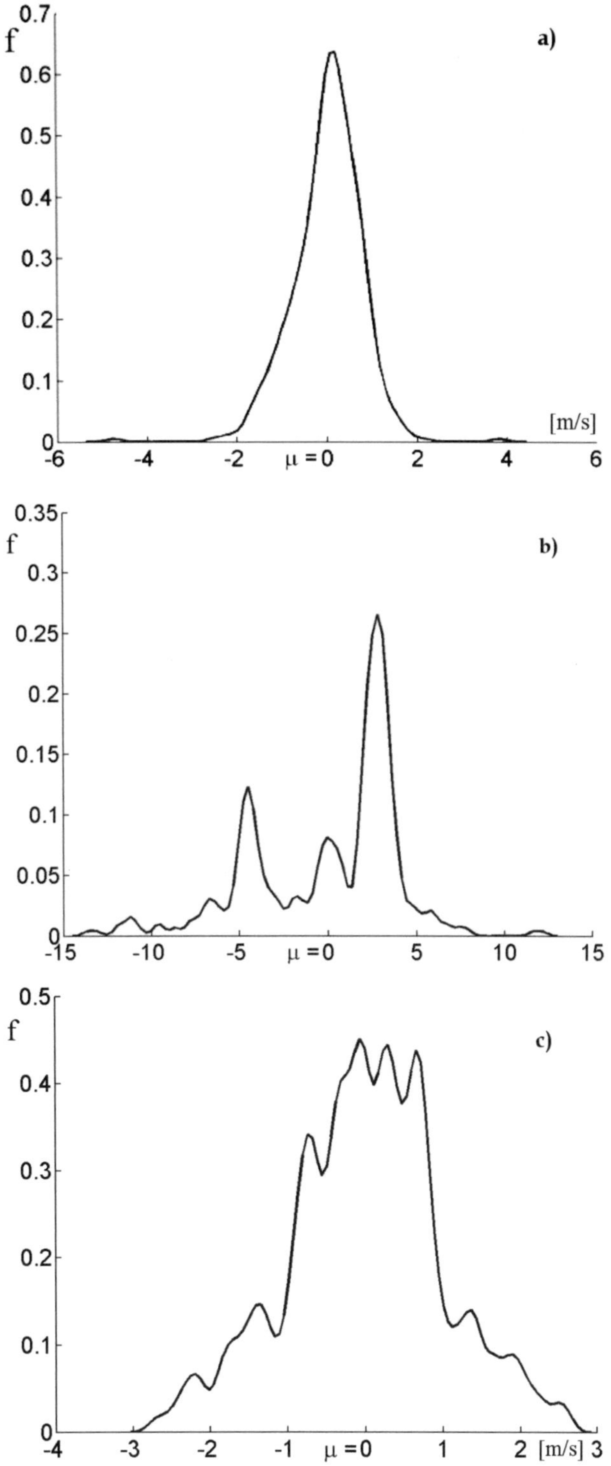

Fig. 5. Examples of probability density distributions of velocity values

and is attributable to both the choice of the sub-pixel estimator and the under-resolved optical sampling of the particle images. So, even though this problem does not affect the meaningfulness of the average velocity value, its identification during a preliminary study could allow to correct the test rig set-up and to increase the quality of the instantaneous flow fields.

Even though the subjective visual analysis of the probability density distributions allowed a preliminary analysis of the measurement repeatability and of the presence of possible experimental problems, to effectively verify the normality of the velocity data distributions in the investigated area, it is necessary to apply a goodness-of-fit test. This test allows to verify the acceptance of the so-called 'null' hypothesis (i.e. if the data follow a specific theoretical distribution) or of the alternative hypothesis (i.e. if the data do not follow the specified distribution). The null hypothesis is rejected with a confidence level α, if the test statistic is greater than a critical value fixed at that confidence level. The greater the difference between the statistic and the critical value is, the greater the probability that the data do not follow the specified distribution.

In literature, several tests with different characteristics and powers are available (Pearson $\chi 2$, Anderson–Darling, Kolmogorov, Cramer–von Mises–Smirnov, and so on). The hypotheses of these tests can be divided into simple ones, when the parameters of the theoretical distribution to be checked are known, and complex ones, when the parameters of the theoretical distribution are determined using the same data sample to be tested. Complex hypotheses are typical of a PIV experimental analysis since the average μ and the standard deviation ε of the theoretical distribution are generally unknown and the average $\overline{C_N}(x,y)$ and the standard deviation $\varepsilon_N(x,y)$ of the sample are used as distribution parameters.

Among the available goodness-of-fit tests, the choice of the most proper one depends on the sampling conditions of the data. Lemeshko et al. (2007) compared the power of several goodness-of-fit tests by means of statistical modelling methods and based the choice on the desired confidence level and on the range of number of samples. For example, for 400 velocity values, as those of the experimental analysis considered as test case, and for a confidence level of 0.05 (5 per cent), the suggested goodness-of-fit test is the Anderson–Darling test.

However, the sampling conditions of an experimental analysis cannot be generalized in ranges and hence the trustworthiness of a goodness-of-fit test must be properly verified with a dedicated simulation. To do this, the test must be applied several times on random Gaussian samples having the same number of values of the experimental sample. The number of test applications has to be great enough to avoid a dependence of the simulation results on it. As this test is not time-expensive, the choice of a very high number of applications, such as 100 000, guarantees its negligible influence on the simulation results.

The trustworthiness of the goodness-of-fit test is verified once the error percentage of its application on Gaussian samples is lower or at most equal to the desired confidence level. In the example considered earlier, the Anderson–Darling test was applied 100 000 times on random Gaussian samples of 400 data, giving an error value of 5 per cent, which is not greater than the desired confidence level. This result demonstrates the applicability of the Anderson–Darling test to the experimental analysis sampling conditions.

Once verified its trustworthiness, the goodness-of-fit test has to be applied to each point of the investigated area so as to obtain a global visualization of the non-Gaussian critical zones of the flow field. Figure 6 reports an example of the Anderson-Darling application to the velocity flow fields determined in the diffuser passage. Cores of non-Gaussian distribution can be clearly identified at the entrance of the diffuser passage, near the blade walls and also in the mean flow (light grey circles in fig. 6). Even though it is always quite difficult to discriminate between fluid-dynamical and experimental problems, these results allow some preliminary hypotheses about the origin of these critical areas. At the diffuser entrance, the position of the impeller blade close to diffuser blade leading edge lets suppose the development of non-periodical turbulent phenomena coming from the impeller discharge, such as impeller blade wakes and/or rotor-stator interaction phenomena.

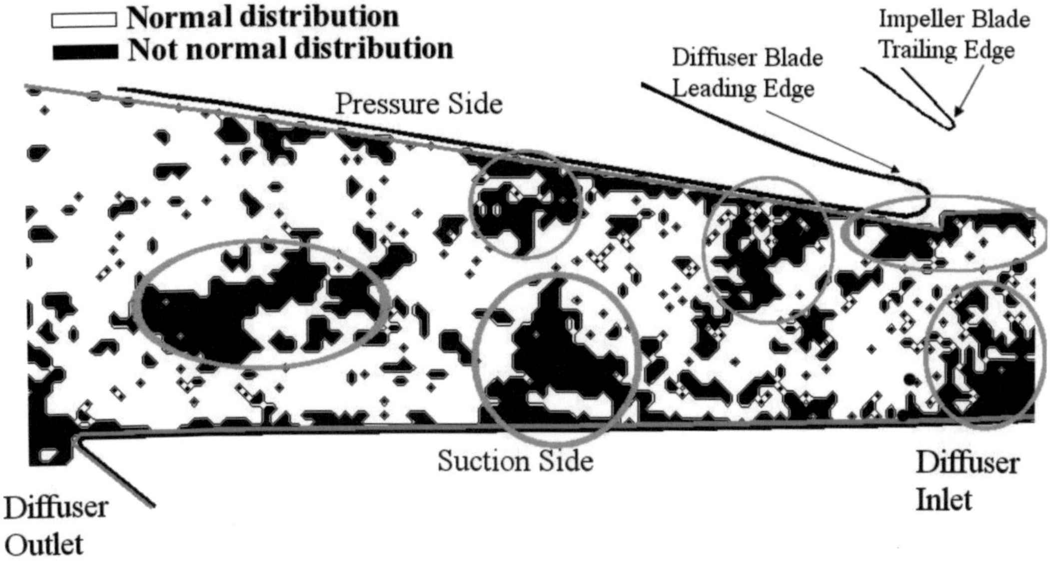

Fig. 6. Example of the results of the Anderson-Darling test applied to the pump diffuser blade passage (the black line represents the blade profile; the dark grey line represents the limit of the mask applied to the grid and the light grey circles are marks to identify the non-Gaussian zones in the figure)

The non-Gaussian cores near the blade sides could be due to possible laser reflections problems or seeding problems, whereas laser reflections can be excluded for the cores in the mean flow because of their distance from the walls. These cores could be probably due to non-periodical phenomena proceeding in the passage, but this hypothesis has to be verified by numerical analysis of the flow field.

2.3 Confidence interval of the measured values

The validation procedure of the experimental results is completed by a measure of the reliability of the averaged flow field, that is obtained by estimation of the confidence interval of the determined averaged velocities, that depends on the effective distribution of the experimental data.

In the hypothesis of normal distribution, the confidence interval of the average velocity $\overline{C_N}(x,y)$ for a confidence level (1-α) is:

$$\left[\overline{C_N} - \frac{\varepsilon_N}{\sqrt{N}}\Phi^{-1}\left(1-\frac{\alpha}{2}\right), \overline{C_N} + \frac{\varepsilon_N}{\sqrt{N}}\Phi^{-1}\left(1-\frac{\alpha}{2}\right)\right] \quad (3)$$

where Φ is the normal cumulative distribution function and ε_N the standard deviation of the average velocity (Montgomery and Runger, 2003). So, in this hypothesis, the maximum error in the estimation of the average velocity is:

$$\frac{\varepsilon_N}{\sqrt{N}}\Phi^{-1}\left(1-\frac{\alpha}{2}\right) \quad (4)$$

When the goodness-of-fit test highlights not-normal distributions of the velocity values, to correctly determine the corresponding confidence interval, the effective distribution of the experimental data should be investigated. However, according to the central limit theorem, the procedure for estimating the confidence interval of normal samples can be also applied, with approximation, to not-normal samples if their dimension is sufficiently great (Montgomery and Runger, 2003). This estimation, even approximate, is useful to critically analyse the meaningfulness of the experimental results and to identify the problematic zones of the investigated area.

Figure 7 reports the distribution of the maxima errors (eq. (4)) that can be made in the evaluation of the averaged velocity in the diffuser blade passage. As it can be seen, the maxima errors are localized near the diffuser blade profiles in the inlet throat of the diffuser

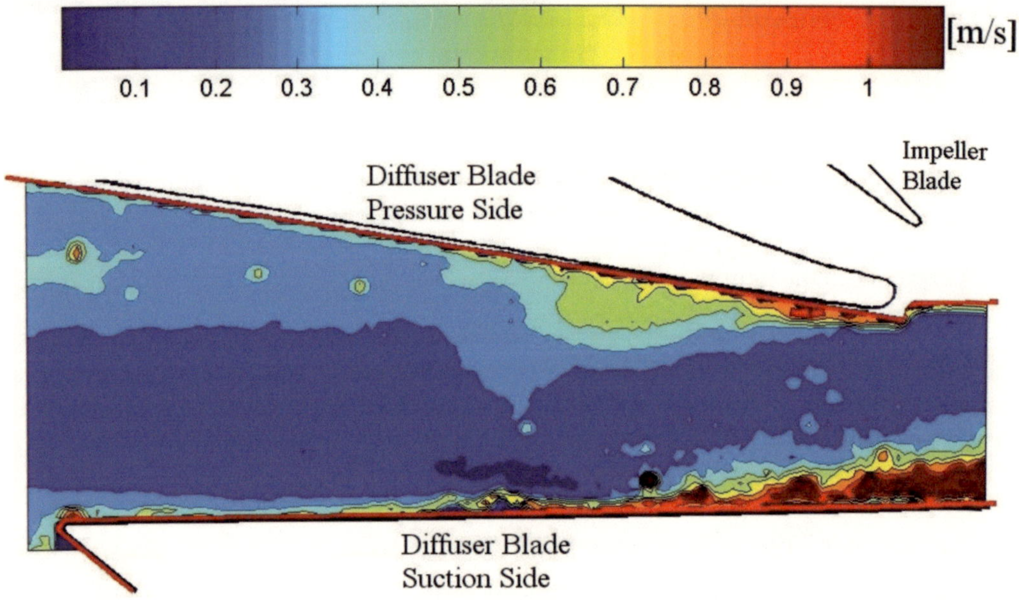

Fig. 7. Example of distribution of the maxima errors of the averaged velocity in the diffuser passage (the black line represents the blade profile; the red lines represents the limit of the mask applied to the grid)

passage and could be attributed to the combination of the boundary-layer development with experimental problems such as reflection or seeding problems and to vortical cores coming from the impeller discharge on the suction side.

The further proof of the possible development of unsteady phenomena could be obtained by the spectral analysis of the velocity signals. Figure 8 reports the fast Fourier transform (FFT)

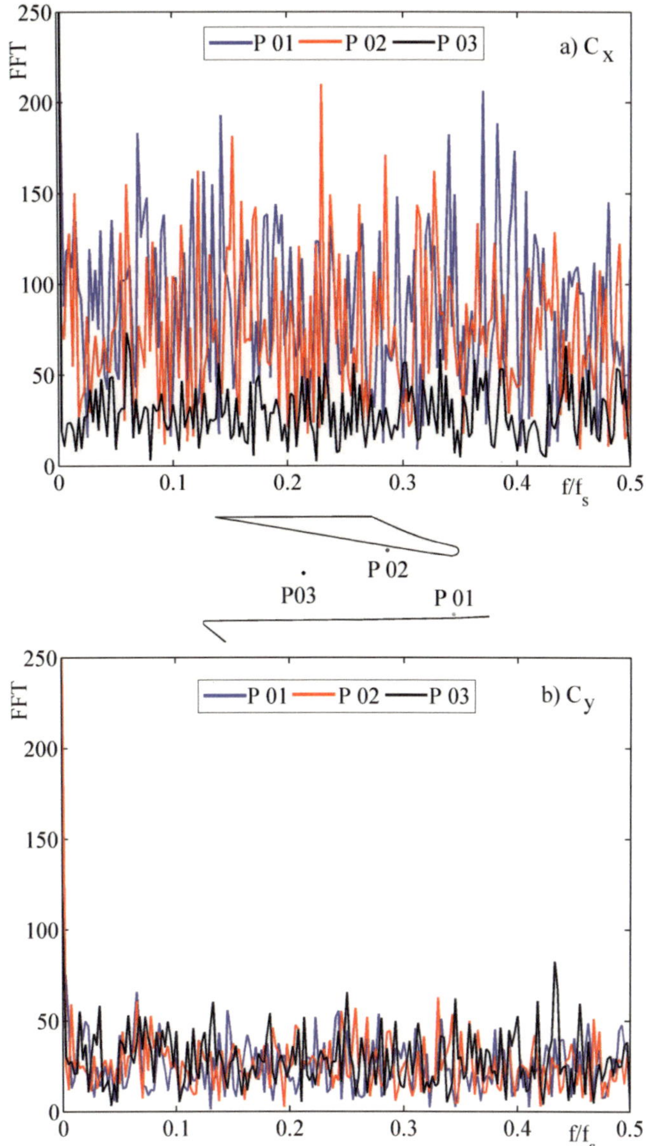

Fig. 8. FFT of the velocity components signals determined in three points of the diffuser passage: at the entrance near the suction side (blue line), near the blade pressure side (red line) and in the second half of the diffuser passage far from the blade profiles (black line). a) Cx b) Cy

of the velocity components signals determined in three significant points of the flow field of fig. 7: near the suction side at the entrance of the diffuser passage (blue line), near the pressure side in the zone of maximum error (red line) and in the second half of the diffuser passage far from the blade profiles (black line). The FFT results are reported as a function of the ratio between the frequency f and the sampling frequency f_s of the data.

Concerning the velocity component in the mean flow direction Cx (fig. 8a), the points near the blade profiles (points 1 and 2) present peaks having amplitudes much greater than those of the point placed in the mean flow, whereas the velocity component in the direction normal to the mean flow Cy presents low FFT peaks for all the three points (fig. 8b). This strengthens the hypothesis of intense unsteady velocity fluctuations near blade profiles, proceeding in the mean flow direction, and hence confirms the results of the previous statistical analysis.

2.4 Comparison between experimental and numerical results: A critical validation

In fluid-dynamical investigations, the experimental results generally represent a significant reference database for the validation of combined numerical analysis models. However, since the experimental analysis could be negatively affected by experimental problems or post-processing limits, the possible discrepancies between experimental and numerical results have to be critically analysed in order to correctly identify the real error sources.

Numerical error sources, such as the grid resolution and the choice of the turbulent model, have to be considered in the comparison, but may not be the only causes. Experimental problems due to the test-rig or to unsteady phenomena could be also possible reasons for discrepancies. In this context, the validation procedure is extremely useful since it allows to identify the problematic zones of the investigated flow field and to appreciate the meaningfulness of the velocity averages for a critical comparison between the numerical and experimental results.

Figure 9 shows a comparison between the experimental results of fig. 7 and the results of a numerical analysis carried out on the same machine at the same operating conditions. Averaged velocity profiles determined in some sections of the diffuser blade passage are compared. The agreement is quite good, but there are some discrepancies near the diffuser blades (y/l=0 and y/l=1). Numerical error sources, such as a low stream-wise grid resolution or an improper choice of the turbulent model have to be considered as possible causes. However, the validation procedure previously applied to the PIV flow fields highlighted a low trustworthiness of the experimental averaged flow field near the blade profiles (fig. 7), indicating that the discrepancies between numerical and experimental results can be also due to experimental limits, such as reflection problems or difficult seeding near the blades.

In a preliminary study, this combined analysis could be also exploited to modify the set-up of the test rig so as to increase the experimental results quality in the problematic zones. For example, the possible reflection problems near the blade profiles of fig. 9 could be reduced modifying the laser configuration (fig. 10).

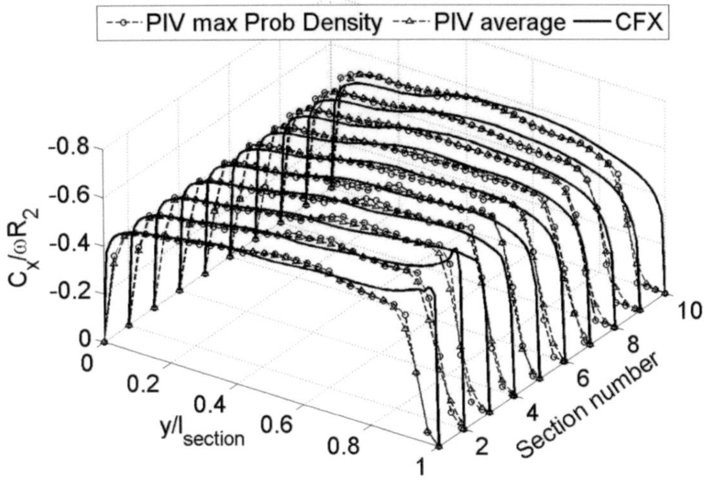

Fig. 9. Comparison between experimental and numerical results: average velocity profiles in some sections of the diffuser passage

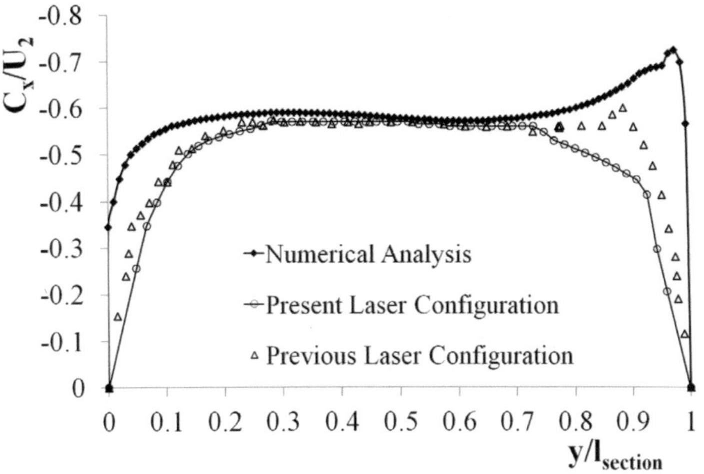

Fig. 10. Effects of the modification of the laser sheet direction on the experimental results quality

3. A new averaging procedure for instabilities visualization

PIV is now a method widely used in the field of Turbomachinery. It has proved its ability to provide useful experimental data for various research topics: rotor stator interaction in radial pumps or fans [Cavazzini et al., 2009; Meakhal & Park 2005; Wuibaut et al., 2002], tip-leakage vortex in axial flow compressors [Voges et al., 2011, Yu & Liu, 2007], swirling flow in hydraulic turbines [Tridon et al., 2010]. Nevertheless, in most cases, PIV was efficiently applied to catch phenomena which were correlated with the impeller rotation: PIV images were taken phase locked with the rotor. Consequently, with this kind of acquisition technique, the measurements were not able to treat phenomena, such as rotating stall or surge, whose frequencies are not constant or simply not linked with the impeller speed.

The recent development of high speed PIV offers new perspective for the application of the PIV technique in Turbomachinery. Van den Braembussche et al. (2010) have recently proposed a original experiment in which the PIV acquisition system was rotating with a simplified rotating machinery. However, this technique does not overcome the problem of studying rotating phenomena whose frequency was not determined before the experiment.

To catch such type of phenomena, an original averaging procedure of the data based on a frequency or time-frequency analysis of a signal characteristic of the phenomenon was developed. The procedure was applied on two different test cases presented below: a constant rotating phenomenon and an intermittent one.

3.1 Experimental set-up of the test cases

The experimental results presented above were obtained in a PIV experimental analysis carried out on the so-called SHF impeller (fig. 11) coupled with a vaneless diffuser. The tests were made in air with a test rig developed for studying the rotor-stator interaction phenomena (fig. 12).

Fig. 11. SHF impeller

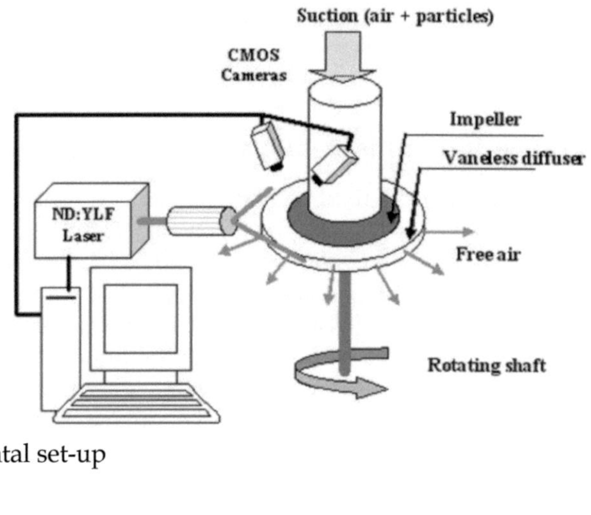

Fig. 12. Experimental set-up

A 2D/3C High Speed PIV combined with pressure transducers was used to study the flow field inside the vaneless diffuser at several flow rates and at three different heights in the hub to shroud direction (0.25, 0.5 and 0.75 of the diffuser width) with an impeller rotation speed of 1200 rpm.

The laser illumination system consists of two independent Nd:YLF laser cavities, each of them producing about 20 mJ per pulse at a pulse frequency of 980 Hz.

Two CMOS cameras (1680 x 930 $pixel^2$), equipped with 50 mm lenses, were properly synchronized with the laser pulses. They were located at a distance of 480 mm from the measurement regions with an angle between the object plane and the image plane of about 45°.

All the details about the experimental set-up, being outside the interest of this work, are not here reported, but can be found in a previous paper [Dazin et al., 2011].

The image treatment was performed by a software developed by the Laboratoire de Mecanique de Lille. The cross-correlation technique was applied to the image pairs with a correlation window size of 32 x 32 $pixels^2$ and an overlapping of 50%, obtaining flow fields of 80 x 120mm^2 and 81 x 125 velocity vectors. The correlation peaks were fitted with a three points Gaussian model. Concerning the stereoscopic reconstruction, the method first proposed by Soloff et al. (1997) was used. A velocity map spanned nearly all the diffuser extension in the radial direction, whereas in the tangential one was covering an angular portion of about 14°.

Each PIV measurement campaign was carried out for a time period of 1.6 seconds, corresponding to 32 impeller revolutions at a rotation speed of 1200 rpm. Since the temporal resolution of the acquisition was of 980 velocity maps per second, the time period of 1.6 second allowed obtaining 1568 consecutives velocity maps, corresponding to about 49 velocity maps per impeller revolution.

As regards the pressure measurements, two Brüel & Kjaer condenser microphones (Type 4135) were placed flush with the diffuser shroud wall at the same radial position (1.05 of diffuser inlet radius r_3) but at different angular position ($\Delta\theta$=75°). The unsteady pressure measurements, acquired with a sampling frequency of 2048 Hz, were properly synchronized with the PIV image acquisition system.

3.2 Constant angular velocity phenomena

At partial load and in particular at 0.26 Q_{des}, previous analyses showed that a rotating instability developed in the vaneless diffuser [Dazin et al., 2008].

The existence of this instability was determined through the analysis of the cross-power spectra of the pressure signals acquired by the two microphones at partial load and at design flow rate (fig. 13) [Dazin et al 2008, 2011]. The spectrum at design flow rate was clearly dominated by the blade passage frequency f_b ($7 \cdot f_{imp}$). At partial load ($Q=0.26Q_{des}$) the frequency spectrum was overcome by several peaks in the frequency band between $0.5f_{imp}$ and $2.0f_{imp}$, particularly by the frequency $f_{ri} = f/f_{imp}=0.84$, that was demonstrated to be the fundamental frequency of a rotating instability composed by three cells rotating around the impeller discharge with an angular velocity equal to 28% of the impeller rotation velocity [Dazin et al 2008].

The identification and visualization of the topology of these instability cells was not immediate since the angular span of one PIV map (about 14° of the whole diffuser) was much smaller than the size of an instability cell (about 75°).

To overcome this limit, a new averaging method was developed so as to combine the PIV velocity maps on the basis of the determined instability precession velocity and to obtain an averaged flow field in a reference frame rotating with the instability.

The knowledge of the instability angular speed was needed to be able to apply the PIV averaging procedure.

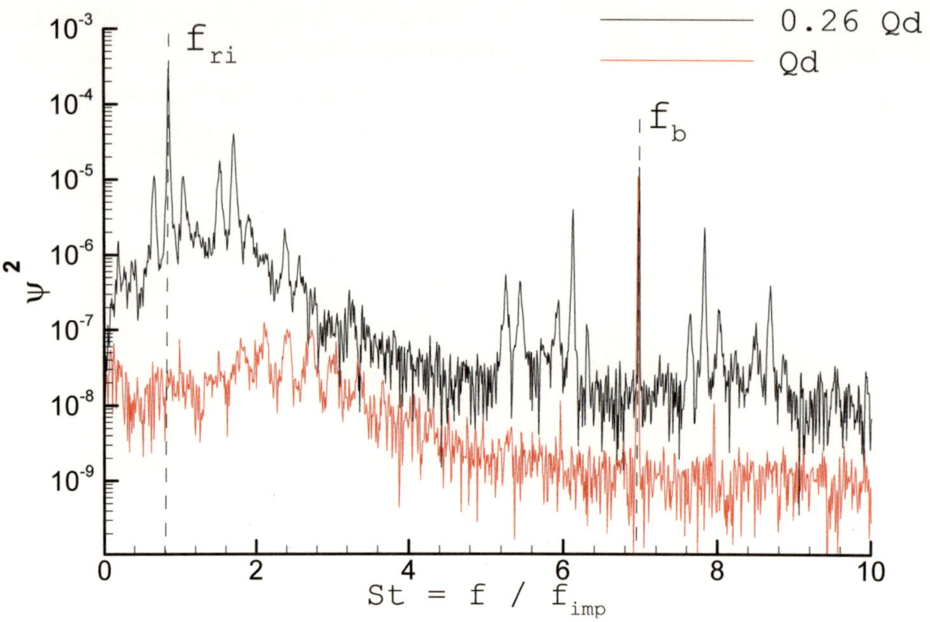

Fig. 13. Cross-power spectra of pressure signals acquired by the microphones at the design flow rate Q_{des} (red line) and at $0.26Q_{des}$ (black line)

First, since the measurements were not synchronized with the instability rotation, the velocity maps could not be exactly superimposed at each impeller revolution. So it was necessary to create a mesh (Fig. 14), having the same dimensions of the diffuser ($0<\theta<360°$, $0.257<r<0.390$ m), to be used as reference grid for the combination of the PIV maps. To have an almost direct correspondence between this mesh and the PIV grid, the size of one cell of the mesh was fixed roughly equal to the size of one cell of the PIV grid.

Then, the first velocity map was bi-linearly interpolated on the new grid, as shown for the tangential velocities in fig. 15a. The velocity values of the mesh were fixed equal to zero (green in the figure) except in the zone corresponding to the first PIV map properly interpolated on the reference grid. Since the reference frame was fixed to rotate with the instability, the second velocity map was added in the new mesh after a rotation of an angle

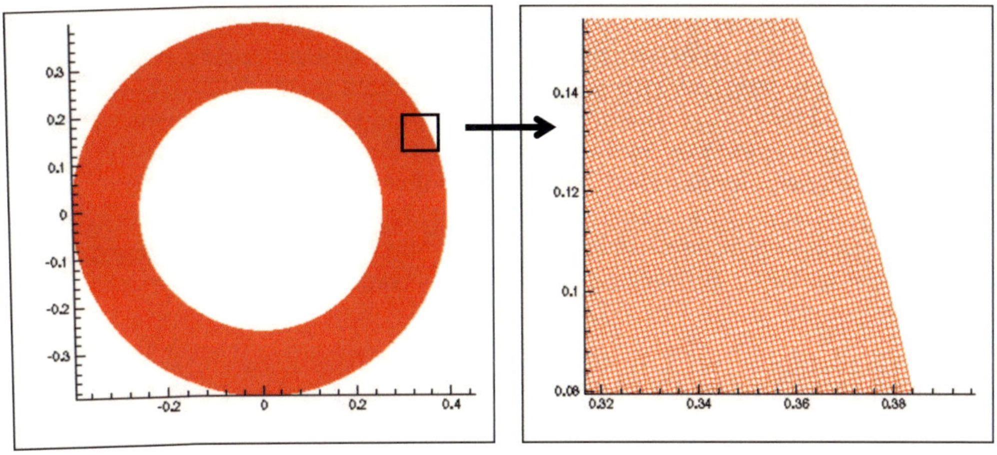

Fig. 14. Mesh used for the averaging procedure.

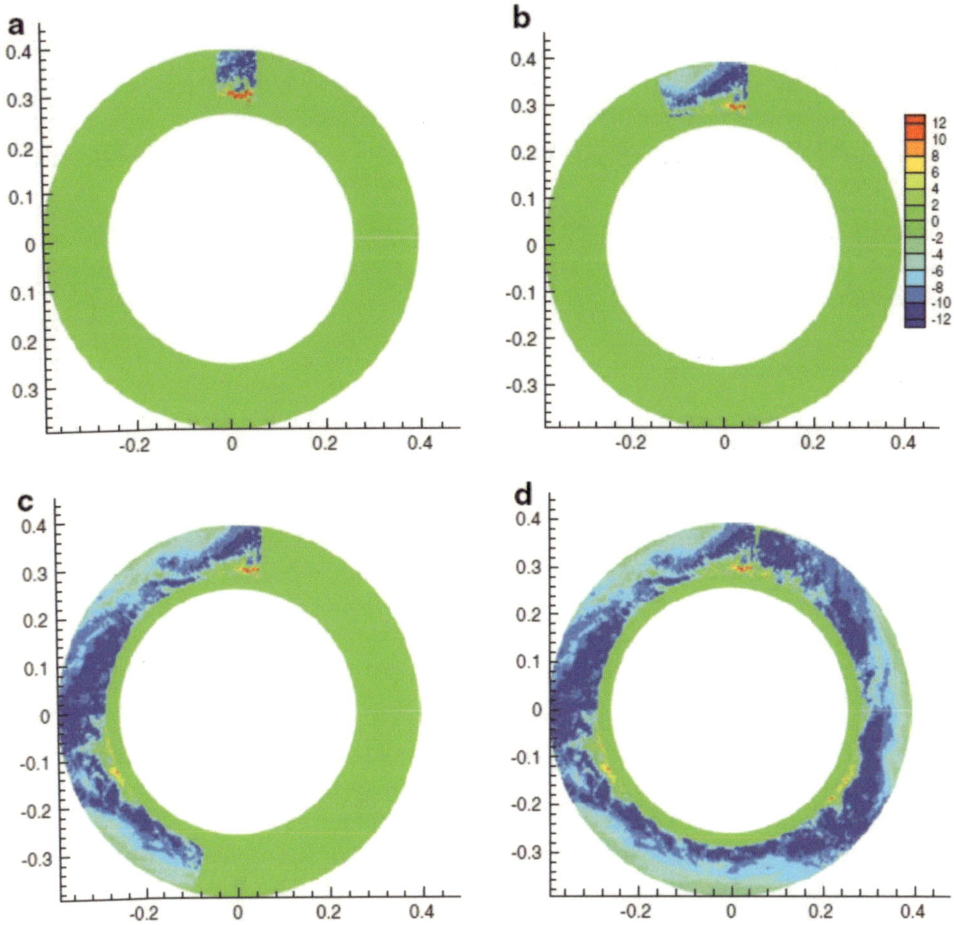

Fig. 15. Averaging computation results after 1, 10, 80 and 175 velocity maps for the tangential velocity component at mid span (in m/s)

equal to the instability velocity multiplied by the sampling period of the PIV measurements. As this second velocity map overlapped the first one, in the overlapping zone the velocity values were properly averaged. This operation was repeated for the following velocity maps till a complete revolution of the instability, corresponding to 175 maps, was made. Afterwards, the maps were averaged with the ones of the previous revolution(s). Examples of the averaging computation results respectively after 10, 80 and 175 velocity maps are reported in fig. 15(b-d). At the end of the procedure, 120 velocity vectors were averaged in each point of the reference grid, obtaining a mean velocity vector. The standard deviation was of the order of 2 m/s and the corresponding 95% confidence interval for each averaged velocity component $\overline{c}_i$ was:

$$[\overline{c}_i \pm 0.4m / s]$$

The procedure described above allowed to obtain averaged flow fields in a reference frame rotating with the instability for the three velocity components (fig. 16). Because of laser sheet reflections on the impeller blades, several instantaneous flow fields were negatively affected at the diffuser inlet by the proximity of the impeller blades. For this reason, the averaged flow fields are presented only for r> 0.3 m. The average flow field of the radial velocity component shows three similar patterns composed of two cores, respectively of inward and outward radial velocities, located near the diffuser outlet (fig 16a). In correspondence to these two cores, a zone of negative tangential velocity is identifiable near the diffuser inlet (fig. 16b) and a zone of slightly positive axial velocity is outlined within the diffuser (fig 16c).

So, the averaging procedure allowed to clearly visualize the topology of the instability rotating in the diffuser and to obtain several information about its fluid-dynamical characteristics.

3.3 Intermittent phenomena

In the same pump configuration at a greater flow rate (0.45 Q_{des}), rotating instabilities were still identified in the diffuser, but resulted to be characterized by two competitive low-frequency modes.

The first mode, which dominated the spectrum, corresponded to an instability composed by two cells rotating at $\omega/\omega_{imp} = 0.28$, whereas the second mode corresponded to an instability composed by three cells rotating at $\omega/\omega_{imp} = 0.26$ [Pavesi et al., 2011]. Moreover, the time-frequency analysis, carried out on the pressure signals, highlighted that these two competitive modes did not exist at the same time but were present intermittently in the diffuser (fig. 17).

Consequently, the averaging procedure defined in §3.2 could not be immediately applied to the PIV results but it was adapted to the new intermittent characteristics of the fluid-dynamical instability. In particular, the results of the time frequency analysis were used to determine the time periods of the acquisition process during which only one mode was dominant. Then, the PIV averaging procedure, described in §3.2, was applied only to the flow fields determined in those time periods characterized by the presence of one mode. In this way, two averaged flow fields corresponding to the two competitive modes were obtained (fig. 18).

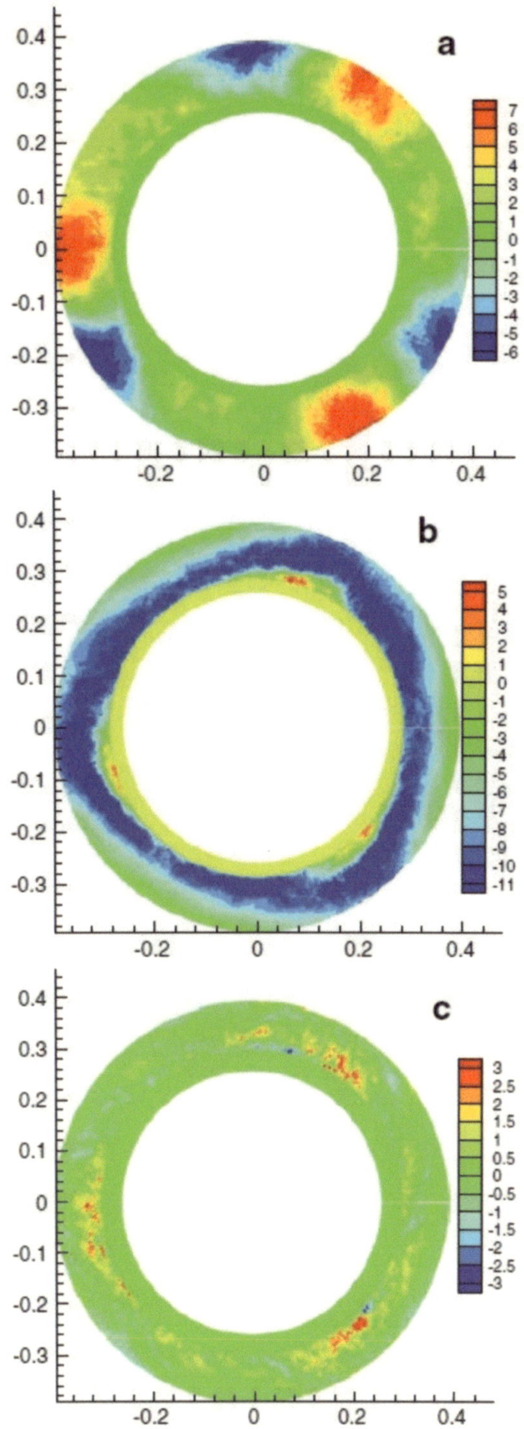

Fig. 16. Results of the averaging procedure: a) radial velocity; b) tangential velocity; c) axial velocity [m/s]

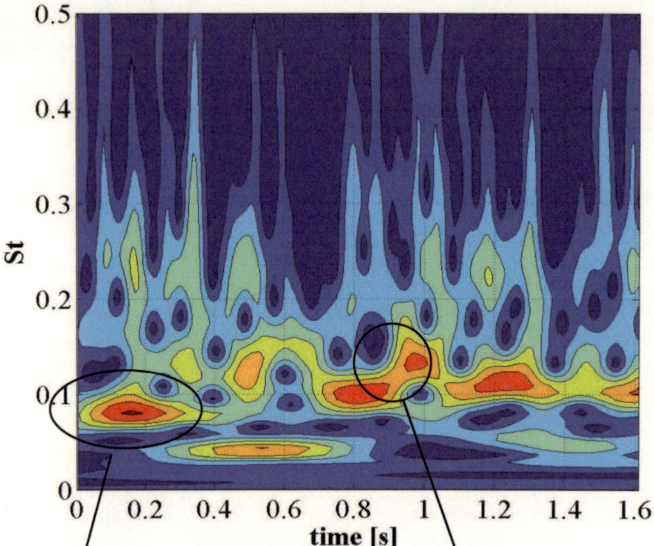

Fig. 17. Detail of the wavelet analysis of the pressure signals acquired at $Q/Q_{des} = 0.45$

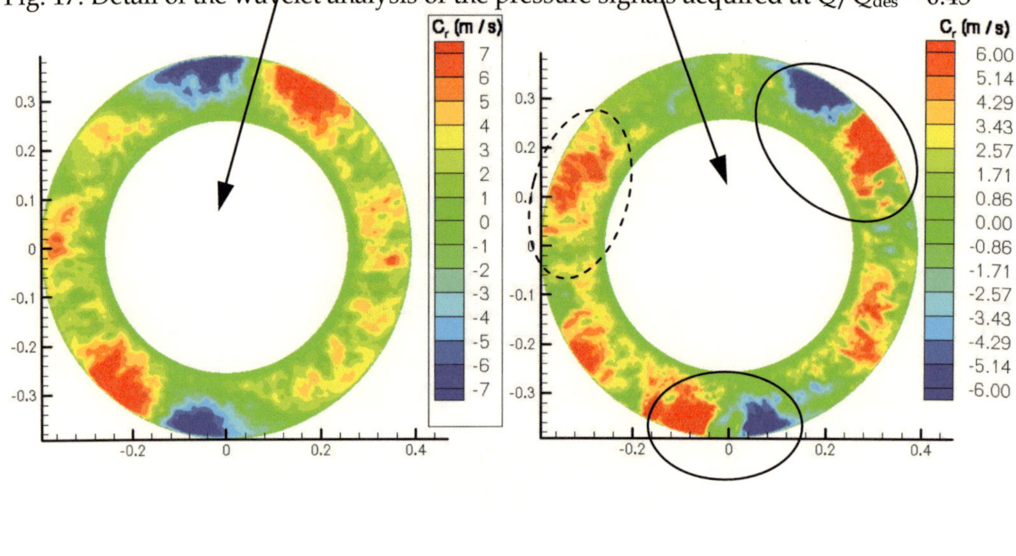

Fig. 18. PIV averaging procedure results for the two modes identified at $Q/Q_{des} = 0.45$

For example, the first mode resulted to be dominant in a time period of about 0.4 s at the beginning of the simultaneous pressure and PIV acquisitions. Consistently with the Fourier spectra analysis, the PIV averaged velocity map (fig. 18a) obtained on this time period presents two instability cells diametrically located, similar to those obtained at the lowest flow rate.

For the second mode, the longer time period identified was of only about 0.1 s. The averaged velocity procedure applied on this time period gives the velocity map plotted on fig 18b. For this mode, the expected number of cells was three, whereas the averaged velocity map presents only two clear cells (surrounded by a solid line). The third cell of this mode is hardly visible (inside the dashed lines), most probably because of a too-short period for the application of the PIV averaging procedure.

4. Conclusions

This work presents two different post-processing procedures suitable to be applied to PIV instantaneous flow fields characterized by the development of unsteady flows.

The first procedure was focused on the PIV experimental accuracy and was aimed at the validation of the averaged flow fields in a PIV analysis. This procedure combines several statistical tools and can be summarized in three main steps:

- a convergence analysis to verify that the number of acquired images allowed to obtain a meaningful averaged flow field
- the analysis of the probability density distribution to verify the repeatability of the measurements and to identify the critical area of the investigated flow field.
- the estimation of the confidence interval to evaluate the maxima errors associated with the determined velocity averages and hence to quantitatively analyze their trustworthiness.

This validation procedure can be considered not only as a necessary critical analysis of the meaningfulness of experimental PIV results, but also as a possible preliminary study for improving the test rig before starting time- and work-intensive measurement campaigns.

The second part of the chapter is focused on the averaging techniques and presents an original averaging procedure of PIV flow fields for the study of unforced unsteadiness. Since the spectral characteristics of the instability and in particular its precession velocity has to be known, the procedure is necessarily combined with a spectral analysis of simultaneously acquired pressure signals.

On the basis of the spectrally determined instability velocity, the PIV flow fields were properly combined and averaged, obtaining an average flow field in the reference frame of the instability to be studied. This result allows to capture and visualize the topology of the phenomenon and to obtain more in-depth information about its fluid-dynamical development and characteristics.

The procedure was also developed and adapted for intermittent instability configurations, characterized by competitive modes alternatively present in the flow field.

5. References

Bowman, A.W. & Azzalini, A. (1997). Applied smoothing techniques for data analysis: the kernel approach with S-plus illustrations, 1997, Oxford University Press, New York.

Cavazzini, G.; Pavesi, G.; Ardizzon, G.; Dupont, P.; Coudert, S.; Caignaert, G. & Bois, G. (2009). Analysis of the Rotor-Stator Interaction in a Radial Flow Pump. *La Houille Blanche, Revue Internationale de l'eau*, n. 5/2009, pp. 141-151.

Christensen, K. T. (2004). The influence of peak-locking errors on turbulence statistics computed from PIV ensembles. *Experiments in Fluids*, vol. 36, pp. 484-497.

Dazin, A.; Coudert, S.; Dupont, P.; Caignaert, G. & Bois, G. (2008). Rotating Instability in the Vaneless Diffuser of a Radial Flow Pump. *Journal of Thermal Science*, vol. 17(4), pp. 368-374.

Dazin, A.; Cavazzini, G.; Pavesi, G.; Dupont, P.; Coudert, S.; Ardizzon, G.; Caignaert, G. & Bois, G. (2011) High-speed stereoscopic PIV study of rotating instabilities in a radial vaneless diffuser. *Experiments in Fluids*, vol. 51 (1), pp. 83-93.

Geveci, M.; Oshkai, P.; Rockwell, D.; Lin, J.-C. & Pollack, M. (2003). Imaging of the self-excited oscillation of flow past a cavity during generation of a flow tone. *Journal of Fluid and Structures*, vol. 18, pp. 665-694

Heinz,O.; Ilyushin, B. & Markovic, D. (2004). Application of a PDF method for the statistical processing of experimental data. *International Journal of Heat and Fluid Flow*, Vol. 25, pp. 864–874.

Lemeshko, B. Y.; Lemeshko, S. B. & Postovalov, S. N. (2007). The power of goodness of fit tests for close alternatives. *Measurement techniques*, vol. 50(2), pp. 132-141.

Meakhail, T. & Park, S.O (2005). A Study of Impeller-Diffuser-Volute Interaction in a Centrifugal Fan, *Journal of Turbomachinery*, vol. 127(1), pp. 84-90.

Meinhart, C. D.; Werely, S. T. & Santiago, J. G. (2000). A PIV algorithm for estimating time-averaged velocity fields. *Journal of Fluids Engineering*, Vol. 122, pp. 285–28

Montgomery, D. C. & Runger, G. C. (2003). Applied statistics and probability for engineers, 2003, John Wiley & Sons, NewYork.

Pavesi, G.; Dazin, A.; Cavazzini, G.; Caignaert, G.; Bois, G. & Ardizzon, G. (2011). Experimental and numerical investigation of unforced unsteadiness in a vaneless radial diffuser. *ETC 9, 9th Conference on Turbomachinery, Fluid Dynamics and Thermodynamics*, Istanbul (Turkey), March 21-25, 2011

Perrin, R.; Cid, E.; Cazin, S.; Sevrain, A.; Braza, M.; Moradei, F. & Harran, G. (2007). Phase-averaged measurements of the turbulence properties in the near wake of a circular cylinder at high Reynolds number by 2C-PIV and 3C-PIV. *Experiments in Fluids*, Vol. 42, pp. 93–109.

Raffel, M.; Kompenhans, J. & Wernert, P. (1995). Investigation of the unsteady flow velocity field above a airfoil pitching under deep dynamic stall conditions. *Experiments in Fluids*, Vol. 19, pp. 103–111.

Raffel, M.; Seelhorst, U.; Willert, C.; Vollmers, H.; Bütefisch, K. A. & Kompenhans, J. (1996). Measurements of vortical structures on a helicopter rotor model in a wind tunnel by LDV and PIV, *Proceedings of the 8th International Symposium on Applications of laser techniques to fluid mechanics*, pp. 14.3.1–14.3.6, Lisbon, Portugal, 1996.

Raffel, M.; Willert, C.; Werely, S. & Kompenhans, J. (2002). Particle image velocimetry – a practical guide, 2007, Springer-Verlag, Berlin.

Schram, C. & Riethmuller, M. L. (2001). Evolution of vortex ring characteristics during pairing in an acoustically excited jet using stroboscopic particle image velocimetry, *Proceedings of the 4th International Symposium on Particle image velocimetry*, paper 1157, Gottingen, Germany, September 17–19, 2001.

Schram, C. & Riethmuller, M. L. (2002). Measurement of vortex ring characteristics during pairing in a forced subsonic air jet. *Experiments in Fluids*, Vol. 33, pp. 879–888.

Soloff, S.M.; Adrian, R.J. & Liu, Z.C. (1997). Distortion compensation for generalized stereoscopic particle image velocimetry. *Measurement Science and Technology*, vol. 8, pp. 1441-1454.

Tridon, T.; Stéphane, B.; Dan Ciocan, G. & Tomas, L. (2010). Experimental analysis of the swirling flow in a Francis turbine draft tube: Focus on radial velocity component determination. *European Journal of Mechanics - B/Fluids*, vol. 29(4), July-August 2010, pp. 321-335

Ullum, U.; Schmidt, J.J.; Larsen, P.S. & McCluskey, D.R. (1997). Temporal evolution of the perturbed and unperturbed flow behind a fence: PIV analysis and comparison with LDAdata, *Proceedings of the 7th International Conference on Laser anemometry and applications*, pp. 809–816, Karlsruhe, Germany, 1997.

Van den Braembussche, R.A.; Prinsier, J. & Di Sante, A. (2010). Experimental and Numerical Investigation of the Flow in Rotating Diverging Channels. *Journal of Thermal Science*, Vol. 19 (2), pp. 115−119

Voges, M.; Willert, C.; Mönig, R.; Müller, M.W & Schiffer, H.P. (2011). The challenge of stereo PIV measurements in the tip gap of a transonic compressor rotor with casing treatment. *Experiments in Fluids*, DOI 10.1007/s00348-011-1061-y

Vogt, A.; Baumann, P.; Gharib, M. & Kompenhans, J. (1996). Investigations of a wing tip vortex in air by means of DPIV. *AIAA* Paper 96-2254

Wernert, P. & Favier, D. (1999). Considerations about the phase-averaging method with application to ELDV and PIV measurements over pitching airfoils, *Experiments in Fluids*, Vol. 27, pp. 473-483

Westerweel, J. (1994). Efficient detection of spurious vectors in particle image velocimetry data, *Experiments in Fluids*, vol. 16, pp. 236-247

Westerweel, J. & Scarano, F. (2005). Universal outlier detection for PIV data, *Experiments in Fluids*, vol. 39, pp. 1096-1100.

Wuibaut, G.; Dupont, P.; Bois, G.; Caignaert, G. & Stanislas, M. (2001). Analysis of flow velocities within the impeller and the vaneless diffuser of a radial flow pump. *Proc. IMechE, Part A: Journal of Power and Energy*, Vol. 215(A6), pp. 801–808, doi: 10.1243/0957650011538938.

Wuibaut, G.; Dupont, P.; Bois, G.; Caignaert, G. & Stanislas, M. (2002). PIV measurements in the impeller and the vaneless diffuser of a radial flow pump in design and off design operating conditions, *Journal of Fluid Engineering, Transactions ASME*, Vol. 124(3), pp. 791–797.

Yao, C. & Pashal, K. (1994). PIV measurements of airfoil wake-flow turbulence statistics and turbulent structures. *AIAA* Paper 94-0085

Yu, X.J. & Liu, B.-J. (2007). Stereoscopic PIV measurement of unsteady flows in an axial compressor stage. *Experimental Thermal and Fluid Science*, vol. 31, pp. 1049–1060

Section 2

PIV Applications

5

PIV Measurements on Oxy-Fuel Burners

Boushaki Toufik[1] and Sautet Jean-Charles[2]
[1]*ICARE-CNRS, Avenue de la Recherche Scientifique, Orléans, University of Orléans,*
[2]*CORIA, CNRS-Université et INSA de Rouen, Saint Etienne du Rouvray*
France

1. Introduction

This chapter concerns the application of the PIV measurements in a semi-industrial combustion system. The emphasis is on oxy-fuel burners with multi-jets. The mixing and the dynamic field for both reacting and non-reacting flows are investigated.

Nowadays, combustion occupies a prominent place to meet the increasing needs of our economic world, in fields as diverse as: the production of electrical energy; the space heating, the development of building materials, the metallurgy; the land and air transport; the synthesis of many chemicals in flames; the production of hydrocarbons from crude oil in refineries, etc. Mastering combustion obtained under as varied conditions requires: a knowledge of more advanced fundamental phenomena governing the reaction processes; a " know-how" to an optimal implementation in terms of energy (efficiency) and in terms of pollution (air pollution, noise).

The evolution of pollution standards and the optimization of combustion chamber performances require a development of new burner types and the improvement of combustion techniques. The industrialists are turning to a new burner generation with separate injection of fuel and oxidant. Design of these burners requires the knowledge of mechanisms controlling the stabilization of flame and the production of pollutants. The present study concerns the control of turbulent natural gas-oxygen flames resulting from burners with aligned separated jets. One possibility to understand the structure and mechanism of flame stabilization more accurately is to analyze the structure of the flow by means of particle image velocimetry (PIV) (Raffel and al., 1999). This method allows direct velocity measurements without and with combustion. The application of PIV in industrial scale flame has been already demonstrated on a 1MW experimental boiler (Honoré et al., 2001).

Oxy-fuel combustion, air is substituted by pure oxygen, is characterized by a higher adiabatic flame temperature, a higher flame velocity, a lower ignition temperature, and a wider flammability range that is the case of combustion with air (Baukal and Gebhart, 1997; GEFGN, 1983; Perthuis, 1983; Ivernal and Marque, 1975). This oxy-fuel combustion allows to have a better thermal efficiency and a better stabilization of flame. Oxy-fuel burners have been adopted in a wide range of industrial furnaces to improve productivity and fuel efficiency, to reduce emissions of pollutants, and, in some applications, to improve product quality and yield or to eliminate the capital and maintenance costs of air preheaters

(Baukhal, 2003). Furthermore, the use of oxy-fuel combustion in separated-jet burners open interesting possibilities in the NOx reduction, and the modularity of flame properties such as stabilization, topology and flame length (Sautet et al., 2006, Boushaki et al., 2007).

Flames from burner with multiple jets have many practical situations. Several studies have been published on the structure and development of non-reacting multiple jets (Krothapalli et al., 1980; Raghunatan et al., 1980; Pani and Dash, 1983; Simonich 1986; Yimer et al., 1996; Moawad et al., 2001). However, the studies of multiple jet flames are mostly limited, for example, flame developing in still air without confinement (Leite et al., 1996) or in a wind tunnel with cross flow (Menon and Gollahali, 1998). Lee et al. (2004) studied the blowout limit considering the interaction of multiple non premixed jet flames and giving a number of variables such as distance between the jets, the number of jets and their arrangements. Lenze et al. (1975) have studied the mutual influence of three and five jet diffusion flame, with town gas and natural gas burners. Their measurements concern concentrations, flame length, and flame width in free and confined multiple flames. The burner configuration used in this work is composed of three round jets, one central jet of natural gas and two lateral jets of pure oxygen.

For the separated jet burner, the principle is based on the geometrical separation of its nozzles. The separation of jets allows high dilution of reactants by combustion products in the combustion chamber. This dilution leads to a lower flame temperature and homogenization of temperature throughout the volume of the flame, and consequently a decline in NOx production. However, the configuration of separated jets, which contributes to the dilution of reactants, may become unfavorable to the stabilization of flame. Indeed, this dilution may be controlled by various burner parameters, such as exit velocities of reactants and the distance between the jets. In previous papers of the authors (Boushaki et al., 2007), the characteristics of flames in burners with separated jets have been studied versus the burner parameters such as exit velocities, separation distance between jets and angle of injection. It is interesting to note that the inclination of jets allows to improve the flame stability and above all reduces significantly the NOx (Boushaki et al. 2008). The present chapter focuses on the dynamic behavior of three jet interactions in more detail by varying angle of the side oxygen jets.

Controlling the flows has been the focus of numerous investigations. The objectives of flow control (passive and active) differ according to the considered industrial application. Among these aims are the improvement of mixing with ambient air (Davis, 1982 ; Denis et al., 1999), the limitation of combustion instabilities (Lang et al., 1987; Candel, 1992), the enhancement of heat transfer of flame (Candel, 1992), the decrease of pollutant emissions (Delabroy, 1998; Demayo, 2002) and the reduction of noise engendered in some combustion chambers (Barrère and Williams, 1968; Strahle, 1978). The two dominant methods of passive flow control include noncircular nozzles (e.g., Ho and Gutmark, 1987; Gutmark and Grinstein, 1999, Gollahalli et al., 1992) or the use of tabs at the nozzle exit (e.g., Bradbury and Khadem, 1975; Ahuja, 1993; Hileman et al., 2003). The active control consists in injecting external energy through actuators .The quality of the control depends directly on the design of the actuators. Some of them are specific to combustion applications but most actuation techniques are encountered in both reactive and non-reactive applications, such as loudspeakers (Bloxsidge et al., 1987, McManus et al., 1993), small jet actuators (Lardeau et al., 2002 ; Faivre and Poinsot, 2004, Boushaki et al., 2009), synthetic jets (Davis, A. Glezer

1999; Tamburello and Amitay, 2008) and flaps (Susuki et al., 1999). It has been proven in the literature that jet actuators have drastic effects on mixing and flow dynamics. In fact, jets actuators are capable to change the flow structure, to act on mixing between the reactants, and thus on the flame characteristics such as stability and flame size, as well as pollutant production. The perpendicular arrangement of tube actuators at the periphery of a main jet can confer to the flow a helical movement (swirl). This kind of swirl with actuators has very significant effects on the flow. Ibrahim et al. (2002) and Faivre and Poinsot (2004) indicated that a radial fluid injection into the main jet enhances mixing with the surrounding air. Béer and Chigier (1972), and Feikema et al. (1990) showed that the addition of swirl significantly changes the aerodynamic pattern and can be used to stabilize the flame. The helical fluid flow creates a recirculation zone, allowing dilution with combustion products, and the decrease of the flame temperature limiting the NOx production (Syred and Béer, 1974; Schmittel et al., 2000; Coghe et al., 2004).

The present chapter reports the results of an experimental investigation of the dynamic field on a burner with 25 kW power composed of 3 jets, one central jet of natural gas and two side jets of pure oxygen. The velocity measurements were carried out using Particle Image Velocimetry (PIV) in both cases of non-reacting flow and reacting flow inside the combustion chamber. Two control systems, one passive and one active, are developed and added to the basic burner to improve the combustion process and to ensure the stabilization of flame and pollutant reductions. The passive control is based on the slope of side oxygen jets towards the central natural gas jet in a triple jet configuration. The active control concerns the use of four small jet actuators, placed tangentially to the exit of the main jets to generate a swirling flow. These actuators are able to strongly modify the flow structure and to act on mixing between the reactants and consequently on the flame behavior.

Nomenclature

d	tube internal diameter, mm	*Greek symbols*	
M	initial velocity ratio ($M = (U^0_{ng} / U^0_{ox})$)	θ	oxygen jet angle, (°)
$\dot{m}$	mass flow rate, kg.s^{-1}	μ	dynamic viscosity, g.m.s^{-1}
r	ratio of volumetric flow rate ($r = \dot{m}_{act} / m_{tot}$)	ρ	gas density, kg.m^{-3}
Re	Reynolds number $(= \rho U^0 d / \mu)$		
S	separation distance between the jets, mm	Subscripts	
SW2J	Burner configuration with 2 jets	ac	actuator
SW3J	Burner configuration with 3 jets	jet	main jet
U^0	nozzle exit velocity, m.s^{-1}	cl	centerline
U	longitudinal mean velocity, m.s^{-1}	lo	lift-off
u′	longitudinal velocity fluctuation ms^{-1}	ng	natural gas jet
V	radial mean velocity, m.s^{-1}	ox	oxygen jet
v′	radial velocity fluctuation, m.s^{-1}	tot	total
x, z	radial and longitudinal coordinate, mm		

2. Burner and operating conditions

2.1 Basic configuration of the burner

The basic configuration of the burner consists of three round tubes, one central of natural gas and two laterals of pure oxygen. A schematic of the oxy-fuel burner apparatus is shown in Fig. 1. This burner can operate with only two jets if the entire oxygen flow rate is injected into a single tube. Fuel and oxidizer flow rates are constant for all experiments to ensure constant power flames of 25 kW ($\dot{m}_{ng} = 556 \times 10^{-3} kg.s^{-1}$, $\dot{m}_{Ox} = 1964 \times 10^{-3}$ kg.s^{-1}). The natural gas ($\rho_{ng} = 0.83$ kg.m^{-3}) flows from the central tube (diameter d_{NG}=6 mm, length 250 mm) and pure oxygen ($\rho_{ox} = 1.35$ kg. m^{-3}) flows from the two side jets (diameter d_{Ox}=6 mm or 8 mm, length 250 mm). The natural gas composition is: 85% CH_4; 9% C_2H_6; 3% C_3H_8; 2%N_2; 1% CO_2; plus traces of higher hydrocarbon species. The flow rate of natural gas is controlled by a regulator of mass flow rate TYLAN RDM 280; the oxygen is regulated by sonic throats calibrated by a classical flowmeter in function of pressure.

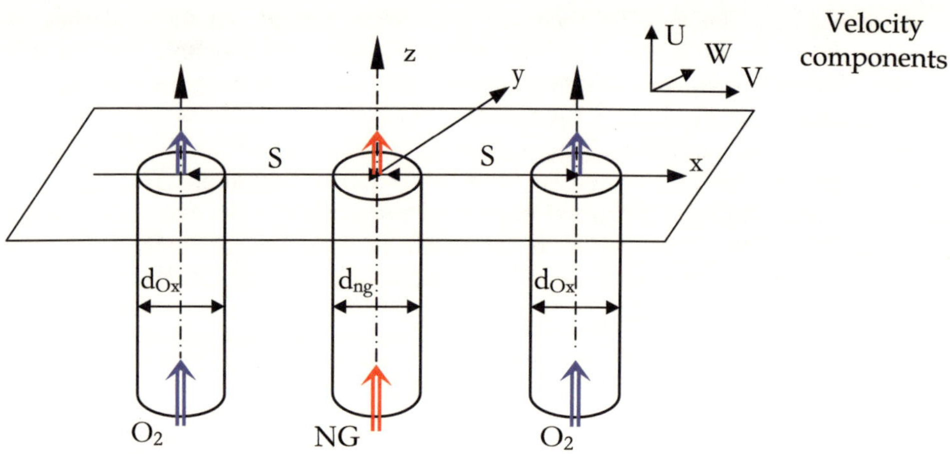

Fig. 1. Schematic diagram of oxy-fuel burner

In the case of reacting flow, the oxy-fuel flames takes place inside a combustion chamber of 1 m-high with square cross section of 60×60 cm. The lateral walls are refractory lined inside and water cooled outside of the combustion chamber. Optical access is provided by quartz windows at many vertical positions of the combustion chamber. The burner is located at the centre of the bottom wall of the combustion chamber (more details, see Boushaki et al. 2009).

2.2 Passive control system

The control technique consists in inclining the side oxygen jets towards the natural gas jet as shown in Fig. 2. The angle of oxygen jets (θ) compared to the vertical direction varies from 0° to 30° (0, 10, 20, 30°). The internal diameters of natural gas and oxygen tubes (d_{ng} and d_{ox}) are both 6 mm. Exit velocities for the natural gas and the oxygen jets without control (θ=0°) inferred from the flow rates are respectively are $U_{NG}^{0} = 23.7$ m/s and $U_{Ox}^{0} = 25.7$ m/s. The separation distance between the jets (S) is fixed at 12 mm. The exit Reynolds numbers of the natural gas jet and the oxygen jet are Re_{ng}= 10772 and Re_{ox}= 10936 respectively.

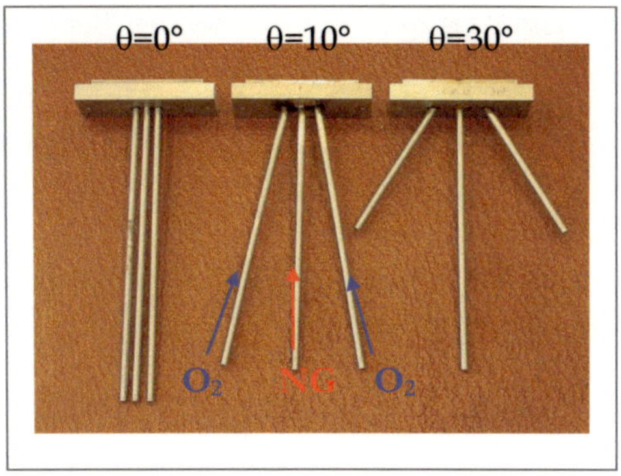

Fig. 2. Photo of burners with angle injection 0°, 10° and 30°

2.3 Active control system

The active control is applied using four small jet actuators arranged slightly upstream the exit of the main jets. Fig. 3 shows the oxy-fuel burner fitted with actuators. Each nozzle is equipped with four small control jets placed tangentially around the main jets. This kind of tangential control generates a swirling motion in the flow, its intensity being controlled by the flow rate in the actuators. The exit diameter of the jet actuators (d_{act}) is 1 mm and their position relative to the main jet exit is z = -1mm. In the present work, the rotating flow intensity is quantified by the ratio of volumetric flow rates of actuators ($\dot{m}_{act}$) and total ($\dot{m}_{tot}$), which varies from 0 to 30%; it is given by:

$$r = \frac{\dot{m}_{act}}{\dot{m}_{tot}} \tag{1}$$

where $\dot{m}_{tot} = \dot{m}_{jet} + \dot{m}_{act}$ and $\dot{m}_{jet}$ is the flow rate of the main jet ($\dot{m}_{tot} = \dot{m}_{NG}$ for the fuel flow rate and $\dot{m}_{tot} = \dot{m}_{Ox}$ for the oxygen flow rate) . The subscripts "tot" and "act" are valuable for both reactants (oxygen and natural gas). For the present configurations, the swirl number (S_n) characterizing rotating flows is an increasing function of the flow rate ratio (r).

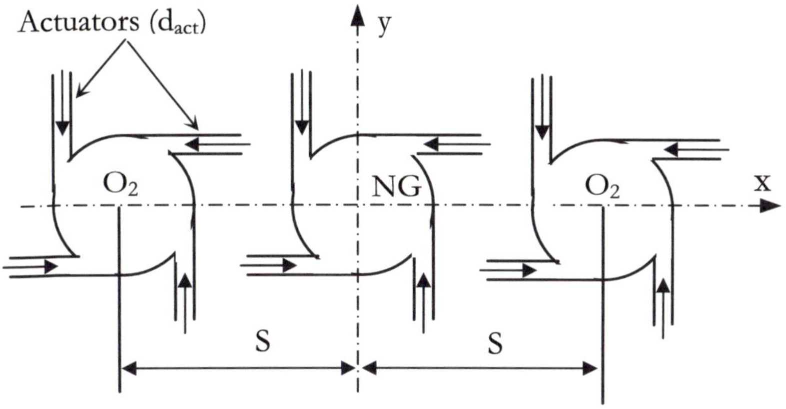

Fig. 3. Schematic view of main nozzles with tangential tube actuators

The swirl number is a dimensionless quantity defined as (Beèr and Chigier, 1972; Sheen et al. 1996):

$$S_n = \frac{G_\varphi}{RG'_x} \tag{2}$$

where G_φ is the axial flux of the tangential momentum, G'_x is the axial flux of axial momentum, and R is the exit radius of the burner nozzle.

$$G_\varphi = \int_0^R (Wr)\rho U 2\pi r\, dr \tag{3}$$

$$G_x = \int_0^R U\rho U 2\pi r\, dr \tag{4}$$

U and W are the longitudinal and tangential components of the velocity respectively.

The corresponding geometrical swirl number, defined following the previous work (Boushaki et al. 2009) as:

$$S_n = \frac{\frac{\dot{m}_{act}}{\dot{m}_{jet}}}{1+\frac{\dot{m}_{act}}{\dot{m}_{jet}}}\left[\frac{2\left(d_{act}^2 + 3R(R-d_{act})\right)}{3R(2R-d_{act})}\right] \tag{5}$$

is calculated from the flow rate of main and actuator jets ($\dot{m}_{jet}$ and $\dot{m}_{act}$), the radius of the main jet and the jet actuators (R and d_{act}). From r=0 to 30%, the swirl number varies from 0 to 0.25 for R=3 mm and to 0.26 for R=4 mm.

The exit parameters of the two studied configurations are listed in Table 1. $U°$, Re and M are the jet exit velocity, Reynolds number and initial velocity ratio (U_{NG}^0 / U_{Ox}^0) respectively. In the presence of jet actuators, the exit velocity of the main jet decreases when the flow rate ratio (r) increases since a portion of the total flow rate is injected into the tube actuators. Conversely, the exit velocity of jet actuators increases with r.

Burner configuration	Number of Jet	S (mm)	Gas	d (mm)	U^0 (m/s)	Re	M= U_{NG}^0 / U_{Ox}^0
SW3J	3	20	Natural gas	d_{ng}=6	23.7	10772	1.64
	1 NG, 2 O_2		Oxygen	d_{ox}=8	14.45	8198	
SW2J	2	20	Natural gas	d_{ng}=6	23.7	10772	0.82
	1 NG, 1 O_2		Oxygen	d_{ox}=8	28.91	16400	

Table 1. Burner configurations and parameters. The notation SW3J means Swirl with three main jets (1NG and $2O_2$)

3. PIV system: Experimental set-up and procedure

Particle Image Velocimetry (PIV) is a laser diagnostic method which has been developed in parallel with Laser Sheet Visualization (LSV)). As with LSV, The PIV is based on the collection of images of Mie scattering of fine particles seeded in the flow. Processing of these particle images provides instantaneous data on two velocity components in a plane crossing the flow. Then, the mean and root-mean-square (rms) velocity fields can be easily deduced from statistical studies of the instantaneous measurements sequences.

Fig. 4 shows a schematic diagram of the PIV system. It includes a laser sheet that illuminates the zone of flow studied, a CCD camera, a PC for data acquisition and a control unit for synchronization. The laser used is double-pulsed Nd-YAG (Big Sky CFR200 Quantel) with a wavelength of 532 nm and a frequency of 10 Hz. Laser energy is adjustable and can be increased up to 150 mJ per pulse with pulse duration of 8 ns. The laser sheet is formed by a first divergent cylinder lens, which spreads out the beam then by second convergent spherical lens, which focuses the sheet. The signal of Mie scattering emitted by particles is collected perpendicularly by a CCD camera FlowMaster of Lavision (12-bit dynamic and 1280×1024 pixels resolution) with a 50 mm lens F/1.2 Nikkon. In reacting flow measurements, to reject the bright luminosity from the oxy-flame, an interference filter (532 nm centre, 3 nm bandwidth) was placed in front of the imaging lens. The time delay between the laser pulses varies from 8 to 20 µs according to the case. For each operation condition, up to 400 pairs of instantaneous images were collected.

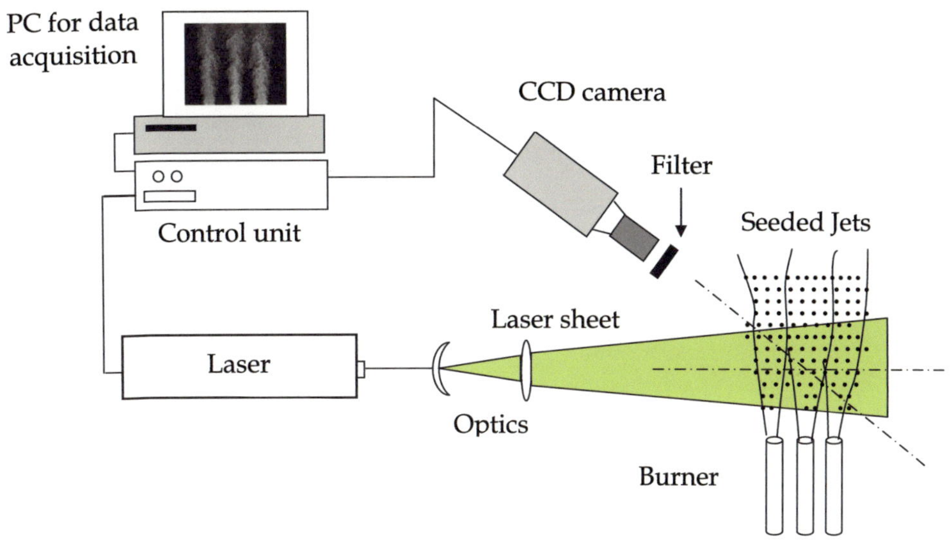

Fig. 4. Schematic view of a PIV system. The CCD camera and the laser sheet are perpendicular

For the non-reacting flow, the jets are seeded with olive oil particles (~ 3 to 4 µm in diameter) whereas for reacting flow they are seeded with zirconium oxide (ZrO_2) particles (~ 5 µm diameter). The reasons for the choice of these ZrO_2 particles are: a high resistance of high temperatures generated by the oxy-combustion (melting point 2715°C), a good refractive index (2.2) which allows a good light scattering, and a little size for a better mixing in the flow. Fig.5 illustrates the two systems of seeding particles. For olive oil particles, the seeding system is an atomizer based on Venturi principle to produce fine particles. For ZrO_2 particles, the system consists of tubes equipped with porous plates, placed at the level of exit gases. The gas passes through the porous medium and drags a certain quantity of particles providing a uniform seeding. Concentration of particles is controlled by valves through the gas flow rate in the line of seeding. A particular attention has been carried in the drying of particles before their injection in the seeding system in order to limit agglomerates. The measurements were conducted with seeding rates relatively low to avoid perturbing the flow. Particles in fact can modify the characteristics of the flow if they are introduced in high quantity, and can even blow out the flame. The criterion assuring a good track of flow by particles is respected here, since the Stokes number, defined as the ratio between the response time of particles and a time characteristic of the flow, obtained in the present experiments is much lower than unity (St <<1). For example, $t_p \approx$0.015ms in the case of oxygen and oxide zirconium particles ($t_p = 2\rho_p r_p^2 / 9\mu_{CH4}$: $\rho_p = 5600\,kg.m^{-3}$, $r_p = 0.5\,\mu m$, $\mu_{CH4} = 20{,}18.10^{-6}\,Pa.s$), with the frequency of 1000Hz, the Stokes number is about 0.015.

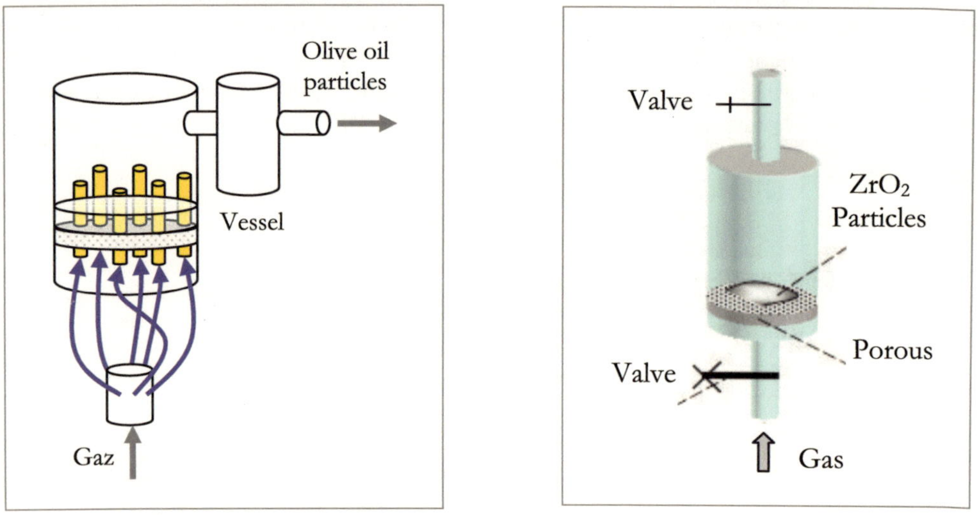

Fig. 5. Seeding systems, for olive oil particles in non-reacting flow (left), for ZrO2 particles in reacting flow (right)

The Davis Lavision software package uses a cross-correlation technique to find the average particle displacement in each subregion (32×32 pixels) of the image. The velocity is found by dividing the particle displacement by the time between laser pulses. The sub-pixel displacement was estimated by means of Gaussian peak of fitting (Lecordier and Trinité, 2003). With a maximum displacement of 8 pixels, this would correspond to less than 2% uncertainty in final velocity measurement. It was necessary to carry out a post processing to detect and correct the aberrant vectors which appear after cross-correlation calculations. Detection of false vectors can be done by the size of the vector. In this case, it is necessary to locate all vectors above a certain threshold of velocity according to the expected results. The direction of vectors can also help to identify false vectors knowing a priori the direction of flow. For that the *allowable vector range*, restricting the filtered vectors to a user specified in units of pixel was performed. A range may be specified for each component of velocity U and V (range of ± value). Any vectors outside this range are removed. In some cases, in particular for instantaneous images, filters to refine the results are used as local median filter or regional median filter based on the neighbouring vectors.

4. PIV measurements on burners with inclined jets

4.1 Instantaneous and mean velocity fields

For the inclined injectors (see Fig. 2), the shape of exit nozzles is elliptical rather than round. Therefore, the profile of velocity for inclined jet at the exit nozzle is quite different to the one of a straight jet. That is why results of the exit velocities are provided; they are very useful for numerical studies. Fig. 6 shows the profiles of mean velocities near the burner exit (z=3 mm) in the non-reacting flow. For the side jets, the longitudinal mean velocity, U, decreases with the angle of the oxygen jets, however, the radial velocity, V, increases as a result of the deflection of injectors. It is noted that from the velocity profiles of PIV measurements, the flow rates at the exit nozzles are calculated and are in very good agreement with the flow rate injected.

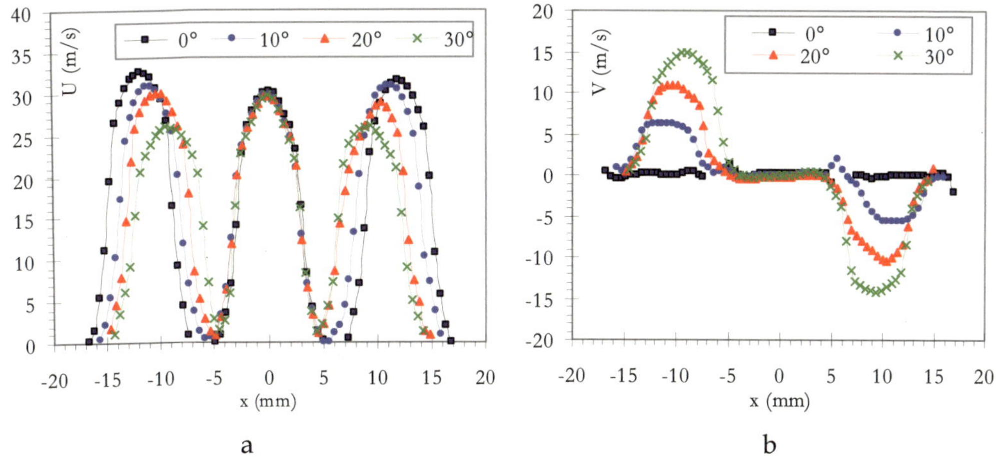

Fig. 6. Radial profiles of longitudinal (a) and radial (b) velocities at exit nozzles (z=3mm) in non-reacting flow for the oxygen jet angle, θ = 0°, 10°, 20° and 30°

Fig. 7 shows an example of instantaneous velocity fields, taken among 400 instantaneous fields, for oxygen jet angles 0° and 30°, with the longitudinal velocity U (along the vertical direction) in color scale. The burner configuration is highly three-dimensional, since some discontinuous aspects of streamlines are observed on the instantaneous velocity fields. However, this aspect is not visible in Fig. 8 because the images averaging remove the discontinuities and high gradients on the velocity distribution. For the central jet, the longitudinal velocity varies weakly and the radial velocity is nearly zero in the near nozzle field. More downstream, the side jets affect the central jet and its radial velocity is no longer zero. In the far field, vortices appeared in the region of the mixing layer between the jet and the ambient air (for θ=30). At the merging zone of three the jets, gradients in the velocity values are noted characterizing the three-dimensionality, particularly when the side jets are inclined. This is due to the transverse flow of jets and the elliptical shape of inclined nozzles, as it was shown in the paper of Gutmark and Grinstein (1999) where the entrainment rate and the mixing for elliptic jets are more significant compared to the round jets. Mean velocity fields, obtained by averaging 400 instantaneous images, in non-reacting flow for the jet angles 0° and 30° are illustrated in Fig.8. These results show that increasing jet angle leads to a decrease of longitudinal velocity and an increase of radial velocity for the side jets.

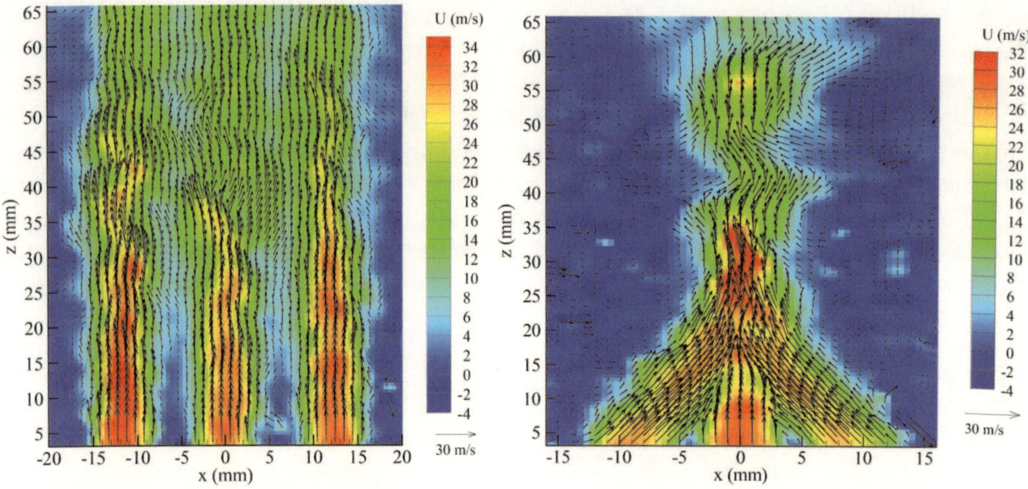

Fig. 7. Example of instantaneous velocity fields for θ = 0° (left) and θ = 30° (right) in non-reacting flow

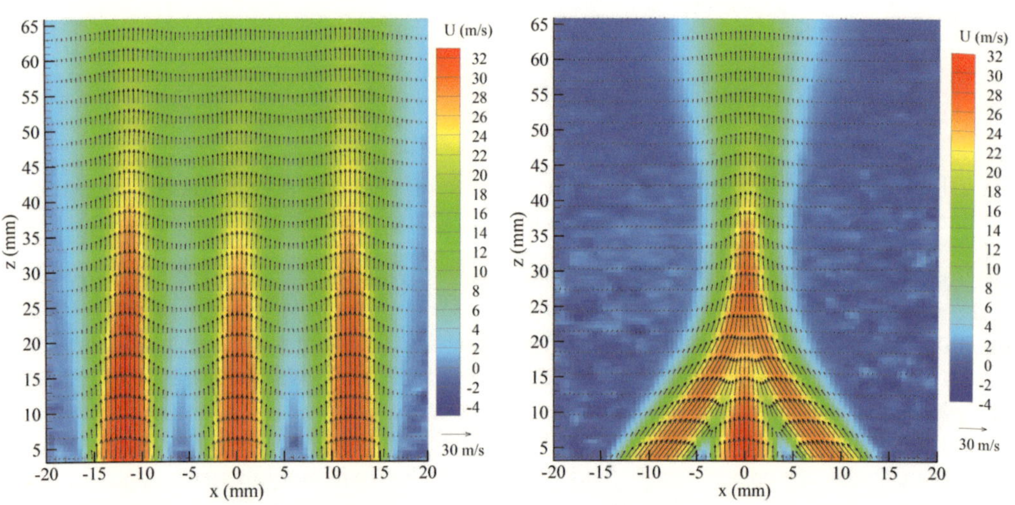

Fig. 8. Mean velocity fields for θ = 0° (left) and θ = 30° (right) in non-reacting flow (with longitudinal velocity in color scale)

From the initial state where θ=0° to inclined states, the structure of the dynamic field changes. As observed in the figure, the mixing of jets is more upstream for the inclined jet configurations. The merging region starts at 15 mm high for the straight jets (θ=0°), and more upstream for inclined jets, 7 mm for θ=10 ° and nearby to the burner for 20° and 30° (z=3 mm). The combined region, in which the velocity profiles combine to form a single jet profile, also starts much more upstream when the jets are angled. An increase of velocities between the jets with increasing jet angle is noted. At the height z=10 mm, the longitudinal velocity is zero between the jets for θ=0°, whereas it is around 12 m/s for θ=20°.

4.2 Radial profiles of mean velocities and fluctuations

Radial profiles of the mean longitudinal velocity (U) at different heights from the burner for the configurations θ = 0°, 10°, 20° and 30° are shown in Fig. 9. For the straight jets, a classical behavior of multiple jets is found for the distribution of longitudinal velocity, maxima in the centre of jets and minima between the jets. In the near burner region (e.g. z=15 mm), the distribution of velocity shows maxima and minima corresponding to the three jets and that

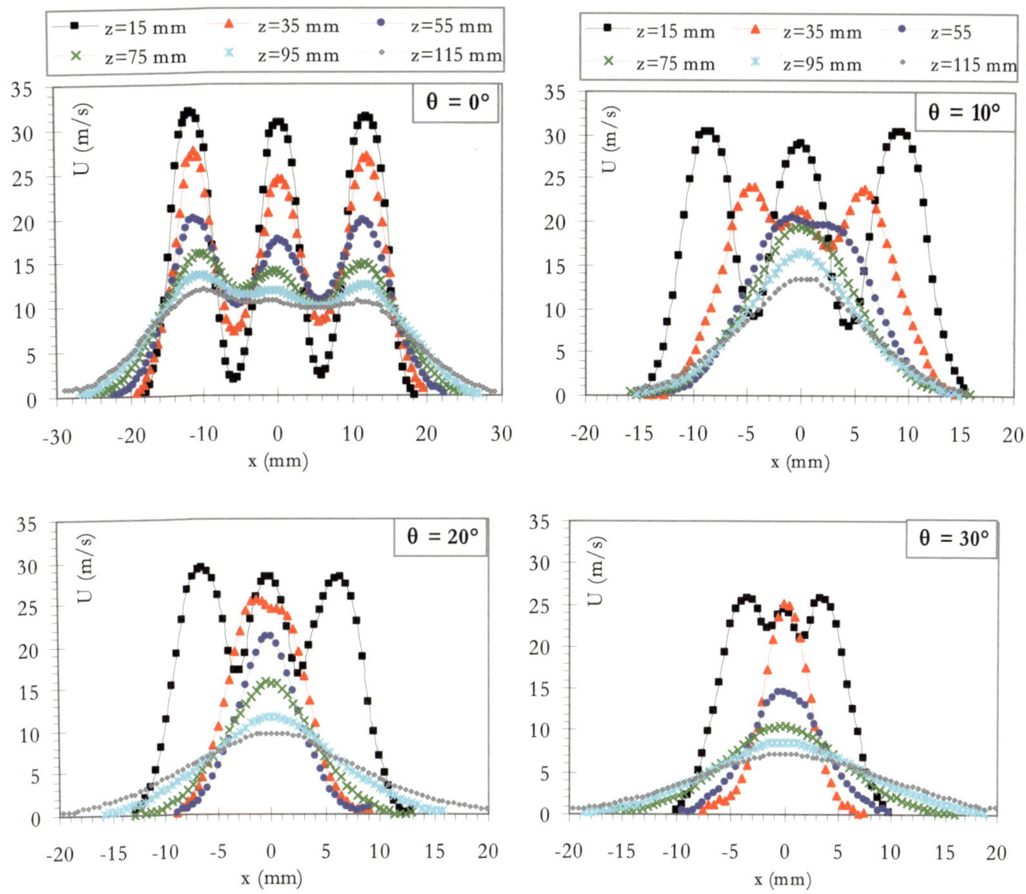

Fig. 9. Radial profiles of mean longitudinal velocity for jet angles 0°, 10°, 20° and 30° in non-reacting flow at different positions from the burner

the maximum velocity decreases when the jet angle increases. More downstream, when inclining the jets, the extreme velocities begin to disappear into a single maximum located along the axis of the center jet. This combined zone of jets, characterizing a single jet, is reached earlier when the jet angle increases.

Without control (θ=0°), the combined zone is not reached even at z=115 mm, while for θ=30° it is already occurred at z=35 mm. With control, in the combined zone it is noted an acceleration of the jet along the axis of the flow. In fact, at z=75 mm, from θ=0° to 10° the centerline longitudinal velocity (U) increases from around 15 to 19 m/s. However, once the combined region is reached, the velocity decreases with the angle of jets. The expansion of the flow decreases with the angle in the region near to the burner, and then increases downstream the flow.

The radial velocity profiles are shown in Fig.10 at different positions from the burner in non-reacting flow. For the straight jets (θ=0°), the radial velocity is low and ranges from -1 to 1 m/s. The deflection of jets leads to an increase of the radial velocity of the oxygen jets, particularly near the burner since at z=15 mm, the maximum value of V varies from 0.6 m/s

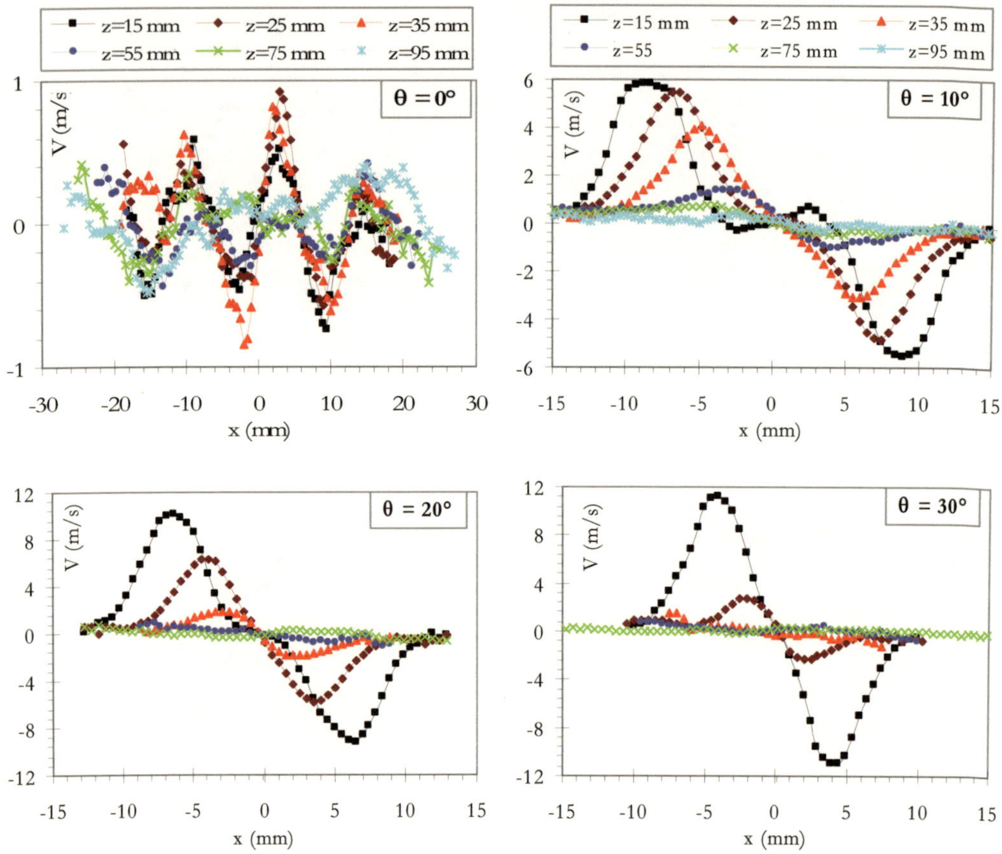

Fig. 10. Radial profiles of mean radial velocity for jet angles 0°, 10°, 20° and 30° in non-reacting flow at different positions from the burner

to 11 m/s when θ varies from 0° to 30°. For the inclined jets the velocity profile is composed of two parts, one positive and one negative with a passage by zero corresponding respectively to the side jets (left and right) and the central jet of natural gas.

Fig. 11 shows the radial distribution of the turbulence intensity (u'/U) with the jet angle at vertical positions z=15 and 75 mm. Results of u'/U highlight the interaction zones between the jets and surrounding air as well as between the jets themselves. In the initial region of flow (z=15mm), four zones of high turbulence are noticed. Two of them take place on the outer region of the side jets, involving a dilution of the oxygen jets by ambient air. The other two zones are located between the jets representing the jets mixing. Near the nozzle exits, the outer zones of turbulence do not seem to be affected by the increase of jet angle, while the inner zones decrease in intensity since the merging region is reached faster by the deflection of jets. Further downstream, when the jets merge, it is found that only the outer zones of turbulence behave as a single jet. At z=95 mm, except for straight jets, it is noted that whatever is the jet angle, the turbulence intensity profile u'/U is similar owing to the complete merging of jets at this position.

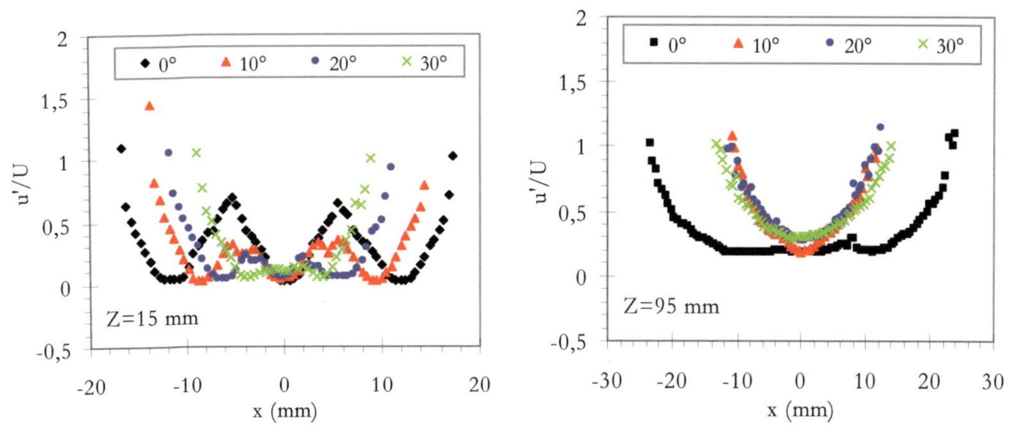

Fig. 11. Turbulence intensity (u'/U) at the vertical positions z=15mm and z=95mm for the jet angles 0°, 10°, 20° and 30° in non reacting flows

4.3 Velocities in oxy-fuel flames

Fig. 12 shows the mean velocity fields in the reacting flow (with combustion) for jet angles 0° and 30°. These vector fields show a significant difference between non-reacting and reacting flow. The first remark concerns the more significant velocities above the stabilization point in the reacting flow. Indeed, for the straight jets (θ=0°), the longitudinal velocity (U) decreases along the flow, however, this decrease is less significant compared to non-reacting flow. For the inclined jets, the longitudinal velocity in combustion keeps higher values even at more significant heights from the burner. The hot environment and the presence of a reaction zone lead to a fast expansion of gases due to the presence of flame, and therefore to an acceleration of the flow. The second remark concerns a greater radial expansion in combustion in particularly above the stabilization zone of the flame.

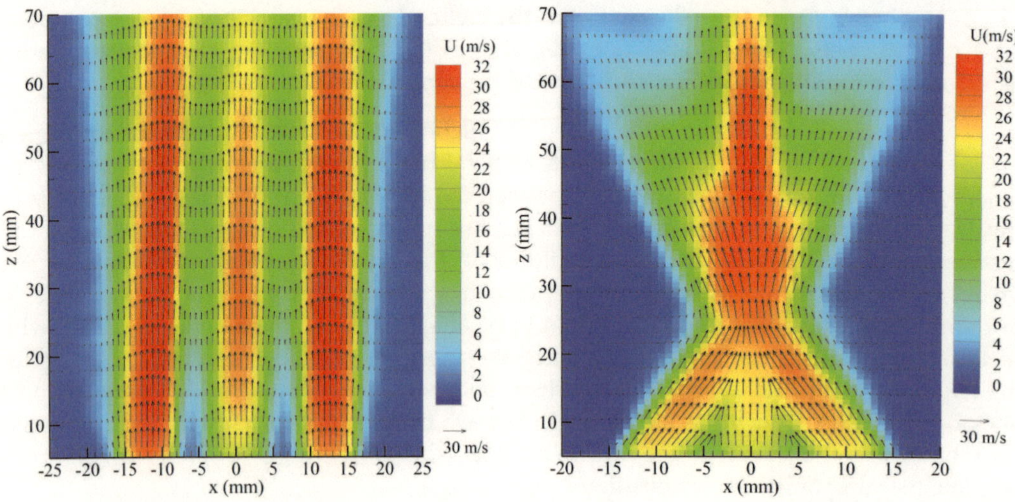

Fig. 12. Mean velocity fields for the oxygen jet angle θ=0° (left) and θ=30° (right) in reacting flow

The longitudinal velocities along centerlines (U_{cl}) in reacting and nonreacting flow are shown in Fig.13. The results concern the central jet of natural gas and one of the two side jets of oxygen for θ=0° and θ=20°. In the case of nonreacting flow, for the straight jets (θ=0°), the centerline velocity (U_{cl}) follows a classical decrease for the central and the side jets: first, a very slight decrease down to 26 mm corresponding to the potential core length of the jet, then a high decay of U_{cl} which corresponds to the merging of mixing layers, and finally a slow decay down to 65 mm for the natural gas jet and 75 mm for the oxygen jet. When the side jets are inclined (θ=20°, nonreacting), a decrease of the length of potential core is noted, and therefore a fast decrease of the longitudinal velocity appears in the first zone of the flow in particular for the central jet. After this first part of the flow, an increase of U_{cl} for the cases

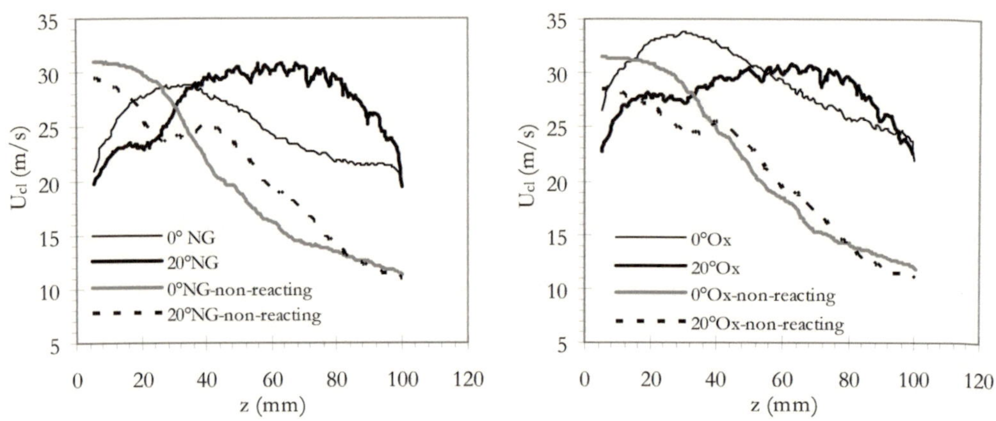

Fig. 13. Mean longitudinal velocity (U_{cl}) along the centerline of the central jet (left) and the side jet (right). A comparison between the reacting and the non reacting flow for the angles θ=0° and θ=20°

θ=20° at z=28 mm is observed. This behavior is attributed to the merging of three jets leading to an acceleration of flow at the beginning of the combined region.

In the case with combustion, the velocity U_{cl} increases with z distance in the first zone up to a maximum, then it conserves a high value, and finally, it slightly decreases along the centerline. The oxy-fuel combustion considerably influences the flow throughout the studied domain. In fact, in comparison with the non-reacting flow in the initial region, the flow is slower for both fluids, and then accelerates under the influence of high flame temperatures. For θ=20°, starting from z=40 mm, the centerline velocity is higher in the case of the reacting flow. After the stabilization point, the flow velocity is higher in the reactive case than that in the non reactive case. This is due both to the rapid expansion of burnt gases which accelerates the flow and to a retardation of mixing due to the presence of a flame.

5. PIV measurements on burners with swirling jets

5.1 Velocity fields in nonreacting flow

In order to examine the effects of the jet actuators on the flow, the case without control (*r*=0) and a case with a control parameter of *r*=0.15 are discussed (*r* represents the ratio of volumetric flow rates of actuators $\dot{m}_{act}$ and total $\dot{m}_{tot}$). Instantaneous velocity fields for the both configurations SW3J and SW2J (see Fig.3 and table 1) in the non reacting flow are shown in Fig. 14 and 16. The mean velocity fields are illustrated in Fig. 15 and 17. The instantaneous or mean fields clearly show the significant effect of the jet actuators on the flow structure. With the activation of actuation, the length of the potential core of the jets decreases and the longitudinal velocity decays more rapidly. This happens in favour of a high jet spreading and an enhancement of mixing, which furthermore leads to the flame stabilization more upstream, as previously indicated. This confirms the results previously reported in the literature on the efficiency of swirl on the mixing between the jet and the surrounding fluid (Syred and Béer, 1974; Faivre and Poinsot, 2004). Without control (r=0),

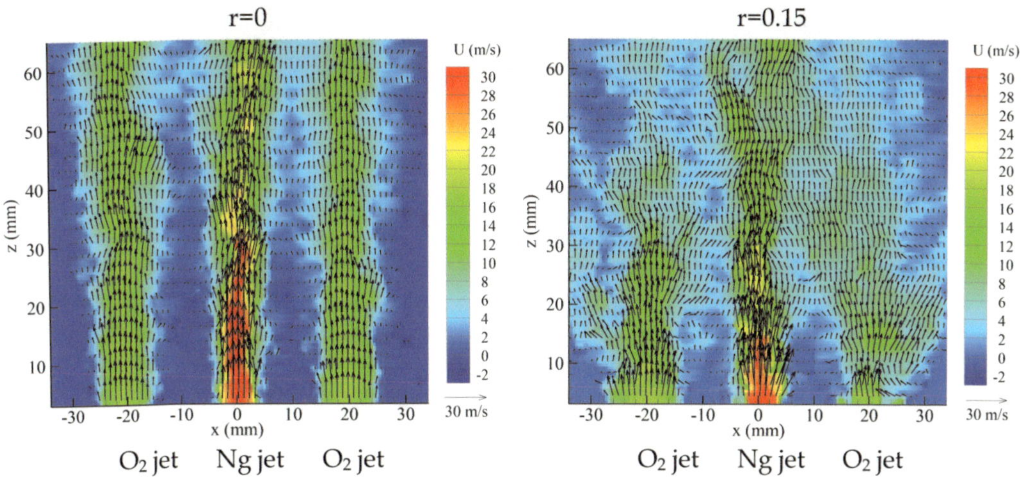

Fig. 14. Samples of instantaneous velocity fields for the flow rate ratio *r*=0 (without control) and *r*=0.15 (with control) in non-reacting flow for the SW3J configuration

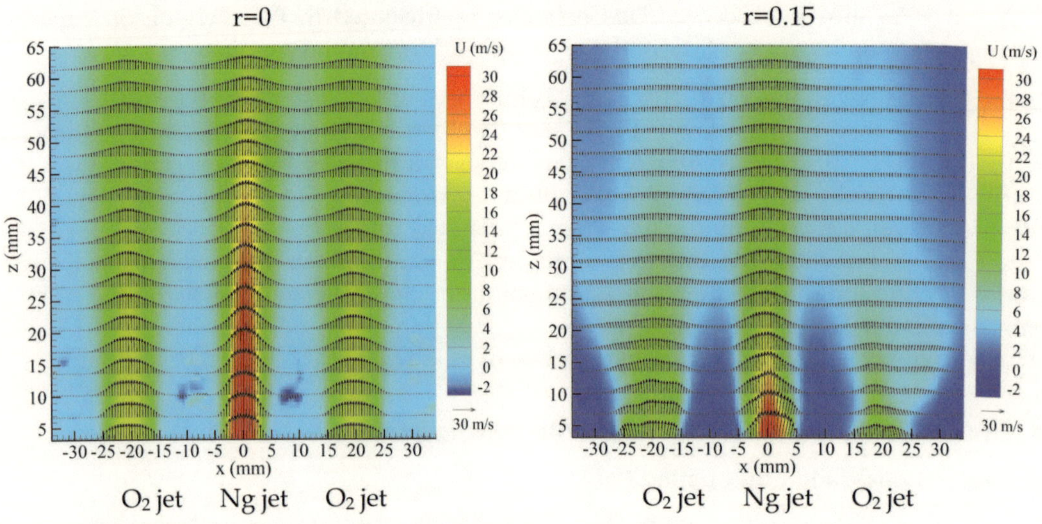

Fig. 15. Mean velocity fields for the flow rate ratio r=0 (without control) and r=0.15 (with control) in non-reacting flow for the SW3J configuration

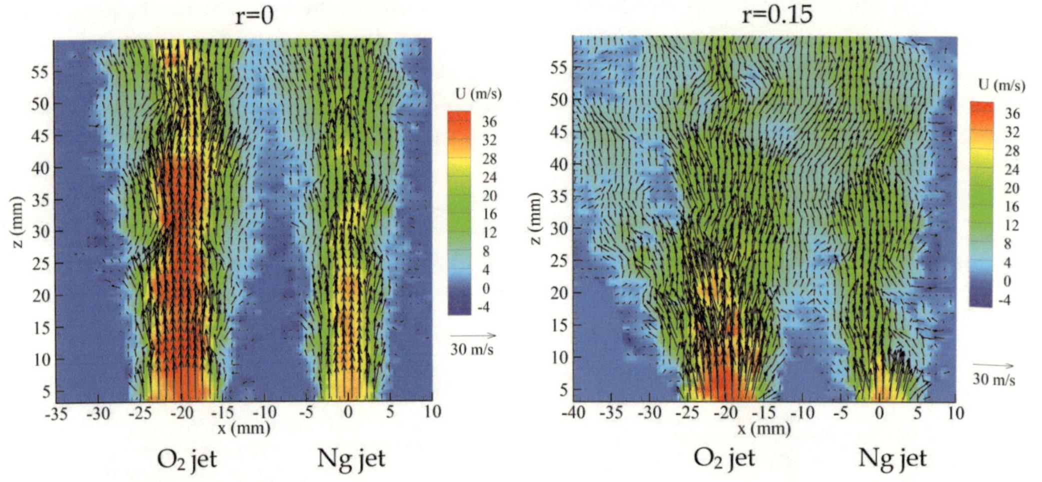

Fig. 16. Samples of instantaneous velocity fields for the flow rate ratio r=0 (without control) and r=0.15 (with control) in non-reacting flow for the SW2J configuration

the jets are well organized with a proper development for each jet in the initial zone of the flow and have a slight velocity variation at the centre. With control (r=0.15), instantaneous flow becomes highly disorganized and velocity gradients become important. In addition, the acceleration and the deflection of velocity vectors clearly appear in the whole field of view. The potential core of the controlled jets is definitely disturbed and its length decreases as a result of the jet actuators. For high swirl intensity ($r > 0.2$), the jets are more affected by the control, highly disorganized and develop also in the transverse plane (x,y); consequently the generated flame would be very oscillating and unstable (see Boushaki et al.2009). It is also noted the strong three-dimensional aspect of the flow, especially when using the jet actuators, since many discontinuities in the velocity contours are observed. The mixing point (where the jets begin to interact) moves upstream in the flow with increasing *r*. In addition, when the distance (S) between the jets decreases, the mixing point gets closer to the burner.

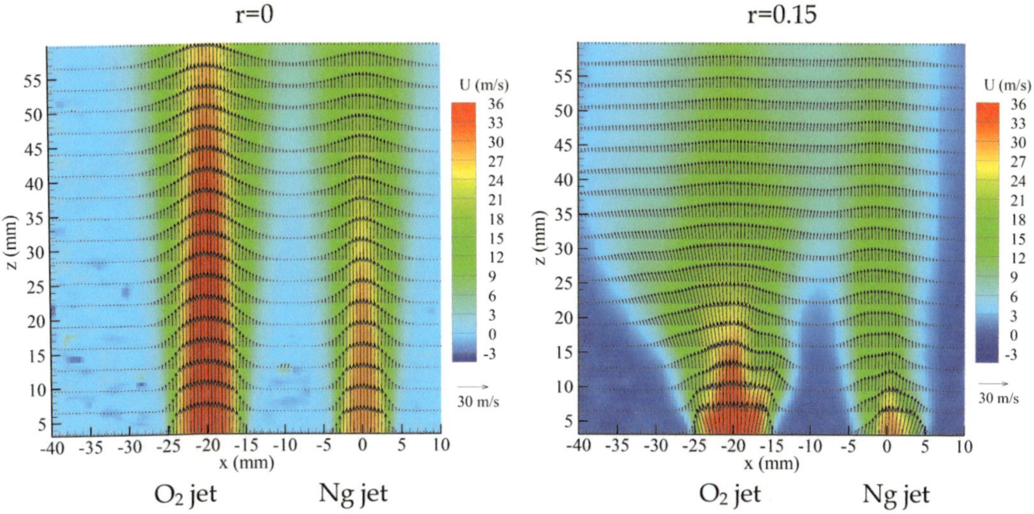

Fig. 17. Mean velocity fields for the flow rate ratio *r*=0 (without control) and *r*=0.15 (with control) in non-reacting flow for the SW2J configuration

Fig.18 shows radial profiles of the longitudinal and radial velocities at different heights from the burner, in the non reacting flow for the SW2J configuration. These profiles are presented for the both cases without control (*r*=0) and with control (*r*=0.15). Without jet actuators, the classical distribution of the longitudinal velocity in a multiple jet configuration is found, with maxima in the centre of the jets and minima between the jets. In the initial zone, each jet follows its own evolution, and then the jets start to interact. The activation of control jets changes the flow behavior and the distribution of jet velocities. The control leads to a faster velocity decay and alters the shape of radial profiles near the burner. Indeed, at z = 35 mm for example, the maximum longitudinal velocity is 34 m/s without control, whereas in the controlled configuration it is 17 m/s. The control also accelerates the longitudinal velocity decay, as expected in as swirling flow. Near the injector exits (z=5 mm), it is observed that the maximum velocity remains high and even slightly higher when *r* increases. It can be explained by the involvement of the tangential velocity component, which increases with the flow rate control and compensates the decrease induced in main jets when *r* increases. This situation was also observed in the work of Faivre and Poinsot (2004) on a single jet. Furthermore, the asymmetry of the velocity profiles in the initial zone of flow induced by the tangential activation of control jets is observed. Far from the burner, this asymmetry however disappears and the profiles become axisymmetric.

The influence of the jet actuators on the flow behaviours also concerns the other velocity components, in particular in the initial zone of the flow. Fig.18.c-d shows that the radial velocity increases in the case of controlled jets. Without control, V is low ($-1 \leq V \leq +1$ m/s), whereas with control the maximum radial velocity can reach 6 m/s for *r*=0.15 and 12 m/s for *r*=0.2 near the burner.

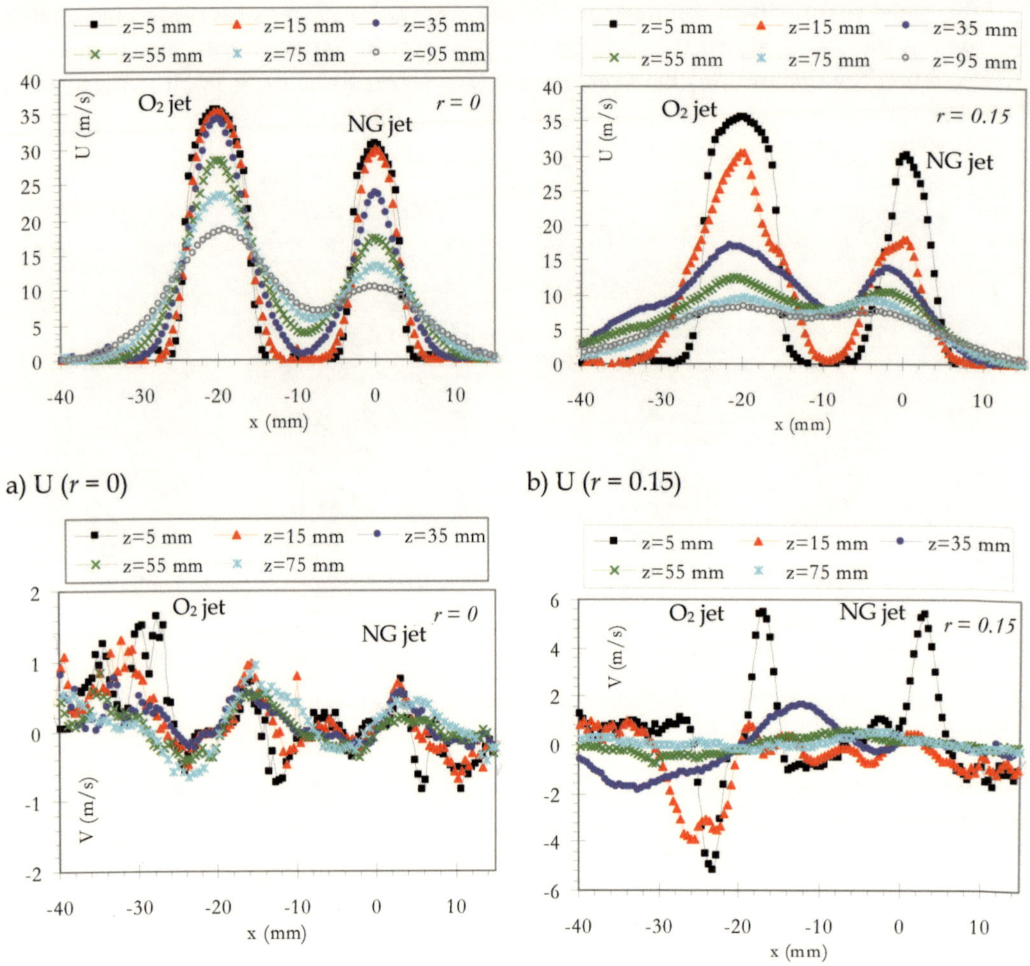

a) U (r = 0) b) U (r = 0.15)

c) V (r = 0) d) V (r = 0.15)

Fig. 18. Radial profiles of mean velocities for the SW2J configuration in the non reacting flow, without control (r=0) and with control (r=0.15); a-b) longitudinal component (U), c-d) radial component (V).

5.2 Axial profiles of mean velocities

Fig.19 shows longitudinal velocities along centrelines of natural gas and oxygen jets for the flow rate ratios r=0, 0.1, 0.15 and 0.2 in the non-reacting flow for the SW2J configuration. Without control, the velocity evolution along the centreline (U_{cl}) is almost similar to the one of a simple jet. First, it can exhibit a plateau corresponding to the potential core of the jet, followed by a pronounced decay of U_{cl} (mixing layers merging), and finally a slow decay. It is noted that the initial plateau is longer for the oxygen jet on account of greater jet diameter (d_{ox}= 8 mm, d_{ng}= 6 mm) with a slight difference in injection velocities of the two fluids.

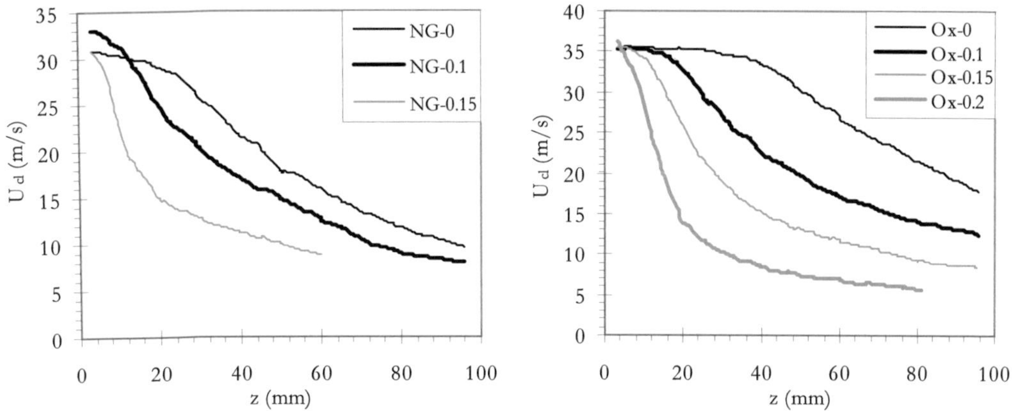

Fig. 19. Mean longitudinal velocity along the centerline of the natural gas jet (left) and the oxygen jet (right) for flow rate ratios (*r*), 0, 0.1, 0.15 and 0.2. Case: configuration SW2J, non-reacting flow

The activation of control jets changes the flow behaviour and the distribution of jet velocities. The decrease of centreline velocity is faster near the burner, and then it further slows downstream. It appears that the greater the flow rate ratio is, the higher the centreline velocity decays. Another important element concerns the potential core of jets, which is strongly affected by the control. Indeed, for *r*=0.1, the length of the potential core is very small compared to the cases without actuators. Between the two cases (*r*=0 and 0.1), the length of the potential core for the oxygen jet decreases from $4d_{ox}$ to $2.25d_{ox}$. Over for *r*=0.1, the plateau of the potential core is completely disappeared and the centres of jets are reached closer to the exit nozzles, which is caused by jet actuators impacting the main jets with high velocities.

5.3 Radials profiles of velocity fluctuations

The radial profiles of rms velocities, u′ (longitudinal velocity fluctuations) and v′ (radial velocity fluctuations) for the SW2J configuration at various heights of the flow with and without control are shown in Fig.20. The results indicate the significant effect of the jet actuators, since the intensity of velocity fluctuations increases in the case *r*=0.15. In the case without actuators, the radial distribution of u′ and v′ with peaks of turbulence is found. The maximum values of u′ are obtained close to the burner, and are of about 6 and 7 m/s for natural gas jet and oxygen jet, respectively (a and c). Further downstream, the intensity of the fluctuations is attenuated in turbulence zones. Indeed, at z=75 mm, u′ is about 3 and 4 m/s in the natural gas and oxygen jets, respectively. In the case with actuators, the values of u′ and v′ are more significant (b and d). In fact, for *r*=0.15, the maximal values of u′ are 10 and 6 m/s for the fuel and oxidizer. For z=15 mm, a high increase of fluctuations appears along the jets axis, affected by the tangential jets. This shows the increase of turbulence zones near the nozzle exits, favouring mixing with the ambient air and between the jets themselves.

The control effect on the fluctuations of radial velocity (v′) is more pronounced as shown in Fig.20.d. The fact of introducing tangentially a portion of jet flow rate tends to influence the initial zone of the flow and induces high radial velocity fluctuations. It is shown that maximum values of v′ are of the order of 4 to 5.5 m/s with actuators (d), instead of 1.5 to 2.5 m/s for the case without jets actuators (c) in the zone close to the burner.

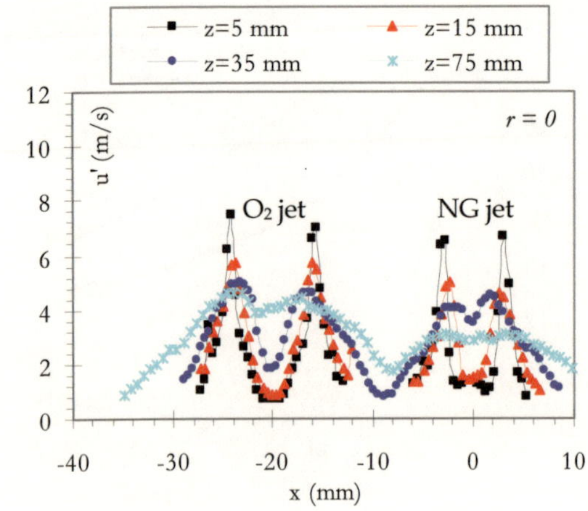

a) u′ (r = 0)

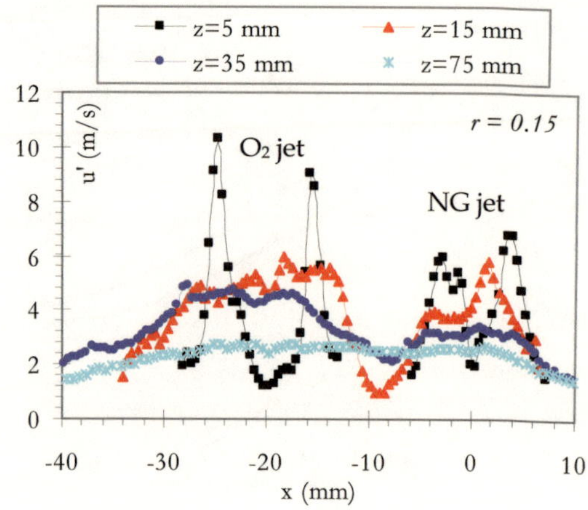

b) u′ (r = 0.15)

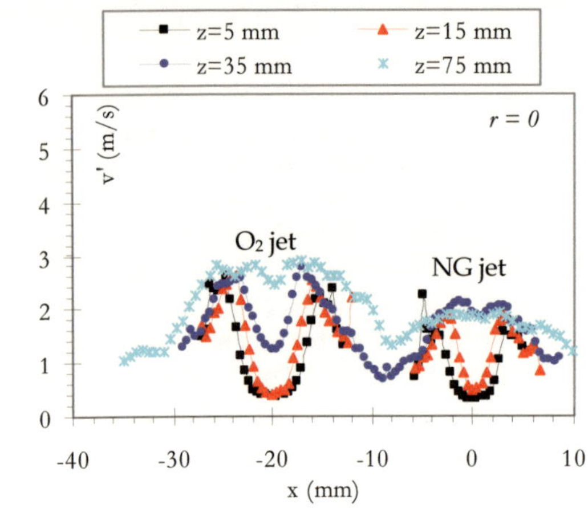

c) v′ (r = 0)

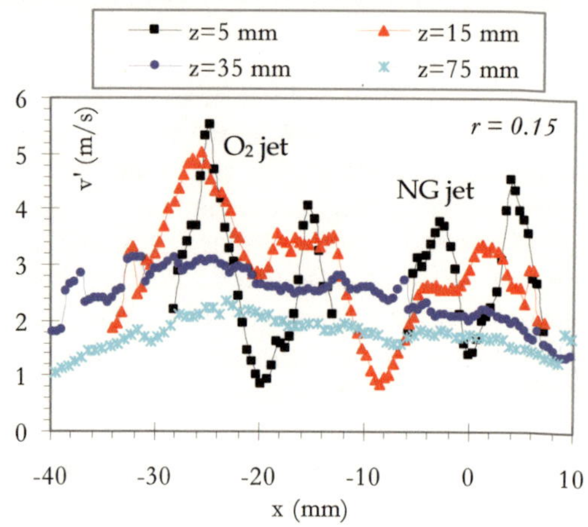

d) v′ (r = 0.15)

Fig. 20. Radial profiles of rms velocities for the SW2J configuration in non reacting flows without control (r=0) and with control (r=0.15); a-b) longitudinal velocity fluctuations (u′), c-d) radial velocity fluctuations (v′)

5.4 Velocity fields in oxy-fuel flames

In this section, the measurements of velocities in oxy-fuel combustion by PIV technique in the furnace are presented. Fig.21 shows the mean velocity fields (with the longitudinal velocity in color scale) for the SW2J configuration without control (r=0) and with control (r=0.15). The radial profiles are shown in Fig.22. These results indicate that the velocity field is also affected by the use of jet actuators. In the presence of jet actuators, a higher decrease and wider spreading of the flow are observed when the flow rate control r increases. The merging and combined zones of the jets occur more and more upstream with increasing r. Therefore, the stabilization point (small squares in pink color in the figure) of the flame moves upstream with the control and locates in the zone inter-jets where the velocity is low. Note that the stabilization point has been deduced from OH* emission measurements (see Bouhaki et al. 2009) and represents the region where the combustion starts. Without actuators (r=0), the flow velocity decreases along the axis but this decrease is slower than in the case of non-reacting flow, in particular for oxygen jet. Between z=5 mm and 95 mm, the maximum velocity of the oxygen jet decreases from 33 to 27 m/s, whereas in non-reacting flow, the maximum velocity passes from 36 to 18 m/s. In non-reactive flow, mixing and turbulence develop faster which generates a faster decrease of velocity. It is shown that in the initial zone of the flow, the maximal velocity is higher in non-reacting flow than in the reacting flow. The velocity profiles in combustion are slightly more flattened and more open, due to the high heat release from oxy-fuel flame. With control, the longitudinal velocity decay and radial spreading are more significant when the flow rate ratio r increases. As in the case without control, the velocity decay is slower with control in the reacting flow compared to the non-reacting flow. Above the stabilization point, the flow keeps a higher velocity owing to the fast expansion of hot gas by the reaction. As an example, for z=55 mm, the maximum velocity of reacting flow is about 21 against 12 m/s in the non-reacting flow. The presence of flame may decrease the entrainment of ambient fluid, which accelerates the flow. This result was observed by Takaji et al. (1981) on a turbulent H_2-N_2/Air flame by comparison between non-reacting and reacting flow.

The enhancement of mixing by the jet actuators leads to a decrease in lift-off heights and a better stability of the flame as shown in Boushaki et al. (2009). However, the flame length decreases with the flow rate of jet control since the mixing is improved by using swirling flow. In practical systems the flame length is an important factor since it defines the distance on which the heat transfer is transmitted. On the other hand, for this control system in the range $0 \leq r \leq 0.2$, the flame length remains relatively higher. Moreover, in this range of low flow rate ratio, globally NOx production decreases when r increases.

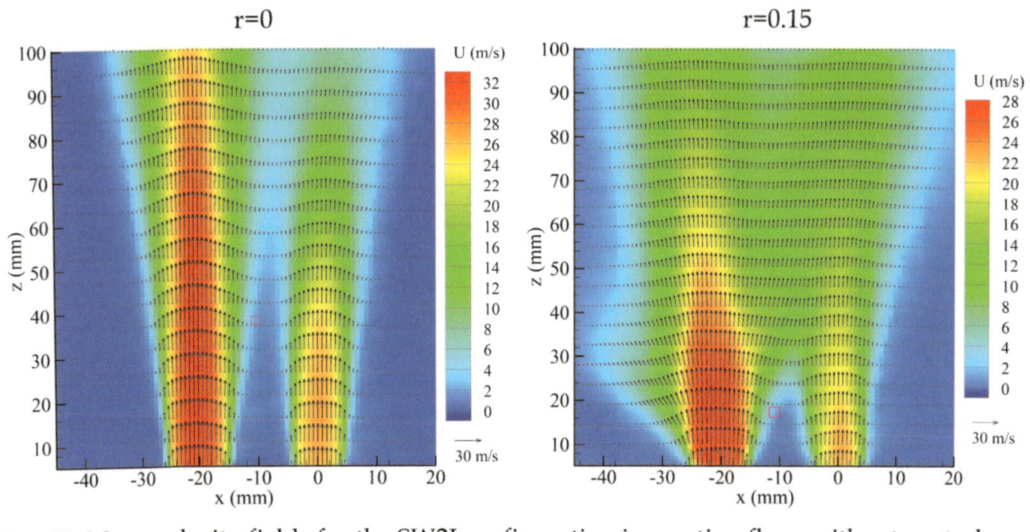

Fig. 21. Mean velocity fields for the SW2J configuration in reacting flow, without control (left) and with control. The point in pink color represents the position of the flame base

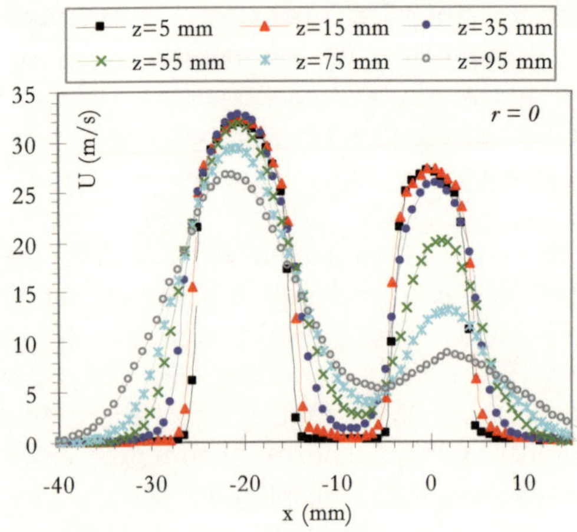

a) U, (r=0)

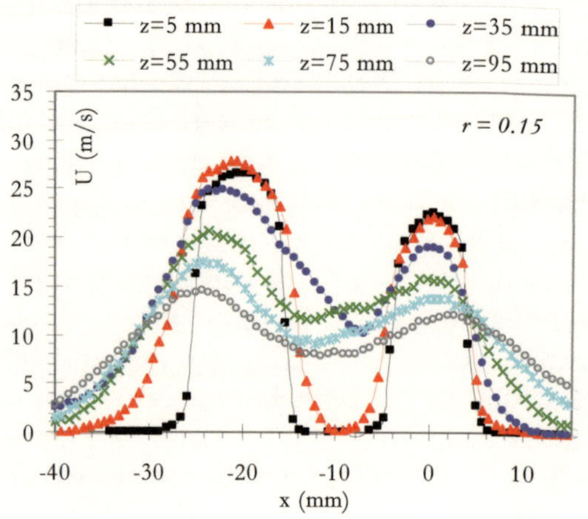

b) U, (r=0.15)

Fig. 22. Radial profiles of mean longitudinal velocity for the SW2 configuration in reacting flow, a) without control (r=0), b) with control (r=0.15).

6. Conclusion

The dynamic study by PIV measurements on oxy-fuel burners with separated jets is investigated in this chapter. Measurements by PIV technique enable the characterization of the behavior and structure of jets and the distribution of velocity fields. Particular attention has been paid to the particle seeding, oil particles for non-reacting configuration and solid ZrO2 particles in combustion. The measurements of the velocity fields are fulfilled on a burner composed of three jets, one central jet of natural gas and two side jets of pure oxygen. Measurements concern the nonreacting flow and the reacting flow in turbulent diffusion flames inside a combustion chamber. Two types of control of flows are developed and applied on the basic configuration of burner. A passive control based on the deflection of jets, and an active control that consists of four small jet actuators, placed tangentially to the exit of the main jets to generate a swirling flow.

Results of passive control show that the inclination of side jets towards the central jet improves the mixing and thus accelerates the merging, and then the combining of jets where velocity profiles become uniform to form a single jet profile. The deflection of injectors induces a radial velocity in the first zone of flow increasing with the jet angle. Along the flow, this velocity decreases since the side jets impact the central jet and the flow forms a single jet. With deflection of jets, the potential core length decreases, and therefore the longitudinal velocity decreases rapidly in the first zone of the flow with the jet angle. The oxy-fuel combustion generates significant changes on the distribution of the dynamic fields. The velocities are higher compared to non-reacting flow, in particular above the stabilization region. This is due both to the rapid expansion of burnt gases which accelerates the flow and a retardation of mixing due to the presence of the flame. Also, a greater radial expansion in reacting flow, particularly above the lift-off position of flame, was observed.

The results about the active control show that the presence of tangential jet actuators appreciably acts on the structure of the flow and consequently on the flame behaviour. The control by jet actuators induces a decrease in the length of potential core of jets, more important spreading of flow and higher longitudinal velocity decay. It is observed that for *r* (flow rate ratio) > 0.1 the potential core disappears completely due to the jet actuation. The center of the jet is reached closer to the exit of nozzles as a result of jet actuators that impact with high velocity the main jets. Moreover, the increase of control flow rate accelerates the merging of the jets and the flow rapidly reaches the characteristics of a single jet. Near the burner, the transverse velocity increases with the control and can reach 10 m/s due to the tangential injection by actuators. Velocity fluctuations (longitudinal and radial) increase with the control, and therefore the layers of turbulence widen significantly and promote mixing with the ambient medium and between the jets themselves. From non controlled jets (r=0) to controlled jets (r>0), the turbulence intensity along the jet axis highly increases, in particular near the exit of jets (from 5 to 25% in the configuration SW2J). Measurements in combustion revealed some changes on the flow velocity fields. The dynamic development of oxy-fuel flame is highly slowed by the presence of the high temperature environment, since the decrease in velocities is slower compared to the nonreactive case. The evolution of the longitudinal velocity shows that even in the presence of control, the values of U_{cl} remain high near the burner, and then decrease slowly along the flow development.

7. Acknowledgment

This work was supported by the CRCD (Centre de Recherche Claude-Delorme) of Air Liquide, Jouy-en-Josas, France. The authors are grateful to Bernard Labegorre for useful discussions.

8. References

Ahuja, K.K. (1993). Mixing enhancement and jet noise reduction through tabs plus ejectors. *AIAA Paper*, 93-4347.

Barrère, M. & Williams, F. (1968). Comparison of Combustion Instabilities found in Various Types of Combustion Chambers, *Proceedings of the Combustion Institute*, 12, pp. 169-181.

Baukal, CE. & Gebhart, B. (1997). Oxygen-enhanced/natural gas flame radiation. *International Journal of Heat and Mass Transfer* 40(11): 2539-2547.

Baukal, CE. (2003). Industrial burners handbook, CRC Press

Beér, J. M. & Chigier N. A. (1972). Combustion Aerodynamics, ed. Krieger, Malabar, Florida.

Bloxsidge, G J; Dowling, A P.; Hooper, N. & and Langhorne, P.J. (1987) Active control of an acoustically driven combustion instability *J. Theor. Appl. Mech.* 6, pp. 161–75.

Boushaki, T.; Sautet, J.C; Salentey, L. & Labegorre, B. (2007)., The behaviour of lifted oxy-fuel flames in burners with separated jets, *International communication in Heat and Mass Transfer*, 34, pp 8-18.

Boushaki, T; Mergheni, M-A. ; Sautet, JC. & Labegorre, B. (2008). Effects of inclined jets on turbulent oxy-flame characteristics in a triple jet burner, *Exp. Therm. Fluid Sci.* 32:1363-1370.

Boushaki, T; Sautet, JC. & Labegorre B. (2009). Control of flames by radial jet actuators in oxy-fuel burners, Combustion and Flame, *Combustion and Flame*, 156, 2043–2055.

Bradbury, L.J.S. & Khadem, A.H. (1975). The distortion of a jet by tabs. *Journal of Fluid Mechanics*, 70, 801–813.

Candel S. (1992). Combustion Instabilities Coupled by Pressure Waves and Their Active Control, *Proceedings of the Combustion Institute* 24, pp. 1277-1296.

Coghe, A.; Solero, G. & Scribano, G. (2004). Recirculation phenomena in a natural gas swirl combustor, *Experimental Thermal and Fluid Science*, 28, pp. 709-714.

Davis, A. & Glezer, A. (1999). Mixing control of fuel jet using synthetic jet technology : velocity field measurements, *37th AIAA Aerospace Sciences Meeting and Exhibit Reno,* NV, AIAA 99-0447, pp. 1-15.

Davis, M.R. (1982). Variable control of jet decay, *AIAA Journal*, 20, pp. 606-609.

Delabroy, O.; Haile, E.; Veynante, D.; Lacas, F. & Candel S. (1996), Réduction de la Production des Oxydes d'Azote (NOx) dans une Flamme de Diffusion à Fioul par Excitation Acoustique, *Revue Générale de Thermique* 35, pp. 475-489.

Demayo, T.N.; Mcdonell, V.G. & Samuelsen, G.S. (2002). Robust Active Control of Combustion Stability and Emissions Performance in a Fuel-staged Natural-gas-fired Industrial Burner, *Proceedings of the Combustion Institute* 29, pp. 131-138.

Denis, S.; Delville, J. ; Garem, J.H. & Bonnet, J.P. (1999). Contrôle de l'expansion d'un jet rond par des jets impactants. *14ème Congrès Français de Mécanique*, pp. 1-6.

Faivre, V. & Poinsot, T. (2004). Experimental and numerical investigations of jet active control for combustion applications, *Journal of Turbulence* 5, 025.

Feikema, D.; Chen, R.H. & Driscoll J.F., Enhancement of flame blowout limits by the use of swirl, *Combustion and Flame*, 80:183-195, (1990).

G.E.F.G.N. (1983). Utilisation de l'oxygène dans les flammes de diffusion de gaz naturel. *Revue Générale de Thermique.*, N° 214, pp: 643-649

Gollahalli, S; Khanna, T. & Prabhu, N. (1992). Diffusion Flames of Gas Jets Issued From Circular and Elliptic Nozzles. *Comb. Sci. Technol*. 86, pp. 267-288.

Gutmark, E.J., Grinstein, F.F., (1999). Flow control with noncircular jets. *Annual Review of Fluid Mechanics* 31, pp. 239–272.

Hileman, J.; Caraballo, E. & Samimy, M. (2003). Simultaneous real-time flow visualizations and acoustic measurements on a tabbed Mach 1.3 jet. *AIAA Paper* 2003-3123.

Ho, C., Gutmark, E.J., (1987). Vortex induction and mass entrainment in small-aspect-ratio elliptical jet. *Journal of Fluid Mechanics* 179, pp. 383–405.

D. Honoré, S. Maurel, A. Quinqueneau, "Particle Image Velocimetry in a semi-industrial 1 MW boiler", *Proceedings of the 4th International Symposium on Particle Image Velocimetry*, Göttingen, Germany, Sept. 11-19, 2001.

Ibrahim, M.; Kunimura, R. & Nakamura, Y. (2002). Mixing enhancement of compressible jets by using unsteady microjets as actuators, *AIAA J*. 40 681–8.

Ivernel, A. & Marque, D. (1975). Flammes industrielles de gaz naturel. Influence de la suroxygénation du comburant. *2nd Europeen Combustion Symposium*, pp 121-128

Krothapalli A.; Bagadanoff D. & Karamchetti K. (1980). Development and structure of a rectangular jet in a multiple jet configuration, *AIAA journal*, vol 18, N° 8, pp 945-950, (1980).

Krothapalli A.; Bagadanoff D. & Karamchetti K. (1980). Development and structure of a rectangular jet in a multiple jet configuration, *AIAA journal*, vol 18, N° 8, pp 945-950, (1980).

Lang, W.; Poinsot, T. & Candel, S. (1987). Active control of combustion instability, *Combustion and Flame*, 70, pp. 281-289.

Lardeau, S.; Lamballais, E. & Bonnet, J.P. (2002). Direct numerical simulation of a jet controlled by fluid injection, *Journal of Turbulence*, 3.

Lecordier B., Trinité M. (2003). Advanced PIV algorithms with image distortion validation and comparison using synthetic images of turbulent flow, *Particle Image Velocimetry: recent improvement*, Springer.

Lee, B-J.; Kim, J-S. & Lee, S. (2004). Enhancement of blow-out limit by the interaction of multiple nonpremixed jet flames, *Combust. Sci. and Tech*, 176, pp. 481-497.

Leite, AOP.; Ferreira, MA. & Carvalho, JA. (1996). An investigation of multiple jet acetylene flames. *International Communications in Heat and Mass Transfer* 23 (7): 959-970.

Lenze, B.; Milano, ME.; & Günther, R. (1975). The mutual influence of multiple jet diffusion flames. *Combust. Sci. and Tech.*, 11:1-8.

McManus, K.R; Poinsot, T. & Candel, S. (1993). A Review of Active Control of Combustion Instabilities, *Progress in Energy and Combustion Science*, 19, pp. 1-29.

Menon, R. & Gollahali, SR. (1998). Combustion characteristics of interacting multiple jets in cross flow. *Combust. Sci. and Tech.*, 60:375-389.

Moawad Ahmed K.; Rajaratnam, N. & Stanley, SJ. (2001). Mixing with multiple circular turbulent jets. *Journal of hydraulic research*, vol. 39, N°2: pp. 163-168.

Pani, B. & Dash, R. (1983). Three dimensional single and multiple jets. *J. Hydr. Engrg*, 109 (2): 254-269.

Perthuis, E. (1983). La combustion industrielle. Editions Technip, Paris

Raffel, M.; Willert, CE. & Kompenhans, J. (1998). Particle Image Velocimetry, A Practical Guide, Springer ed. ISBN-10: 3540636838.

Raghunatan, S. & Reid, I.M. (1981). A study of multiple jet. *AIAA paper* 19, 124-127

Sautet, JC; Boushaki, T.; Salentey, L. & Labegorre, B. (2006). Oxy-combustion properties of interacting separated jets. *Combust. Sci. and Tech.*, 178: 2075-2096.

Schmittel, P.; Günther, B.; Lenze, B.; Leuckel, W. & Bockhorn, H. (2000). Turbulent swirling flames: experimental investigation of the flow field and formation of nitrogen oxide, *28th Symposium (International) on Combustion*, pp 303-309.

Sheen H.J; Chen, W.J.; Jeng, S.Y.; Huang, T.L. (1996). Correlation of swirl number for a radial-type swirl generator. *Experimental Thermal and Fluid Science* 1996; 12: pp 444-451.

Simonich JC (1986) Isolated and interacting round parallel heated jets. AIAA paper 86-0281

Strahle, W.C. (1978). Combustion Noise. *Pro. Energy Combust. Sci.*, 4, pp. 157-176.

Susuki, H; Kasagi, N. & Susuki, Y. (1999). Active control of an axisymmetric jet with an intelligent nozzle *Proc. 1st Int. Symp. on Turbulent Shear Flow (Santa Barbara)*, pp 665–70.

Syred N. & Béer J.M. (1974). Combustion in swirling flows : A review, *Combustion and Flame*, 23:143-201.

Takagi, T., Shin, H.D. & Ishio, A. (1981). Properties of turbulence in turbulent diffusion flames, *Combustion and Flame*, 40:121-140.

Tamburello, DA. & Amitay, M. (2008). Manipulation of an Axisymmetric Jet by a Single Synthetic Jet Actuator. *International Journal of Heat and Fluid Flow* 29, pp. 967-984.

Yimer, I.; Becker, HA. & Grandmaison, EW. (1996). Development of flow from multiple jet burner, *A.I.C.H.E Journal* 74, pp. 840-851.

6

Characterization of the Bidirectional Vortex Using Particle Image Velocimetry

Brian A. Maicke and Joseph Majdalani
University of Tennessee Space Institute
United States of America

1. Introduction

The practical applications and theoretical difficulties associated with swirl-dominated flow fields have turned them into a source of constant scientific inquiry. Although some modern research efforts can be traced back to the piecewise formulation developed by Rankine (1858), an understanding of fundamental aspects of vortex dynamics has been demonstrated as far back as in ancient Greece (Vatistas, 2009). In fact, a multitude of naturally occurring phenomena such as hurricanes and tornadoes (Penner, 1972), as well as celestial events such as galactic jets are often modeled as vortex-dominated spiraling motions (Kirshner, 2004; Königl, 1986).

Confined vortex modeling involves small-scale helical motions that can be affected by the presence of viscous resistance. In naturally occurring vortices, an inviscid model is often sufficient as the natural length scales of the vortex substantially exceed the dimensions of the fluid regions influenced by viscous effects. However, in a confined vortex environment, the length scales are greatly reduced and therefore accounting for viscous effects becomes vital, especially near the axis of rotation.

A wealth of literature exists on confined vortex studies. For example, the efficiency of cyclone separators has been the focus of an investigation by ter Linden (1949). Bloor & Ingham (1987) have also introduced an incompressible formulation for a conical separator in spherical coordinates. In addition to these practical applications, the confined vortex possesses important academic value. As far as stability of unidirectional vortices is concerned, Rusak et al. (1998) describe the evolution of a perturbed vortex in a pipe in an attempt to characterize axisymmetric vortex breakdown. This work is further extended by Rusak & Lee (2004) to include compressible vortices. The intention of these studies is to not only characterize confined vortex breakdown, but to also extend the mechanisms entailed in generalized vortex breakdown to unconfined vortices.

1.1 Experimental studies

While only few analytical models have been proposed for describing the various swirl-dominated solutions of a confined vortex, there exists a significant body of literature that is devoted to experimental investigations. These studies can be roughly separated depending on the methods employed in their data collection: probes, Laser Doppler Velocimetry (LDV), and Particle Image Velocimetry (PIV).

1.1.1 Probes

Within the context of cyclone separators, the experimental study by Smith (1962b) employs a glass tube filled with smoke particles to capture the general structure of a confined swirling flow. Smith also utilizes a special slender probe stretched across the diameter of the chamber to determine the magnitude and direction of the velocity in the cyclone. This setup allows the measurement of the axial and tangential components of the velocity. In this case, the radial velocity is assumed to be so small that it can be inferred from continuity. In a companion paper, Smith (1962a) combines analytical methods with experimental measurements to characterize the dynamics and possible instabilities that occur in a confined vortex.

In a later investigation into the behavior of cyclone combustors, Vatistas et al. (1986) conduct a similar experiment in which a prismatic pitot tube is used to capture the velocity and pressure maps within a cyclonic chamber. These researchers compare their findings to an experimentally correlated inviscid model and find what may be essentially viewed as a favorable agreement. Their study highlights a key realization in confined vortex modeling, namely, that swirl velocity variations in the axial direction are so small that they may be ignored (see Ogawa, 1984; Reydon & Gauvin, 1981, among others). This simplification is common in the analytical modeling of vortices.

Furthermore, these studies provide early insights into the conditions arising in a confined vortex; however, some deficiencies must be noted. Even with proper calibration, the minimally intrusive probes can introduce disturbances into the flow, and these, in turn, can lead to potentially misleading results. This is especially important when investigating dynamic effects such as vortex instability.

1.1.2 Laser Doppler Velocimetry

Improvements in technology give rise to increasingly sophisticated experimental techniques that help to provide valuable information regarding confined swirl velocities without the intrusion invariably present with even the smallest probes. For instance, LDV minimizes flow disruptions by seeding the fluid domain with particles and then using a focused laser to scatter light off those particles. The interference patterns are then correlated to velocity measurements obtained in localized regions. Subsequently, the corresponding subvolumes are summed together to reconstruct the overall velocity profile of a given flow pattern.

Hoekstra et al. (1999) take an increasingly common approach of pairing a CFD solution with LDV measurements to validate their proposed turbulence models. Their experimental setup uses a back-scatter LDV to collect the axial and tangential velocity profiles in small volumes and these are then combined and correlated to provide an overall velocity profile. In this effort, however, the turbulent cross-correlation of the LDV measurements is found to be problematic because of the finite wall thickness of the cyclonic chamber which, in itself, can cause refraction and dissimilar levels of distortion based on the spatial location within the chamber. Without proper accounting for these optical disparities, a perfect correlation between the acquired signal and the flow profile will be difficult to realize. The quality of the seeding in the core region also proves to be an issue, as the natural motion in the cyclone tends to separate particles from the flow.

Hu et al. (2005) conduct a similar study for industrial-size cyclone separators. Whereas Hoekstra et al. (1999) focus on the separation section of the cyclone, Hu et al. (2005)

consider the full geometry of the separator including the inlet, hoppers, and other supporting hardware. Moreover, their experimental investigation is accompanied by a turbulent computational solution. In addition to verifying turbulent vortex models, their study aims at improving the prediction of cyclone efficiency.

Along similar lines, an investigation into the turbulent kinetic energy of a confined vortex is reported in a forward scatter LDV study by Yan et al. (2000). In this work, data collected at a wide range of Reynolds numbers is used to validate their empirically derived solutions. These particular models rely on scaling laws to reduce the problem's dependence in each case to one or two key parameters and these tend to involve some combination of the inlet flow rate and the contraction ratio.

1.1.3 Particle Image Velocimetry

PIV is another minimally intrusive technique that will be discussed in the remainder of this chapter. Much like LDV, PIV employs particle seeding to collect velocity measurements. The primary difference between the two techniques stands in how the data is gathered. Whereas LDV relies on two focused beams to generate interference patterns, PIV uses optics to create a laser sheet that illuminates a plane in the chamber. High-speed cameras are then utilized to capture images of the illuminated particles at two closely spaced intervals such that a net velocity profile may be deduced from the cross-correlation of these images.

By way of comparison, both PIV and LDV methods are used by Sousa (2008) to determine the velocity field that accompanies vortex breakdown in a closed container. Sousa finds that accurate measurements may be acquired using either method; he also reports several challenges that may be associated with PIV techniques. The fully three-dimensional nature of the flow field can lead to a decrease in correlation accuracy as seed particles move normal to the light sheet. Sousa accounts for this factor by shortening the duration between laser pulses and by slightly thickening the laser sheet to increase the chances that the particles of interest will remain in the area of investigation.

In the spirit of improvement, Zhang & Hugo (2006) use a stereoscopic PIV setup to investigate the vortex motion in a pipe. Stereoscopic PIV can be used to capture the fully three-dimensional flow field; however, it requires an additional high speed camera and more elaborate calibration skills to ensure that both cameras will target the same area. This obviously leads to an increase in post-experimental processing as the images from two cameras have to be analyzed for each exposure, effectively doubling the amount of data acquired. Finally, Zhang & Hugo (2006) develop and implement an improved calibration technique to reduce the optical distortion caused by refraction through the fluid and the curved chamber wall.

1.2 Classical solutions

While a large body of literature may be reported on vortical motions, the segment devoted to analytical modeling remains much smaller in comparison. For this reason, a brief introduction to some of these classical models will follow.

In examining the classic models for describing swirl-dominated flowfields, it is helpful to further discriminate between solutions by specifying whether the attendant equations allow for unidirectional or multidirectional swirl. Unidirectional models are characterized by a one-directional axial velocity, whereas multidirectional motions exhibit a reverse flow character in which the axial velocity switches polarity. In practical applications, multidirectional flows are reduced to a bidirectional profile in which the axial velocity reverses only once.

1.2.1 Unidirectional models

One of the earliest unidirectional models is provided by Rankine (1858). Accordingly, the swirl velocity is formulated as a simple piecewise solution that is radially dependent. A normalized representation of this model may be written as

$$\frac{\bar{u}_\theta}{(\bar{u}_\theta)_{\max}} = \begin{cases} \dfrac{\bar{r}}{\delta_c} & \bar{r} \le \delta_c \\ \dfrac{\delta_c}{\bar{r}} & \bar{r} > \delta_c \end{cases} \quad \text{or} \quad u_\theta = \begin{cases} r & r \le 1 \\ r^{-1} & r > 1 \end{cases} \tag{1}$$

where overbars denote dimensional quantities and the radius is referenced to δ_c, the distance from the origin to the point at which $\bar{u}_\theta$ reaches its peak value, $(\bar{u}_\theta)_{\max}$. The profile consists of a forced core that exhibits solid body rotation. The outer vortex varies with the inverse of the radius, so the swirl velocity diminishes away from the axis of rotation. Rankine's model is often used as a baseline solution for comparison or initialization owing to its simplicity. Nonetheless, Eq. (1) is somewhat limited in that the optimal matching location, δ_c, cannot be specified a priori, but requires data for anchoring. Moreover, the model is not differentiable at the matching location, which may be undesirable in subsequent analysis.

The Lamb-Oseen vortex consists of another solution that incorporates a time-dependent decay of the vortex motion (Wendt, 2001). This makes the model particularly suitable for capturing the behavior of wing-tip vortices. The dimensional representation of its swirl velocity may be expressed as

$$\bar{u}_\theta(\bar{r}, \bar{t}) = \frac{\Gamma}{2\pi\bar{r}}\left[1 - \exp\left(-\frac{\bar{r}^2}{\delta^2}\right)\right] \tag{2}$$

Here Γ refers to the circulation and $\delta = 2\sqrt{\nu\bar{t}}$, to the characteristic radius which is dependent on time, $\bar{t}$, and the kinematic viscosity, ν. Equation (2) starts as a potential vortex, behaving as $1/\bar{r}$ away from the centerline before smoothly switching to a linear dependence on $\bar{r}$ in the forced vortex core evolving around $\bar{r} = 0$. As time elapses, the vortex decays exponentially.

The Burgers-Rott vortex (Burgers, 1948) is similar in form to the Lamb-Oseen profile with two notable exceptions. First, rather than a time-dependent decay, the exponential function here is governed by the suction parameter, A. Secondly, the Burgers-Rott vortex possesses well-defined relations for the axial and radial velocities. It can be written as

$$\bar{u}_\theta(\bar{r}) = \frac{\Gamma}{2\pi\bar{r}}\left[1 - \exp\left(-\frac{\bar{r}^2}{\delta^2}\right)\right]$$
$$\bar{u}_r(\bar{r}) = -A\bar{r}; \qquad \bar{u}_z(\bar{z}) = 2A\bar{z} \tag{3}$$

where $\delta = \sqrt{2\nu/A}$. The presence of an axial velocity and a suction parameter has proven useful in applications related to the modeling of thunderstorms.

1.2.2 Bidirectional solutions

In what concerns bidirectional behavior, an unbounded bipolar solution to the Navier–Stokes equations is provided by Sullivan (1959). For what is essentially a two-celled vortex, Sullivan's inner region exhibits a descending axial velocity coupled with an outward radial motion. Conversely, the outer cell flows inwardly and up. The model itself can be written in an integral representation using

$$\begin{cases} \bar{u}_\theta(\bar{r}) = \dfrac{\Gamma}{2\pi\bar{r}}\dfrac{1}{H(\infty)}H\left(\dfrac{\bar{r}^2}{\delta^2}\right); \quad \delta = \sqrt{2\nu/A} \\ H(x) = \displaystyle\int_0^x e^{f(t)}\mathrm{d}t; \quad f(t) = -t + 3\int_0^t \left(1 - e^{-y}\right)\frac{\mathrm{d}y}{y} \end{cases} \tag{4}$$

As before, A denotes the suction strength and ν, the kinematic viscosity. The corresponding axial and radial components may be expressed as

$$\bar{u}_z(\bar{r},\bar{z}) = 2A\bar{z}\left[1 - 3\exp\left(-\frac{\bar{r}^2}{\delta^2}\right)\right]; \quad \bar{u}_r(\bar{r}) = -A\bar{r} + \frac{6\nu}{\bar{r}}\left[1 - \exp\left(-\frac{\bar{r}^2}{\delta^2}\right)\right] \tag{5}$$

It is the combination of the axial and radial velocities that makes the Sullivan vortex a suitable candidate for modeling tornadoes (Wu, 1986).

A comparison of the above-mentioned swirl velocities is presented in Fig. 1. In all cases, the equations are normalized such that their peak velocities occur at a dimensionless radius of one. This is accomplished by dividing the radius by δ_c, which is the distance from the axis of rotation to the point where the maximum swirl velocity occurs. Traditionally, a diameter of $2\delta_c$ may be used to define the thickness of the forced viscous core. While all of the models capture similar trends, a significant amount of variability in the profiles may be seen. The Rankine solution displays an abrupt change in behavior at the peak velocity. The remaining models exhibit smooth contours, with the Sullivan profile concentrating the swirl velocity to a narrower region than that of Burgers-Rott. Although not depicted, the Lamb-Oseen velocity becomes identical to that of Burgers-Rott especially that time dependence is not featured in this figure.

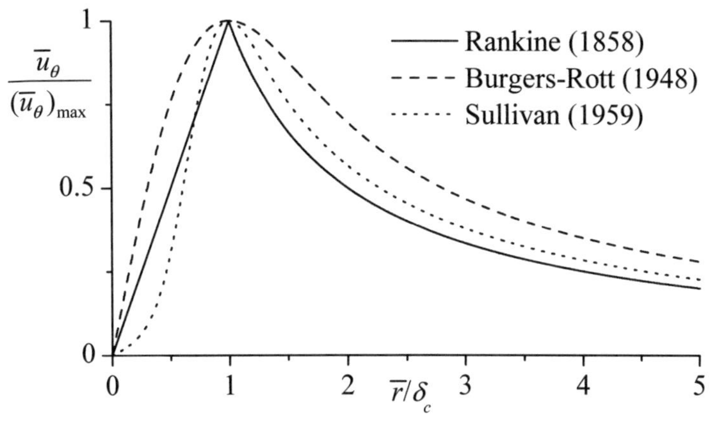

Fig. 1. A comparison of selected swirl velocity models, rescaled so that the peak velocity occurs at $r = 1$.

In work related to cyclone separators, a study by Bloor & Ingham (1987) leads to one of the most frequently cited models. The resulting inviscid solution arises in the context of a conical cyclone (see Fig. 2). Bloor and Ingham solve the Bragg-Hawthorne equation in spherical coordinates to the extent of producing a stream function of the form

$$\psi = \sigma R^2 \left\{ \left[\csc^2\alpha + \ln\left(\tan\frac{\alpha}{2}\right) - \csc\alpha\cot\alpha \right] \sin^2\phi - \sin^2\phi \ln\left(\tan\frac{\phi}{2}\right) + \cos\phi - 1 \right\} \tag{6}$$

Equation (6) translates into the following velocity components

$$u_R = 2\sigma R \left\{ \left[\csc^2\alpha + \ln\left(\tan\frac{\alpha}{2}\right) - \csc\alpha\cot\alpha \right] \cos\phi - \cos\phi \ln\left(\tan\frac{\phi}{2}\right) - 1 \right\} \tag{7}$$

$$u_\phi = \frac{2\psi}{R^2\sin\phi}; \quad u_\theta = \frac{1}{R\sin\phi}\sqrt{1 - \frac{\bar{Q}_i^2\sigma\psi}{\pi^2 a^4 U^2}} \tag{8}$$

Here $\bar{Q}_i$ denotes the volumetric flow rate through the cyclone, U and W stand for the average swirl and axial velocities at the entrance, α represents the taper angle of the cyclone, and σ refers to the dimensionless swirl parameter described by

$$\sigma = \frac{\pi a_0^2 U^2}{\bar{Q}_i W} \tag{9}$$

Here a_0 is the outer radius of the virtual inlet (see Fig. 2). Equations (6)– (9) constitute an improvement on previous work (Bloor & Ingham, 1973), where use of the Polhausen technique leads to a solution that is insensitive to injection conditions. It should be noted that Eqs. (6)– (9) represent a corrected form of the Bloor–Ingham solution according to Barber & Majdalani (2009).

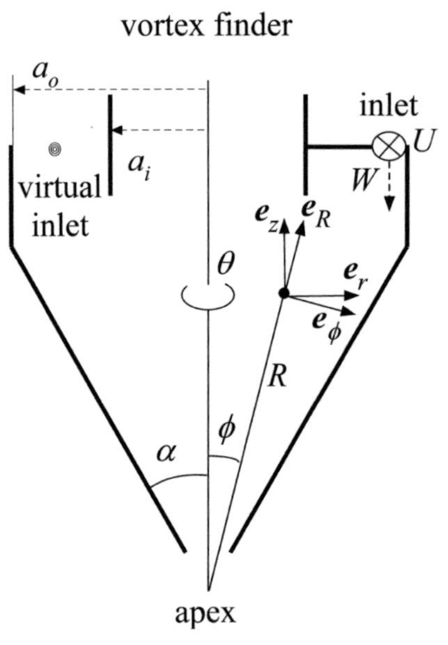

Fig. 2. Bloor-Ingham solution domain and geometry.

Moving beyond the Bloor–Ingham approximation, Vyas & Majdalani (2006) introduce a bidirectional model with a reversing axial character. Their complex-lamellar solution, which constitutes the basis for the upcoming analysis, seeks to describe the bulk gaseous motion in the Vortex Combustion Cold-Wall Chamber (VCCWC) developed by Chiaverini et al. (2002). In this vortex-driven engine, the swirling motion of the oxidizer insulates the sidewalls against thermal loading, thus leading to a substantial reduction in engine weight. The swirling motion also has a mitigating impact on pressure oscillations in the chamber as shown by Batterson & Majdalani (2011a;b). The mathematical character of this application is described in the following section.

The remainder of this chapter is focused on the synergistic combination of analytical and experimental methods in the context of confined vortex flows. First, two distinct analytical frameworks are devised with the intent of modeling the swirl velocity of a confined vortex. These models introduce control parameters that guide the construction of an experimental apparatus. A PIV setup is then used to capture the swirl velocity in the vortex chamber. Finally, the data is compared to the analytical frameworks in an effort to reconcile between theory and experiment.

2. Analytical formulations

The vortex chamber, shown in Fig. 3, is modeled as a right circular cylinder of radius a and length L. The origin of the coordinate system is fixed at the center of the inert headwall, while the aft section is partially left open with a radius of b. The radial and axial coordinates are denoted by $\bar{r}$ and $\bar{z}$, respectively, while the characteristic geometric parameters are the outflow fraction $\beta = b/a$ and the aspect ratio $l = L/a$.

A single phase, non-reactive fluid is injected tangentially at $\bar{z} = L$. The fluid spirals up the outer annular region towards the headwall, thus forming an outer vortex (see Fig. 3). Once the fluid reaches the top of the chamber, it reverses axial direction and funnels out of the chamber as an inner vortex in the region $0 < \bar{r} < b$. These two vortices are separated by a spinning, non-translating layer called the 'mantle.' This fluid interface materializes at $\bar{r} = b$. Mathematically, these conditions translate into the following boundary conditions:

$$\bar{r} = a,\ \bar{z} = L,\ \bar{u}_\theta = U \qquad \text{tangential velocity at entry} \qquad (10)$$

$$\bar{z} = 0,\ \forall \bar{r},\ \bar{u}_z = 0 \qquad \text{impervious headwall} \qquad (11)$$

$$\bar{r} = 0,\ \forall \bar{z},\ \bar{u}_r = 0 \qquad \text{no flow across centerline} \qquad (12)$$

$$\bar{r} = a,\ \forall \bar{z},\ \bar{u}_r = 0 \qquad \text{impervious sidewall} \qquad (13)$$

$$2\pi \int_b^a \bar{u}_z(\bar{r}, L)\bar{r}\mathrm{d}\bar{r} = \bar{Q}_i \qquad \text{axial outflow matching tangential source} \qquad (14)$$

$$\bar{r} = 0,\ \forall \bar{z},\ \bar{u}_\theta = 0 \qquad \text{forced vortex center} \qquad (15)$$

$$\bar{r} = a,\ \bar{z} < L,\ \bar{u}_\theta = 0 \qquad \text{no slip condition at the sidewall} \qquad (16)$$

The last two conditions are only used when accounting for viscous effects in the chamber. The $\bar{Q}_i$ term represents the inlet volumetric flow rate and $\bar{Q}_0$ is the outlet volumetric flow rate, calculated from the integral in (14).

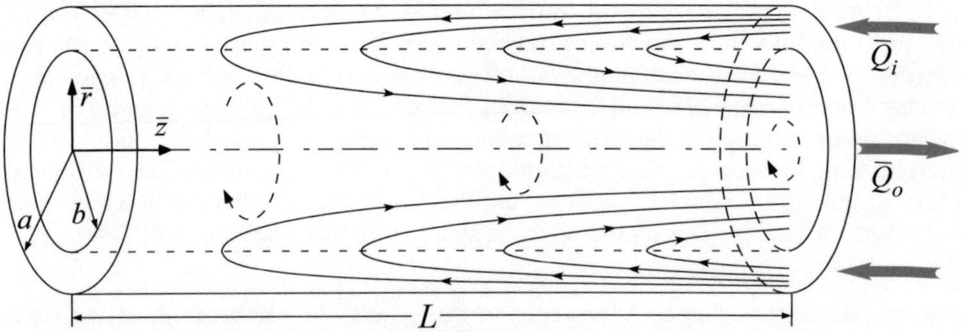

Fig. 3. Sketch of the bidirectional vortex chamber and corresponding coordinate system.

To facilitate comparisons to existing models, it is beneficial to use a non-dimensional form of the governing equations. This is achieved by setting

$$r = \frac{\bar{r}}{a};\ z = \frac{\bar{z}}{a};\ \delta = \frac{\bar{\delta}}{a};\ \nabla = a\bar{\nabla}; \tag{17}$$

$$u_r = \frac{\bar{u}_r}{U};\ u_\theta = \frac{\bar{u}_\theta}{U};\ u_z = \frac{\bar{u}_z}{U};\ p = \frac{\bar{p}}{\rho U^2};\ Q_i = \frac{\bar{Q}_i}{Ua^2} = \frac{A_i}{a^2} \tag{18}$$

Here all spatial variables are normalized by the chamber radius and the velocities by the injection velocity. The variable δ represents the characteristic length scale used in the viscous analysis.

2.1 Laminar core model

The laminar core model is a solution to the Navier–Stokes equations using a stream function approximation. An inviscid base flow is obtained initially, followed by a boundary layer type correction at the centerline. In modeling confined vortices, it is common to assume that the flow is both axisymmetric and that the swirl velocity is axially invariant (see Leibovich, 1984). The equations for the inviscid flow, after applying these assumptions, reduce to:

$$\frac{1}{r}\frac{\partial(ru_r)}{\partial r} + \frac{\partial u_z}{\partial z} = 0 \tag{19}$$

$$u_r\frac{\partial u_r}{\partial r} + u_z\frac{\partial u_r}{\partial z} - \frac{u_\theta^2}{r} = -\frac{1}{\rho}\frac{\partial p}{\partial r} \tag{20}$$

$$u_r\frac{\partial u_\theta}{\partial r} + \frac{u_r u_\theta}{r} = 0 \tag{21}$$

$$u_r\frac{\partial u_z}{\partial r} + u_z\frac{\partial u_z}{\partial z} = -\frac{1}{\rho}\frac{\partial p}{\partial z} \tag{22}$$

Equation (21) clearly shows the tangential velocity as fully decoupled from the radial and axial components such that it can be directly retrieved. Further application of the boundary condition from (10) gives

$$u_\theta = \frac{1}{r} \tag{23}$$

This free vortex form is common to many swirl-dominated inviscid flows. However, to realistically model the bidirectional vortex, viscous corrections will be required especially at the centerline.

With the swirl velocity successfully decoupled from the remaining two velocities, the Stokes stream function may be introduced using

$$u_r = -\frac{1}{r}\frac{\partial \psi}{\partial z}; \; u_z = \frac{1}{r}\frac{\partial \psi}{\partial r} \tag{24}$$

Applying (24) and the vorticity transport equation (for additional detail see Vyas & Majdalani, 2006) leads to the following relation in ψ, namely,

$$\frac{\partial^2 \psi}{\partial z^2} + \frac{\partial^2 \psi}{\partial r^2} - \frac{1}{r}\frac{\partial \psi}{\partial r} + C^2 r^2 \psi = 0 \tag{25}$$

Equation (25) can be solved via separation of variables. After applying the boundary conditions, the solution becomes

$$\psi = \kappa z \sin(\pi r^2) \tag{26}$$

where κ is a constant determined from the mass conservation boundary condition in (14)

$$\kappa = Q_i/(2\pi l) \tag{27}$$

The stream function provides the axial and radial velocity profiles which are key quantities in the calculation of the pressure. They may be expressed as

$$u_r = -\kappa \sin(\pi r^2)/r; \qquad u_z = 2\pi\kappa z \cos(\pi r^2) \tag{28}$$

With the choice of normalization being based on the average tangential speed at entry, the small size of κ ensures that these two quantities remain of secondary importance relative to u_θ.

The irrotational form in (23) accurately describes an inviscid vortex. However, for a confined vortex, some deficiencies arise, notably the singularity at the centerline where the velocity approaches infinity, and at the sidewall where slippage occurs. Bloor & Ingham (1987) note that the unbounded behavior at the centerline appears to be a common characteristic of inviscid models to the extent of becoming archetypical (cf. Leibovich, 1984). To treat this core singularity, the second-order viscous terms in the θ-momentum equation must be considered. Retention of these terms leads to

$$u_r \frac{\partial u_\theta}{\partial r} + \frac{u_r u_\theta}{r} = \frac{1}{Re}\frac{\partial}{\partial r}\left[\frac{1}{r}\frac{\partial (r u_\theta)}{\partial r}\right]; \; Re \equiv \frac{\rho U a}{\mu} \tag{29}$$

where Re represents the mean flow Reynolds number, μ, the dynamic viscosity, and ρ, the density. In this particular case, Re is of the order of 10^5. Thus, a perturbation parameter may be formed by setting $\varepsilon \equiv 1/Re$. For simplicity, the dimensionless angular momentum is consolidated into a single variable, $\xi \equiv r u_\theta$; this turns Eq. (29) into

$$\varepsilon \frac{\mathrm{d}}{\mathrm{d}r}\left(\frac{1}{r}\frac{\mathrm{d}\xi}{\mathrm{d}r}\right) - \frac{u_r}{r}\frac{\mathrm{d}\xi}{\mathrm{d}r} = 0 \quad \text{or} \quad \varepsilon \frac{\mathrm{d}}{\mathrm{d}r}\left(\frac{1}{r}\frac{\mathrm{d}\xi}{\mathrm{d}r}\right) + \frac{\kappa \sin(\pi r^2)}{r^2}\frac{\mathrm{d}\xi}{\mathrm{d}r} = 0 \tag{30}$$

By converting the independent coordinate using $\eta \equiv \pi r^2$, Eq. (30) simplifies into

$$\frac{\varepsilon}{\kappa}\frac{d^2\xi}{d\eta^2} + \frac{\sin\eta}{2\eta}\frac{d\xi}{d\eta} = 0 \tag{31}$$

In order to bring the swirl velocity to zero along the chamber axis, one must account for the rapid changes caused by the local emergence of viscous stresses (see Fig. 4). To do so, one may introduce the slowly varying scale, $s \equiv \eta/\delta(\varepsilon)$. The stretching transformation maps the region of non-uniformity about the centerline to an interval of order unity. Applying the transformation and linearizing the equation near the core where $s \approx 0$ yields

$$\frac{\varepsilon}{\kappa\delta}\frac{d^2\xi}{ds^2} + \frac{1}{2}\left[1 - \frac{\delta^2 s^2}{3!} + O(\delta^2 s^2)\right]\frac{d\xi}{ds} = 0 \tag{32}$$

The diffusive and convective terms strike a balance near the core. This occurs in (32) when $\delta \sim \varepsilon/\kappa$. Having identified the distinguished limit, the core boundary layer equation becomes

$$\frac{d^2\xi^{(i)}}{ds^2} + \frac{1}{2}\frac{d\xi^{(i)}}{ds} = 0 \tag{33}$$

where the superscript (i) stands for the inner, near-core approximation. Using a standard perturbation series, successive viscous corrections can be determined to any desired order. For this chapter, a one-term inner solution is sought. Hence, by integrating (33) and insisting on a forced vortex near the core, one retrieves

$$\xi^{(i)} = C_0\left[\exp\left(-\tfrac{1}{2}s - 1\right)\right] \tag{34}$$

The remaining constant may be determined through matching with the outer expansion. Using Prandtl's matching principle, the outer limit of the inner solution may be equated to the inner limit of the free vortex. This process results in a composite solution that is valid everywhere except in the close vicinity of the sidewall. After transforming back to the original coordinate, the composite inner solution collapses into

$$\xi^{(ci)} = \bar{C}\left[1 - \exp\left(-\tfrac{1}{4}Vr^2\right)\right] \tag{35}$$

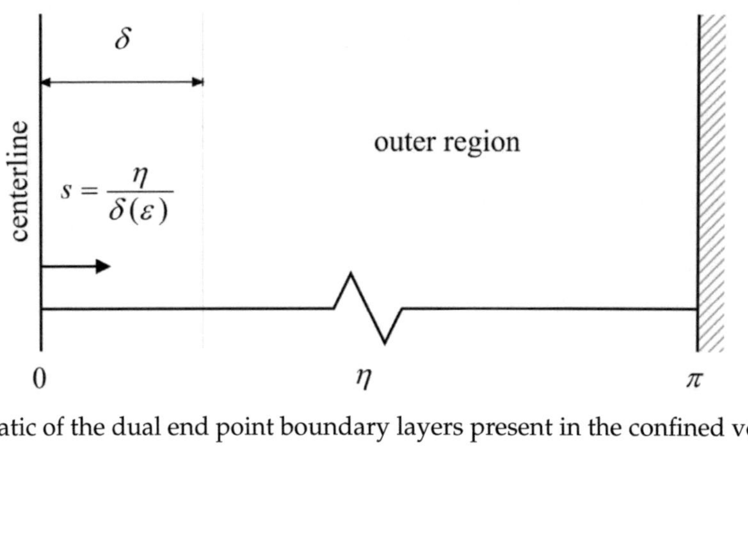

Fig. 4. Schematic of the dual end point boundary layers present in the confined vortex.

where V stands for the vortex Reynolds number,

$$V \equiv \frac{1}{\varepsilon \sigma l} = \frac{Re}{\sigma}\frac{a}{L} = \frac{\rho U A_i}{\mu L} = \frac{\dot{m}_i}{\mu L} \tag{36}$$

For a complete physical description of V, the reader may consult with Majdalani & Chiaverini (2009). The injection mass flow rate appears as $\dot{m}_i$ while the density and viscosity are given by ρ and μ. The outer constant $\bar{C}$ may be procured from the tangential injection condition. This implies

$$\zeta^{(ci)}(1) = 1 \text{ or } \bar{C} \simeq 1 \tag{37}$$

and so, when suitably combined, one has

$$u_\theta \simeq \frac{1}{r}\left[1 - \exp\left(-\tfrac{1}{4}Vr^2\right)\right] \tag{38}$$

Equation (38) provides a laminar representation of the swirl velocity in the bidirectional vortex chamber in which slip is permitted at the sidewall. To reconcile between this model and experimental data, additional steps are required as it will be shown below.

Before leaving this subject, it should be noted that a similar asymptotic analysis can be used to obtain the sidewall boundary layer correction. However, given the scope of this chapter, the attendant analysis is omitted. This is partly due to the fidelity of the PIV procedure and the data near the sidewall being too coarse to resolve the wall boundary layer. The additional complexity engendered by the sidewall correction is hence disregarded.

2.2 Constant shear stress model

The constant shear stress model provides a piecewise swirl velocity solution that may be used to model confined vortex motions. The basis for the model is that a free vortex of the $1/r$ type develops away from the core of the vortex; in contrast, equilibrium is maintained near the core between the shear and pressure terms. This balance leads to a model that is equally valid for both laminar and turbulent conditions. It is important to note that the corresponding flow is not turbulent per se as there are no unsteady effects included, but rather a mean velocity profile that can be used to represent the bulk flow field in a turbulent regime.

The justification for this approach can be seen mathematically from the conservation of momentum, namely,

$$(\boldsymbol{U} \cdot \nabla)\boldsymbol{U} = -\nabla p + \nabla \cdot \boldsymbol{\tau} \tag{39}$$

At the centerline, the tangential velocity vanishes and this leaves the pressure and shear stress terms available to balance each other (Townsend, 1976). The flow under consideration for this work has a zero tangential pressure gradient; therefore, the shear stress in the tangential direction may be assumed to have a constant value. The equation for the dominant shear stress becomes

$$\tau_{r\theta} = \epsilon\left[\frac{1}{r}\frac{\partial u}{\partial \theta} + r\frac{\partial}{\partial r}\left(\frac{v}{r}\right)\right] \tag{40}$$

where ϵ is the same viscous parameter $1/Re$. Because the flow is axisymmetric, the θ derivative is eliminated such that

$$\epsilon\left(\frac{\partial v}{\partial r} - \frac{v}{r}\right) = C_1 \tag{41}$$

The traditional forced vortex model can be recovered by setting the constant equal to zero; nonetheless, the model examined here will retain the general constant. After integration, the inner swirl velocity becomes

$$u_\theta^{(i)} = r\left[\frac{C_1}{\epsilon}\ln(r) + C_2\right] \tag{42}$$

It may be interesting to note that each of the two undetermined constants, C_1 and C_2 , has a clear physical meaning: while the first relates to the swirl strength of the velocity component generating the stress, the second corresponds to the swirl strength of a flow undergoing solid body rotation. The two undetermined constants can be manipulated to match the inner solution with the outer, free vortex expression at their intersection point. This is achieved by equating the velocity and its derivative to the outer vortex at a specific matching radius. However, since the matching radius is not known a priori, it must be carefully specified. For the moment, the matching point X is left arbitrary. The equation to match the velocities at X translates into

$$X\frac{C_1}{\epsilon}\ln(X) + XC_2 = \frac{1}{X} \tag{43}$$

Equation (43) represents an effort to match the inner solution from (42) to the outer, free vortex solution. The same procedure can be used on the derivatives to provide

$$\frac{C_1}{\epsilon}\left[1 + \ln(X)\right] + C_2 = \frac{1}{X^2} \tag{44}$$

After solving (43) and (44) for (C_1, C_2) and substituting back into (42), the result may be expressed as

$$u_\theta = \begin{cases} \dfrac{r}{X^2}\left[1 - \ln\left(\dfrac{r^2}{X^2}\right)\right]; & r \leq X \\ \dfrac{1}{r}; & r > X \end{cases} \tag{45}$$

At this point in the analysis, the value of X is still unknown although it produces a smooth function due to the matching conditions imposed through (43) and (44). The actual value of X can be determined in a number of different ways, the foremost being a comparison to experimental data.

3. Experimental setup

With the analytical models for the confined vortex determined, it is possible to design a meaningful experiment through which collections of data can be obtained and compared directly to their theoretical predictions. Having identified the vortex Reynolds number, V, as the key parameter that governs the shape of the swirl velocity, any experimental setup must be able to incorporate a number of different values for V. When limited to a single working fluid, two ways exist for inducing variations in V: changing the aspect ratio or the injection velocity. Rom (2006) uses this information to design a PIV experiment through which the character of the bidirectional vortex is carefully investigated.

To provide the necessary geometric flexibility, a modular chamber is used that allows for a variety of vortex Reynolds numbers. The chamber itself is made from a quartz cylinder and top plate. Four different chamber lengths provide a range of aspect ratios from 2.8 to 8.8. The bottom plate is acrylic and can be modified for a number of injection conditions. Additionally, the base of the chamber is retrofitted with a simple convergent nozzle with a throat diameter of 1.27 cm and an inlet radius of curvature of 0.635 cm. Figure 5 showcases the modular chamber.

The working fluid for the PIV experiment is gaseous nitrogen. A number of configurable injectors are used to provide injection pressure drops that range from 10% to 30% of the chamber pressure. This, in turn, generates an increasing set of injection velocities. The nitrogen flow is seeded by a Corona Integrated Technologies Colt4 smoke generator. The process creates liquid particles of 0.2 μm diameter.

The imaging system consists of a 250 mJ/pulse Nd:Yag laser to illuminate the seed particles. For this study, a typical pulse separation of 1 μs yields adequate resolution for the anticipated swirl velocities. The laser is focused through a series of adjustable optical devices to produce a sheet in the r–θ plane that can be positioned at three different axial locations. The particle images are captured by a LaVision 1280 x 1024 Flowmaster 3 camera at 1 μs increments which are spaced out over a thirty second run time. The images are then cross-correlated to deduce the swirl velocity at three axial locations in the chamber. A schematic of the setup is provided in Fig. 6.

The cross-correlation is carried out with LaVision DaVis 6.2 software. The two images, separated by 1 μs, are analyzed through deformed interrogation windows, initially 64 x 64 pixels in size. The windows decrease to 16 x 16 pixels during successive passes over the image. For each experimental configuration, ten sets of images are acquired and correlated to ten velocity fields.

In a related study by Rom et al. (2004), the PIV apparatus is supplemented with a modified end cap that is fitted with pressure taps. The taps are spaced at intervals of 15% of the radius with two additional taps at $r = 0.9$ and 0.967 to capture the near wall behavior. The pressure measurements provide an additional avenue to verify the analytical approximations.

3.1 Trial overview

The modularity of the experimental apparatus will permit for a widely varying range of trials. For all trials, the radius of the chamber is kept fixed at 1.27 cm. According to this arrangement, the smallest aspect ratio allows sampling at the midpoint of the chamber only. Table 1(a) provides the geometric variability of the test chamber.

In addition to the geometric variability, the injectors are configurable for three separate injection pressure drops. This is achieved by varying the available port area of the incoming fluid. All trials are conducted using eight equally spaced tangential injection ports. Details of the port construction are available in Table 1(b).

4. Results

After a successful trial, the ten pairs of image files are cross-correlated with DaVis 6.2 to produce a vector field for the swirl velocity, represented by a 128 x 160 matrix. Further data analysis is furnished via Matlab scripts which act upon the matrix exported by the DaVis software. The scripts average the ten raw velocity magnitudes at each axial location such that a radial profile of the swirl velocity may be reconstructed. The swirl velocity profile is an ideal candidate for comparison to the analytical models developed in Sec. 2.

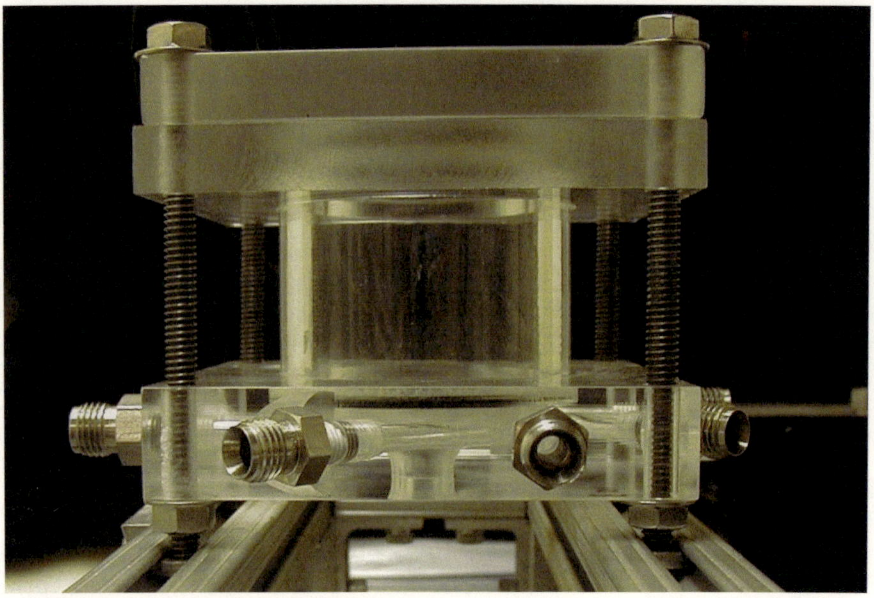

Fig. 5. Modular vortex chamber for the PIV experiment.

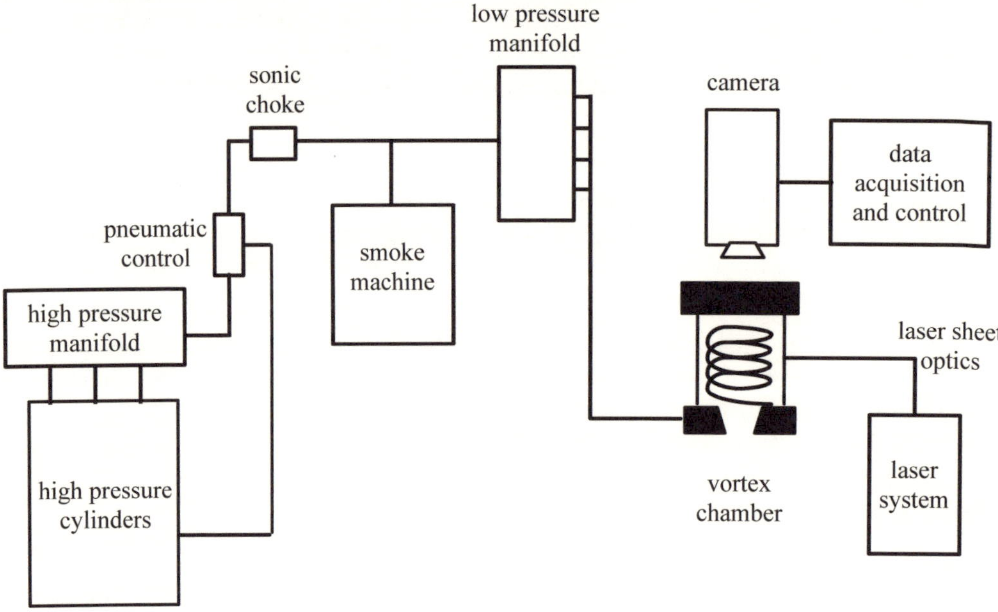

Fig. 6. Schematic of the PIV setup.

(a) Geometric configurations

Length (cm)	Aspect Ratio	Axial Sampling Locations (z/L)
3.56	2.80	0.5
6.10	4.80	0.2 0.5 0.7
8.64	6.80	0.2 0.5 0.7
11.2	8.82	0.2 0.5 0.7

(b) Injector configurations

Δp (% of p_c)	Port Dia. (cm)	Aggregate Injection Area (cm^2)
10	0.605	2.299
20	0.500	1.571
30	0.442	1.228

Table 1. Available configurations for experimental trials.

Parameter	Case 1	Case 2	Case 3	Case 4	Case 5	Case 6
Injector Pressure Drop, $\Delta\bar{p}$ (kPa)	27.6	55.2	82.8	55.2	27.6	55.2
Aspect Ratio, L	2.4	2.4	4.4	3.4	4.4	4.4
Average Injection Speed, U (m/s)	68.73	77.72	89.61	88.78	74.81	88.31
Modified Swirl Number, $\sigma = a^2/A_i$	2.81	4.10	5.26	4.10	2.81	4.10
Inflow Parameter, $\kappa = 1/(2\pi\sigma L)$	0.0239	0.0164	0.0069	0.0114	0.0129	0.0088
Vortex Reynolds Number, $V = \dot{m}_i/(L_0\mu)$	47 150	36 540	30 160	29 150	27 650	22 370

Table 2. Operational parameters for the PIV experiments.

Six cases are chosen from the available trials for comparison with the laminar core and constant shear stress models. Data fidelity is the primary consideration in selection, as many of the other trials exhibit an increasing amount of scatter to the data, especially in the core region. Of the six cases considered, the first three are chosen to develop the empirically based correlations needed to match the data to the analytical approximation. The remaining three cases are held in reserve, to provide verification of the corrected models. The experimental parameters for these test cases are furnished in Table 2.

4.1 Laminar core correlation

In order to compare the analytical models to the experimental data, some additional effort is required. For the laminar core case, a simple least-squares regression will permit the vortex Reynolds number to be tuned to the experimental conditions. This correlation is necessary as the flow conditions in the experiment are beyond the range of validity of the laminar model.

V_t	X_0	X	$50/\sqrt{V_t}$	ℓ_t
47 150	49.04	0.243	0.230	150.3
36 540	49.63	0.267	0.262	154.1
30 160	50.67	0.314	0.288	151.0

Table 3. Least-squares parameters for laminar core and constant shear stress frameworks.

However, since the experimental vortex Reynolds number is also known, the test results can be linked to the analytical vortex Reynolds number through an empirically based viscosity ratio of the form $\ell_t = \mu/\mu_t$, with μ_t denoting a turbulent viscosity. In this manner, the vortex Reynolds number definition may be modified such that

$$V = \frac{\rho U A_i}{\mu L} = \frac{\rho U A_i}{\mu_t \ell_t L} = \frac{V_t}{\ell_t} \tag{46}$$

An average eddy viscosity may be calculated with this correlation that will properly match the laminar, analytical model to the high-speed experimental counterpart.

The standard least-squares regression scheme is used to determine the laminar vortex Reynolds number that best fits the model to the data. This value is then used to calculate the corresponding ℓ_t that will rectify the laminar and experimental vortex Reynolds numbers. The three trials provide 879 data points for the regression, producing an average $\ell_t = 151.8$. Results of the least-squares regression runs for the individual trials are given in Table 3. Note that the three sets of experiments yield rather consistent values of $\ell_t = 150$, 154, and 151, respectively.

4.2 Constant shear stress correlation

The constant shear stress model requires a more elaborate analysis to achieve the same sort of comparison. First, the model itself consists of a piecewise function and the optimization parameter appears in both the solution and the boundary. To properly match the model to the experiment, a modified least-squares routine is developed using an iterative process that is capable of accounting for the moving boundary in the optimization.

4.2.1 Piecewise least-squares regression

The piecewise least-squares code contains several distinct components. The first element is a rewrite of the standard least-squares technique in a manner to incorporate the piecewise nature of the function into the derivative calculations. The function returns the optimized parameter, in this case the matching radius, X. The second function is simply a truncation function that adjusts the data set to reflect the new value of the optimization parameter. Finally, a control function loops over the data set, calling the least-squares function and comparing the new radius to the previous trial, X, until a satisfactory tolerance is reached, such as 0.0001 in this case. For the reader's convenience, a flow chart of the numerical procedure is provided in Fig. 7.

This iterative procedure is necessary because of the nature of this particular piecewise solution. For most piecewise solutions, a standard least-squares algorithm is sufficient. However, in this case, the optimization parameter coincides with the matching radius that determines the boundary between the inner and outer solutions. As a result, the optimization space changes every time that a new radius is calculated. The iterative approach continues to calculate new values of X until the difference between successive radii falls below a user-specified tolerance. Since the data comprises a limited set of discrete points, the solution converges rapidly. The final radius is checked against neighboring values to ensure that the solution is in fact fully optimized.

4.2.2 Viscous correlation

Another fundamental issue associated with the shear stress framework is that the model is not written in terms of the vortex Reynolds number, V. The laminar core analysis highlights the importance of the vortex Reynolds number as it incorporates the geometry of the chamber, l, as well as the swirl dominated nature of the flow through the modified swirl number, σ. This parameter clearly determines the behavior of the swirl velocity near the core. To determine the vortex Reynolds number dependence, the constant shear stress model is matched to the laminar core solution such that both representations will return the same maximum velocity. Mathematically, this requires taking

$$\left.\frac{\mathrm{d}u_\theta}{\mathrm{d}r}\right|_{r=r_{\max}} = 0 \tag{47}$$

Solving Eq. (47) gives $r_{\max} = X/\sqrt{e}$. The same procedure applied to the laminar model yields

$$r_{\max} = \sqrt{\frac{2}{V}\left[-1 - 2\mathrm{pln}\left(-1, -\frac{1}{2}e^{-1/2}\right)\right]} \simeq \frac{2.242}{\sqrt{V}} \tag{48}$$

With the maximum velocity locations known, these values may be substituted back into their respective expressions for velocity and then equated to produce

$$\frac{2}{Xe^{1/2}} = 1 - e^{\frac{1}{2}+\mathrm{pln}\left(-1,-\frac{1}{2}e^{-1/2}\right)}\sqrt{\frac{-V}{2\left[1+2\mathrm{pln}\left(-1,-\frac{1}{2}e^{-1/2}\right)\right]}} \tag{49}$$

Equation (49) may be solved for X in terms of V to obtain

$$X = \frac{2}{\sqrt{eV}}\frac{\sqrt{-2\left[1+2\mathrm{pln}\left(-1,-\frac{1}{2}e^{-\frac{1}{2}}\right)\right]}}{1-\exp\left[\frac{1}{2}+\mathrm{pln}\left(-1,-\frac{1}{2}e^{-\frac{1}{2}}\right)\right]} \simeq \frac{3.802}{\sqrt{V}} \tag{50}$$

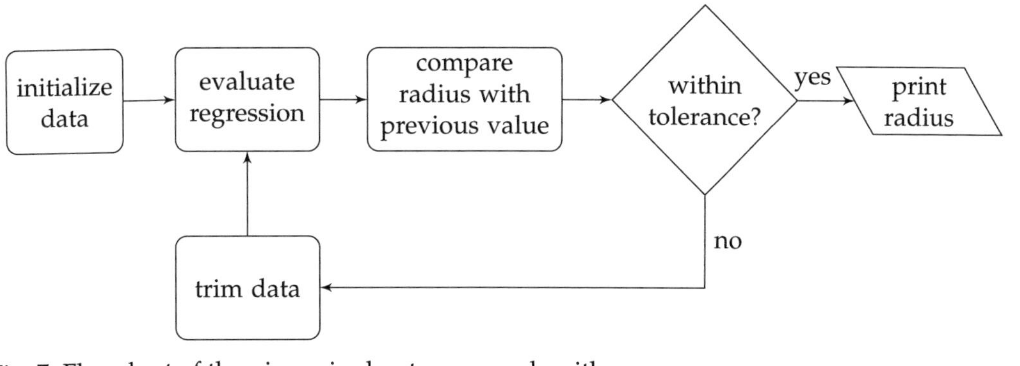

Fig. 7. Flowchart of the piecewise least-squares algorithm.

This solution serves a twofold purpose. First, it demonstrates how the constant shear stress model can be matched to existing data as illustrated in Fig. 8. For additional matching paradigms the reader is directed to Maicke & Majdalani (2009). Secondly, it provides an expected form of the relationship between V_t and the matching radius. The iterative least-squares method uses the relationship

$$X = \frac{X_0}{\sqrt{V_t}} \tag{51}$$

The algorithm optimizes the constant X_0 to produce a matching radius for each trial that incorporates the experimental vortex Reynolds number. The calculated values are shown in Table 3. Note that an average value of approximately 50 provides very good agreement with the experimental data.

4.3 Graphical comparison

Using $X_0 = 50$ and $\ell_t = 151.8$, a comparison is drawn in Figs. 9 and 10 between theory and experiment. The measurements collected in each trial correspond to the data acquired at three axial locations in the chamber, specifically $z = 0.2, 0.5$, and 0.7. The velocity plots in Fig. 9 display the collection of data used in the least-squares analysis. Consequently, good agreement is expected and achieved in all three figures. Figure 10 showcases the reserve data sets that were not employed in the regression algorithm, but rather saved to test the accuracy of the models at various vortex Reynolds numbers. As with the first three trials, the data in the outer, free vortex region is seen to be tightly grouped to the extent of faithfully following the shape predicted by both analytical models. Evidently, the agreement in the core region may be seen to be less appreciable. The additional scatter observed in the data is due to PIV limitations that will be discussed in Sec. 5. Nonetheless, the models do capture the general trends in the core. To accommodate the scatter, the fits for both theoretical models are weighted such that they provide an envelope in the core region, rather than a strict regression fit through the data. This weighting is chosen specifically because of the increased scatter in the core region that artificially lowers the predicted velocities.

It may be interesting to note the ability of the laminar core model with a turbulent eddy viscosity to duplicate the essential features of the flow. This behavior is consistent with

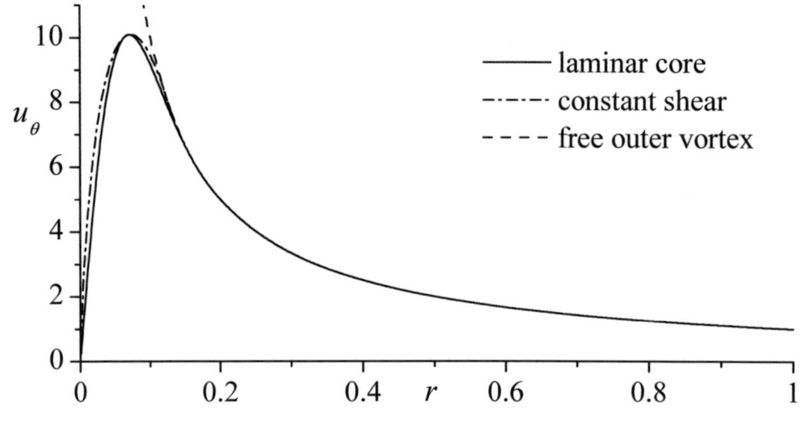

Fig. 8. Peak velocity matched models shown at $V = 1000$.

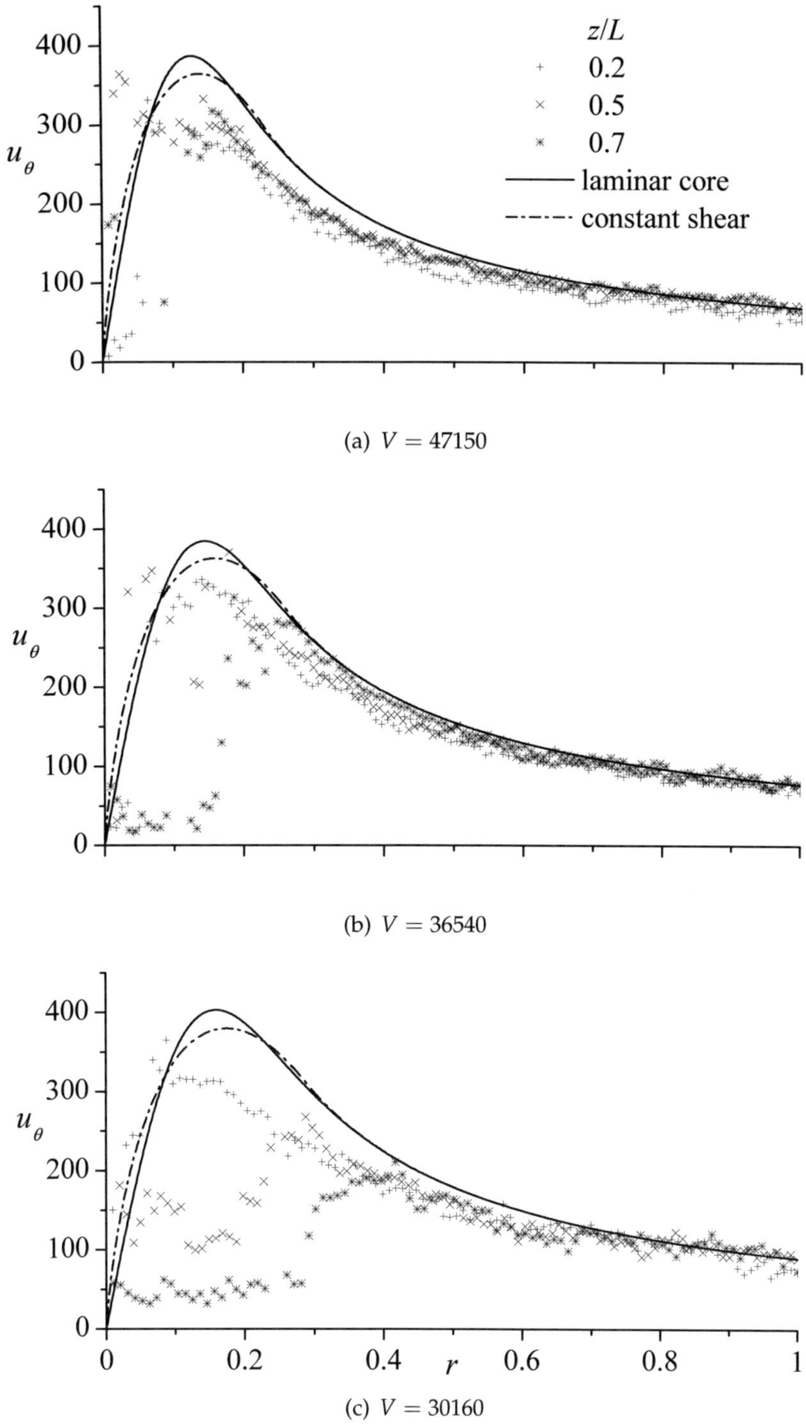

Fig. 9. Laminar and constant shear stress models compared to experimental data from Rom (2006) that is used in the least-squares regression.

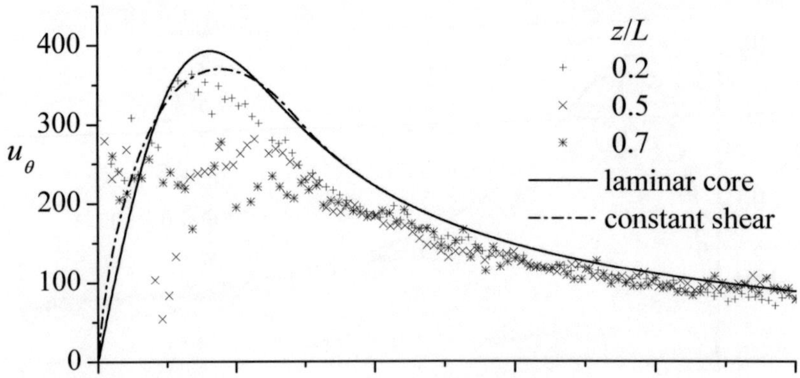

(a) $V = 29150$

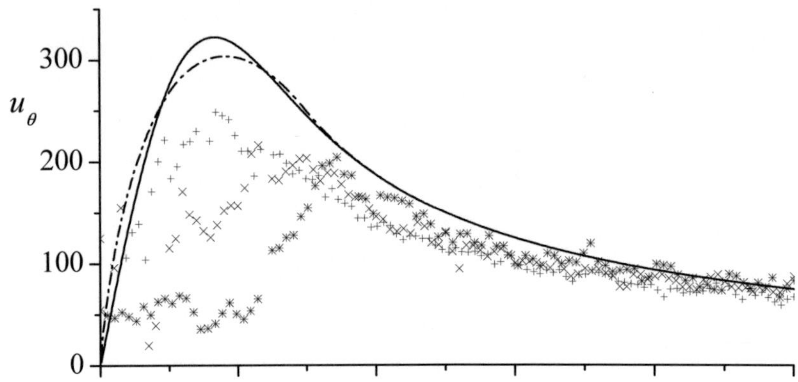

(b) $V = 27650$

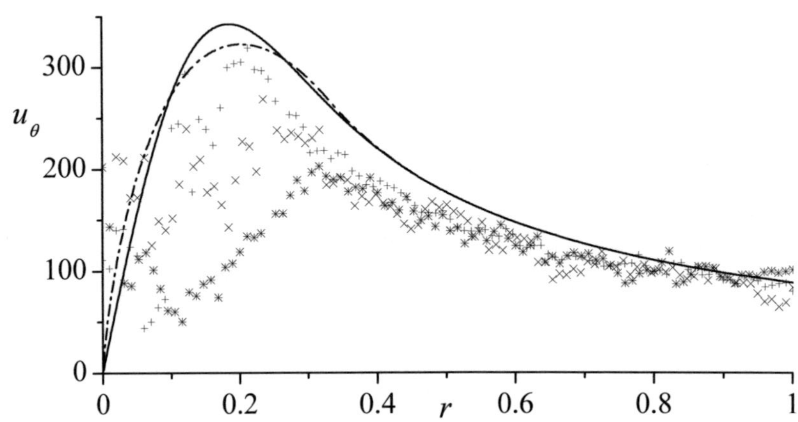

(c) $V = 22370$

Fig. 10. Laminar and constant shear stress models compared to experimental data from Rom (2006) that is reserved for verification.

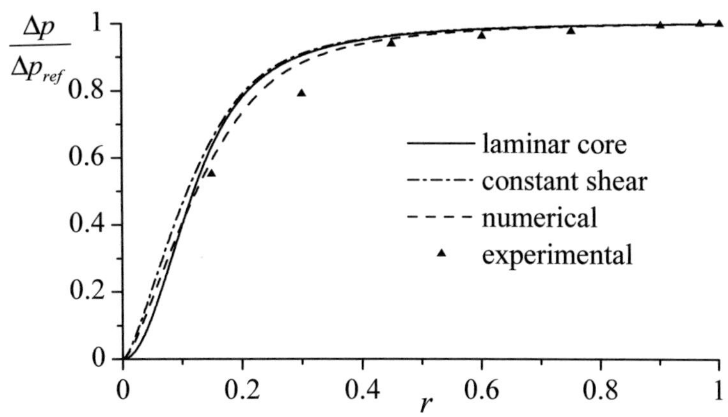

Fig. 11. Comparison of laminar core and constant shear stress models to experimental and numerical pressure data for $V_t = 47150$.

the results of several experimental studies in which vortices are shown to exhibit an approximately constant angular velocity core with low turbulence levels to the extent of appearing as nearly laminar. With turbulent diffusion being restricted to an annular region about the laminar core, it is not surprising to see that the free vortex approximation in the outer domain continues to hold.

In addition to the swirl velocity distribution, the pressure measurements from Rom et al. (2004) may be used to verify the pressure predictions associated with the existing theoretical models. For the laminar core approximation, the equation of interest is

$$\Delta p = -2\pi^2\kappa^2 z^2 \\ -\tfrac{1}{2}r^{-2}\left[\left(1-e^{-\frac{1}{4}Vr^2}\right)^2+\kappa^2\sin^2\left(\pi r^2\right)\right]+\tfrac{1}{4}V\left[\mathrm{Ei}\left(-\tfrac{1}{4}Vr^2\right)-\mathrm{Ei}\left(-\tfrac{1}{2}Vr^2\right)\right] \quad (52)$$

where $\mathrm{Ei}(x)$ is the second exponential integral function. For the constant shear stress model, the piecewise relation for the pressure drop becomes

$$\Delta p = \begin{cases} \frac{1}{2\kappa^4 r^2}\left(r^4\left\{5+\ln\left(\frac{r^4}{X^4}\right)\left[\ln\left(\frac{r}{X}\right)-2\right]\right\}-\kappa^2 X^4\sin^2(\pi r^2)\right) \\ \quad +\frac{1}{2}-\frac{3}{X^2}-\kappa^2\left[4\pi^2 z+\frac{\sin^2(\pi r^2)}{2r^2}\right]; & r \le X \\ \frac{1}{2}-\frac{1}{2r^2}\left[1+\kappa^2\sin^2(\pi r^2)\right]-\kappa^2\left[4\pi^2 z+\frac{\sin^2(\pi r^2)}{2r^2}\right]; & r > X \end{cases} \quad (53)$$

For more detail on the pressure relations, the reader is referred to Majdalani & Chiaverini (2009) and Maicke & Majdalani (2009).

Figure 11 depicts a normalized form of Δp in the vortex chamber. Note that along with the experimental data, a numerical solution from Murray et al. (2004) is presented for the same conditions. All data sets are normalized to a reference value at the sidewall. The agreement displayed is encouraging especially near the wall region. Both analytical models exhibit

	V_t								
	47 150			36 540			30 160		
	r_{cc}	σ_e	$\Delta E_t\%$	r_{cc}	σ_e	$\Delta E_t\%$	r_{cc}	σ_e	$\Delta E_t\%$
laminar core	0.887	0.592	4.91	0.962	0.276	1.47	0.870	0.391	3.45
constant shear	0.900	0.558	4.36	0.968	0.253	1.23	0.880	0.376	3.19

Table 4. Statistical parameters for the regression of the laminar and constant shear stress models.

a slight overprediction of the pressure, but follow the experimental trend admirably. The correlation of the theory with the numerical simulation appears to be excellent, with only slight deviations near the core region.

4.4 Statistical verification

To objectively compare the accuracy of the two models, several statistical parameters may be evaluated. By comparing correlation coefficients, r_{cc}, standard errors, σ_e, and total relative errors, ΔE_t, the constant shear-based model seems to provide a slightly better fit to the data than the modified laminar distribution. The standard and total relative errors are calculated from

$$\sigma_e = \frac{1}{\sqrt{n-1}} \sqrt{\sum_{i=1}^{n} [\hat{u}_\theta(r_i) - u_\theta(r_i)]^2} \tag{54}$$

and

$$\Delta E_t = \sum_{i=1}^{n} [\hat{u}_\theta(r_i) - u_\theta(r_i)]^2 / \sum_{i=1}^{n} \hat{u}_\theta^2(r_i) \tag{55}$$

where n and $\hat{u}_\theta$ denote the number of data points and the measured velocity at r_i, the radius of the ith data point. The standard error of the estimate quantifies the spread of data about the regression line, much like the standard deviation that measures the spread about a mean value. As shown in Table 4, the total relative error falls under 3.2, 1.3, and 4.4% for the three cases associated with the constant shear stress approach. The corresponding experimental correlation coefficients are calculated at 0.88, 0.97 and 0.90, respectively. When the modified laminar core technique is used, the relative errors slightly increase to 3.5, 1.5, and 5.0%, with an equally minute reduction in r_{cc}.

The graphical and statistical verifications highlight the effectiveness of the theoretical approach that is pursued. The agreement between all cases, even those held in reserve, is particularly satisfying. However, additional experiments are necessary to both confirm the correlations developed here and to extend their usefulness to a wider variety of engine configurations and injection conditions.

5. Challenges and limitations

While the data provided by the PIV experiments proves invaluable in the verification of theoretical models, the technique is not without its challenges. In this section, the challenges and limitations of PIV applications to confined vortex flows are discussed.

In most cases, the PIV technique constitutes a powerful non-intrusive method for measuring velocities. However, as demonstrated in the previous sections, the vortical flow field can be difficult to capture in the region of maximum swirl, even with properly sized particles. In the core region, the particles exhibit increased drag as the swirl velocity reaches its peak. The particles also display a tendency to separate via centrifugal entrainment. Visually, this manifests itself as a reduction in the seed particles near the flow centerline. Furthermore, the core of the chamber happens to be the region of highest axial velocity; thus, the particles that are tagged in the first exposure may have traveled below the laser sheet by the time the second image is taken. This scenario can also lead to decreased confidence in the correlated data in the core region (see also Hu et al., 2005).

The verification process in this chapter is concentrated on a single aspect of the theoretical models, namely, the tangential velocity. To fully characterize the analytical approximations, the axial and radial velocities should be measured alongside the pressure. Practically, this cannot be accomplished in one trial with the current experimental configuration as the camera cannot track the fluid motion perpendicularly to the field of view. A stereoscopic PIV setup can overcome this deficiency albeit with increased cost in both equipment and procedural time as the pair of cameras require careful calibration in order to properly account for the velocities along the third axis. For this particular application, Laser Doppler Velocimetry may prove to be a better choice. In addition to providing velocity measurements in all three directions, a properly configured LDV experiment can produce data at more axial locations in a single setup. While not specific to PIV, acquiring the necessary pressure measurements may require yet another setup in which the clear headwall module can be replaced with a separate apparatus containing pressure taps.

Finally, the PIV approach cannot provide an accurate assessment of flow motion over the entire chamber. In the vicinity of the nozzle, accurate measurements are difficult to obtain as the fluid is restricted to a narrower cross-section; moreover, the character of the flow changes abruptly as it transitions from a swirl-dominated to a predominantly axial motion. In this region PIV is not sufficient to accurately resolve the velocity field; alternative techniques such as X-ray radiography may be required to more suitably investigate this region experimentally.

6. Conclusion

This chapter focuses on the use of PIV measurements in conjunction with two analytical models to describe the flow field in a bidirectional vortex chamber. Although the primary motivation for this work is to better understand the gas dynamics within a vortex-fired liquid rocket engine known as the Vortex Combustion Cold-Wall Chamber, the principal attributes associated with the resulting cyclonic motion have industrial, geophysical, meteorological, and astrophysical applications that fall beyond the propulsive dimension.

Using a cold-flow environment as a basis for this work, the findings indicate that both PIV and theoretical frameworks converge in predicting the presence of a forced vortex core region followed by a free, irrotational vortex tail that are characteristic of not only bidirectional vortices, but of unidirectional swirl-dominated flows as well. Moreover, despite the underlying restriction to non-reactive conditions, the essential features captured here seem to be somewhat representative of those reported in combustion chambers in which chemical reactions are prominent (Fang et al., 2004). It appears that the addition of high intensity swirl leads to a robust cyclonic motion that tends to retain its fundamental structure even in the presence of chemical kinetics. Nonetheless, more effort in modeling reactive and multiphase flows in cyclonic chambers will be critically important to further illuminate the particular mechanisms and parameters that control the performance of the VCCWC engine over a wide range of physical properties and geometric scales.

At this stage, the need to advance both computational and experimental capabilities in the modeling of reactive mixtures in swirl-dominated flows in general and cyclonic chambers in particular cannot be overrated. In what concerns the latter, it would be advantageous if a standard set of universal group parameters can be defined against which data acquired experimentally or numerically can be properly correlated. Thus guided by rigorous mathematical formulations that are derived from first principles, a basic set of dimensionless groupings such as the vortex Reynolds number, swirl parameter, inflow parameter, and chamber aspect ratio can be used consistently by various researchers while undertaking wide-range parametric studies. Presently, several different expressions of the swirl and vortex Reynolds numbers exist in the literature and this may be attributed to the dissimilar approaches used in defining them, be it rationalization, order-of-magnitude scaling, or conjecture. It would be strongly recommended that these are replaced by their corresponding forms that emerge naturally in analytical solutions that are obtained directly from first principles. The use of standard forms of these parameters will be essential before investing in new equipment and undertaking full-range laboratory or numerical experiments.

Novel investigations will be required to open up new lines of research inquiry and address several fundamental questions that remain unanswered. For example, it is still unclear what conditions will lead to the formation of a coherent cyclone, to vortex breakdown in a cyclonic chamber, or to the precession of the core vortex. It is also uncertain what techniques will be best suited to capture such behavior. The same may be said of acoustic and/or combustion instability of vortex-fired engines. Preliminary studies using the biglobal stability approach by Batterson & Majdalani (2011a;b) show that stability is promoted with successive increases in swirl. Nonetheless, their study is limited to a narrow range of chamber aspect ratios and the hydrodynamic instability effects only represent one of the factors that must be accounted for in studying engine stability. The sidewall and headwall boundary layers, which were omitted in this chapter, pose another challenge that future examinations are hoped to overcome. Resolving the viscous stresses along the chamber wall will be essential not only to improve the velocity field description, but also to make any sort of formulation of a Nusselt number possible. This effort will have to be carried out using a three-pronged approach that leverages analytical, computational, and experimental techniques.

In hindsight, the availability of a laminar core flow solution to describe the bidirectional vortex has proven helpful in the design of a successful PIV experiment. The converse is also true given that the production of experimental data has been quintessential in refining the analytical approximations. The need to investigate the existence of more accurate, multi-dimensional, or higher-order helical solutions is evident. So far the analytical framework has provided several dimensionless parameters, including the vortex Reynolds number which is critical in controlling the flow behavior and providing guidance to minimize the number of experimental trials. The availability of a core shear stress model has also led to an iterative least-squares method that is specifically developed to enable comparisons between piecewise functions and experiments. To mitigate the bias in the unavoidable scatter in PIV data, the measurements have been carefully averaged and distilled in the process of calculating the turbulent eddy viscosity, μ_t, and the corresponding viscosity ratio, $\ell_t = \mu_t/\mu \simeq 152$, for the simulated vortex chamber. This effort epitomizes theory and experiment working hand-in-hand. In the absence of the PIV data, it would have been virtually impossible to predict the effective vortex Reynolds number, $V = V_t/\ell_t$, which will reduce the overshoot in the swirl velocity of the laminar core flow model. It is gratifying that the data generated from the PIV technique leads to a consistent agreement with the theoretical predictions despite the decreased fidelity in the core region.

Clearly, Particle Image Velocimetry seems to offer a vital resource in the quantification of confined vortex flows. Though not perfect, the technique captures the structure of the swirl velocity in the bidirectional vortex engine quite satisfactorily. By knowing the limitations of the method, specifically the likelihood of increased drag on particles near the peak flow velocity, it may be possible to refine the analytical models to the extent of better predicting their peak swirl velocities measured experimentally. Another aspect that must be brought into perspective is that, unlike the case of an unbounded vortex, the axial and radial velocities associated with confined vortices can have appreciable contributions in some regions of the flow domain. Then given that most studies to date emphasize the swirl velocity to the exclusion of the remaining components, more research is warranted to fully characterize the three-dimensional nature of the bidirectional vortex. This in turn will require a concerted effort between theoretical techniques and more advanced stereoscopic PIV or LDV procedures.

Because stereoscopic PIV relies on a rigorous procedure to capture the multidimensionality of complex flows, the addition of a second camera will be required to permit investigators to deduce particle velocities normal to the laser sheet. At the outset, all three components of velocity can be acquired at a given experimental section. With proper equipment, the light sheet can traverse the entire chamber to the extent of reproducing a comprehensive map of the vortex structure. In cases where the added expense of a stereoscopic apparatus proves impractical, a traditional PIV setup can still be employed to compile the necessary results, albeit with multiple trials.

An LDV technique may also be used in reconstructing the three-dimensional structure of the velocity field. In fact, Sousa (2008) relies on LDV measurements in his confined vortex study. However, a concern is raised in his investigation regarding the gains in spatial resolution that are offset by a substantially prolonged measurement time. In practice, each of these laser-based methods serves a different overall purpose. Whereas the PIV technique offers superb capabilities in capturing large scale velocity structures, LDV yields localized point measurements of fluid velocities. It can thus be seen that PIV may be the preferred method for visualizing large scale flow patterns such as those arising in cyclonic chambers.

In closing, it may be helpful to re-emphasize the tight balance between theory and experiment that stands behind this work. All too often, an investigation is steered towards one aspect of research to the detriment or neglect of the other. Without the theory to guide the proper design of experiments, more trials would have been necessitated to fully characterize the swirl velocity. Conversely, without the experimental values, the theoretical model could not have been refined and validated, nor could the turbulent eddy viscosity correlation been obtained. Pushing the borders of scientific inquiry certainly requires both theory and experiment to be pursued diligently and in unison.

7. Acknowledgments

This material is based on work supported partly by the National Science Foundation, and partly by the University of Tennessee Space Institute and the H. H. Arnold Chair of Excellence in Advanced Propulsion. The authors thank Dr. Martin J. Chiaverini of Orbital Technologies Corporation (ORBITEC) for providing the experimental data featured in this chapter.

8. References

Barber, T. A. & Majdalani, J. (2009). Exact Eulerian solution of the conical bidirectional vortex, *45th AIAA/ASME/SAE/ASEE Joint Propulsion Conference and Exhibit*, Denver, CO. AIAA Paper 2009–5306.

Batterson, J. & Majdalani, J. (2011a). Biglobal instability of the bidirectional vortex. part 1: Formulation, *47th AIAA/ASME/SAE/ASEE Joint Propulsion Conference and Exhibit*, San Diego, CA. AIAA Paper 2011-5648.

Batterson, J. & Majdalani, J. (2011b). Biglobal instability of the bidirectional vortex. part 2: Complex lamellar and Beltramian motions, *47th AIAA/ASME/SAE/ASEE Joint Propulsion Conference and Exhibit*, San Diego, CA. AIAA Paper 2011-5649.

Bloor, M. I. G. & Ingham, D. B. (1973). Theoretical investigation of the flow in a conical hydrocyclone, *Transactions of the Institution of Chemical Engineers* 51(1): 36–41.

Bloor, M. I. G. & Ingham, D. B. (1987). The flow in industrial cyclones, *Journal of Fluid Mechanics* 178: 507–519.
URL: *http://dx.doi.org/10.1017/S0022112087001344*

Burgers, J. M. (1948). A mathematical model illustrating the theory of turbulence, *Advances in Applied Mechanics* 1: 171–196.

Chiaverini, M. J., Malecki, M. J., Sauer, J. A. & Knuth, W. H. (2002). Vortex combustion chamber development for future liquid rocket engine applications, *38th AIAA/ASME/SAE/ASEE Joint Propulsion Conference and Exhibit*, Indianapolis, IN. AIAA Paper 2002–4149.

Fang, D., Majdalani, J. & Chiaverini, M. J. (2004). Hot flow model of the vortex cold wall liquid rocket, *40th AIAA/ASME/SAE/ASEE Join Propulsion Conference and Exhibit*. AIAA Paper 2004–3676.

Hoekstra, A. J., Derksen, J. J. & van den Akker, H. E. A. (1999). An experimental and numerical study of turbulent swirling flow in gas cyclones, *Chemical Engineering Science* 54(13): 2055–2065.
URL: *http://dx.doi.org/10.1016/S0009-2509(98)00373-X*

Hu, L. Y., Zhou, L. X., Zhang, J. & Shi, M. X. (2005). Studies of strongly swirling flows in the full space of a volute cyclone separator, *AIChE Journal* 51(3): 740–749.
URL: *http://dx.doi.org/10.1002/aic.10354*

Kirshner, R. P. (2004). *The Extravagant Universe: Exploding Stars, Dark Energy, and the Accelerating Cosmos*, Princeton University Press, Princeton, New Jersey.

Königl, A. (1986). Stellar and galactic jets: Theoretical issues, *Canadian Journal of Physics* 64: 362–368.
URL: *http://dx.doi.org/10.1139/p86-063*

Leibovich, S. (1984). Vortex stability and breakdown: Survey and extension, *AIAA Journal* 22(9): 1192–1206.
URL: *http://dx.doi.org/10.2514/3.8761*

Maicke, B. A. & Majdalani, J. (2009). A constant shear stress core flow model of the bidirectional vortex, *Proceedings of the Royal Society, A* 465: 915–935.
URL: *http://dx.doi.org/10.1098/rspa.2008.0342*

Majdalani, J. & Chiaverini, M. J. (2009). On steady rotational cyclonic flows: The viscous bidirectional vortex, *Physics of Fluids* 21: 103603–15.
URL: *http://dx.doi.org/10.1063/1.3247186*

Murray, A. L., Gudgen, A. J., Chiaverini, M. J., Sauer, J. A. & Knuth, W. H. (2004). Numerical code development for simulating gel propellant combustion processes, *52nd JANNAF Propulsion Meeting*, Las Vegas, NV. JANNAF Paper 2004-0115.

Ogawa, A. (1984). Estimation of the collection efficiencies of the three types of the cyclone dust collectors from the standpoint of the flow pattern in the cylindrical cyclone dust collectors, *Bulletin of the JSME* 27(223): 64–69.

Penner, S. S. (1972). Elementary considerations of the fluid mechanics of tornadoes and hurricanes, *Acta Astronautica* 17: 351–362.

Rankine, W. J. M. (1858). *A Manual of Applied Mechanics*, 9th edn, C. Griffin and Co., London, UK.

Reydon, R. F. & Gauvin, W. H. (1981). Theoretical and experimental studies of confined vortex flow, *The Canadian Journal of Chemical Engineering* 59: 14–23.
URL: *http://dx.doi.org/10.1002/cjce.5450590102*

Rom, C. (2006). *Flow field and near nozzle fuel spray characterizations for a cold flowing vortex engine*, Master's thesis, University of Wisconsin-Madison, Madison, WI.

Rom, C. J., Anderson, M. H. & Chiaverini, M. J. (2004). Cold flow analysis of a vortex chamber engine for gelled propellant combustor applications, *40th AIAA/ASME/SAE/ASEE Joint Propulsion Conference and Exhibit*, Fort Lauderdale, FL. AIAA Paper 2004-3359.

Rusak, Z. & Lee, J. H. (2004). On the stability of a compressible axisymmetric rotating flow in a pipe, *Journal of Fluid Mechanics* 501: 25–42.
URL: *http://dx.doi.org/10.1017/S0022112003006906*

Rusak, Z., Wang, S. & Whiting, C. H. (1998). The evolution of a perturbed vortex in a pipe to axisymmetric vortex breakdown, *Journal of Fluid Mechanics* 366(1): 211–237.

Smith, J. L. (1962a). An analysis of the vortex flow in the cyclone separator, *Journal of Basic Engineering* 84(4): 609–618.

Smith, J. L. (1962b). An experimental study of the vortex in the cyclone separator, *Journal of Basic Engineering* 84(4): 602–608.

Sousa, J. M. M. (2008). Steady vortex breakdown in swirling flow inside a closed container: Numerical simulations, PIV and LDV measurements, *Open Mechanical Engineering Journal* 2: 69–74.
URL: *http://dx.doi.org/10.2174/1874155X00802010069*

Sullivan, R. D. (1959). A two-cell vortex solution of the Navier-Stokes equations, *Journal of the Aerospace Sciences* 26(11): 767–768.

ter Linden, A. J. (1949). Investigations into cyclone dust collectors, *Proceedings of the Institution of Mechanical Engineers* 160: 233–251.

Townsend, A. A. (1976). *The Structure of Turbulent Shear Flow*, Cambridge University Press.

Vatistas, G. H. (2009). Vorticies in Homer's Odysssey–A scientific approach, *Science and Technology in Homeric Epics*, History of Mechanism and Machine Science, Springer Netherlands, pp. 67–75.
URL: *http://dx.doi.org/10.1007/978-1-4020-8784-4_4*

Vatistas, G. H., Lin, S. & Kwok, C. K. (1986). Theoretical and experimental studies on vortex chamber flows, *AIAA Journal* 24(4): 635–642.
URL: *http://dx.doi.org/10.2514/3.9319*

Vyas, A. & Majdalani, J. (2006). Exact solution of the bidirectional vortex, *AIAA Journal* 44(10): 2208–2216.
URL: *http://dx.doi.org/10.2514/1.14872*

Wendt, B. J. (2001). Initial circulation and peak vorticity behavior of vortices shed from airfoil vortex generators, *Technical Report NASA/CR–2001-211144*, NASA.

Wu, J. Z. (1986). Conical turbulent swirling vortex with variable eddy viscosity, *Proceedings of the Royal Society of London, Series A* 403(1825): 235–268.
URL: *http://dx.doi.org/10.1098/rspa.1986.0011*

Yan, L., Vatistas, G. H. & Lin, S. (2000). Experimental studies on turbulence kinetic energy in confined vortex flows, *Journal of Thermal Science* 9(1): 10–22.
URL: *http://dx.doi.org/10.1007/s11630-000-0040-z*

Zhang, Z. & Hugo, R. J. (2006). Stereo particle image velocimetry applied to a vortex pipe flow, *Experiments in Fluids* 40: 333–346.
URL: *http://dx.doi.org/10.1007/s00348-005-0071-z*

7

Application of the Particle Image Velocimetry to the Couette-Taylor Flow

Innocent Mutabazi[1], Nizar Abcha[2],
Olivier Crumeyrolle[1] and Alexander Ezersky[2]
[1]*LOMC, UMR 6294, CNRS-Université du Havre 53, rue Prony, Le Havre Cedex,*
[2]*M2C, UMR 6143, CNRS-University of Caen-Basse Normandie,*
France

1. Introduction

For longtime, the investigation of flow regimes has been achieved using fluorescent particles or anisotropic reflective particles. Fluorescent particles are suitable for open flows (Van Dyke, 1982) such as flows in channels (Peerhossaini *et al.*, 1988) or flows behind a cylinder (to visualize Benard-von Karman street) (Provansal *et al.*, 1986; Mutabazi *et al.*, 2006). For closed flows such as flows in a rectangular cavity or in an annular cylindrical rotating cavity, fluorescent particles rapidly color the entire flow and no flow structure can be caught. Anisotropic reflective particles (aluminium, iriodin or Kalliroscope flakes) are more convenient for detection of the flow structures (Taylor, 1923; Andereck *et al.*, 1986; Coles, 1965; Matisse *et al.*, 1984; Dominguez-Lerma *et al.* , 1985; Thoroddsen *et al.* 1999). A laser light is used to illuminate the flow cross-sections and to detect the flow structure in the axial, radial and azimuthal directions. The motion of the seeded particles in a fluid gives a qualitative picture of flows which can be used to develop appropriate theoretical models. The development of chaotic models of fluid flows (Rayleigh-Bénard convection, Couette-Taylor flow or plane Couette flow) has benefited from observations using visualizations techniques (Bergé *et al.*, 1994). Using appropriate signal processing techniques such as space-time diagrams and complex demodulation, it is possible to obtain spatio-temporal evolution of the flows (Bot *et al.*, 2000). In order to obtain quantitative data on velocity fields, different velocimetry techniques have been developed such as Laser Doppler Velocimetry (LDV, Durst *et al.*, 1976, Jensen 2004), Ultrasound Doppler Velocimetry (UDV, Takeda *et al.*, 1994) and Particle Image Velocimetry (PIV, Jensen, 2004). Nowadays, there is a lot of literature on velocimetry techniques the development of which is beyond the scope of this chapter, some of them and their applications are described in this volume. Each velocimetry technique has its advantages and own limitations depending on the flow system under consideration. For example, in the case of the Couette-Taylor flow, the LDV (Ahlers *et al.*, 1986) gives time averaged velocity in a point, the UDV measures a velocity profile along a chosen line in the flow and the PIV gives a velocity field in a limited flow cross section. The Couette-Taylor system is composed of a flow in the gap between two coaxial differential rotating cylinders. This system represents a good hydrodynamic prototype for the study of the transition to turbulence in closed systems. The experimental results obtained from this system have led to the development of powerful theoretical models for the transition to chaos (Chossat *et al* . (1994)). Beside the theoretical interpretation of patterns observed in the Couette-Taylor system, many theoretical attempts have been made to connect flow quantitative properties and visualized structures by anisotropic particles (Matisse *et al.* , 1984; Savas (1985), Gauthier *et al.*, (1998)). The application of PIV in the Couette-Taylor system with a fixed outer cylinder was first performed by Wereley & Lueptow (Wereley *et al.*, 1994, 1998) followed later by few authors. The question of correlation between velocimetry data and qualitative structure given by anisotropic reflective particles in the Couette-Taylor flow was addressed only recently (Gauthier *et al.*, 1998; Abcha *et al.*, 2008). However, many questions connected with the interpretation of results obtained by different techniques have not been answered. A special attention is paid to some of these unresolved problems.

This chapter illustrates how the PIV technique can be applied to the Couette -Taylor system. Two special cases are described: 1) flow patterns obtained when the outer cylinder is fixed while the inner is rotating; 2) flow patterns achieved when both cylinders are in contra-rotation. A detailed comparison between PIV and visualisation by anisotropic reflective particles will be provided for illustration of the complementarity between these two techniques. The chapter is organized as follows. The experimental setup and procedure are presented in the next section. Section 3 is devoted to the flow visualization by Kalliroscope particles and the space-time diagram technique. In section 4, the description of PIV and its adaptation to the Couette-Taylor flow are described. Section 5 contains results for flow regimes when the outer cylinder is fixed (Taylor Vortex Flow (TVF) and Wavy Vortex Flow (WVF)). Section 6 gives results for spiral vortex flow when both cylinders are sufficiently counter-rotating. Section 7 summarizes the content of the chapter.

List of symbols

a	Inner cylinder radius, cm
b	Outer cylinder radius, cm
d	Size gap, cm
g	Gravity acceleration
L	Cylinder length, cm
Re	Reynolds number
Ta	Taylor number
n	Optical refraction index
vs	Sedimentation velocity
V_r	Radial velocity component, m/s
u	Dimensionless radial velocity
V_z	Axial velocity component, m/s
w	Dimensionless axial velocity
Δt_{res}	Residence time, s
Abs	Absolute value
CCF	Circular Couette Flow
TVF	Taylor Vortex Flow
WVF	Wavy Vortex Flow
MWVF	Modulated Wavy Vortex Flow
TTVF	Turbulent Taylor Vortex Flow
SVF	Spiral Vortex Flow

OC	Outer cylinder
IC	Inner cylinder
Γ	Aspect ratio
η	Radius ratio
Ω	Angular velocity, rad/s
ξ	Dimensionless radial coordinate
ζ	Dimensionless axial coordinate
ν	Kinematic viscosity, m^2/s
ρ	Fluid density, g/cm^3
λ	Pattern wavelength

2. Experimental apparatus

The experimental system consists of two vertical coaxial cylinders, immersed in a large square Plexiglas box filled with water in order to maintain a controlled temperature (Fig. 1). The square box allows to minimize distortion effects of refraction due to curvature of the outer cylinder during optical measurements. The inner cylinder made of aluminium has a radius a = 4 cm, the outer cylinder made of glass has a radius b = 5 cm, the gap between the cylinders is $d = b\text{-}a$ = 1 cm and the working height is L = 45.9 cm. Therefore the radius ratio $\eta = a/b = 0.8$ and the aspect ratio is $\Gamma = L/d = 45.9$. Such an aspect ratio is large enough to avoid the end effects; the flow system is considered as an extended system. The gap is filled with a deionized water for which $\nu = 9.8.10^{-3} cm^2/s$ at the temperature T = 21.2°C. Its size has been chosen in order to obtain a good resolution in the radial direction.

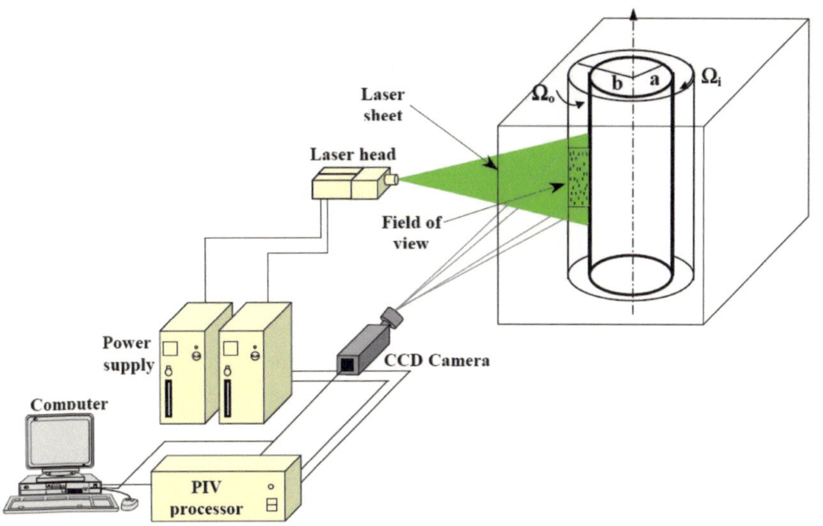

Fig. 1. Experimental apparatus: scheme of visualization and data acquisition system

The cylinders are rotated by two servomotors at controlled angular rotation frequencies Ω_i and Ω_o. The control parameters are the Reynolds numbers defined for each cylinder : $Re_i = \Omega_i ad/\nu$ and $Re_o = \Omega_o bd/\nu$ for the inner and outer cylinder respectively. When $Re_0 = 0$, the Taylor number is often preferred: $Ta = Re_i (d/a)^{1/2}$. Three cameras have been implemented on the experimental table: a linear CCD camera of 1024 pixels that records the light reflected by anisotropic reflective particles along a line parallel to cylinder axis. The second camera is a 2-d IEEE1394 camera (A641f, Basler) that is used to record the flow motion in the (r,z) plane; this record allows a better investigation of the flow in the radial direction. The third camera is a CCD camera (Kodak) with 1034x779 pixels for PIV data recording. All cameras were connected to a computer for data recording and processing.

3. Visualization by Kalliroscope particles and space-time diagrams

Particles added to the flow must have controlled characteristics such as size, distribution, and concentration. These particles must be small enough to be good flow tracers and large enough to scatter sufficient light for imaging. In the Couette-Taylor flow, the commonly used particles are Kalliroscope flakes of typical size of 30 μm x 6 μm x 0.07 μm (Matisse *et al.* 1984) with a relatively large reflective optical index $n = 1.85$ and a density of $\rho' = 1.62$ g/cm^3. A concentration of 1% to 2% reflective particles is added to water to realize a Kalliroscope AQ1000 suspension, 2% per volume of which was added to the working solution. The sedimentation of these particles remains negligible in horizontal or vertical configurations if the experiment lasts less than 10 hours (Matisse *et al.*, 1984) because their sedimentation velocity is $v_s = 2\, a^2 g(\rho' - \rho)/(9\nu\rho) = 2.8\ 10^{-5}$cm/s. The time scales related to the particle motions (transient, rotation and diffusion) were discussed in detail by Gauthier *et al.* (1998). These particles do not modify significantly the flow viscosity and no non-Newtonian effect was detected as far as small concentrations ($c < 5\%$) are used (Dominguez-Lerma *et al.*, 1985). The choice of the concentration of 2% was done to ensure the best contrast in the flow. The values of the control parameters (Re_o, Re_i) were determined within a precision of 2%.

Increasing values of the control parameters leads to the occurrence of different patterns in the Couette-Taylor flow depending of whether both cylinders rotate or only the inner cylinder is rotating (Fig. 2, 3). A whole state diagram of flow regimes in the Couette-Taylor system has been established by Andereck (Andereck *et al.*, 1986) for a configuration with radius ratio $\eta = 0.883$ and aspect ratios Γ ranging from 20 to 48. When the outer cylinder is fixed and the inner Reynolds number Re_i is increased, the transition sequence is the following : Circular Couette Flow (CCF) bifurcates to Taylor Vortex Flow (TVF) which is formed of axisymmetric stationary vortices, then to Wavy Vortex Flow (WVF) oscillating in the azimuthal and axial directions with a frequency f and an azimuthal wavenumber m; the later bifurcates to Modulated Wavy Vortex Flow (MWVF) characterized by two incommensurate frequencies. The ultimate state is the Turbulent Taylor Vortex Flow (TTVF) iin which large scale vortices of the size of the gap and small vortices of different scales coexist. In case of counter-rotating cylinders, the bifurcation of the circular Couette flow leads to spiral vortex flow (SVF) composed of helical vortices travelling in axial and azimuthal directions, followed by interpenetrating spirals then by wavy spirals and modulated waves before transition to turbulence. Interpenetrating spirals, wavy spirals and modulated waves are characterized by incommensurate frequencies. Using a He-Ne Laser sheet (whose wavelength is 632 nm, one millimetre wide beam, spread by a cylindrical lens), it was possible to visualize the cross section of the flow in the r-z plane. Fig. 3 gives the cross section of regimes observed in the Couette-Taylor system for different values of the control parameters. Linear CCD camera of 1024 pixels was used to record a reflected light intensity $I(z)$. Records were performed at regular time intervals along a line in the centre of the flow

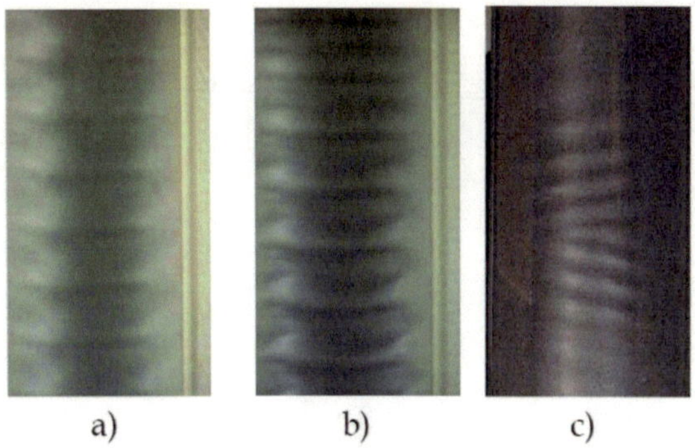

Fig. 2. Pictures of flow regimes in the Couette-Taylor system : a) TVF, b) WVF, c) SVF with a sink.

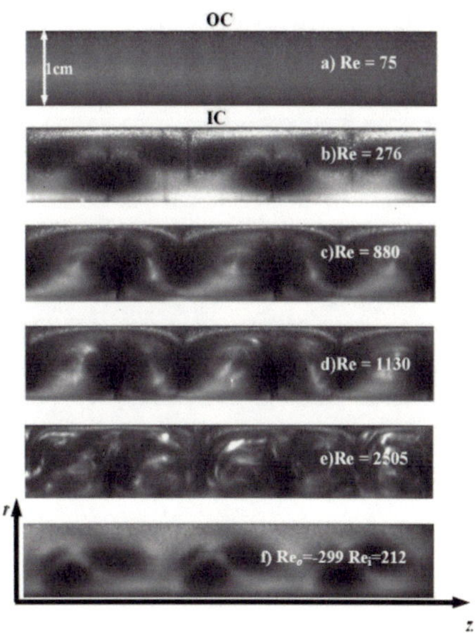

Fig. 3. Cross-section of : a) CCF, b) TVF, c) WVF, d) MWVF, e) TTVF, f) SVF. OC and IC stand for outer and inner cylinder respectively.

cross-section ($r = a + d/2$), parallel to the cylindrical axis over a length of 27.8 cm in the central part of the flow system. The intensity was sampled over a linear range of 256 values, displayed in gray levels at regular time intervals in order to produce space-time diagrams $I(z,t)$ of the pattern which exhibits the temporal and spatial evolution of vortices (Fig. 4a). The radial variation of intensity $I(r)$ was recorded using a 2-d IEEE1394 camera, and then sampled at regular time intervals to obtain the space-time diagram $I(r,t)$. Examples are illustrated (Fig. 4) for wavy vortex flow and in Figure 5 for spiral vortex flow.

A cross-section of spiral vortex flow is shown in Fig. 5-a; its space-time diagram in the axial and radial directions are shown respectively in Fig. 5-b and Fig. 6. The right and left travelling spirals merge into a single point called "sink" at $z = z_0$ (Fig. 5-a). Using a 2-D Fast Fourier Transform (FFT), it is possible to obtain the axial wavenumbers and the frequencies

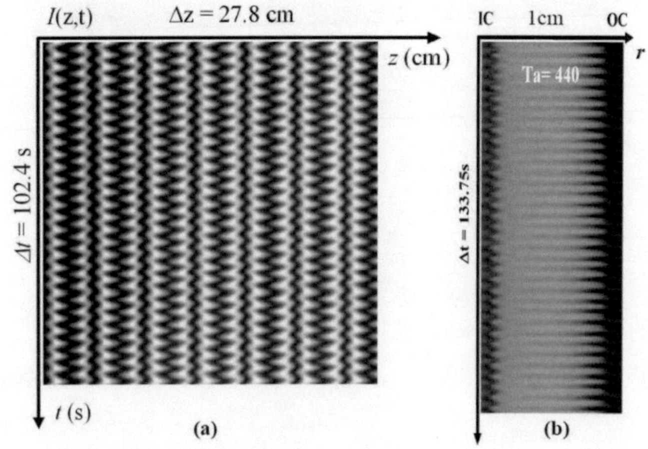

Fig. 4. Space-time diagrams for Wavy Vortex Flow (Re_o = 0; Re_i = 880) cross section of which is shown in Fig. 2 : (a) axial distribution *I(z,t)*; (b) radial distribution *I(r,t)*.

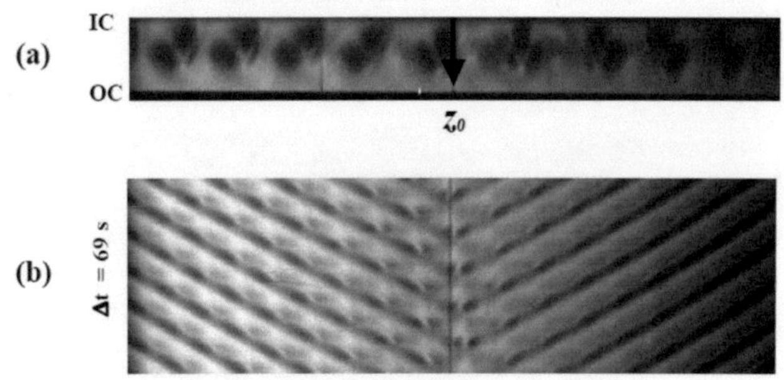

Fig. 5. Spiral pattern for Re_o = - 230 and Re_i = 174 just above the critical value (Re_{ic} = 160, ε = 0.0875): (a) cross-section of flow; (b) space- time diagrams *I(z,t)* taken in the mid-gap position (x = 0.5) over the axial length Δz = 13.8 cm

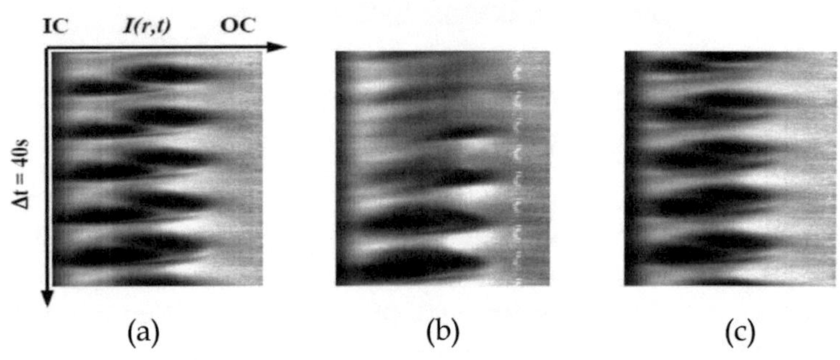

Fig. 6. Space- time diagrams *I(r,t)* of the spiral pattern : (a) before the sink; (b) in the sink; (c) after the sink.

of oscillations of the patterns. The FFT can be complemented by the complex demodulation technique (Bot *et al.* 2000) in order to determine the physical properties like phase, amplitude, frequency, and wavenumber of more complex patterns (in the presence of localized defects, sinks or sources) . These techniques were applied to the spiral pattern of

Figure 5-b. Analysis of temporal and spatial spectra show that the left and right spirals have different frequencies and wavenumbers: $f_R = 0.166$ Hz $\neq f_L = 0.163$ Hz and $k_R = 5.21$ cm$^{-1} \neq k_L =$ 5.06 cm^{-1}. The standard deviations on measured frequencies and wavenumbers are $\Delta f = \pm$ 0.002 Hz and $\Delta k = \pm$ 0.04 cm^{-1} respectively.

The azimuthal wavenumber of the spiral flow can be determined by measuring the inclination angle θ of the spirals and extracting the value of m from the formula $\theta = \arctan(m\lambda / (2\pi R)) = \arctan(m / (kR))$ where λ is the wavelength and $R = (a+b)/2$ is the mean radius. For $Re_o = -230$ and $Re_i = 174$ the obtained values are $m = 2$ and $\theta = 4.88°$ while for $Re_o = -251$ and $Re_i = 202$: $m = 2$ and $\theta = 4.85°$. Therefore, the space-time diagrams of the spiral pattern in Figure 4 can be represented by the following signal [Cross *et al.*, 1993]:

$$I(r,z,t) = F(r)\left\{ \mathrm{Re}\left[A(z,t)e^{i\left(\omega_R t - k_R z\right)} + B(z,t)e^{i\left(\omega_L t + k_L z\right)}\right]e^{-im\theta} \right\} \qquad (1)$$

where A and B are the amplitudes of right-handed and left-handed spirals respectively, $\omega_R = 2\pi f_R$, $\omega_L = 2\pi f_L$ are the corresponding frequencies, k_R and k_L the corresponding axial wavenumbers, m their azimuthal wavenumber and c.c. stands for complex conjugate. The "structure function" $F(r)$ characterizes the radial dependence of the spiral pattern and vanishes at the cylindrical surfaces: $F(r = a) = F(r = b) = 0$ (Fig. 7). The amplitudes $A(z,t)$ and $B(z,t)$ satisfy the complex Ginzburg-Landau equations (Cross *et al.*, 1993):

$$\tau_0\left(\frac{\partial A}{\partial t} + s\frac{\partial A}{\partial z}\right) = \varepsilon(1+ic_0)A + \xi_0^2(1+ic_1)\frac{\partial^2 A}{\partial z^2} - g(1+ic_2)|A|^2 A - \delta(1+ic_3)|B|^2 A \qquad (2a)$$

$$\tau_0\left(\frac{\partial B}{\partial t} - s\frac{\partial B}{\partial z}\right) = \varepsilon(1+ic_0)B + \xi_0^2(1+ic_1)\frac{\partial^2 B}{\partial z^2} - g(1+ic_2)|B|^2 B - \delta(1+ic_3)|A|^2 B \qquad (2b)$$

where τ_0, ξ_0 represent the characteristic time and characteristic length of perturbations, s their group velocity, ε is the criticality, g the Landau constant of nonlinear saturation, δ is the coupling constant of left and right travelling spirals, c_i are the dispersion coefficients. These coefficients can be determined either numerically (Demay *et al.* , 1984; Tagg *et al.* 1990) or experimentally (Goharzadeh *et al.*, 2010); their values depend on the control parameters

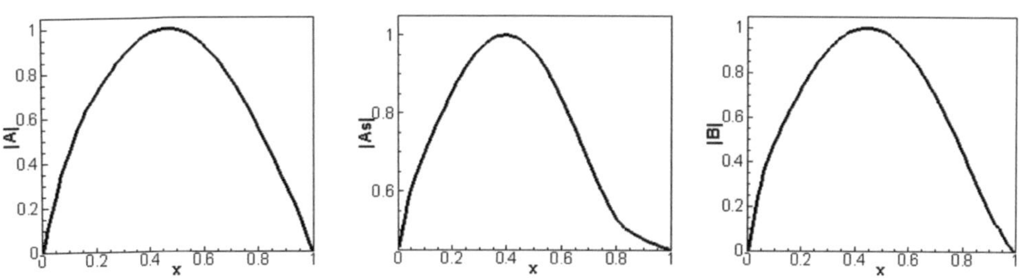

(a) (b) (c)

Fig. 7. Structure function $F(r)$ of the time-averaged amplitude of the pattern $I(r,t)$ for $Re_o = -230$ and $Re_i = 174$: (a) before the sink; (b) in the sink; (c) after the sink.

Re_i and Re_o. The patterns shown in Fig. 5 and time-averaged amplitude profiles of which are shown in Fig. 8 correspond to the case when $\delta > 1$, i.e. the wave coupling is destructive. The sink corresponds to the intersection of two amplitudes solutions given by $A(z) = B(z)$.

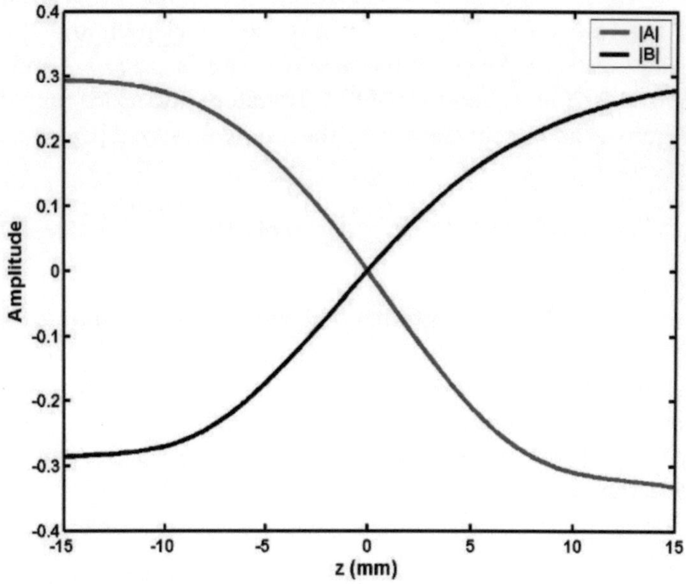

Fig. 8. Spatial distribution of the time-averaged amplitude of the right and left spiral, for Re_o = -230 and Re_i = 174 in the neighbourhood of the sink localized in z_0 = 64 mm.

The fit of experimental data with the theoretical curves (Fig.8) $A(z) = A_0 \tanh[(z - z_0)/\xi]$ and $B(z) = A_0 \tanh[-(z - z_0)/\xi]$ gives A_0 = 0.3, ξ = 7.14 mm = 0.714 d. The value of the coefficient ξ_0 in the equations (2) is given by $\xi_0 = \xi\,\varepsilon^{1/2} = 2{,}11$ mm. This value is in a good agreement with theoretical values $\xi_0/d \approx 0.2$ (Tagg *et al.*, 1990]. The structure functions in Fig. 7 show that the source weakly affect the radial flow.

4. Description of the PIV and its adaptation to the Couette-Taylor flow

The technique of space-time diagrams does not provide quantitative data on the velocity or vorticity fields that are important for the estimate of energy or momentum transfer in the different regimes. Thus it is necessary to perform particle image velocimetry in order to get more quantitative data. For PIV measurements, the working fluid was seeded with spherical glass particles of diameter 8-11 μm and density ρ = 1.6 g/cm^3, with a concentration of about 1ppm. The diffusion time of such particles is $\tau_D = 3\pi\rho v d_p^3 / 4k_B T \approx 500\ s$ where d_p is the particle diameter. The particle Reynolds number $Re_p = Ud_p/\nu < 0.1$ or equivalently their Stokes number $St = (\rho_p d_p/\rho\, d)Re_p < 10^{-3}$ so that they are assumed to follow the flow streamlines, i.e. they are good tracers of the flow. Here U is the characteristic velocity of the particle. The PIV system consists of two Nd-

YAG Laser sources, a MasterPiv processor (from Tecflow) and a CCD camera (Kodak) with 1034x779 pixels. The time delay between two Laser pulses varies from 0.5 to 25 ms, depending on the values of Reynolds numbers Re_i and Re_o. The flow in the test area of the plane (r,z) is visualized with a thin light sheet that illuminates the glass particles, the positions of which can be recorded at short time intervals. To obtain velocity field, 195 pairs of images of size 1034 x 779 pixels were recorded. Each image of a pair was sampled into windows of 32x32 pixels^2 with a recovering of 50%. The velocity fields were computed using the intercorrelation function, which is implemented in the software "Corelia-V2IP" (Tecflow). The PIV measurements were performed in the CCF, in the TVF and WVF regimes in order to calibrate our data acquisition system and to fit data available in the literature for these regimes (Wereley *et al.* 1994, Wereley *et al.* 1998, Abcha *et al.* 2008).

In the circular Couette flow, the spherical glass particles are uniformly distributed (Fig. 9a) while in the TVF and WVF, after 10 hours, the particles have migrated towards the vortex cores where the radial velocity vanishes (Fig. 9b). The PIV allows to visualize velocity and vorticity fields in the cross section (r,z). The results of the complete process is illustrated by 2D velocity fields of Fig. 10a, b. The inflow (arrow (2)) and outflow (arrow (3)) are clearly evidenced in the case of the TVF and WVF. The measured radial and axial velocity components $V_r(r)$, $V_z(r)$ at a given axial position z or at a given radial position r are plotted in Fig. 10c-h in scaled units.

The radial and axial velocity components have been fitted by a polynomial function satisfying the non-slip condition at the cylindrical walls $r = a$ and $r = b$. The velocity data are scaled by the inner cylinder velocity $a\Omega_i$ as follows : $u = V_r/a\Omega_i$ $w = V_z/a\Omega_i$. The lengths are scaled by the gap size, the radial position becomes $\xi = (r - a)/d$ and the axial coordinate is $\zeta = z/d$. In order to plot their profiles, time-averaged velocity components were computed in the axial and radial directions (Fig.10). The instantaneous velocity components can be superposed chronologically at regular time intervals in order to obtain space-time diagrams in both direction (z,t) and (r,t) (Fig. 11). The resulting diagrams are colour-coded as in Abcha *et al.* 2008.

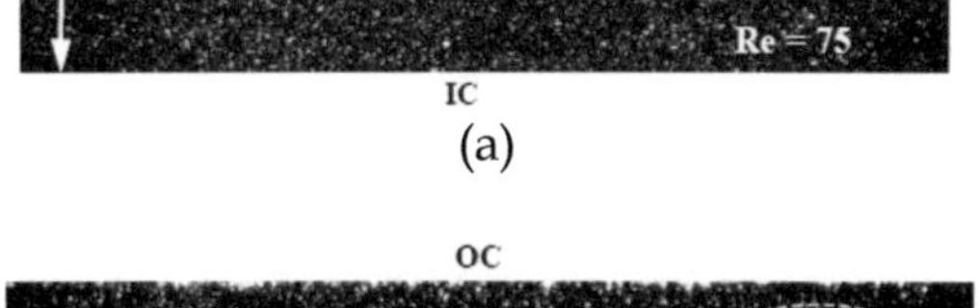

(a)

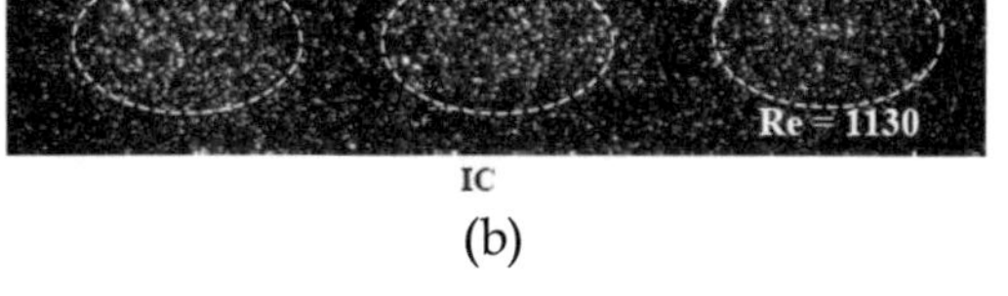

(b)

Fig. 9. Cross section of flow visualized with glass particles for: a) the CCF ($Ta = 37.5$), b) the WVF ($Ta = 565$).

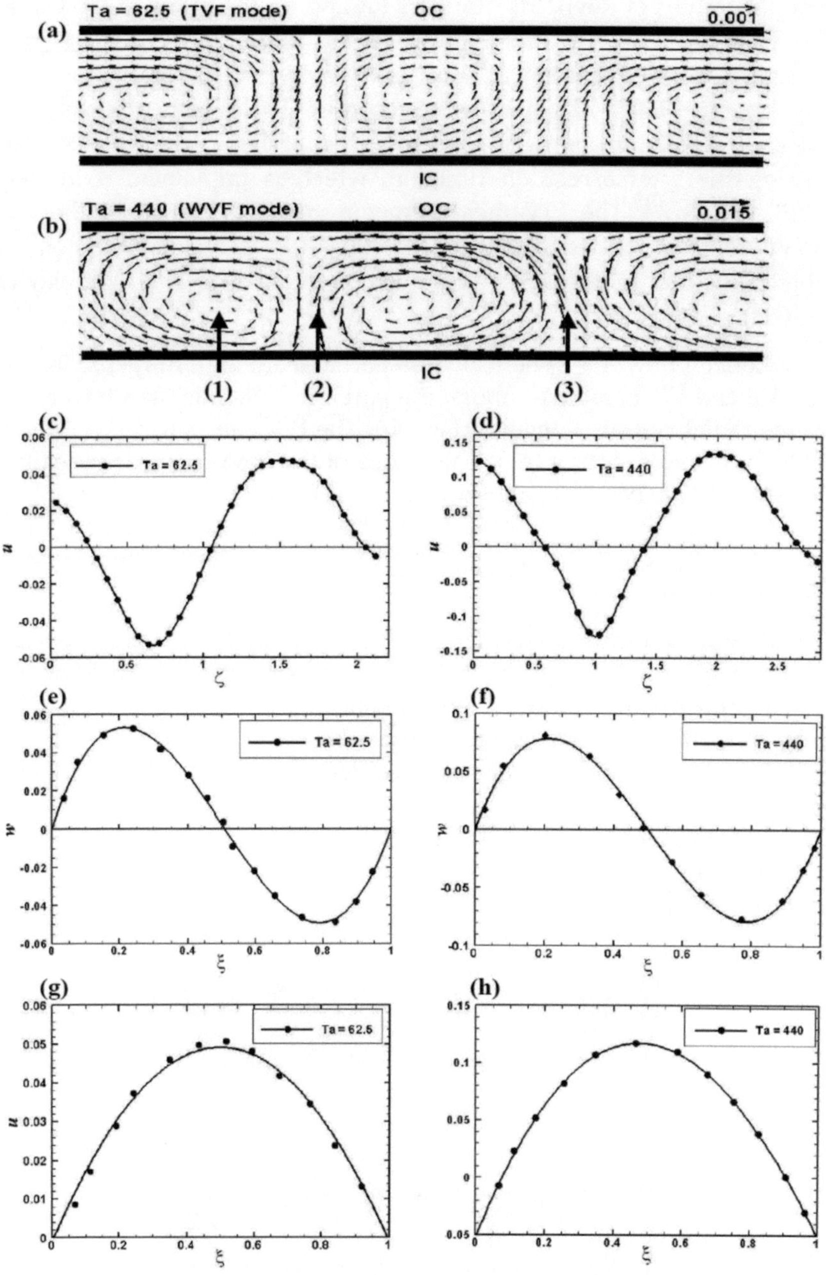

Fig. 10. a); b) Velocity field from PIV measurement; c-d) axial variation of velocity component $u(\zeta)$ at the midgap ($\xi = 0.5$). Radial variation of velocity components: e-f) axial component $w(\xi)$ in the vortex core (arrow (1)): and g-h) radial component $u(\xi)$ at outflow (arrow (3)) for TVF (Ta = 62.5), WVF (Ta = 440) .

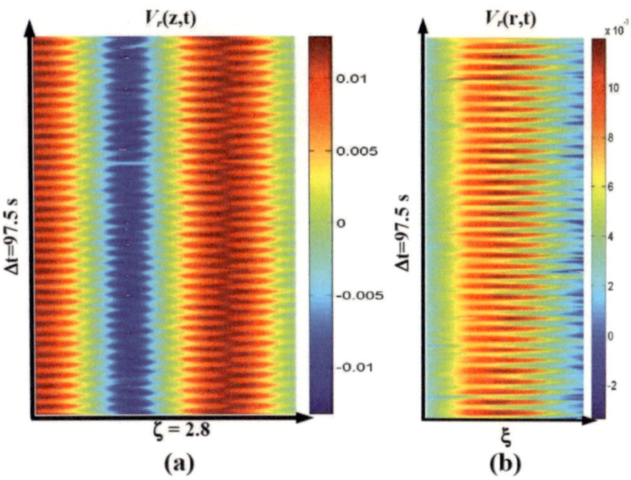

Fig. 11. Space-time diagrams for $Ta = 440$: a) $V_r(z,t)$ taken at the midgap $x = 0.5$, b) $V_r(r,t)$ at outflow position. The red colour corresponds to positive values and blue to negative values of the velocity.

5. Spatio-temporal structure of Taylor vortex flow and wavy vortex flow

5.1 Velocity fields

The instantaneous velocity fields of the Taylor vortex flow in the radial-axial plane, just above the transition to supercritical flow at $Re_i = 125$ for 4 records ($t_{i+1} = t_i + 0.5s$) are shown in Fig. 12. These velocity fields illustrate the dynamics of the TVF: a regime composed of stationary counter-rotating vortices characterized by spatial periodicity equal to twice the size of the gap. The traditional flow visualization of wavy vortex flow by observing the motion of small particles at the outer cylinder suggests that the vortices passing a point on

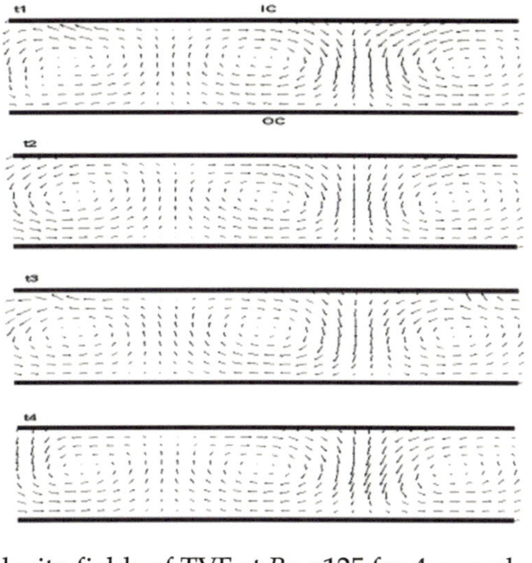

Fig. 12. Instantaneous velocity fields of TVF at Re_i =125 for 4 records ($t_{i+1}=t_i+0.5s$).

the outer cylinder oscillate axially. The PIV permits to visualize the significant transfer of fluid between adjacent vortices in the time and to check the time-dependent theory of shift-and-reflective symmetry in the vortex using the models of wavy vortex flow (Marcus *et al.* 1984). Although the axial motion of the vortices is evidently based on the location of the vortex centres, marked by diamonds (Fig.13), the significant transfer of fluid between adjacent vortices indicates that vortex cells are not independent. The transfer of fluid occurs in a cyclic fashion with a particular vortex gaining fluid from adjacent vortices and then losing fluid to adjacent vortices.

The cycle can be described most easily with reference to the center vortex of « vortex 0 » in Fig. 13. The cycle begins by the frame (i) and ends by the frame (vii). During the cycle the fluid moves from the inner part of the left-hand vortex flowing into the middle vortex and toward the outer cylinder. Simultaneously, fluid from the center vortex moves into the right-hand vortex and toward the inner cylinder. The flow out of the right-hand side of the middle vortex shifts as it is shown in frame (iii), so that now the middle vortex is gaining fluid from the left-hand vortex without losing any fluid. An inward flow from the right-hand vortex also feeds fluid into the middle vortex (see frame (iv)). Frames from (v) to (vii) demonstrate the reversed process beginning with flow around the inner side of the middle vortex from right to left, followed by flow out of the middle vortex to the left, then flow out

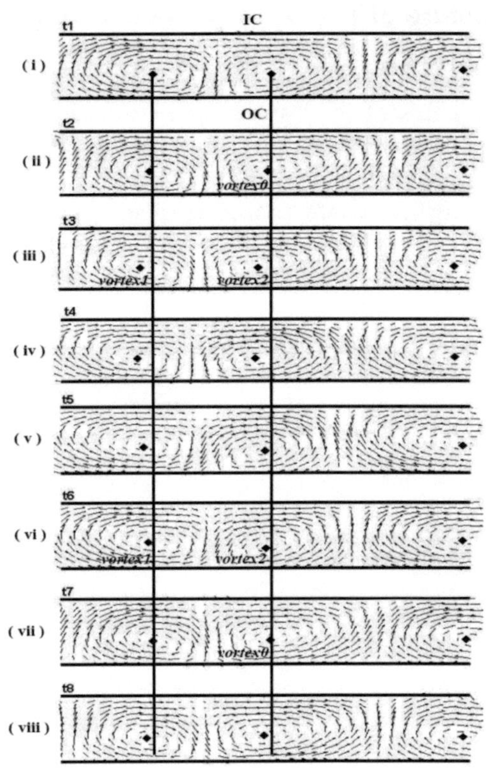

Fig. 13. The instantaneous velocity fields of WVF at $Re_i = 880$ for 8 records ($t_{i+1}=t_i+0.5s$), the time progresses from top to bottom through one complete cycle of an azimuthal wave passing the measurement plane

to the left and right, and finally to the right only. These observations led to the conclusion that the middle vortex is losing the same amount of fluid during the second half of the cycle as it gained in the first half. The second half of the cycle (frames (v)-(vii)) appears identical to the first half of the cycle (frames (i)-(iv)) by a rule of reflective symmetry. For example, the vortex 2 in the field (vi) is a reflection of vortex 1 in the field (iii) with reversal of flow direction, that is to say that the flow of the left vortex from to the middle of the field (iii) became the flow of the middle vortex to the left the vortex of field (vi). This is called ''shift-and-reflect'' symmetry (Marcus, 1984).

5.2 Other hydrodynamic fields

From velocity fields (Fig. 12, 13), different quantities of the flow perturbations in the cross section (r,z) can be computed for each regime (Fig. 14):

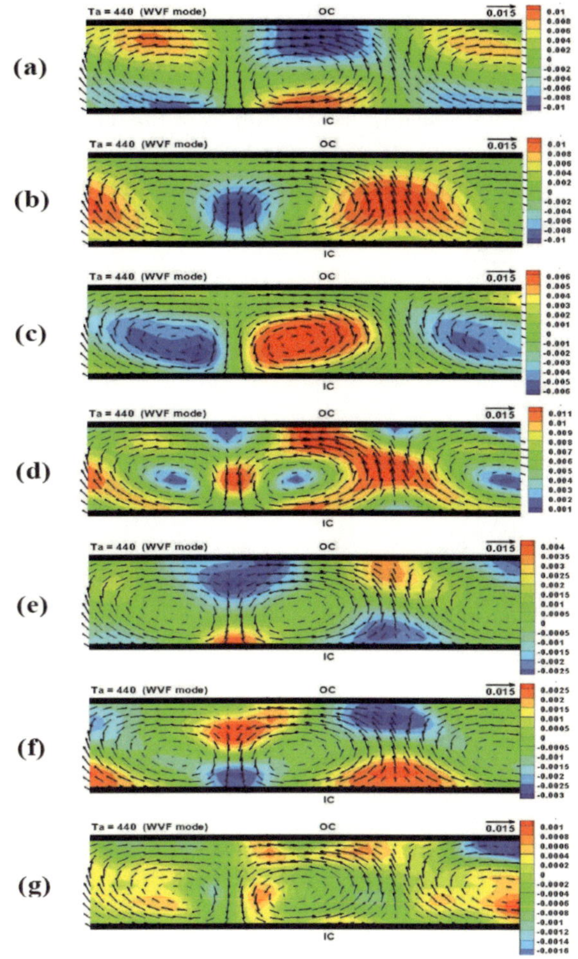

Fig. 14. Cross-section (r,z) of hydrodynamics fields for WVF on the background of vector velocity field: a) Axial velocity; b) radial velocity; c) vorticity; d) kinetic energy; e) axial elongation; f) radial elongation; g) shear rate.

- the azimuthal vorticity component ω_θ and the kinetic energy E:

$$\omega_\theta = (\partial V_r / \partial z - \partial V_z / \partial r) \; ; \; E = (V_r^2 + V_z^2)/2 \tag{3}$$

- three components of the shear rate tensor:

$$\dot{\varepsilon}_{rr} = \partial V_r / \partial r \; ; \; \dot{\varepsilon}_{zz} = \partial V_z / \partial z \; ; \; \dot{\varepsilon}_{rz} = (\partial V_r / \partial z + \partial V_z / \partial r)/2 \tag{4}$$

The vorticity fields and velocity components show that inflow and outflow are almost symmetric in the Taylor vortex flow (Fig. 10a 10c) while they are dissymmetric in the wavy vortex flow (Fig. 10b 10d) because of the oscillations of the separatrix.

5.3 Space-time dependence of velocity profiles

In order to have the most complete information on dynamics of vector velocity field, records of instantaneous profiles of both axial and radial velocity components were superimposed chronologically at regular time intervals (Fig. 15, 16) with color code as in Figure 11. For example, Fig. 15 illustrates the space-time diagram of radial $V_r(z,t)$ and axial velocity $V_z(z,t)$ of TVF (Re_i = 125). The red colour corresponds to the outflow and the blue colour to the inflow. In Fig. 16, the space-time diagram of radial $V_r(r,t)$ and axial velocity $V_z(r,t)$ of WVF (Re_i = 880), were the red colour corresponds to anti-clockwise vortex core and the blue colour to clockwise vortex core.

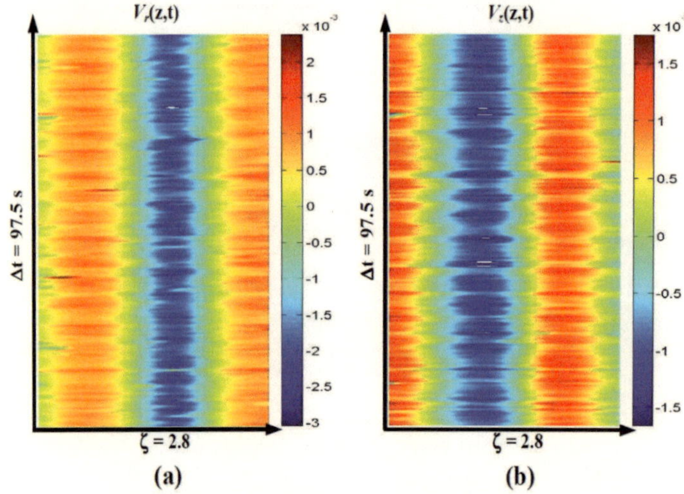

Fig. 15. Space-time diagrams of velocity components for TVF (Re_i=125): a) V_r(z,t) maeasured at the midgap ξ = 0.5 and b) V_z(z,t) near the outer cylinder at ξ = 0.75.

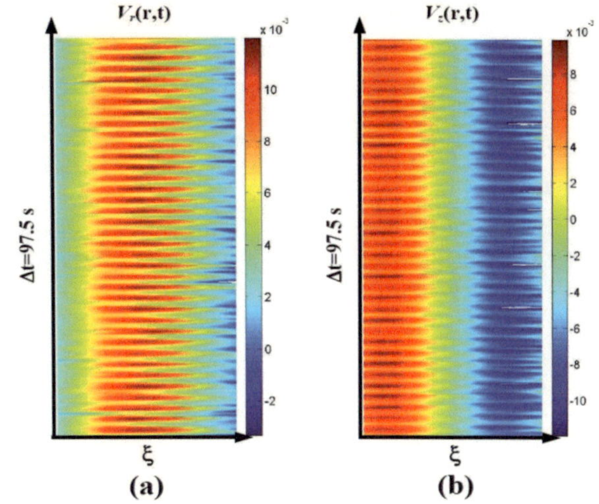

Fig. 16. Space-time diagrams for WVF (Re_i = 880): (a) V_r(r,t) at outflow position, (b) $V_z(r,t)$ in the core of anticlockwise vortex.

5.4 Intensity of light reflected by Kalliroscope *vs.* velocity component

Fig. 17a and Fig. 18a compare the space-time diagrams obtained from PIV measurements (Fig. 11) with those obtained from flow visualization by Kalliroscope flakes in both directions for a wavy vortex regime at Re_i = 880. Unlike space time diagrams of velocities components, space-time diagrams obtained from the reflected light intensity do not give any information about the flow direction (upward or downward for $I(z,t)$, inward vs. outward for $I(r,t)$). That is why for comparison, the absolute values of the velocity components obtained by PIV are used.

At first glance, it was realized that the space-time diagrams obtained by Kalliroscope flakes are very similar with those of the radial velocity component $V_r(z,t)$ and $V_r(r,t)$. Fig. 17b and Fig. 18b illustrate the time-average profiles in the axial and radial directions. These plots highlight the fact that Kalliroscope particles give a signature of the radial velocity component measured in the centre of the gap (ξ = 0.5). The minima and maxima are reached for identical axial positions. A similar correspondence is obtained with the envelopes of the space-time diagrams in the axial and radial directions and leads to the same conclusion (Fig. 17b and Fig. 18b): a perfectly identical evolution in the annular space, a maximum reached in the middle of the gap and minima at the walls of two cylinders. Moreover, the absolute value of the radial velocity vanishes in the vortex core while it reaches the maximum in the outflow and in the inflow. The reflected light intensity vanishes in the vortex core because of the weak motion of Kalliroscope flakes. In the inflow and outflow where the Kalliroscope flakes are faster in the radial direction, the intensity is much larger than in the other parts of the flow.

Recent numerical simulations (Gauthier *et al.* 1998) have shown that the Kalliroscope or iriodin particles may be related to the radial velocity component but no measurements were provided to sustain these arguments. The relaxation time τ of the Kalliroscope flakes is about 0.01 T_p, where the precession time $T_p \sim d/V_r \sim d/V_z \approx$ 2s for the TVF and $T_p \approx$ 1s for WVF. The time scale of the Brownian orientation in a water flow is about 100 s at room temperature (Savas, 1985); it is large enough compared to other time scales of our experiment

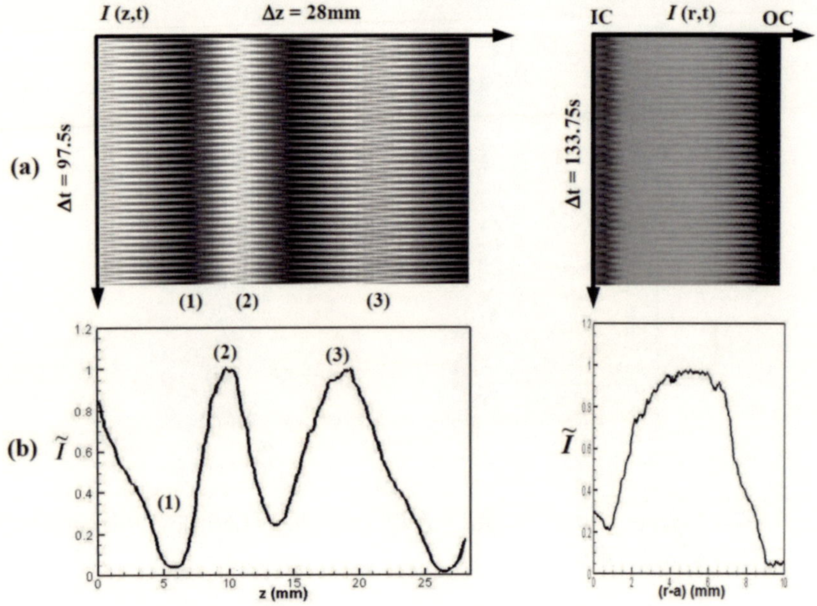

Fig. 17. a) Space-time diagrams of the intensity distribution in the axial $I(z,t)$ and radial $I(r,t)$ direction for Re_i = 880. b) Radial profile and axial profile of light reflected intensity $\tilde{I} = I / I_{\max}$ taken at $\xi = 0.5$. (1) : vortex core, (2): inflow, (3): outflow.

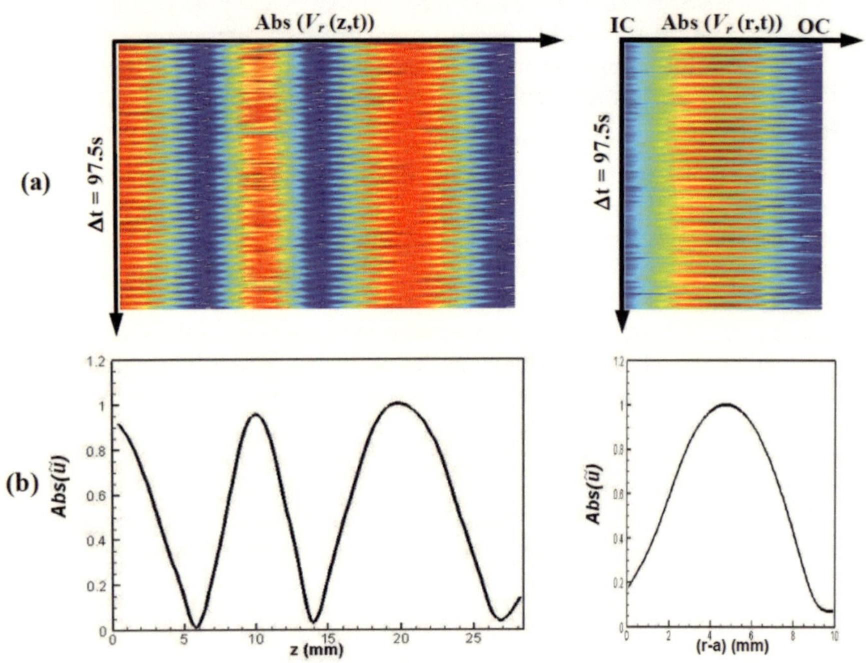

Fig. 18. a) Space-time diagrams of the absolute value of radial velocity component for Re_i = 880, b) Radial and axial profiles of the absolute value $\tilde{u} = V_r / V_{r\max}$ of the radial velocity component measured at $\xi = 0.5$.

so that the Brownian motion can be neglected. The comparison of the space-time diagrams obtained from flow visualization and PIV measurements performed for different flow regimes confirms that in the case of the fixed outer cylinder the reflective particles in the flow give information on the radial velocity component. Therefore, the commonly admitted conjecture that the reflective particles give information on the shear rate (Savas 1985) is in contradiction with the quantitative results. In fact, Fig. 19 shows profiles of different flow properties in the axial and radial direction. None of them has a similar behaviour as the reflected light intensity profile (Fig. 17b, 18b). These results give a more precise content on the fact the small anisotropic particles align with the flow streamlines (Savas 1985, Gauthier *et al.* 1998, Matisse *et al.* 1984) by giving the precision on the velocity component which bears these alignment.

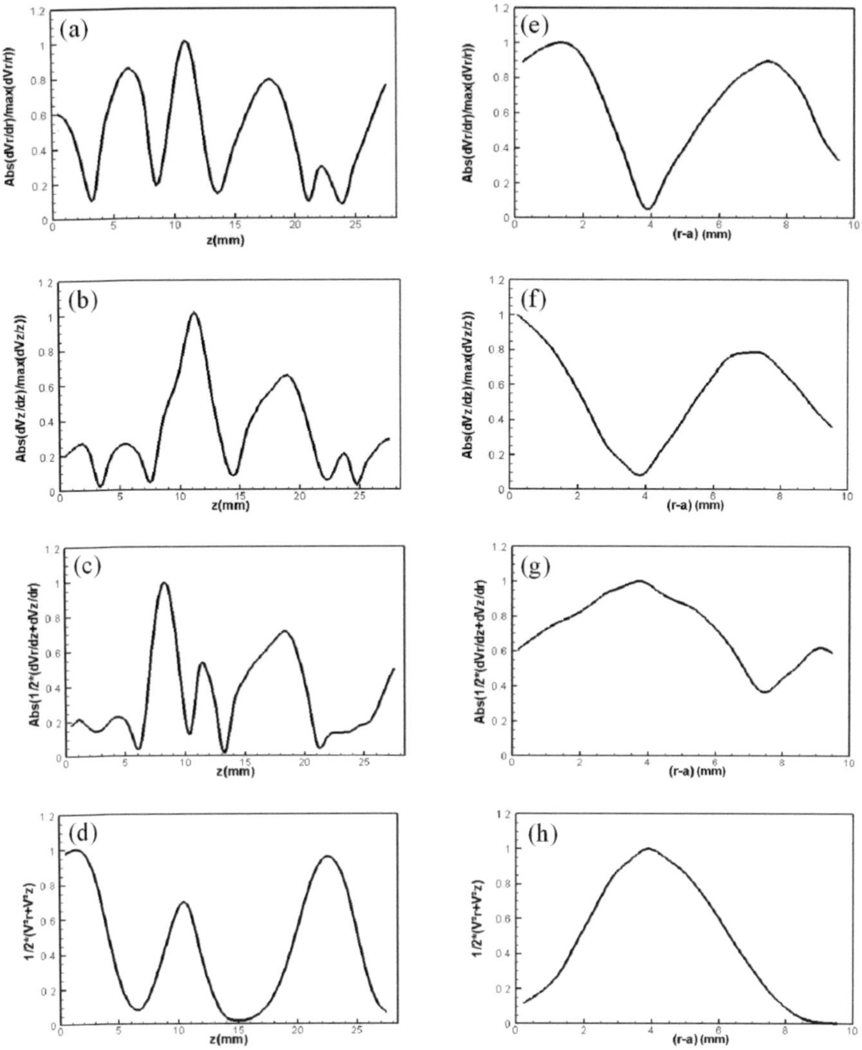

Fig. 19. Axial profiles of the absolute values of flow characteristics measured at $\xi = 0.5$: a) $\dot{\varepsilon}_{rr}$, b) $\dot{\varepsilon}_{zz}$, c) $\dot{\varepsilon}_{rz}$ and d) kinetic energy E; Radial profiles of the absolute values of flow characteristics measured in the outflow: e) $\dot{\varepsilon}_{rr}$, f) $\dot{\varepsilon}_{zz}$, g) $\dot{\varepsilon}_{rz}$ and h) kinetic energy E.

One should mention that our results were verified for TVF, WVF and MWVF in which the radial velocity component has a magnitude larger than that of the axial component. For turbulent Taylor vortex flow (TTVF), no conclusive observation has been made (Fig. 20). Following the results from (Gauthier et *al.* 1998) one would expect the applicability of the present results to pre-turbulent patterns observed in flow between differentially rotating discs (Cros *et al.* 2002).

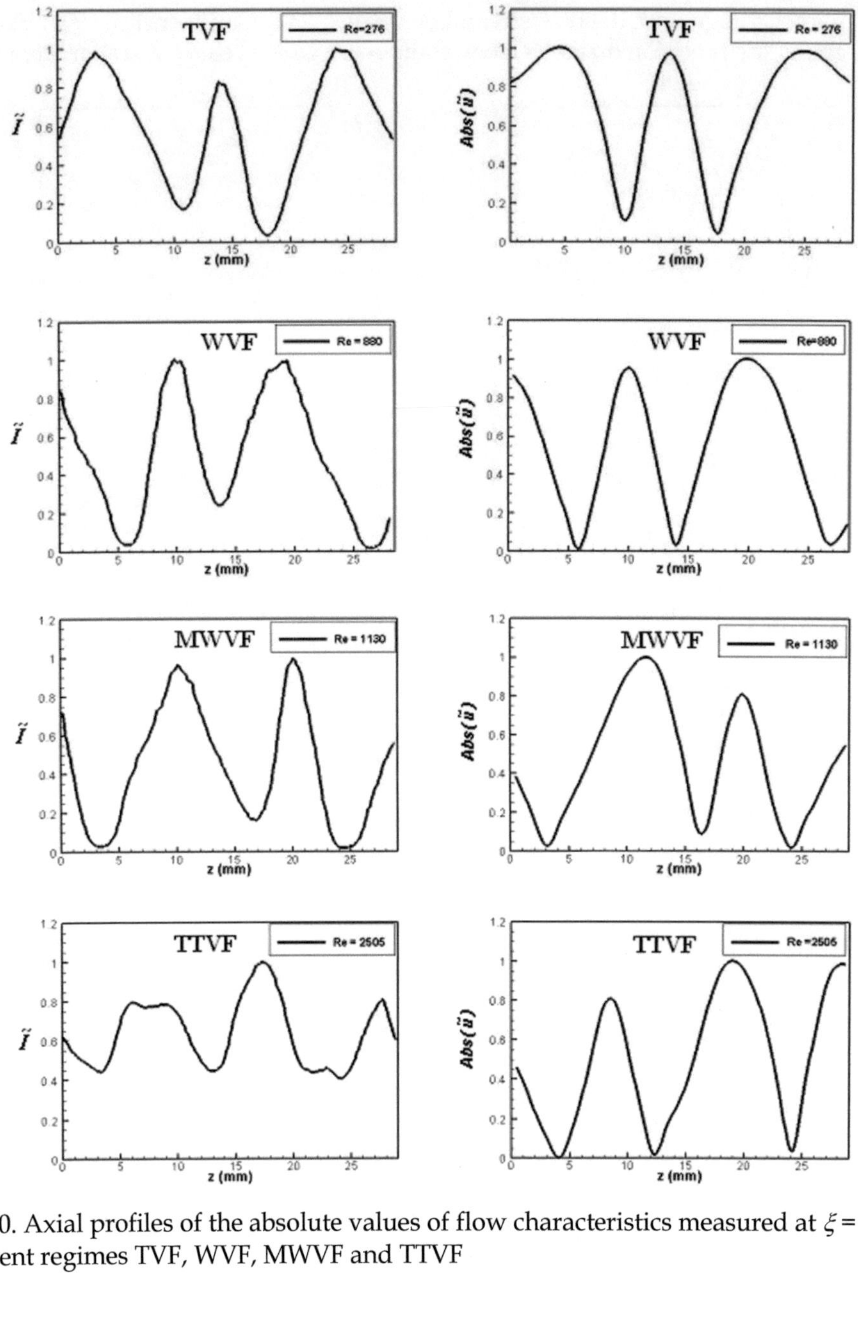

Fig. 20. Axial profiles of the absolute values of flow characteristics measured at $\xi = 0.5$ for different regimes TVF, WVF, MWVF and TTVF

6. Spatio-temporal structure of spiral vortex flow

6.1 PIV velocity measurements

The velocity and vorticity fields of the spiral vortex flow (SVF) for Re_o = -299 and Re_i = 212 in the radial-axial plane (r,z) of the flow are shown in Fig. 21. The measurement zone is located in the lower part of the system ($\zeta \in [14,17]$) from the bottom. The instantaneous velocity fields are regular and show very well the axial motion of the vortices.

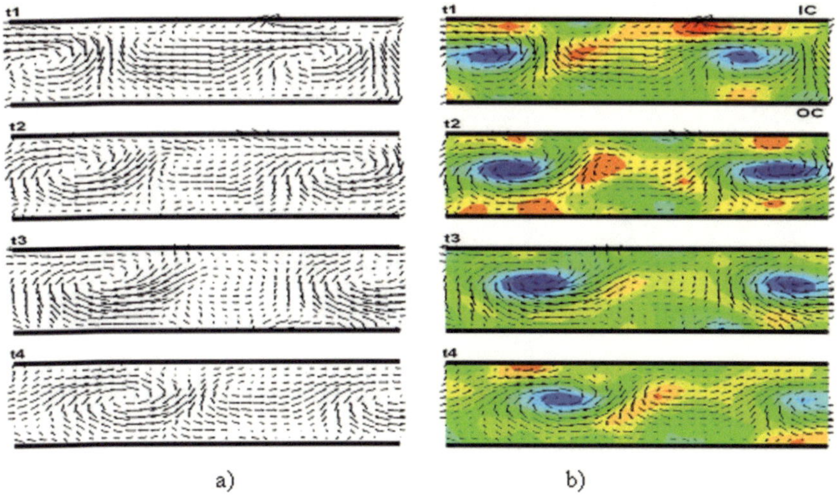

Fig. 21. The instantaneous velocity (arrows) and vorticity fields of the SVF for 4 records ($t_{i+1} = t_i$+0.5s). The color varies from blue (minimal negative vorticity) to red (maximal positive vorticity).

The space-time diagrams of velocity component $V_r(z,t)$ and $V_z(r,t)$ (Fig. 22c-d) confirmed the result from visualization using Kalliroscope flakes that the Taylor spiral vortex pattern is composed of a pair of vortices which propagate along and around the inner cylinder Fig. 22 a-b. Moreover it was revealed that the separatrix between two vortices in a spiral are inclined as in numerical simulations (Ezersky *et al.*, 2010). The radial velocity vanishes in the vortex core while its amplitude is maximal in the outflow and in the inflow. There exists an asymmetry between the inflow and outflow which is well pronounced for the spiral flow (Fig. 23a). In the radial direction, the axial velocity $w(\xi)$ is characterized by an asymmetry in the radial direction: it vanishes at $\xi \approx 0.39$ (Fig. 23d). Similarly the profile of the radial velocity $u(\xi)$ presents an asymmetry, and reaches a maximum around $\xi_0 \approx 0.4$, meaning that the spiral core is located in the region near the inner cylinder (Fig. 23b). The application of the Rayleigh circulation criterion for counter-rotating cylinders shows that the potentially unstable zone is located between $\xi = 0$ and the nodal surface $\xi_0 = \sqrt{(\eta^2 - \mu)/(1-\mu)}$, where $\mu = \eta \mathrm{Re}_o / \mathrm{Re}_i$. In our experiment with $\eta = 0.8$, $\xi_0 = 0.38$ for $\mu = -1.13$ and $\xi_0 = 0.42$ for $\mu = -0.99$. This indicates that the centrifugal instability in case of counter-rotating cylinders is penetrative instability (i.e. it invades the potentially stable zone near the outer cylinder). Using the formula (3-5), the meridional kinetic energy, the radial and axial elongations and the shear rate for the spiral vortex pattern have been computed (Fig. 24).

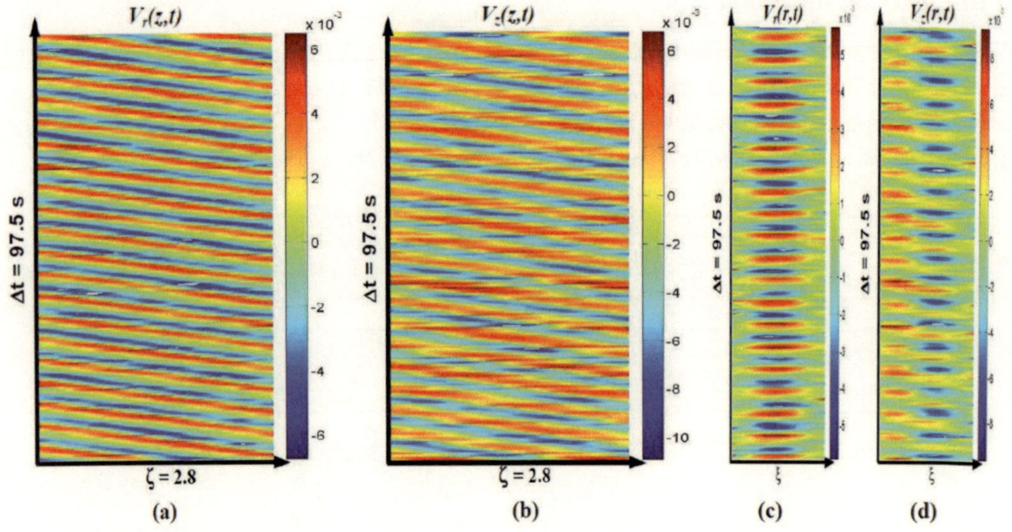

Fig. 22. Space-time diagrams of velocity components for Re_o = -299 and Re_i = 212 : a) $V_r(z,t)$, b) $V_z(z,t)$, c) V_r(z,t), d) $V_z(x,t)$.

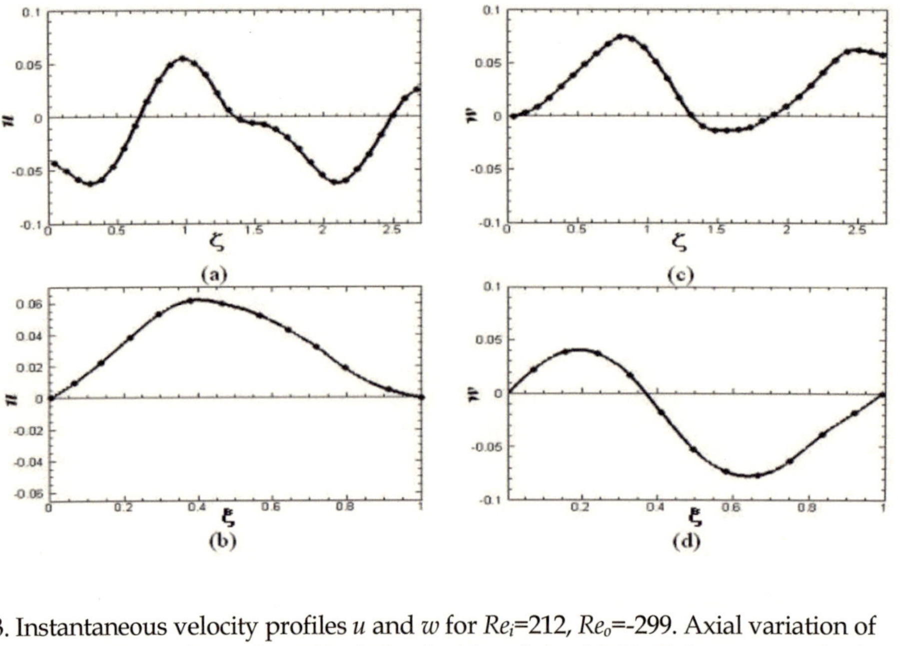

Fig. 23. Instantaneous velocity profiles u and w for Re_i=212, Re_o=-299. Axial variation of velocity component at the midgap (ξ = 0.5) : a) $u(\zeta)$ and c) $w(\zeta)$. Radial variation of velocity components: b) Radial component $u(\xi)$ at outflow, d) axial component $w(\xi)$ in the vortex core.

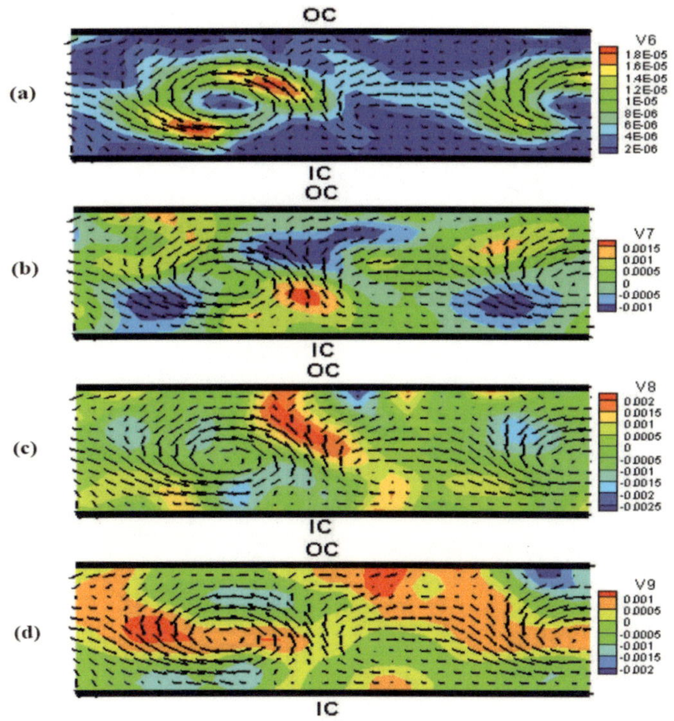

Fig. 24. Cross-section (*r,z*) of hydrodynamics fields for SVF : a) kinetic energy *E*; b) axial elongation $\dot{\varepsilon}_{zz}$ c) radial elongation $\dot{\varepsilon}_{rr}$ d) shear rate ε_{rz} .

6.2 Intensity of light reflected by Kalliroscope *vs.* velocity component

Comparison of the space-time diagrams, reveals a strong similarity between diagrams obtained by Kalliroscope flakes and those of the axial velocity component $V_z(z,t)$ and $V_z(r,t)$. Fig. 25a and Fig. 25b illustrate the time-average profiles in the axial and radial directions respectively of light reflected intensity and axial velocity component. Fig. 25c demonstrates discrepancies between radial velocity and intensity *I* for this case. The minima and maxima of intensity *I* are observed at approximately the same coordinates as minima and maxima of axial velocity V_z. It should be noted that the absolute value of the axial velocity vanishes in the vortex core while it reaches the maximum in the outflow and in the inflow. The reflected light intensity vanishes in the vortex core because of the weak motion of Kalliroscope flakes.

These plots highlight the fact that Kalliroscope particles give a signature of the axial velocity component measured in axial and radial direction.

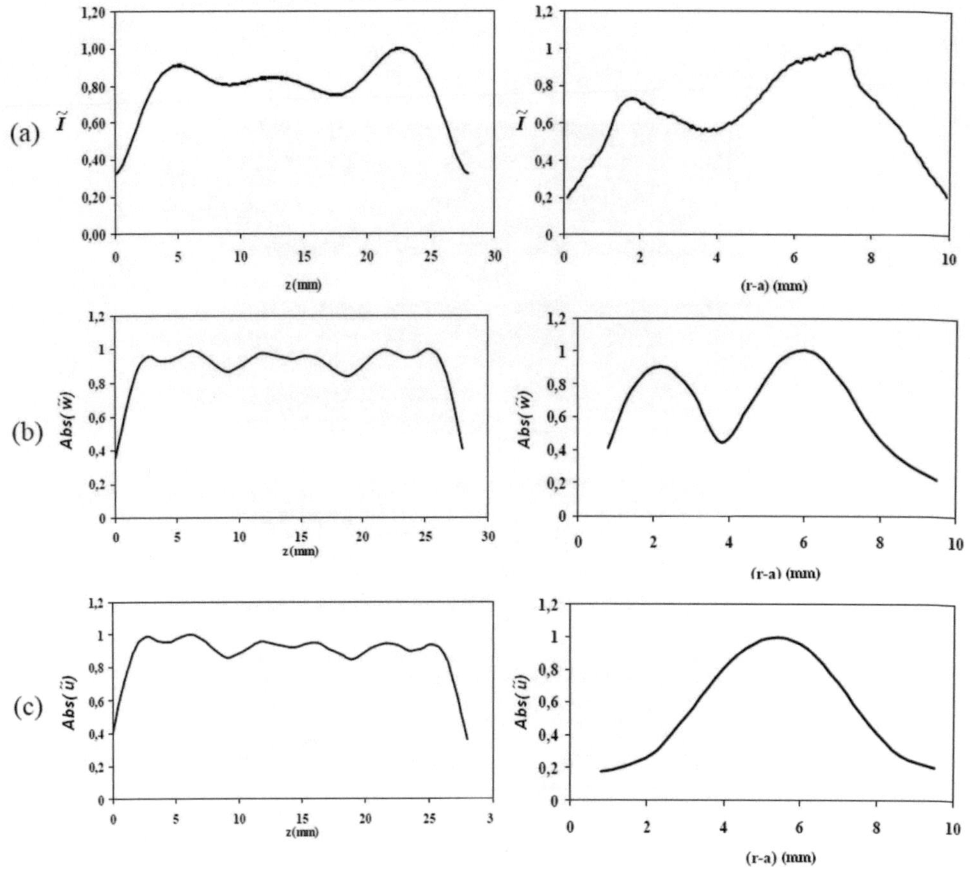

Fig. 25. a) Radial profile and axial profile of light reflected intensity $\tilde{I}$ taken at $\xi = 0.5$ b) Radial and axial profiles of the absolute value of the axial velocity component $\tilde{w} = V_z / V_{z\max}$ measured at $\xi = 0.5$; c) Radial and axial profiles of the absolute value $\tilde{u}$ of the radial velocity component measured at $\xi = 0.5$

6.3 The velocity field in the vicinity of defects

When the Reynolds number Re_i is increased , the pattern of the spiral vortex flow becomes unstable and spatio-temporal defects appear as a result of vortex merging (annihilation event) or of splitting of a vortex (creation event). The creation and annihilation events appear randomly in the pattern; they are due to long wavelength modulations. The velocity field was determined in the neighborhood of the spatio-temporal defects for Re_i = 227 and Re_o = -299. The defect was localized as point where amplitude of velocity field is closed to zero $V_z^2 + V_r^2 \approx 0$ and phase of the field has nonzero circulation around this point in the plane (*z*,*t*) (Fig. 26a-b black ellipses near t_d = 42s and z_d = 12 mm). A special attention was focused on the spatiotemporal behavior of radial and axial velocity components across the defect. Fig. 27a shows the temporal evolution of the axial and radial velocity components at the position of the defect and Fig. 27b shows the velocity profile taken at the collision time.

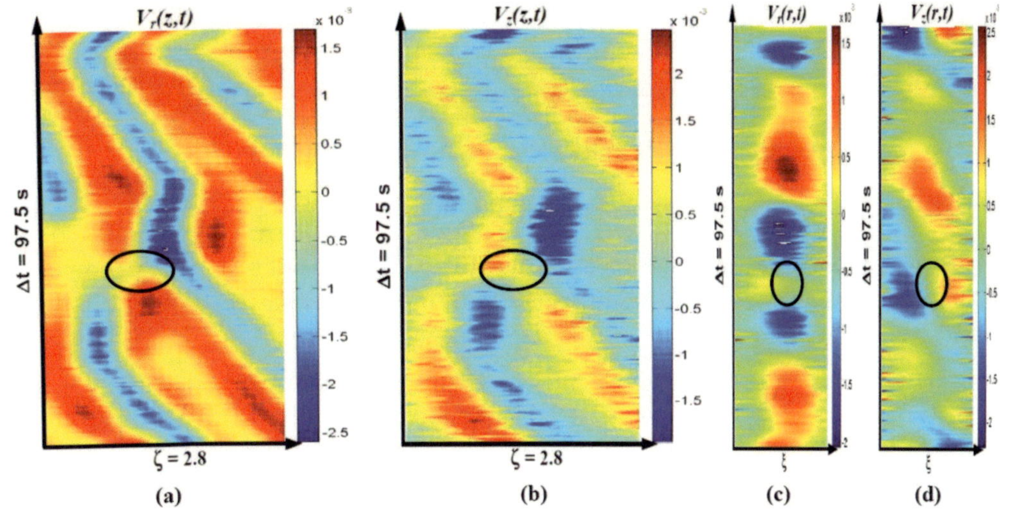

Fig. 26. Space-time diagrams of velocity components in the neighbourhood of the defect for Re_i=227 , Re_o=-299 : a) V_r(z,t) , b) V_z(z,t), c) V_r(r,t) and d) V_z(r,t).

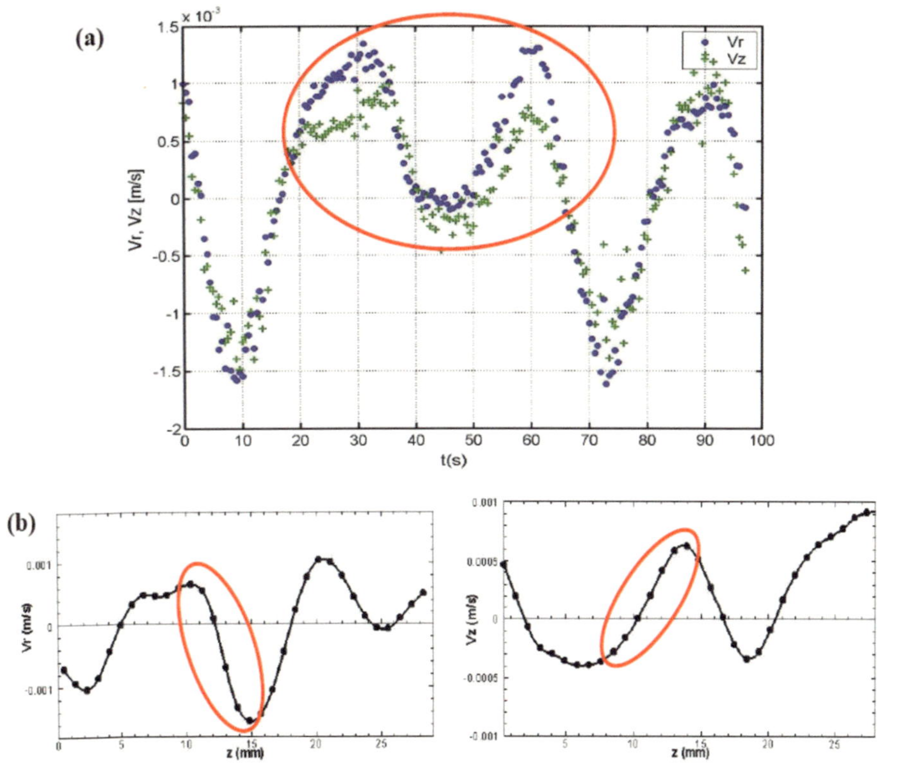

Fig. 27. Velocity profiles through the defect at the mid-gap (ξ = 0.5), for Re_i = 227 and Re_o = -299: a) The temporal evolution of radial and axial velocity components, b) Spatial evolution of radial and axial velocity components.

In the neighborhood of the defect, the temporal variation of both velocity components follows a parabolic law (Fig. 27a):

$$V_r(t) \approx \alpha(t - t_d)^2; \quad V_z(t) \approx \beta(t - t_d)^2 \tag{5}$$

while the spatial evolution is linear in the neighborhood of the defect (Fig. 27b):

$$V_r(z) \approx a(z - z_d) \quad ; \; V_z(z) \approx b(z - z_d) \tag{6}$$

The coefficients of the best fit are given in the Table. These results are in agreement with the solutions of the Ginzburg-Landau equation near a defect as was shown in Ezersky et *al.*, 2010.

Coefficient	α	β	a	b
Best fit value	0.01	0.07	-0.62	0.21

Table 1. Best fit coefficients of the temporal and spatial evolution of the velocity field in the neighbourhood of a defect.

7. Summary

This chapter has made a focus on the correspondence between the intensity of reflected light by particles and the velocity components in the meridional plane (*r,z*). When the outer cylinder is fixed, there is a correspondence between radial velocity component and the intensity of light reflected by anisotropic particles. This result has confirmed recent numerical simulations [Gauthier 1998]. When cylinders are counter-rotating, the intensity of light reflected by anisotropic particles is related to the axial velocity component. To investigate all the aspects of the transition to turbulence in closed or open flows, visualization by particle seeding and velocimetry techniques (LDV, UDV, PIV)are very complementary as they permit to access to different flow characteristics. In fact, the reflective flakes allow to access to flow properties on a large spatial extent and for a long time, and it is possible to evidence the spatio-temporal evolution to turbulence in each directions. For example the study of defects (sources, sinks, dislocations, ...) has facilitated the application of the Ginzburg-Landau model to the study of stationary and time-dependent patterns, the transition to chaos or weak turbulence has been characterized using results from visualization. The velocimetry techniques allow therefore to access to physical quantities needed in the model of turbulence (kinetic energy, rate of strain, vorticity, momentum,...) which are useful for validation of theoretical models for example in computing structure coefficients of the statistical distributions. The choice of appropriate technique depends on flow under consideration. In some cases, visualization by anisotropic particles is more preferable than LDV or PIV technique. Besides its simplicity and low cost, it is possible to visualise larger spatial extent, and to record long time data and therefore obtain a better power spectra. The problem of correlation of data obtained from anisotropic particles and velocimetry data is far from being solved, it represents a big challenge for experimental research in Hydrodynamics. Although some similarities between these two methods were found in the case of Couette-Taylor flow patterns, there are many fundamental questions that are far from being resolved:

1. How image brightness depends on particle concentration or particle orientation?
2. How concentration or orientation depends on velocity field characteristics?
3. Intensity of image brightness is a two dimensional field. How to project three dimensional field of particle concentration or orientation on two dimensional plane?

The answers to these questions will enable researchers in Hydrodynamics to understand spatio-temporal structures of closed flows by comparing results obtained from different techniques (velocimetry, visualization) and numerical simulations.

8. Acknowledgements

This work has been benefited from a financial support from the CPER-Haute-Normandie under the program THETE.

9. References

Abcha N.; Latrache N.; Dumouchel F. & Mutabazi I.(2008). Qualitative relation between reflected light intensity by Kalliroscope flakes and velocity field in the Couette-Taylor flow system, *Experiments in Fluids* Vol. 45, p.85.

Ahlers G.D.; Cannell D.S.; Dominguez-Lerma M.A.& Heinrichs R. (1986). Wavenumber selection and Eckhaus instability in Couette-Taylor flow, *Physica D* Vol. 23, p.202.

Andereck C. D.; Liu S.S. & Swinney H.L. (1986). Flow regimes in a circular Couette system with independently rotating cylinders, *Journal of Fluid Mechanics* Vol.164, p.155.

Bergé P., Pomeau Y. & Vidal Ch. (1984). L'ordre dans le chaos, Hermann, Paris.

Bot P. & Mutabazi I. (2000). Dynamics of spatio-temporal defects in the Taylor-Dean system, *European Physics Journal* B Vol. 13, p.141.

Chossat P. & Iooss G. (1994). *The Couette-Taylor Problem*, Applied Mathematical Science 102, Springer-Verlag, New-York.

Coles D. (1965). Transition in circular Couette flow, *Journal of Fluid Mechanics* Vol.21, p.385.

Cros A. & Le Gal P.(2002). Spatiotemporal intermittency in the torsional Couette flow between a rotating and a stationary disk, *Physics of Fluids* Vol.14, p.3755.

Cross M. C. & Hohenberg P. C. (1993). Pattern Formation Out of Equilibrium, *Review of Modern Physiscs* Vol. 65, p.851.

Demay Y. & Iooss G. (1984). Calcul des solutions bifurquées pour le problème de Couette-Taylor avec les deux cylindres en rotation, *Journal de Mécanique Théorique et Appliquée*, Numéro spécial p.193.

Dominguez-Lerma M.A.; Ahlers G. & Cannell D.S. (1985). Effects of Kalliroscope flow visualization particles on rotating Couette-Taylor flow, *Physics of Fluids* Vol. 28, p.1204.

Durst F. & Whitelaw J.H. (1976). *Principles and Practice of Laser-Doppler Anemometry*, Academic Press, New York.

Egbers Ch. & Pfister G. (eds) (2000). *Physics of Rotating Fluids*, Lectures Notes in Physics 549, Springer Berlin.

Ezersky A.B.; Abcha N. & Mutabazi I. (2010). The structure of spatio-temporal defects in a spiral pattern in the Couette-Taylor flow. *Physics Letters A* Vol.374, p. 3297.

Gauthier G., Gondret P. & Rabaud M. (1998). Motions of anisotropic particles : Application to visualization of three-dimensional flows, *Physics of Fluids* Vol. 10, p.2147.

Goharzadeh A. & Mutabazi I. (2010). Measurement of coefficients of the Ginzburg-Landau equation for patterns of Taylor spirals, *Physical Review E* Vol. 82, p.016306.

Hoffmann C.; Lücke M. & Pinter A. (2005). Spiral vortices traveling between two rotating defects in the Taylor-Couette system, *Physical Review E* Vol.72, p. 056311.

Jensen K. D. (2004). *Journal of the Brazilian Society of Mechanical Science & Engineering*, Vol. XXVI(4), p. 401.

Langford W. F.; Tagg R.; Kostelich E. J.; Swinney H. L. & Golubitsky M. (1988). Primary instabilities and bicriticality in flow between counter-rotating cylinders, *Physics of Fluids* Vol.31, p. 776.

Marcus P. S. (1984). Simulation of Taylor-Couette flow, Part 2: Numerical results for wavy-vortex flow with one travelling wave. *Journal of Fluid Mechanics* Vol. 146, p.65.

Matisse P. & Gorman M. (1984). Neutrally buoyant anisotropic particles for flow visualization, *Physics of Fluids* Vol.27, p.759.

Mutabazi I.; Wesfreid J.E. & Guyon E. (2006). *Dynamics of Spatio-Temporal Cellular Structures*, Springer, New York.

Peerhossaini H. & Wesfreid J.E. (1988). On the inner structure of streamwise Görtler vortices, *International Journal of Heat and Fluid Flow* Vol. 9, p. 12.

Provansal M.; Mathis C. & Boyer L. (1987). Bénard-von Karman instability: transient and forced regimes, *Journal of Fluid Mechanics* Vol. 182, p.1.

Savas Ö (1985). On flow visualization using reflective flakes, *Journal of Fluid Mechanics* Vol. 152, pp.235.

Takeda Y.; Fischer W.E. & Sakakibara J. (1994). Decomposition of the modulated waves in rotating Couette system, *Science* Vol. 263, p.502.

Taylor G.I. (1923). Stability of a viscous liquid contained between two rotating cylinders, *Philosophical Transactions of the Royal Society A* Vol. 223, p.289.

Thoroddsen ST; Bauer JM (1999). Qualitative flow visualization using colored lights and reflective flakes, *Phys. Fluids* Vol. 11, p.1702.

Van Dyke M. (1982). *An Album of Fluid Motion*, Parabolic, Stanford.

Wereley S.T. & Lueptow R.M. (1994). Azimuthal velocity in supercritical circular Couette flow. *Experiments in Fluids* Vol.18, pp.1.

Wereley S.T. & Lueptow R.M. (1998). Spatio-temporal character of non wavy and wavy Taylor-Couette flow, *Journal of Fluid Mechanics* Vol. 364, p.59.

8

Rheo-Particle Image Velocimetry for the Analysis of the Flow of Polymer Melts

José Pérez-González[1], Benjamín M. Marín-Santibáñez[2], Francisco Rodríguez- González[3] and José G. González-Santos[4]

[1]*Laboratorio de Reología, Escuela Superior de Física y Matemáticas, Instituto Politécnico Nacional,*

[2]*Sección de Estudios de Posgrado e Investigación, Escuela Superior de Ingeniería Química e Industrias Extractivas, Instituto Politécnico Nacional,*

[3]*Departamento de Biotecnología, Centro de Desarrollo de Productos Bióticos, Instituto Politécnico Nacional,*

[4]*Departamento de Matemáticas, Escuela Superior de Física y Matemáticas, Instituto Politécnico Nacional,*

México

1. Introduction

The knowledge of the flow kinematics of polymer melts is relevant for basic rheology as well as for polymer processing, particularly for the design of molds and extrusion dies. Nevertheless, the analysis of the flow behavior of polymer melts has been typically performed by using rheometrical (mechanical) measurements and numerical simulation. In spite of the large amount of publications in the field, few works have been dedicated to the analysis of the underlying flow kinematics by using velocimetry techniques. This may in part be due to the difficulties to implement velocimetry techniques during the processing of polymer melts at high temperatures and high pressures.

With the advent of modern technologies that permit the efficient measurement of the velocity of particles seeded in a fluid, it has been possible to obtain velocity maps in fluids, which provide precise details of their flow kinematics. The analysis of flow fields has been mainly done by using optical techniques like laser Doppler velocimetry (LDV), particle tracking velocimetry (PTV) and particle image velocimetry (PIV), which are powerful non-invasive techniques to describe the flow kinematics in transparent fluids. However, while LDV is a single-point measurement technique and PTV requires the tracking of individual particles, PIV is a whole-field method that allows for the determination of instantaneous velocity maps of a flow region. This last approach, of common use in fluid mechanics, has been gradually implemented for the analysis of the flow behavior of polymer melts. Major limitations for the use of PIV in polymer melts rely on the design and adaptation of transparent dies and molds capable to withstand the high temperatures and pressures characteristic of polymer processing operations.

In this contribution, the extrusion of polymer melts is analyzed by using simultaneous rheometrical and two-dimensional particle image velocimetry measurements (2D PIV), or what has been called Rheo-PIV. First, the fundamentals of rheology and rheometry, shear flow and pressure driven flows are introduced. Then, a brief description of flow instabilities and slip in polymer melts is presented, followed by a short account of the work carried out in the study of the kinematics of Poiseuille flows of polymer melts by using optical velocimetry techniques. The geometry and materials of construction of the dies used for velocimetry measurements are highlighted and problems of actual interest in the field, as flow instabilities and slip at solid boundaries, are particularly addressed. Another section includes the basics of the PIV technique, as well as a discussion of some algorithms relevant for calculation of velocity vectors near solid boundaries and high shear gradients regions, which are very common in the flow of polymer melts. Finally, we present some applications of the PIV technique to the study of the stable and unstable Poiseuille flow under slip and no-slip boundary conditions of molten polyolefins with great practical importance, namely, low-density polyethylene (LDPE), polypropylene (PP), high-density polyethylene (HDPE) and linear low-density polyethylene (LLDPE).

2. Rheology

Rheology is the science of deformation and flow of matter. Following this general definition, one might think that rheology is used to analyze any type of material. In practice, however, rheology has been mainly devoted to the study of fluids containing large molecules (macromolecules) or suspended particles, and fluids having a structure, also known as complex fluids. The flow behavior of these fluids cannot be described by the Newton's law of viscosity and are then called non-Newtonian.

Rheometers are based on shear and shear-free flows. Shear-free flows occur when there is not contact of the fluid with solid walls, for instance, during the film blowing or during the elongation of a polymer melt filament. Contrarily, shear flows may be generated by a moving surface into contact with the fluid or by an applied pressure gradient (Fig. 1). For example, that generated in a capillary rheometer.

2.1 The capillary rheometer

Most polymer processing operations and fluid transport take place at high shear rates. In such cases, the rheological characterization of the fluids is performed by using rheometers based on Poiseuille flows, since rotational rheometers are limited to low shear rates. The capillary rheometer is the most used for this purpose; it has a great practical importance since it is found in different polymer processing operations, particularly in extrusion dies and mold runners.

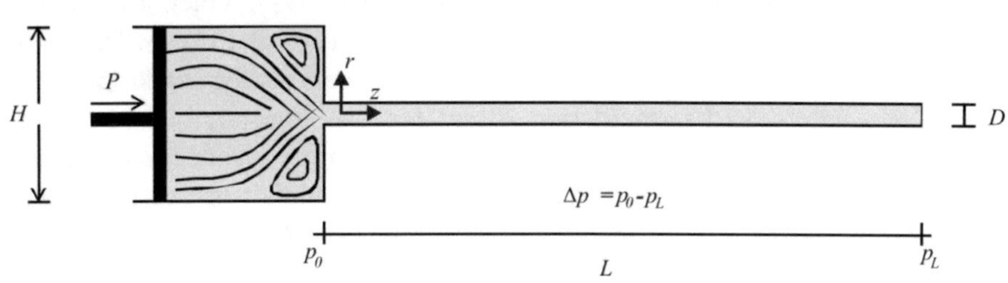

Fig. 1. Schematic representation of a capillary rheometer

The capillary rheometer may be operated at constant pressure or constant flow rate. It consists of a reservoir that holds the fluid to be characterized and a capillary through which the fluid is forced to flow by an imposed pressure (see Fig. 1). The data obtained from this rheometer are the pressure drop (Δp) between capillary ends and the volumetric flow rate (Q). The flow variables, wall shear stress (τ_w) and shear rate ($\dot{\gamma}$), are obtained by solving the momentum and mass conservation equations, respectively, along with a constitutive equation for the fluid (Bird et al., 1977). These equations are solved for isothermal conditions considering the following boundary conditions:

a. Laminar flow
b. Incompressible fluid
c. Steady and well developed flow
d. No-slip at the capillary wall

The assumption of zero velocity at a solid boundary, also known as the no-slip condition, has serious implications from the basic and practical point of view. This condition is generally satisfied for the flow of Newtonian fluids, but is not necessarily valid for some non-Newtonian ones, for example, entangled polymer melts. An investigation of the validity of such condition for the extrusion flow of polymer melts is precisely one of the objectives of this work. Considering the previous assumptions, the shear stress is given by:

$$\tau_{rz}(r) = \tau_w \frac{r}{R} \tag{1}$$

$$\tau_w = \frac{R\Delta p}{2L} \tag{2}$$

where τ_w is the wall shear stress, τ_{rz} is the radial dependent shear stress, r and z are the cylindrical coordinates, R and L are the capillary radius and length, respectively. In practice, the pressure drop (Δp) is not linear between capillary ends due to rearrangements of the velocity profiles at the inlet and outlet region. Then, a correction for end effects is usually applied to the pressure drop (Bagley, 1957). The analysis of such a correction is beyond the scope of this work, but its details may be found elsewhere (de Vargas et al., 1995).

For a Newtonian fluid the constitutive equation $\tau = \mu \frac{\partial v_z}{\partial r}$ is included in the equation of conservation of momentum, which results in a parabolic velocity profile, Eq. 3 (see Fig. 2a), and an expression for the shear rate at the capillary wall given by Eq. 4:

$$v_z(r) = \frac{\Delta p R^2}{4\mu L}\left[1 - \left(\frac{r}{R}\right)^2\right] \tag{3}$$

$$\dot{\gamma}_w = \frac{4Q}{\pi R^3} \tag{4}$$

where Q is the volumetric flow rate given by:

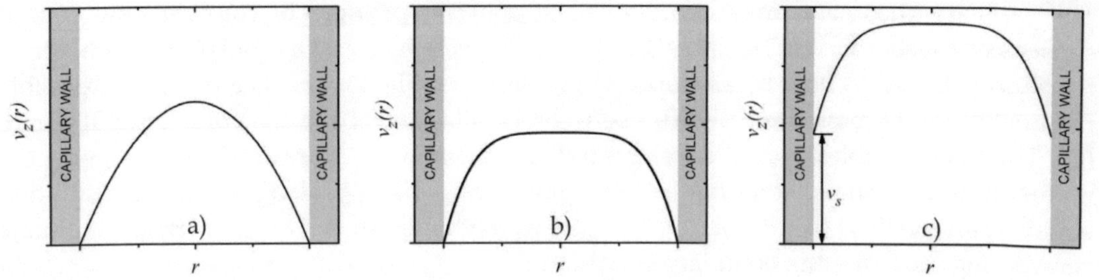

Fig. 2. Velocity profiles: a) Newtonian fluid; b) Non-Newtonian fluid; c) Non-Newtonian fluid with slip at the wall (v_s)

$$Q = 2\pi \int_0^R v_z(r) r dr \tag{5}$$

The fluid viscosity (μ), which is defined as the ratio $\mu = \tau_w / \dot{\gamma}_w$, is given by the well known Hagen-Poiseuille equation:

$$\mu = \frac{\pi \Delta p R^4}{8LQ} \tag{6}$$

For non-Newtonian fluids, the viscosity (η) is dependent on the shear rate, i. e., $\eta = \eta(\dot{\gamma})$ and a commonly used constitutive equation to describe the viscous behavior is the power-law model, $\tau = m\dot{\gamma}^n$, where m and n are the consistency and shear thinning index, respectively. By using this model in solving the motion equations, the fluid velocity (see Fig. 2b) and the corresponding shear rate at the capillary wall are given by:

$$v_z(r) = \frac{nR}{1+n}\left(\frac{R\Delta p}{2mL}\right)^{\frac{1}{n}}\left[1-\left(\frac{r}{R}\right)^{\frac{n+1}{n}}\right] \tag{7}$$

$$\dot{\gamma}_w = \dot{\gamma}_{app} \frac{3n+1}{4n} \tag{8}$$

Where $\dot{\gamma}_{app}$ is given by Eq. 4 and it is referred to as the apparent shear rate when used for non-Newtonian fluids. Eq. 8 is used to calculate the true shear rate and it is obtained after using the Rabinowitsch's correction (Bird et al., 1977).

2.2 Flow instabilities and slip in polymer melts

Flow instabilities restrain rheometrical measurements and limit productivity in polymer processing operations. Unstable flow is obviously unsteady and may produce distortions of extruded materials (see Fig. 3a). Due to these facts, a great deal of work has been devoted to understand and explain the origin of flow instabilities. Most of the research done in the field in the last decades has been thoroughly reviewed by several authors, being some of the most influential reviews those by Petrie and Denn (1976), Denn (1990) and Denn (2001).

Amongst the different flow instabilities, the stick-slip phenomenon occurring in the extrusion of entangled linear polymers melts has received particular attention. This phenomenon starts at a critical shear stress (τ_c) and manifests itself as periodic oscillations of the pressure drop and volumetric flow rate under controlled flow rate experiments. Such oscillations are related to dynamic transitions, from stick to slip, of the boundary condition at a solid wall and produce bamboo-like distortions on the extrudates (Fig. 3a-ii).

The flow curve for a polymer that exhibits the stick-slip instability is discontinuous and is commonly divided into three regions (Fig. 3b). The flow is stable in the first region for $\tau < \tau_c$ (low shear rate branch), unstable for $\tau = \tau_c$ (stick-slip region), and assumed to be stable again for $\tau > \tau_c$ (high shear rate branch). Slip may occur or not in the low shear rate branch depending on the molecular characteristics of the polymer, whereas it is characteristic of the high shear rate branch.

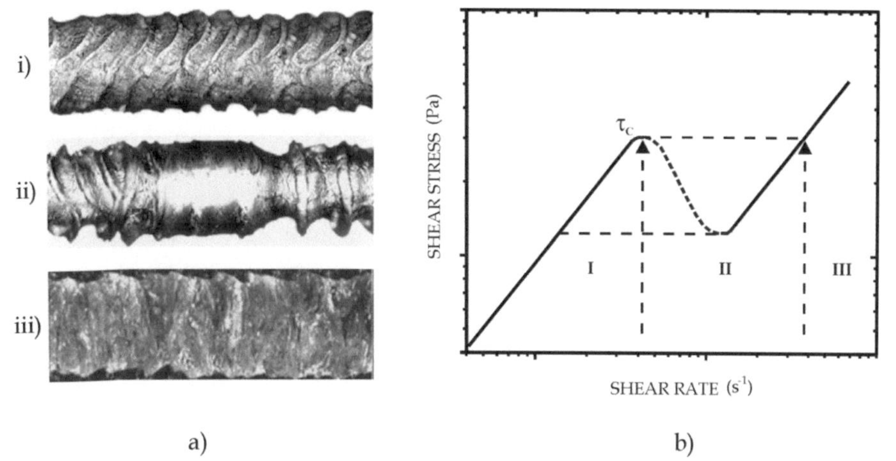

Fig. 3. a) Extrudate distortions: i) Sharkskin, ii) Bamboo, and iii) Gross melt fracture. b) Non-monotonic flow curve.

The assumption of zero relative velocity between a fluid and a solid wall, or "no-slip" condition, has been a matter of debate for a long time. Nowadays, it is widely accepted that such condition may not be satisfied during the flow of entangled polymer melts (see Fig. 2c). The calculation of the slip velocity (v_s), for a given shear stress, has been done by using the Mooney's method (1931), which is a phenomenological correction representing the contribution of slip to the experimentally measured flow rate. According to this, v_s is given by:

$$v_s = \frac{D(\dot{\gamma}_s - \dot{\gamma}_{f-s})}{8}\bigg|_{\tau_w} \tag{9}$$

Where $\dot{\gamma}_s$ and $\dot{\gamma}_{f-s}$ represent the shear rates with slip and free of slip, respectively, for a given τ_w. Following this scheme, v_s and $\dot{\gamma}_{f-s}$ are calculated from the slope and ordinate to the origin, respectively, of a linear plot of $\dot{\gamma}_{app}$ versus $1/D$ for any given shear stress. The dependence of v_s on the shear stress has been modeled by a linear Navier's condition $v_s = \eta b \tau_w$, where b is a parameter with dimension of length that depends on the fluid-solid pair into contact (Denn, 2008). Another model for v_s, more often used in numerical simulations with polymer melts, is a power-law, $v_s = k\tau_w{}^p$, where k and p also depend on the fluid-solid pair.

3. Flow kinematics in polymer melts

Despite the large amount of publications about the flow of polymer melts, the experimental description of their kinematics has received scarce attention. Limitations rely on the adaptation of transparent flow geometries capable to withstand high pressures and high temperatures. Following, a short account of the work done on the analysis of the kinematics of polymer melts by using optical techniques is presented.

Early works on the flow kinematics of polymer melts were mainly focused on visualization by using streak photography (Han, 2007), which mainly provides qualitative information on the characteristics of the flow field. In contrast, velocimetry techniques provide not only qualitative, but also quantitative information about the flow field. Accordingly to Mackley and Moore (1986), the pioneer work in the description of the flow kinematics in polymer melts by using velocimetry techniques goes back to the work by Kramer and Meissner, who reported the measurements of velocity profiles for LDPE flowing through and abrupt contraction in a rectangular duct by using LDV. Mackley and Moore (1986) measured the velocity profiles of HDPE during the steady state flow in a slit die with glass windows by using LDV and reported normalized velocity profiles almost insensitive to temperature and flow rate. Piau et al. (1995) carried out LDV measurements in a polybutadiene flowing through a metal slit channel with glass windows and observed slip, characterized by a nearly plug flow, when the die surface was fluorinated. Münstedt and coworkers have made an intensive use of the LDV technique to analyze the flow of polymer melts. Schmidt et al. (1999) analyzed the flow of LDPE through a metal slit channel with a glass wall. These authors obtained the velocity along the channel and measured the length for fully developed flow. Also, they obtained the velocity profiles in the slit and calculated the viscosity from a velocity profile. Later, Wassner et al. (1999) analyzed the secondary entry flow of LDPE in the same channel by using LDV. In a subsequent paper Münstedt et al. (2000) reported what is probably the first detailed description of the velocity profiles during the slit flow of HDPE under stable and unstable conditions by using LDV. Even though their work was limited to a narrow shear rate range, these authors described the typical characteristics of the velocity profiles in the three regimes of the unstable flow curve (see Fig. 3b). A more detailed description of the stick-slip kinematics of HDPE in a slit die, also by using LDV, was provided by Robert et al. (2004). These researchers reported that slip was not homogeneous across the die, then, from a numerical computation, they suggested that the measured slip velocities were of the same order of magnitude as those measured in a capillary rheometer. Combeaud et al. (2007) analyzed the flow of polystyrene under stable and unstable flow conditions at the entry of a slit channel by using LDV. These authors reported periodic oscillation of the velocity corresponding to instabilities and volume distortions in the extruded materials.

In all this set of reports, which are based on single-point measurements, the measurement region (of micrometric size) was generally located by micrometrical displacements in a direction normal to the flow to cover the whole region of interest. This may be very time consuming and impose limitations for the analysis of fast changing or unsteady flow. On the other hand, slit channels with flat walls have been preferred for measurements, since they do not introduce optical distortions. However, the shear rate in a slit is dependent onto two directions across the channel, so a full description of the flow field requires measurements into two directions, being the slit central plane the proper measurement place to obtain meaningful rheometrical results. In addition, slit channels are typically long and require the use of high pressures to induce flow, which limits the range of shear rates to explore.

Migler et al. (2001) developed a rheo-optics technique to visualize how polymer processing additives (PPA) eliminate sharkskin in a metallocene LLDPE (mLLDPE). These authors measured tracking velocity profiles near the wall and the coating process of PPA in a sapphire capillary, and showed evidence of slip in the coated die and adhesion when this was uncoated. These authors used polydimethysiloxane between the die and an external cube to minimize refractive index variations and to avoid optical distortions. Nigen et al. (2003) studied the entry flow of polybutadiene and its relation with flow instabilities in channel flow by using PIV. Polybutadiene is a low melting point polymer that may be extruded at room temperature, so their analysis does not require special fittings for temperature control. Mitsoulis et al. (2005) analyzed the flow of branched polypropylene in a quartz capillary by visualization combined with laser speckle velocimetry and simulation. These authors showed the formation of vortices in the contraction region and suggested their variation in size with increasing shear rate to be related to the presence of slip in the die. The reported velocity profiles, however, do not exhibit slip. Fournier et al. (2009) characterized the extrusion flow of polycarbonate and polystyrene in a rectangular die with quartz windows by using PIV. The measured velocity profiles in this case, showed good agreement with numerical simulations that considered the influence of slip at the die wall.

The use of capillaries has been generally avoided in studies of the kinematics of polymer melts mainly because of optical distortions introduced by their curvature. However, capillary flow is very often found in rheometry and polymer processing operations. Then, the description of the flow kinematics in such geometry was required. Recently, Rodríguez-González et al. (2009) described for the first time the flow kinematics of HDPE in a glass capillary die by using PIV. These authors corroborated the main results by Münstedt et al. (2000) and Robert et al. (2004) and showed that the velocity profiles cannot become plug-like in the presence of shear thinning in the melt. In a subsequent work (2010) these authors provided clear evidence of alternating behavior between full adhesion and slip in the unstable stick-slip regime of mLLDPE.

4. Fundamentals of PIV

Particle image velocimetry (PIV) has been developed from the early 1980's and can be applied to virtually any kind of flow, as long as the fluid is transparent to enable the imaging the suspended particles. This technique has been developed rapidly over the last three decades and the main findings have been reviewed by Adrian (1991) and Raffel et al. (2007). PIV is a non-intrusive technique commonly used to obtain instantaneous measurements of the velocity vectors in a flow (velocity maps). For two-dimensional PIV, the observation plane is illuminated by a laser light sheet, where two consecutive images of particles seeded in the fluid are obtained. Each image is divided into subsections called interrogation areas, and the statistical displacement of the seeding particles between corresponding areas of the two images gives the displacement vector. The resulting displacement vector is divided by the time elapsed between the two consecutive images, Δt, to obtain the velocity vector. The seeding particles need to be small enough to follow the flow with minimal drag, but sufficiently large to scatter light to obtain a good particle image. The success of a good analysis is greater when the interrogation areas contain about *8-10* particle images.

Almost all algorithms for estimation of the displacement of a group of tracer particles use cross-correlation techniques. One of the most popular methods is the cross-correlation of two frames of singly exposed recordings. This is a robust method that uses the fast Fourier transform (FFT) to evaluate the cross-correlation coefficient in the interrogation windows. The cross-correlation between the two consecutive images is given by:

$$R(x,y) = \sum_{i=-M}^{M} \sum_{j=-N}^{N} f(i,j) g(i+x, j+y) \tag{10}$$

for $x=0,\pm1,\pm2,\ldots\pm(N-1)$ and $y=0,\pm1,\pm2,\ldots\pm(M-1)$. $f(i,j)$ and $g(i,j)$ are the gray level functions at the position (i,j) of the images taken at time t and $t+\Delta t$, respectively. NxM pixels is the size of the interrogation area. By applying this operation for a range of (x,y) shifts a correlation plane of size $(2N-1)X(2M-1)$ is obtained. The location of the first order intensity peaks in this plane is directly proportional to the mean displacement.

There are two ways to compute Eq. 10; the first is by the evaluation of the cross-correlation function directly by using sum of products. In this case the interrogation windows can have different sizes. Although this approach also reduces the sources of errors, it is rarely used due to the computational cost. The second approach to compute Eq. 10 is via the correlation theorem, which states that the correlation of two functions is equal to the inverse Fourier transform of the complex conjugate product of their Fourier transform. The two dimensional FFT transform is efficiently implemented for an image (2D array of values of the gray scale) using the fast Fourier transform, which reduces the computation from $O(N^2)$ to $O(N^2 log_2 N)$, for a NxN image size. When the FFT is used, it is generally assumed that the data are periodic; this means that the image sample repeats itself in x and y directions. Besides the most common implementation of the FFT needs the interrogation areas are squares of the same size; NxN, with N a power of 2. There may be two systematic errors in the cross-correlation calculation. One of them is the aliasing; since the input data are considered to be periodic a correlation peak will be folded back in the correlation plane when the data contains a signal exceeding half of the sample size ($N/2$) and appears on the opposite place. Aliasing may be reduced by either increasing the size of the interrogation areas or by decreasing the laser pulse delay. Another consequence of the periodicity of the correlation data is that the correlation estimation is biased; when the shift magnitude increases less data are actually correlated with each other and their contribution to the actual correlation is limited. Computation speed requirements favor in most cases FFT-based correlation, although the aliasing effect and bias error can be present.

The cross-correlation algorithm may produce several peaks and it is necessary to obtain the position and magnitude of the strongest one. Several algorithms to determine the displacement, from the correlation data at subpixel level, have been proposed by Raffel et al. (2007). Three alternatives are the most common; i) the peak position is approximated by the centroid of the cross-correlation function in the vicinity of the data peak. ii) the digital cross-correlation function, in the vicinity of the peak, is approximated by a known continuous function. The coefficients of the fitting function are found by least squares. The new peak position is taken where the fitting function is a maximum. A parabolic or Gaussian fitting function is used. iii) Finally, the digital auto-correlation around the peak location is computed in a refined grid using an interpolation scheme.

On the other hand, the cross-correlation method only produces an approximation to the velocity field and may be necessary to apply a post-processing to eliminate the wrong vectors (so-called outliers). Detection of either a valid or a spurious displacement depends on the number and spatial distribution of the particle-image pairs inside the interrogation area. In practice, a minimum of four particle-image pairs is required to obtain an unambiguous measurement of the displacement.

4.1 Adaptive second-order accuracy method

Most PIV algorithms use the simple forward difference interrogation (FDI) scheme to calculate the velocity. The velocity at time t is calculated from the particle images recorded at times t and $t+\Delta t$, by using the forward finite difference (Raffel et al., 2007). This approximation is accurate to order Δt and it can be improved by using a central difference interrogation (CDI) scheme, which is accurate to order $(\Delta t)^2$.

Wereley and Meinhart (2001) suggested a PIV technique that performs better than the conventional PIV. The main part of this technique is a central difference approximation of the flow velocity. An adaptive interrogation region-shifting algorithm is used to implement the central difference approximation. Adaptive shifting algorithms also have the advantage of helping to eliminate the velocity bias error. The adaptive central difference interrogation technique has the following advantages over the forward difference approximation with or without adaptation. This technique performs better near flow boundaries or in the presence of velocity gradients because the spatial shift between interrogation areas is based on the local and not on the global velocity. The adaptive method automatically reduces the spatial shift between interrogation areas in regions where the particle displacement is small. In addition, the adaptive CDI technique is more accurate, especially at large time delays between camera exposures and it provides a temporally symmetric view of the flow.

4.2 Data validation

The post-processing of the PIV data is an extensive area and the procedures applied to the vector field depend on the application. At least two stages are necessary for any application; validation and replacement of the incorrect data. When the number of outliers is small their validation can be done by visual inspection and the wrong velocities may be deleted in an interactive way. The spurious vectors are visually recognized by their deviation with respect to the neighbor vectors. Such analysis becomes prohibitive when a great number of recordings have to be evaluated. Then, several automatic procedures have been proposed to validate the raw data. An algorithm that can reject most of the outliers due to noise in the cross-correlation is based on the assumption that for real flow fields the vector difference between the neighboring velocity vectors is small. Typically, a $3x3$ neighborhood is used and eight neighbors are used. The vector velocity is rejected if the magnitude of the difference between the vector and the average over its neighbors is greater than certain threshold. Raffel et al. (2007) used this validation procedure to detect automatically the invalid vectors from the velocity map above an airfoil. Other data validation techniques are found in the literature, however, no general method can be offered for the problem of data validation in PIV. After removing the outliers, it is necessary to replace the missing data. This may be done, for instance, by using bilinear (or bicubic) interpolation from the valid neighbor vectors.

4.3 Velocity bias

Several authors have proposed methods to reduce or to eliminate the velocity bias, which is produced by using equally-sized interrogation areas in the first and the second images. Westerwell (1994) showed that the error bias can be completely eliminated by dividing the image correlation function by the areas correlation function. Keane and Adrian (1992) suggested reduction of the bias by doing the second interrogation area larger than the first, so all particle images in the first interrogation area will likely be contained in the second. They also proposed that spatially-shifting the second interrogation area by an integer part of the displacement will substantially reduce the bias error. Haung et al. (1997) proposed an efficient algorithm to eliminate velocity bias by renormalizing the values of the correlation function in the neighborhood of the peak location prior to calculation of the subpixel peak location. These techniques improve, in general, the PIV results. However, the improvement is limited near to flow boundaries.

To reduce the errors of the displacement estimation, iterative methods have been developed which use a shift of the interrogation areas where the offset can be an integer number of pixels or a fraction (sub-pixel accuracy). If the displacement of the particles of the first interrogation area, between t and $t+\Delta t$, can be estimated, then the interrogation area from the second image can be matched via a relative offset. Scarano and Riethmuller (1999) proposed a method to obtain the displacement and optimize it within an iterative process. Since the components of the predicted displacement, $\Delta s=(\Delta x,\Delta y)$ are an integer number of pixels, their method does not require to use interpolation.

5. Experimental methods

5.1 Materials and rheometry

Molten polyolefins were analyzed in this work, namely, LDPE, PP, HDPE, and mLLDPE, respectively. All the polyethylenes are reported as free of additives that might screen interactions at the die wall and PP is an industrial grade polymer. Their main characteristics are given in Table 1.

Polymer	Melt Index (*g/10min*)	M_w (*g/mol*)	ρ (*g/cm³*)	T_m (°C)	T_{Exp} (°C)	Supplier
LDPE	1.5	---	0.922	115	190	Aldrich
PP	1.8	---	0.9	151	200	Basell
HDPE	0.25	125000	0.950	130	180	Aldrich
mLLDPE	0.8	93200	0.880	60	190	Aldrich

Table 1. Polymer characteristics and experimental conditions

A fluoropolymer polymer processing additive (FPPA, Dynamar™ FX-9613) at a concentration of *0.1 wt.%* was used to produce slip at the interface between the LLDPE melt and the capillary wall. Experiments were carried out at the temperatures given in Table 1 under continuous extrusion with a Brabender single screw extruder of *0.019* m in diameter and length to diameter ratio of *25/1*. The pressure drop (Δp) between capillary ends was measured with a Dynisco™ pressure transducer, whose voltage signal was sent to an independent computer, via an USB data acquisition board, in order to follow the pressure evolution while evaluating the flow kinematics. Pressure data were acquired at a rate of *100* points/s. The volumetric flow rate (Q) was determined by collecting and measuring the ejected mass as a function of time.

Measurements were performed with a capillary die ($D = 0.0017\ m$) made up of borosilicate glass, with an entry angle of *180º* and $L/D=20$; no corrections for end effects were performed. This type of glass capillaries has more than *90%* of light transmissibility, high resistance to wear and high dimensional stability under several processing conditions.

A fixture made up of stainless steel was adapted to the extruder die head in order to support a capillary die as shown in Fig. 4a. The fixture has two pairs of perpendicular windows, one pair was used to pass the laser light sheet through the flow region and the other for visualization. Image distortions produced by the curved geometry of the capillary were eliminated by using an aberration corrector, which was made up of a small rectangular prism with glass walls containing a fluid as shown in Fig. 4b. For this purpose, the refractive index (n) of the borosilicate glass capillary ($n_{glass}=1.43$) was closely matched by filling the prism with glycerol ($n_{gly}=1.47$). In order to perform the PIV measurements, one part of the capillary corresponding to a length of *15D* was kept inside the extruder die head at controlled temperature, and the other part of the capillary was inside the aberration corrector, in which the temperature was continuously monitored and supplied with pre-heated glycerol.

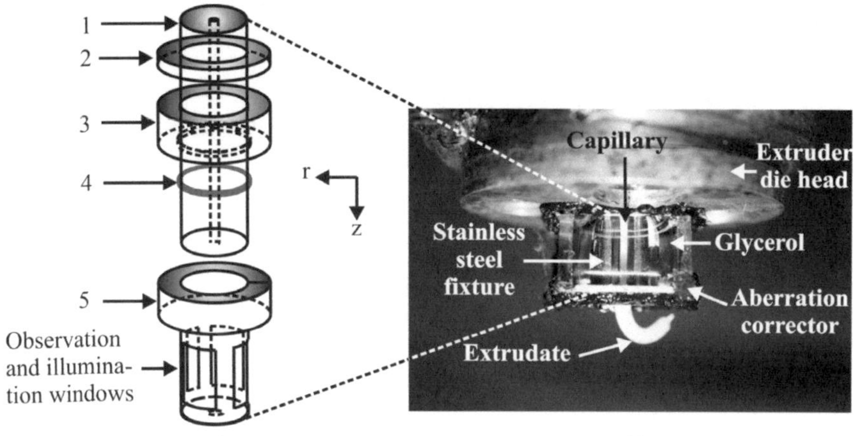

Fig. 4. a) Stainless steel fixture to support glass capillary in the extruder die head: 1) borosilicate glass die, 2) cooper ring, 3) stainless steel ring, 4) high-temperature o-ring, 5) stainless steel fixture. b) Aberration corrector made up of a small rectangular prism with glass wall filled with glycerol.

5.2 PIV measurements

The study of the flow kinematics in the capillary was performed with a two dimensional (2D) PIV Dantec Dynamics system as shown in Fig. 5. The PIV system consists of a high speed and high sensitivity HiSense MKII CCD camera of *1.35 Mega-pixels*, two coupled Nd:YAG lasers of *50 mJ* with $\lambda = 532\ nm$ and the Dantec Dynamic Studio 2.1 software. The light sheet was reduced in thickness up to less than *200 μm* by using a biconvex lens with *0.05 m* of focal distance, and then sent through the center plane of the capillary by using a prism oriented at *45°* relative to the original direction of the laser beam. The prism was mounted on a rail carrier and the center plane of the capillary was found by horizontal displacements of the prism up to see the longest chord on the image plane. The particles used were solid copper spheres < *10 μm* in diameter (Aldrich 32,6453) at a concentration of *0.5 wt.%*. This amount of particles is not expected to affect the rheological behavior of the polymer. Using the Einstein relation for spherical particles in a fluid, the increase in the viscosity due to the presence of copper particles was calculated to be less than *0.1%*, which is negligible.

Fig. 5. PIV set up: 1) Single screw extruder, 2) Extruder die head with the stainless steel fixture and the aberration corrector as shown in Fig. 6, 3) Nd:YAG lasers, 4) high-speed and high -sensitivity CCD camera with a continuously-focusable video microscope, 5) biconvex lens, 6) prism, 7) pressure transducer.

An InfiniVar™ continuously-focusable video microscope CFM-2/S was attached to the CCD camera in order to increase the spatial resolution. The depth of field of the microscope at the position of observation was measured as *40 μm*. A small depth of field reduces the true observation volume and the error introduced by out of plane particles. This fact is particularly important in experiments in which the shear stress is not homogeneous, since the wider the light sheet, the bigger the variation of the shear stress in the observation volume. Thus, the variation of the shear stress in the region associated with the depth of field of the microscope was only *5%* of τ_w.

The images taken by the PIV system covered an area of *0.00171 m x 0.0032 m* and were centered at an axial position *z=17D* downstream from the contraction. Series of fifty image pairs were obtained for each flow condition. For stable conditions, the frequency was *6.1 Hz* and all the image pairs were interrogated and ensemble averaged to obtain a single velocity map. This was not made for unstable regimes where the velocity oscillates with time. In such a case, two frequencies were used, *2.0* and *6.1 Hz*, for slow and fast changes, respectively. The axial velocity component as a function of the radial position (velocity profile) was obtained by averaging the profiles in a map.

According to discussion in subsection 4.1 an adaptive correlation algorithm with a central difference approximation was used to calculate the velocity vectors. This technique has been proved to be more accurate than conventional PIV algorithms when measuring near flow boundaries or in the presence of velocity gradients (Wereley & Meinhart, 2001), which is the case during the evaluation of velocity vectors in the neighborhood of the capillary wall. Since fully developed flow is unidirectional, each interrogation area was chosen as a long rectangle in the flow direction of *256 pixels* long and *16 pixels* wide (*642 μm x 40 μm*, radial and axial direction, respectively), with an overlap of *50%* in both axis. With this width of the interrogation area and the particle size used for seeding, the closest distance to the capillary wall at which measures were made was *40 μm*. Further approach to the wall would require the seeding with smaller particles. Finally, data validation was performed by using a moving average filter.

6. Results and discussion

6.1 Analysis of stable flow conditions

6.1.1 Low-density polyethylene

The rheometrical flow curve for the LDPE is shown in Fig. 6a, along with the one obtained from the integration of the velocity profiles according to Eq. 7. Note that both curves are well fitted by the power-law relationship in the apparent shear rate range studied (see the equation inserted in Fig. 6a). Validation of the PIV measurements is performed via its comparison with rheometrical data. In this case, the data obtained from the velocity profiles agree well with the rheometrical ones; the maximum difference in the volumetric flow rates obtained by using the two methods was *6.5%* at most (the average difference was around *3.4%*), which shows the reliability of the PIV technique to describe the behavior of the polymer melt in capillary flow. The origins of the differences between the two methods are likely found in the location of the laser light sheet with respect to the real central plane of the capillary and its thickness, as well as in the uncertainty of the flow rate measurements.

Figure 6b shows the velocity maps in the capillary for different flow conditions. It is clear that almost all vectors in each map are parallel to the flow direction, which shows that the flow in the observation region was unidirectional for the different apparent shear rates studied, with a velocity field simply given by $v_z=v_z(r)$, as it is expected for a fully developed shear flow.

The PIV velocity profiles for different apparent shear rates are shown in Figs. 7a-b along with the profiles calculated with Eq. 7. Observe that the velocity profiles in the capillary are symmetric with respect to the flow direction and that they are very well matched by Eq. 7. Also, the standard deviation of the time average of fifty profiles, which is represented by the error bars, shows variations below *5%*, which indicates that the flow was steady. It is interesting to note that all the velocity profiles in Figs. 7a-b extrapolate to zero value at the capillary wall, indicating the absence of slip. This result agrees with the well known fact that branched polyethylene melts do not exhibit slip.

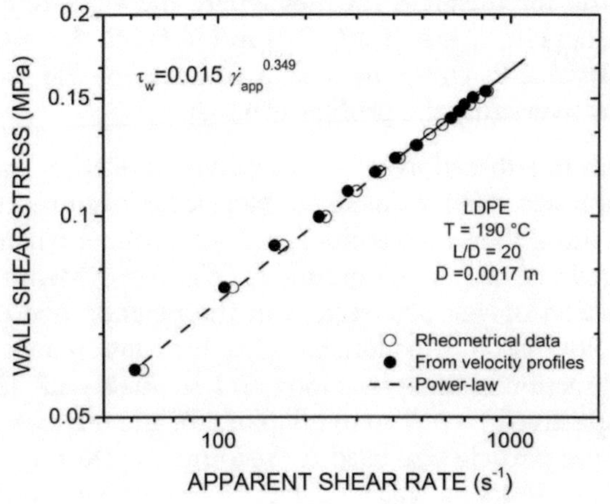

a)

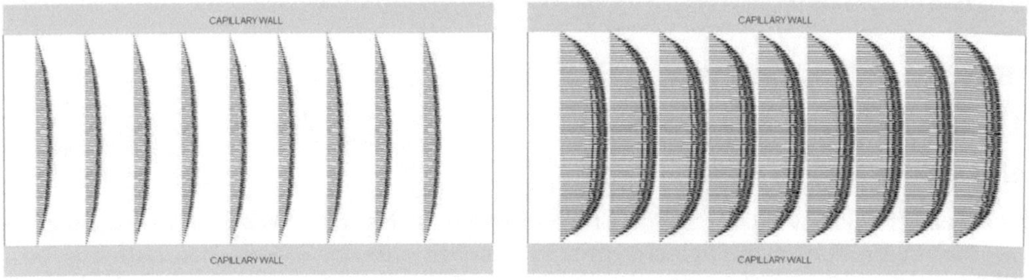

54.78 1/s 834.31 1/s

b)

Fig. 6. a) Flow curve for LDPE obtained from rheometrical measurements and by the integration of the velocity profiles. b) Velocity maps in the capillary for different flow conditions.

6.1.1.1 Determination of the flow and viscosity curves from velocity profiles

Due to the radial distribution of the shear stress in capillaries (Eq. 1), i.e., $\tau=\tau(r)$, the true flow curve for a polymer may be obtained from the velocity profiles if a determination of the true shear rate as a function of the radial position is performed. There is a range of shear rates and stresses in a capillary for a given flow rate, namely, $0\leq \dot{\gamma}_r \leq \dot{\gamma}_R$ and $0\leq\tau_r\leq\tau_w$, which enables one to obtain the flow curve from the velocity profiles and the measured wall shear stress. The local shear rate may be calculated from the numerical derivative of the velocity profiles with respect to the radial position. A central difference approximation was used in this work. Meanwhile, the corresponding shear stress was calculated from the measured τ_w by using Eq. 1.

The flow curve calculated by using the velocity profiles corresponding to *54.78, 112.31, 167.1, 359.79* and *834.31* s^{-1}, respectively, is displayed in Fig. 8a along with that obtained by applying the Rabinowitsch's correction to the rheometrical data. The segments of the flow curve reconstructed from the velocity profiles are represented with different symbols for

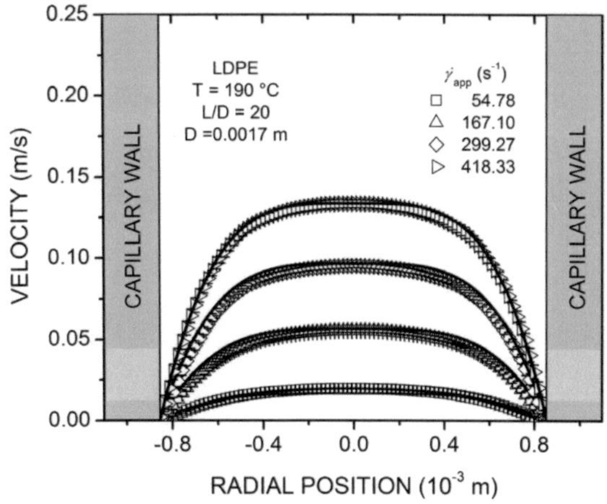

a)

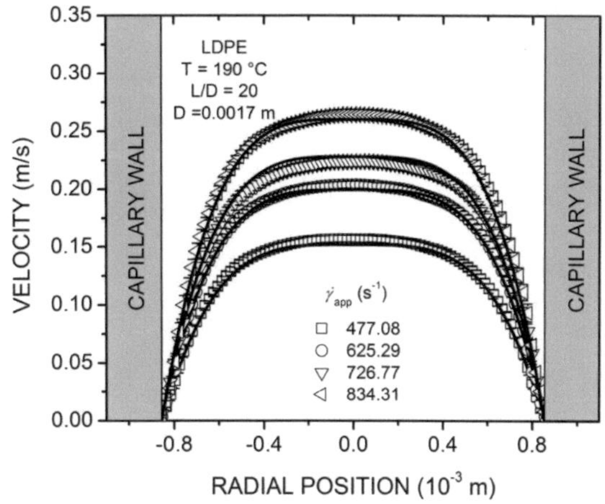

b)

Fig. 7. Velocity profiles for different apparent shear rates for LDPE at a) low and b) high shear rate regimes. Continuous lines represent the power-law model (Eq. 7).

clarity. It is clear that the data obtained from the velocity profiles follow the trend of the rheometrical ones and extend well into the transition between the Newtonian and power-law regions, which is clearly observed in the shear viscosity curve shown in Fig. 8b. The PIV data extend to lower shear rate values than the rheometrical $\dot{\gamma}_w$. This is a valuable fact, since allows the analysis of the low shear rate region that is not accessible by using the macroscopic physical quantities provided by a single capillary. Instead, a rotational rheometer or a capillary of bigger diameter must be used to study the low shear rate behavior.

The wider shear rate range covered by PIV data in Figs. 8a-b permits the fitting of the viscous behavior of the LDPE, $\eta = \eta(\dot{\gamma})$, by a more realistic constitutive equation including the response at low shear rates, as for example the Carreau´s model (Bird et al., 1977).

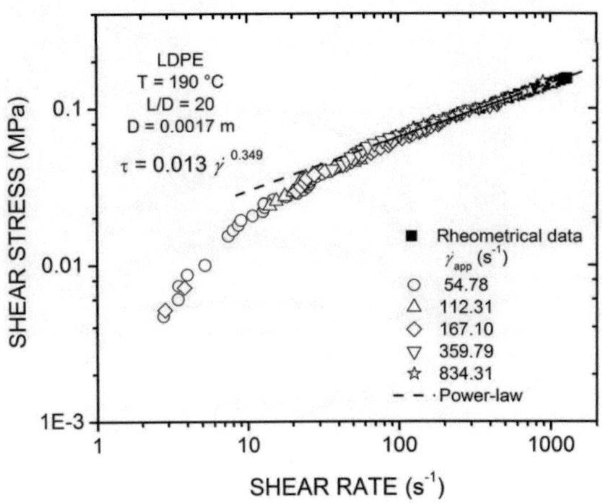

a)

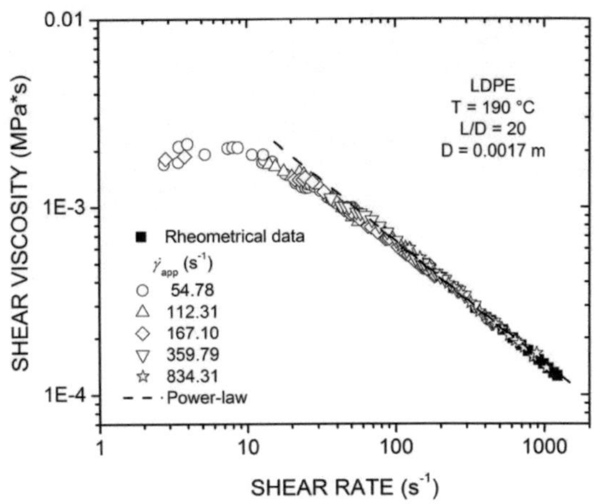

b)

Fig. 8. a) True flow curve and b) viscosity curve for LDPE obtained from local shear rate data. Dashed and continuous lines indicate the power-law and Carreau's model fit, respectively.

6.1.2 Polypropylene

PP shows a similar behavior to the observed for LDPE. The flow curve in Fig. 9a may also be well fitted by a power-law at shear stresses prior to the onset of gross melt fracture, as well as the velocity profiles in Fig 9b. Akin to the LDPE, all the velocity maps (not shown here) agree with a fully developed flow, meanwhile the profiles appear symmetric and extrapolate to zero value at the capillary wall, indicating the absence of slip.

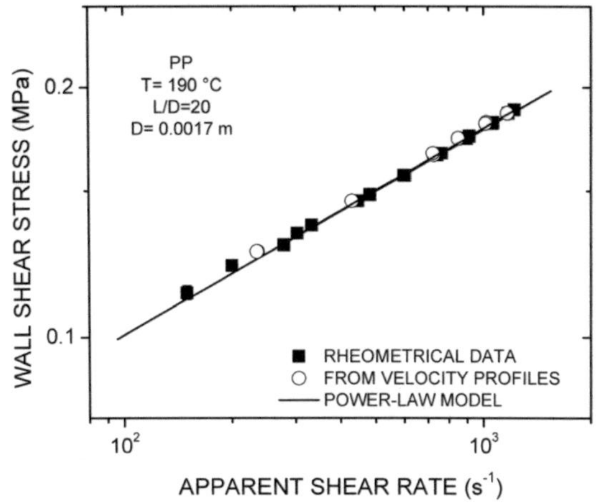

a)

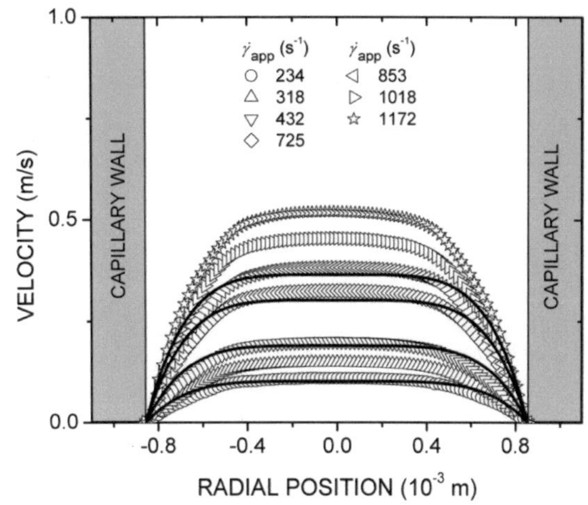

b)

Fig. 9. a) Flow curve for PP obtained from rheometrical measurements and by the integration of the velocity profiles. b) Velocity profiles in the capillary for different flow conditions.

6.2 Analysis of unstable flow conditions

6.2.1 High-density polyethylene

The feature of the PIV technique to produce instantaneous velocity maps, allows for the detection of rapid variations in a region of the flow field. Thus, the PIV technique permits, for example, an accurate description of the flow kinematics under unstable flow conditions occurring in some polymer processing operations. This is particularly the case for HDPE and LLDPE, which are known to exhibit extrusion instabilities like those described in section 2.2 (see Fig. 3a).

The flow curve for HDPE shown in Fig. 10a displays the typical three regions (I-III) of a non-monotonic flow curve (see Fig. 3b). Region I corresponds to a stable flow regime characterized by a power-law behavior. The stick-slip instability appears in region II; vertical bars represent the amplitude of the pressure oscillations. Region III begins at an apparent shear rate of *652* s^{-1} and is characterized by the presence of lower amplitude and much faster pressure oscillations than those in region II, which might be related to the onset of the gross melt fracture regime.

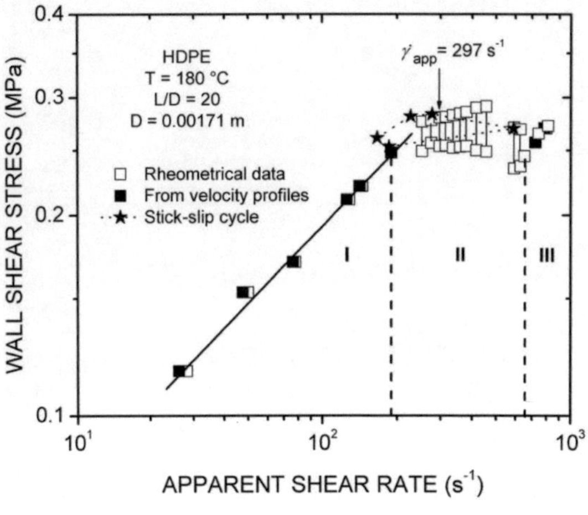

a)

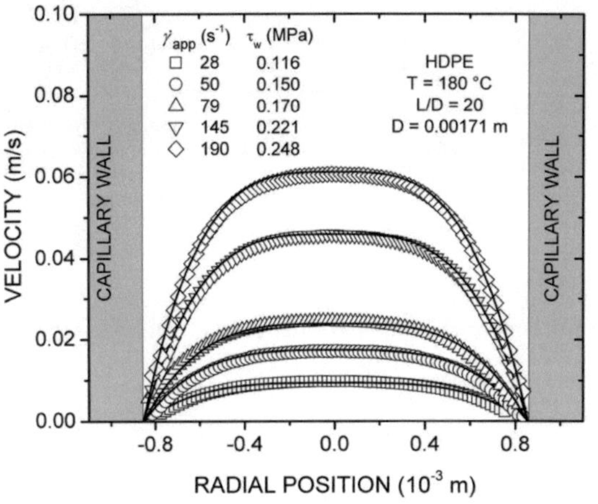

b)

Fig. 10. a) Flow curve for HDPE obtained from rheometrical measurements and by the integration of the velocity profiles. b) Velocity profiles in the capillary for different flow conditions. Reprinted with permission from Rodríguez-González et al., *Chemical Engineering Science*, Vol. 64, No. 22, (November 2009), pp. 4675-4683, ISSN 0003-2509. Copyright Elsevier.

The velocity profiles for the different apparent shear rates in region I are shown in Fig. 10b along with the profiles calculated by using Eq. 7. Also in this case, the flow rate data obtained from the integration of the velocity profiles are included in the flow curve in Fig.

10b. Again, the velocity profiles are very well described by the power-law relationship and there is a good agreement between the rheometrical and PIV data in region I. The velocity profiles in this flow regime do not show slip, even though it is known that HDPE may slip at the die wall. This may be explained by a relatively low molecular weight of the polymer. The magnitude of slip depends on the polymer molecular weight. The higher the molecular weight, the larger the slip velocity.

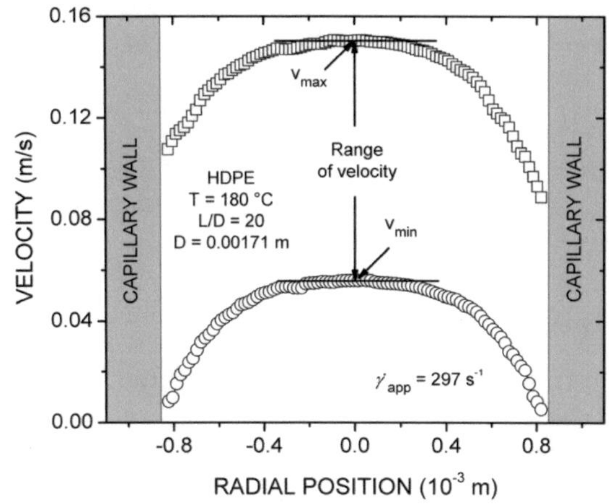

a)

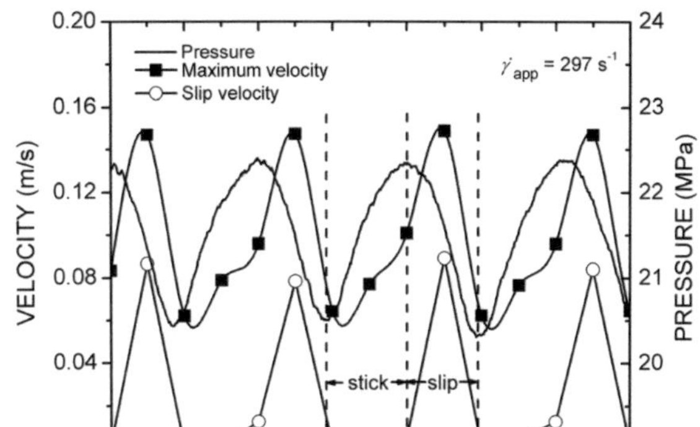

b)

Fig. 11. a) Minimum and maximum velocity profiles during one pressure oscillation at an apparent shear rate of 297 s^{-1}. b) Evolution of pressure, maximum velocity and slip velocity for the same apparent shear rate. Reprinted with permission from Rodríguez-González et al., *Chemical Engineering Science*, Vol. 64, No. 22, (November 2009), pp. 4675-4683, ISSN 0003-2509. Copyright Elsevier.

The pressure signal and the flow maps were recorded simultaneously in the stick-slip regime. The minimum and maximum velocity profiles during one pressure oscillation at an apparent shear rate of *297* s^{-1} are shown in Fig. 11a. There is a large difference between the maxima of the two profiles, v_{min} and v_{max}, and their values during one oscillation change from v_{min}=*0.055 m/s* up to v_{max}=*0.151 m/s*. Similar variations in velocity are typical of other apparent shear rates in the stick-slip regime. In addition, the velocity profiles in Fig. 11a show that the boundary conditions change from stick to slip at the capillary wall, in agreement with the term, stick-slip, used to describe this oscillating phenomenon.

Figure 11b shows the evolution of pressure together with the maximum velocity and slip velocity an apparent shear rate of *297* s^{-1}. Note that both, the maximum velocity and the slip velocity, increase continuously during the stick part of the oscillations. In contrast, at the beginning of the slip part of the cycle both velocities rise steeply, reach a maximum and then decrease abruptly up to a minimum at the end of the cycle.

In the high shear rate branch, pressure oscillations decreased in amplitude and became much faster than those in region II. The high frequency of the oscillations in this case did not permit the recording of consecutive velocity maps to describe one full cycle with good resolution (see Rodríguez-González et al, 2009).

6.2.2 Linear low-density polyethylene

The flow curves for the mLLDPE with and without additive are shown in Fig. 12. The flow curve corresponding to the pure polymer resembles that of Fig. 10a; only regions and I-II were explored. Region I corresponds to a stable flow regime that cannot be well fitted by a simple power-law. Instead, a more complex constitutive equation seems to describe the melt behavior. The stick-slip instability appears in region II, the amplitude of pressure oscillations are represented by double points.

Figure 12a also includes the flow curve obtained for the mLLDPE containing additive. In this case, an increase in the flow rate with respect to the pure polymer is evident, as well as the absence of the stick-slip regime. This increase results from interfacial slip between the polymer melt and FPPA. It is noteworthy here that the flow curve for the mLLDPE containing additive also deviates from a power-law and should be fitted by a more complex constitutive equation (Rodríguez-González et al., 2010).

v_s was calculated for the mLLDPE+FPPA system by comparing the data from both flow curves in Fig. 12a using Eq. 9, To use this equation, only stable data from the pure polymer before the stick-slip were considered, assuming them as free of slip. The slip velocity as a function of the shear stress is shown in Fig. 12b. The slip velocity increases with the shear stress and follows a power-law behavior at low shear stresses, but the trend changes as the shear stress is further increased. This change of the rate of increase of the slip velocity along with the shear stress has been attributed to shear thinning of the melt (Pérez-González & de Vargas, 2002).

Figure 13 shows the velocity profiles for the different apparent shear rates in region I. According to the discussion in the previous section, a single power-law relationship was not appropriate to describe all of these profiles. Note that the velocity profiles in this flow regime extrapolate to a zero value at the die wall (no slip), which supports the assumption made in using Eq. 9.

The main results for this mLLDPE agree with those described for HDPE in the previous section. However, an issue to highlight in the analysis of the stick-slip flow of this polymer is the appearance of non-homogeneous slip (Fig. 14a), i.e., the simultaneous appearance of

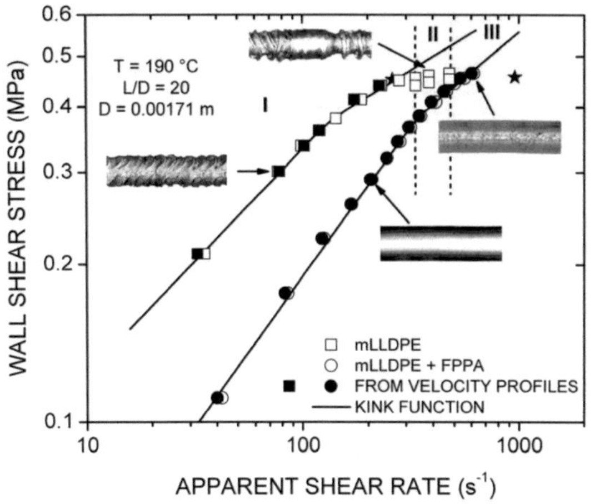

a)

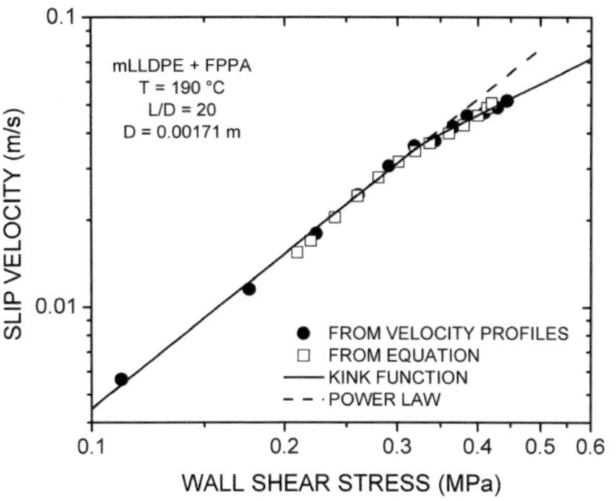

b)

Fig. 12. a) Flow curves for the mLLDPE with and without FPPA. b) Slip velocity as a function of wall shear stress calculated by using Eq. 9 and from velocity profiles. Continuous lines represent kink functions (Shaw, 2007). Reprinted with permission from Rodríguez-González et al. (2010). *Rheologica Acta*, Vol. 49, No. 2, (February 2010), pp. 145-154, ISSN 0035-4511. Copyright Springer-Verlag.

regions with and without slip at the die wall. This new characteristic of the stick-slip flow was recently discovered by Rodríguez-González et al. (2009) in HDPE and it is nicely visualized in the photograph of the extrudate in Fig. 14b. It is noteworthy that these peculiar details of the stick-slip flow may only be detected by instantaneous measurements of the flow field. The detection of this feature of the stick-slip flow is not possible by one point measurements, like in LDV in previous works, but it is allowed by the instantaneous recording of the PIV maps. In fact, the existence of slip during the oscillations can not be assured by LVD, unless a full velocity profile across the plane of interest is constructed.

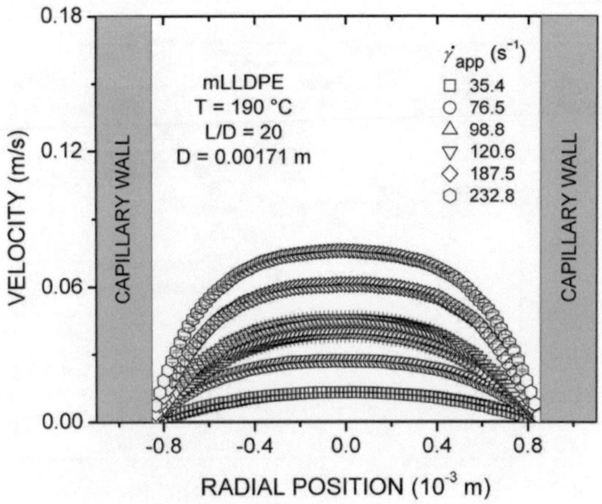

Fig. 13. Velocity profiles for the different apparent shear rates in region I of the mLLDPE flow curve. Reprinted with permission from Rodríguez-González et al. (2010). *Rheologica Acta*, Vol. 49, No. 2, (February 2010), pp. 145-154, ISSN 0035-4511. Copyright Springer-Verlag.

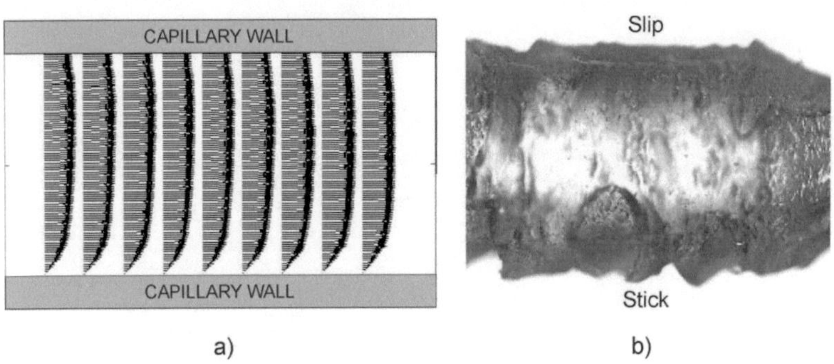

Fig. 14. Non-homogeneous slip characterized by a) velocity map and b) extrudate. Reprinted with permission from Rodríguez-González et al. (2010). *Rheologica Acta*, Vol. 49, No. 2, (February 2010), pp. 145-154, ISSN 0035-4511. Copyright Springer-Verlag.

Figures 15a-b show the velocity profiles obtained for the mLLDPE+FPPA. In contrast to the observed for the pure polymer, the velocity profiles in Figs. 15a-b exhibit a non-zero velocity at the die wall, whose magnitude increases along with the apparent shear rate (shear stress). The slip velocity values obtained from the extrapolation of the velocity profiles at the die wall are plotted in Fig. 12b along with those calculated from the rheometrical data by using Eq. 9. The agreement between both sets of data is remarkable, which provides a direct proof for the validity of Eq. 9.

The relationship between the slip velocity and wall shear stress in Fig. 12b clearly deviates from the power-law behavior at a shear stress above *0.30 MPa*. Then, a more realistic equation, that could be used in numerical calculations, may be obtained by fitting the data to a continuous "kink" function (Shaw, 2007):

$$\log v_s = -1.4276 + 1.7716(\log \tau_w - 5.5202) + 0.0078(1.11 - 1.7716)\ln\left\{1 + \exp\left[\frac{\log \tau_w - 5.5202}{0.0078}\right]\right\} \quad (11)$$

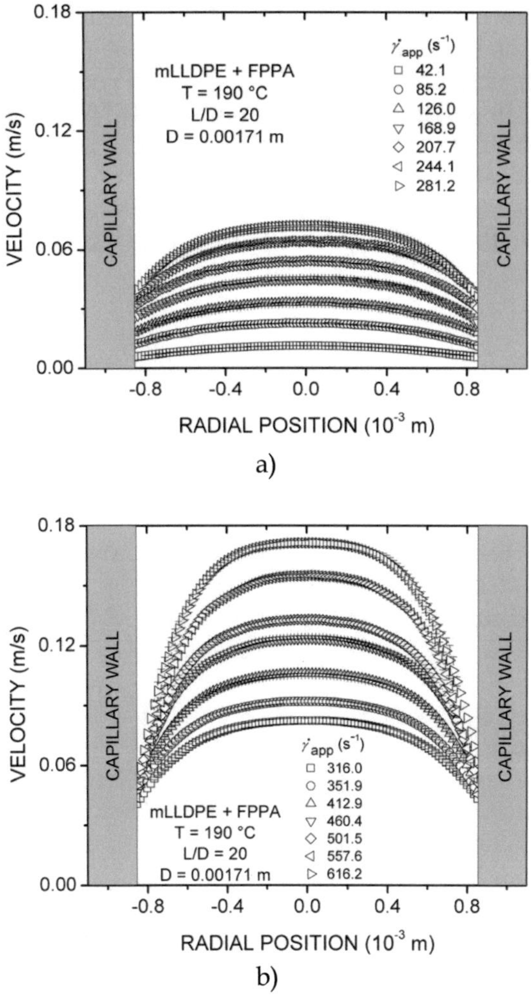

Fig. 15. Velocity profiles obtained for the mLLDPE under strong slip conditions at a) low and b) high shear rates. Reprinted with permission from Rodríguez-González et al. (2010). *Rheologica Acta*, Vol. 49, No. 2, (February 2010), pp. 145-154, ISSN 0035-4511. Copyright Springer-Verlag.

A comparison of the slip velocity calculated by using the kink function (Eq. 11) and a power-law (see Fig. 12b) at τ_w= *0.443 MPa* leads to an overestimation of *23%* in v_s when using the power-law model. Considering the trend in Fig. 12b for the slip velocity, the error introduced by using a power-law model for this polymer will become even more significant at higher shear stresses.

7. Conclusion

The extrusion of polyolefins of significant practical importance, namely, LDPE, HDPE, PP and LLDPE, was analyzed in this work by using simultaneous rheometrical and two-dimensional particle image velocimetry measurements (2D PIV), or what has been called

Rheo-PIV. The results in this work show that PIV is a reliable tool to describe the flow kinematics of this sort of fluids under stable and unstable flow conditions, including the flow with slip at the die wall. Moreover, the use of the PIV technique enables the detection of characteristics of the flow that are not distinguished by using other optical velocimetry techniques like PTV or LDV. The main limitation in the use of PIV for the analysis of polymer processing operation is the need to adapt transparent molds and dies able to withstand high temperatures and pressures. In spite of this, in general, the use of the PIV has a promising future in the study of polymer melts and other complex fluids.

8. Acknowledgment

This work was supported by SIP-IPN (20111119). J. P.-G. and J. G. G.-S. are COFFA-EDI fellows. B. M. M.-S. had a fellowship under the CONACyT's program for the *Apoyos Complementarios para la Consolidación Institucional de Grupos de Investigación* (Sol. 147970).

9. References

Adrian, R. J. (1991). Particle-imaging techniques for experimental fluid mechanics, *Annual Review of Fluid Mechanics*, Vol. 23, (January 1991), pp. 261-304, ISSN 0066-4189.

Bagley, E. B. (1957). End corrections in the capillary flow of polyethylene, *Journal of Applied Physics*, Vol. 28, No. 5, (May 1957), pp. 624-628, ISSN 0021-8979.

Bird, R. B.; Armstrong, R. C. & Hassager, O. (1977). *Dynamics of polymeric liquids, Fluid Mechanics*, Vol. 1, John Wiley & Sons, ISBN 0-471-07375-X, United Sates of America.

Combeaud, C.; Vergnes, B.; Merten, A. & Hertel, D. (2007). Volume defects during extrusion of polystyrene investigated by flow induced birefringence and laser-Doppler velocimetry, *Journal of Non-Newtonian Fluid Mechanics*, Vol. 145, No. 2-3, (September 2007), pp. 69-77, ISSN 0377-0257.

Denn, M. M. (1990). Issues in Viscoelastic Fluid Mechanics, *Annual Review of Fluid Mechanics*, Vol. 22, (January 1990), pp. 13-32, ISSN: 0066-4189.

Denn, M. M. (2001). Extrusion instabilities and wall slip, *Annual Review of Fluid Mechanics*, Vol. 33, (January 2001), pp. 265-287, ISSN 0066-4189.

Denn, M. M. (2008). *Polymer melt processing*. (1st Edition). Cambridge University Press, ISBN 978-0-521-089969-7, United States of America.

de Vargas, L.; Pérez-González, J. & Romero-Barenque, J. (1995). Evaluation of end effects in capillary rheometers for solutions of flexible polymers, *Journal of Rheology*, Vol. 39, No. 1, (January 1995), pp. 125-137, ISSN 0148-6055.

Fournier, J. E.; Lacrampe, M. F. & Krawczak, P. (2009). Characterization of extrusion flow using particle image velocimetry, *eXPRESS Polymer Letters*, Vol. 3, No. 9, (September 2009), pp. 569-578, ISSN 1788-618X.

Han, C. D. (2007). *Rheology and processing of polymeric materials*. Vol. II. (1st Edition). Oxford University Press, ISBN 978-0-19-518783-0, United States of America

Huang, H.; Dabiri, D. & Gharib, M. (1997). On errors of digital particle image velocimetry, *Measurement Science Technology*, Vol. 8, No. 12, (December 1997), pp. 1427-1440, ISSN 0957-0233.

Keane, R. D. & Adrian, R. J. (1992). Theory of cross-correlation analysis of PIV images, *Applied Scientific Research*, Vol. 49, No. 3, (July 1992), pp. 191-215, ISSN 0003-6994.

Mackley, M. R. & Moore, I. P. T. (1986). Experimental velocity distribution measurements of high density polyethylene flowing into and within a slit, *Journal of Non-Newtonian Fluid Mechanics*, Vol. 21, No. 3, pp. 337-358, ISSN 0377-0257.

Migler, K. B.; Lavallée, C.; Dillon, M. P.; Woods, S. S. & Gettinger, C. L. (2001). Visualizing the elimination of sharkskin through fluoropolymer additives: Coating and polymer–polymer slippage, *Journal of Rheology*, Vol. 45, No. 2, (March 2001), pp. 565-581, ISSN 0148-6055.

Mitsoulis, E.; Kazatchkov, I. B. & Hatzikiriakos, S. G. (2005). The effect of slip in the flow of a branched PP melt: experiments and simulations, *Rheologica Acta*, Vol. 44, No. 4, (April 2005), pp. 418-426, ISSN 0035-4511.

Mooney, M. (1931). Explicit formulas for slip and fluidity, *Journal of Rheology*, Vol. 2, No. 2, (April 1931), pp. 210-222, ISSN 0148-6055.

Münstedt, H.; Schmidt, M. & Wassner, E. (2000). Stick and slip phenomena during extrusion of polyethylene melts as investigated by laser-Doppler velocimetry, *Journal of Rheology*, Vol. 44, No. 2, (March 2000), pp. 413-427, ISSN 0148-6055.

Nigen, S.; El Kissi, N.; Piau, J. M. & Sadun, S. (2003). Velocity field for polymer melts extrusion using particle image velocimetry: Stable and unstable flow regimes, *Journal of Non-Newtonian Fluid Mechanics*, Vol. 112, No. 2-3, (June 2003), pp. 177-202, ISSN 0377-0257.

Petrie, C. J. S. & Denn, M. M. (1976). Instabilities in polymer processing, *A.I.Ch.E. Journal*, Vol. 22, No. 2, (March 1976), pp. 209-236, ISSN 1547-5905.

Pérez-González, J. & de Vargas, L. (2002). Quantification of the slip phenomenon and the effect of shear thinning in the capillary flow of linear polyethylenes, *Polymer Engineering and Science*, Vol. 42, No. 6, (June 2002), pp. 1231-1237, ISSN 1548-2634.

Piau, J. M.; Kissi, N. & Mezghani, A. (1995). Slip flow of polybutadiene through fluorinated dies, *Journal of Non-Newtonian Fluid Mechanics*, Vol. 59, No. 1, (August 1995), pp. 11-30, ISSN 0377-0257.

Raffel, M.; Willer, C. E.; Wereley, S. T. & Kompehhans, J. (2007). *Particle Image Velocimetry. A Practical Guide*. (2nd edition), Springer-Verlag Berlin Hidelberg, ISBN 978-3-540-72307-3, Germany.

Robert, L.; Demay, L. & Vergnes, B. (2004). Stick-slip flow of high density polyethylene in a transparent slit die investigated by laser Doppler velocimetry, *Rheologica Acta*, Vol. 43, No. 1, (February 2004), pp. 89-98, ISSN 0035-4511.

Rodríguez-González, F.; Pérez-González, J.; de Vargas, L. & Marín-Santibáñez, B. M. (2010). Rheo-PIV analysis of the slip flow of a metallocene linear low-density polyethylene melt, *Rheologica Acta*, Vol. 49, No. 2, (February 2010), pp. 145-154, ISSN 0035-4511.

Rodríguez-González, F.; Pérez-González, J.; Marín-Santibáñez, B. M. & de Vargas, L. (2009). Kinematics of the stick–slip capillary flow of high-density polyethylene, *Chemical Engineering Science*, Vol. 64, No. 22, (November 2009), pp. 4675-4683, ISSN 0003-2509.

Scarano, F. & Riethmuller, M. L. (1999). Iterative multigrid approach in PIV image processing wiyh discrete window offset, *Experiments in Fluids*, Vol. 26, No. 6, (May 1999), pp. 513-523, ISSN 0723-4864.

Schmidt, M.; Wassner, E. & Münstedt, H. (1999). Setup and Test of a Laser Doppler Velocimeter for Investigations of Flow Behaviour of Polymer Melts. *Mechanics of*

Time-Dependent Materials, Vol. 3, No. 4, (December 1999), pp. 371-393, ISSN 1385-2000.

Shaw, M. T. (2007). Detection of multiple flow regimes in capillary flow at low shear stress, *Journal of Rheology,* Vol. 51, No. 6, (November 2007), pp. 1303-1318, ISSN 0148-6055.

Wassner, E.; Schmidt, M. & Münstedt, H. (1999). Entry flow of a PE-LD melt into a slit die: An experimental study by laser-Doppler velocimetry, *Journal of Rheology,* Vol. 43, No. 6, (November 1999), pp. 1339-1353, ISSN: 0148-6055.

Wereley, S. T. & Meinhart, C. D. (2001). Second-order accurate particle image velocimetry, *Experiments in Fluids,* Vol. 31, No. 3, (September 2001), pp. 258-268, ISSN 0723-4864.

Westerweel, J. (1994). Efficient detection of spurious vectors in particle image velocimetry data, *Experiments in Fluids,* Vol. 16, No. 3-4, (February 1994), pp. 236-247, ISSN 0723-4864.

9

PIV as a Complement to LDA in the Study of an Unsteady Oscillating Turbulent Flow

Chong Y. Wong*, Graham J. Nathan, Richard M Kelso
University of Adelaide,
**Now at CSIRO Process Science and Engineering,*
Australia

1. Introduction

In the last decade, particle image velocimetry (PIV) has become a standard laser diagnostic tool in numerous fluid mechanics laboratories worldwide. At first glance, it appears straight-forward to set up an off-the-shelf system for a quick investigation of the flow of interest. However, unless additional effort is taken to understand the flow, results from 'quick investigations' may lead to limited and sometimes spurious interpretation of the physical flow phenomena. This is especially so for highly three-dimensional and turbulent flows. This chapter examines the efficacy of phase-averaged particle image velocimetry results in assessing the physical phenomena occurring in highly periodic flows and how they complement results from phase-averaged laser Doppler anemometry (LDA) and surface flow visualisation techniques. A specific case study will be presented to demonstrate the complementary nature of these techniques.

The selected flow, a fluidic precessing jet, is a turbulent and highly three-dimensional jet that is used as a fluid mixing device in a combustion lance, or "burner", in rotary kilns. It has been found to achieve low-NO_x (Oxides of Nitrogen; a type of Greenhouse gas) emissions as a gas-fired burner, developed by researchers at the University of Adelaide (Luxton & Nathan, 1988). The flow field lowers flame temperatures by reducing flame strain, which enhances soot formation (Nathan et al., 2006). This reduces NO_x emissions by 30-40% in typical cement kilns (Manias & Nathan, 1994) compared with conventional kiln burners (free jet burners), and its enhanced radiation heat transfer also improves the product quality (Manias & Nathan, 1994) and output (Videgar, 1997) of the clinker in rotary cement kilns. Specific fuel savings of approximately 3-6% were typically reported (Videgar, 1997). The patented burner (hereinafter called the 'fluidic precessing jet' or FPJ nozzle) is commercially known as the Gyrotherm™ and is based on a geometrically simple nozzle configuration.

Although the FPJ nozzle is simple in design, the flow within and emerging from the nozzle is unsteady and highly three-dimensional (Fig. 1). Several experimental FPJ nozzles were developed to study the fundamental characteristics of the precessing jet. Some of the experiments employed classical flow visualisation techniques, such as particle-tracing (using glass beads or gas bubbles), shadowgraph, smoke, cotton tufts, coloured dyes and China-clay surface flow visualisations (Nathan, 1988), as well as quantitative methods including yaw-probe meters, hot-wire anemometers, pressure probes, an entrainment shroud (Nathan, 1988), planar laser induced fluorescence (PLIF), OH-PLIF and particle image velocimetry (PIV) (Newbold, 1997; Nobes, 1997). Nevertheless, many details of the flow structure eluded researchers for some two decades and it is only recently that its phase-averaged flow structure has been revealed through the use of laser diagnostic techniques such as phase-averaged PIV and LDA, complemented by a classical surface flow visualisation technique (Wong, Nathan & Kelso, 2008).

The chapter presents a summary both of the flow field itself and of the approaches used to investigate it. Section 2 provides details of what is known about the fluidic precessing jet flow and explains how a qualitative understandaing of the flow was used to develop a methodology for a systematic investigation of it. Finally section 3 discusses the key results and constructs a qualitative image of the flow based on these results.

2. A case study of an unsteady oscillating turbulent flow

This section presents a case study in the use of PIV to investigate a fluidic precessing jet (FPJ). As discussed earlier, the FPJ produces an unsteady, oscillating turbulent flow with unique fluid mechanical features. In order to apply any laser diagnostic technique, it is necessary to understand the characteristic features of the flow so that an appropriate experimental strategy can be planned.

2.1 Characteristic of a fluidic precessing jet

The FPJ nozzle, introduced above, comprises a cylindrical chamber with a diameter D_1 and a length L (~$3D_1$) with a small axisymmetric inlet (d=15.79mm) at one end and an exit lip (D_2) at the other (Fig. 3). Referring to Fig. 1, the inlet flow forms a central jet which in the jet precession mode, reattaches non-preferentially onto the curved wall of the chamber (Nathan, 1988). As a result of flow instabilities setup within the chamber, the flow precesses around the chamber wall (Nathan & Luxton, 1992). A region of swirl is formed at the upstream end of the chamber, approximately $x/L \leq 1/6$. A larger recirculation region is also observed to feed the swirling region, originating from near to the lip. The fluid in the swirling flow is found to comprise contributions both from the recirculated fluid from the main jet and a typically smaller contribution by induced ambient fluid (Nathan, 1988; Nathan, Hill & Luxton, 1998; Parham, 2000). As the jet exits, it does not completely occupy the exit plane (Nathan & Luxton, 1992). The exit flow is then directed through a large angle (typically θ=45° from the nozzle axis) towards the axis and across the face of the nozzle outlet. The emerging flow is also highly three-dimensional and the precession extends for several chamber diameters downstream from the exit plane (Nathan, 1988).

2.1.1 Mode switching nature

Nathan, Hill and Luxton (1998) reported on the mode-switching behaviour of the precessing jet. They identified two major flow modes: an axial-jet mode and a precessing-jet mode. The corresponding conjectured flow patterns for the flow in either mode are also given in Nathan, Hill and Luxton (1998). When in the axial jet mode, the inlet jet emerges from the exit plane without significantly attaching to the inner walls of the chamber. A symmetric region of recirculating flow on the upper and lower sides of the jet is clearly observed

(Nathan, Hill & Luxton, 1998). Hill (1992) reported that the axial jet mode is greatly suppressed by use of a centrebody positioned near to the exit of the chamber. The presence of a centrebody blocks the path of the axial jet. Consequently, some of the flow becomes re-directed back into the nozzle while the majority of the flow by-passes the centrebody and emerges in an asymmetric fashion from the exit plane resulting in the precessing-jet mode. Therefore, in order to increase the probability of the precessing jet mode for all the phase-averaging experiments, a centrebody arrangement is used in the investigation reported here.

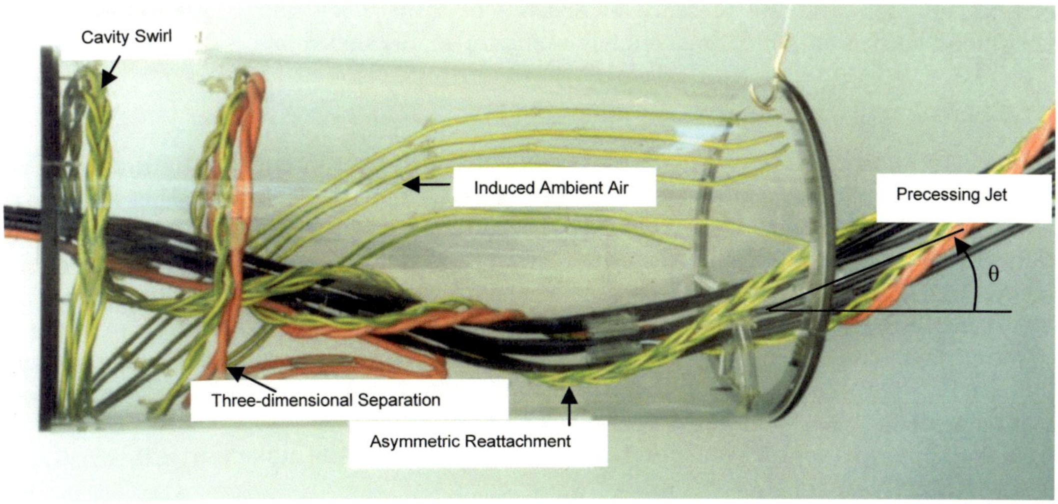

Fig. 1. Perspex-wire model of the conjectured flow topology from flow visualisation studies (reprinted with permission and adapted from Nathan, 1988).

Dependence on inlet flow conditions

Nathan and Luxton (1992) reported that a precessional instability occurs in an axisymmetric duct when the expansion ratio ($E=D_1/d$) of a sudden expansion at its inlet is sufficiently large (ie. $E>5$). In addition, they noted that the chamber's length-to-diameter ratio must be about $L/D_1=2.7$ and that the Reynolds number based on the inlet conditions be sufficiently high, with the precessional instability becoming dominant above a critical value ($Re_d>20{,}000$) for precession with a deflected jet to occur. Hill, Nathan and Luxton (1995) investigated, in water, the flow through an axisymmetric large sudden expansion into a long downstream duct. They found that jet precession can be generally described by the axial momentum (M), duct diameter (D_1) and the fluid properties (i.e. density and kinematic viscosity) of the inlet fluid. In quantifying their results for duct diameters of D_1=60mm and D_1=140mm for varying expansion ratios, using a video camera with a framing rate of 30Hz, they visually counted (frame by frame) the frequency of precession by noting the oscillations in the seeded jet in water. Their inlet flow was seeded with 0.6mm diameter neutral-density polystyrene beads. The minimum precession cycle count was N_p=3 for E=3.75 at Re_d=4400 while the maximum count recorded was N_p=107 for E=14.2 at Re_d=56500. The small sample size, used for the low Re_d number experiments, meant that the result was not statistically conclusive. Despite the measurement difficulties, especially for the low Re_d flows, a useful relationship between expansion ratio E and Reynolds number Re_d was established for suddenly-expanded flows into long ducts.

2.1.2 Jet precession and its dependence on nozzle configuration

Mi and Nathan (2000) conducted a parametric study to determine the influence on precession frequency of the systematic variation of key dimensional parameters. Their single stationary hot-wire study concluded that jet precession frequency was mainly a function of the chamber's length-to-diameter, lip-to-diameter, and centrebody-position-to-diameter ratios, further supporting the study conducted earlier by Nathan and Luxton (1992). Other factors included the size of the centrebody (CB), the size of the exit lip, the presence or absence of a CB, and the presence or absence of a lip. They noted that different inlet conditions, such as an orifice or a contraction inlet, also influenced precession frequency.

2.1.3 Variation of exit angles in the emerging precessing jet

Nathan (1988), using flow visualization, documented that the exit angle of the emerging jet (Fig. 1) can vary from θ=30° to θ=60° relative to the geometric centreline. This adds even more variability to the flow and renders measurements in the flow more difficult. The influence of the jet exit angle on the downstream flow has been examined in further detail by Nobes (1997) who used a mechanically-rotated precessing jet (MPJ). He found that increasing the jet exit angle increases the spread of the jet helix and this results in an increase in the initial mixing rate in the jet. For every given angle, there is a critical Strouhal number, above which the jet converges to the axis of rotation and, below which, it does not. For an exit angle of 45°, $St_{p,cr} \sim 0.008$ (Mi and Nathan, 2005), where $St = f_p\, d_e \,/\, u_e$, where f_p is the frequency of precession, d_e is the nozzle exit diameter and u_e is the bulk mean exit velocity.

2.1.4 Three-dimensional nature of flow

The flow reversals within the FPJ chamber are caused by the adverse pressure gradient downstream from the expansion face and the large expansion (Nathan, Hill & Luxton, 1998). However, this reverse flow is not only turbulent, it also exhibits the large-scale coherent precession motion. This distinguishes it from the relatively steady asymmetry that has been observed in channel flows with similar expansion ratios (Ouwa, Watanabe & Asawo, 1981). The investigation of this three-dimensional oscillation in absolute flow direction requires a measurement technique that is able to differentiate direction in the flow velocity, as well as other standard requirements, including adequate spatial resolution and the need to minimise disturbing the flow field of interest. This rules out the use of stationary hot-wire anemometers and pressure probes, which are not able to resolve such turbulent flows reliably. Flying hot-wires, although able to resolve such unsteady flows, are very complex and are difficult to apply to flows within confined cavities. Laser diagnostic tools such as LDA and PIV are considered to be ideal because they are non-intrusive, do not suffer from directional ambiguity when measuring flow velocities (Durst, Melling & Whitelaw, 1981) and can be applied to flows within cavities when adequate optical access is provided.

2.1.5 Bi-directional azimuthal direction

In addition to the oscillation in axial direction of the flow with each cycle, the azimuthal direction of the entire flow can change. Nathan and Luxton (1992), while conducting surface flow visualisation experiments in an FPJ chamber, reported that "for no apparent reason, the surface flow patterns were occasionally destroyed for a short time and then reappeared".

They added that following the reappearance of the flow patterns the precession was reported to sometimes reverse direction in the azimuthal direction. That observation was similarly reported in the numerical simulations of Guo (2000) who explained that the change in precession direction was the result of the phase interaction between the pressure gradient driving the precession and the jet momentum.

2.2 Experimental methodology

Summarising the previous section, the emerging precessing flow is highly three-dimensional and unsteady, with both the exit angle and precession frequency exhibiting considerable cycle-to-cycle variation. To complicate matters, the precession direction and flow mode also change intermittently with time. These flow variations make the flow challenging to study. Laser-based techniques such as LDA and PIV, which were chosen to study the flow field due to their minimally-intrusive nature and capacity to resolve the flow direction, nevertheless have a limited dynamic range. Hence, the simple application of these techniques to such a flow will lead to large uncertainties and also would not provide much information of the 'instantaneous' or phase-averaged structure of the flow. To resolve the flow in a way that accounts for the variations in flow mode, exit angle, precession direction and phase, requires the use of additional flow sensors to condition, or select, the measurements obtained with these laser-based techniques. This type of conditioning allows "phase averaged" data to be recorded, as used successfully by Fick, Griffiths and O'Doherty (1997) to study the precessing vortex core in swirl burners and by Fernandes and Heitor (1998) to measure oscillating flames. Triggering of a PIV data collection system by pressure probes has also been used by Fick, Griffiths and O'Doherty (1997) to study the precessing vortex core near the exit of a swirl burner. A key element to the success of that measurement was the predictability of the precession direction of the PVC. The naturally-excited fluidic precessing jet studied here not only precesses, but is known to change direction intermittently for the Chamber-Lip (Ch-L) configuration (Nathan, 1988). Frequent directional changes in the emerging flow for a Chamber-Lip-Centrebody (Ch-L-CB) configuration were also observed in the LDA tangential velocity measurements (Wong, Nathan & O'Doherty, 2004). It is also numerically predicted by Guo, Langrish and Fletcher (2001) in a long pipe downstream from a sudden expansion inlet. However, those techniques did not require the resolution of precession direction since the directions of these oscillations are predetermined by the physical geometries of the burners. Hence their triggering devices required only one phase sensor. In investigating the unsteady FPJ flow, both precession direction and phase information should be accounted for to provide a suitable trigger to resolve the phase-averaged structure of the jet.

2.2.1 Geometric configuration of the fluidic precessing jet nozzles

The configuration of the FPJ nozzles investigated here is based on the specifications proposed by Hill, Nathan and Luxton (1992) and Hill (1992) for reliable jet precession. A total of nine configurations, each of which give rise to a precessing jet flow, were studied by Wong, Nathan & O'Doherty (2004). Three different inlet conditions were combined with three alternative ways of arranging the FPJ chamber (Wong, Nathan & O'Doherty, 2004). Of these nine, the Ch-L-CB configuration was chosen for further studies because its inlet flow is uniform and well-defined, and the configuration provides reliable jet precession. Details of these are provided below.

2.2.2 Experimental arrangement

Fig. 2 shows the apparatus used in the experiment, while Fig. 3 shows the geometry of the devices and the coordinate system used for the PIV experiments. A compressor with an operating pressure of up to 650 kPa was used to deliver conditioned and compressed air to the experimental nozzle. The compressed air was regulated to maintain a constant flow rate and was divided into three sub-streams. This was done to provide a flow through the nozzle, while also providing separate seeding both to the nozzle flow and to the entrained air. This avoids the bias that would occur were only the nozzle fluid to be seeded. The first sub-stream was fed into a 6-jet particle generator (TSI Model 9306), while the second sub-stream was diverted to a bypass valve. The two streams were re-combined at the exit of the 6-jet particle generator and transported via a flexible hose into the brass section used to condition the flow. The by-pass arrangement allows the particle generator to function optimally, whilst providing a large air flow rate to the FPJ chamber. The third sub-stream was fed into an in-house-built nozzle particle generator system.

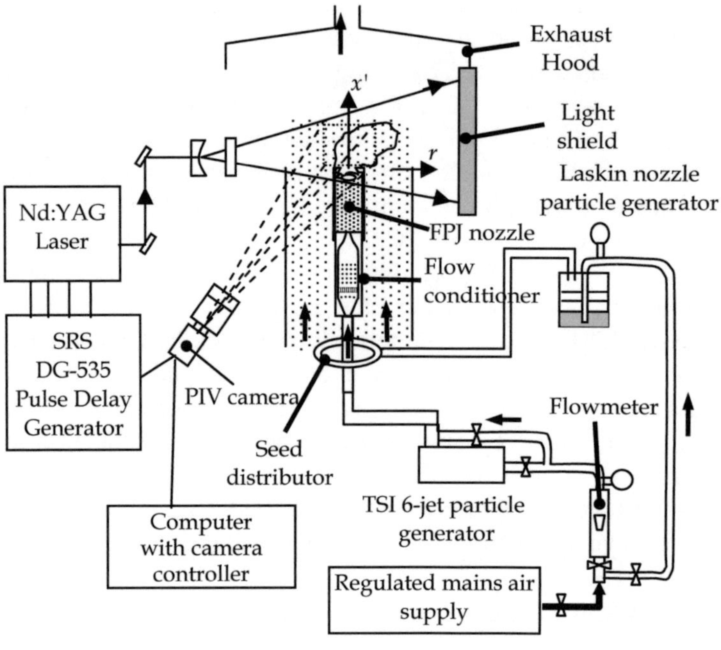

Fig. 2. Experimental arrangement used for PIV experiments. Note that for *x'*-*r* plane experiments, light sheet plane is normal to camera plane as shown. For y-z plane experiments, the camera is directed at an inclined mirror positioned downstream from the FPJ exit plane (not shown here).

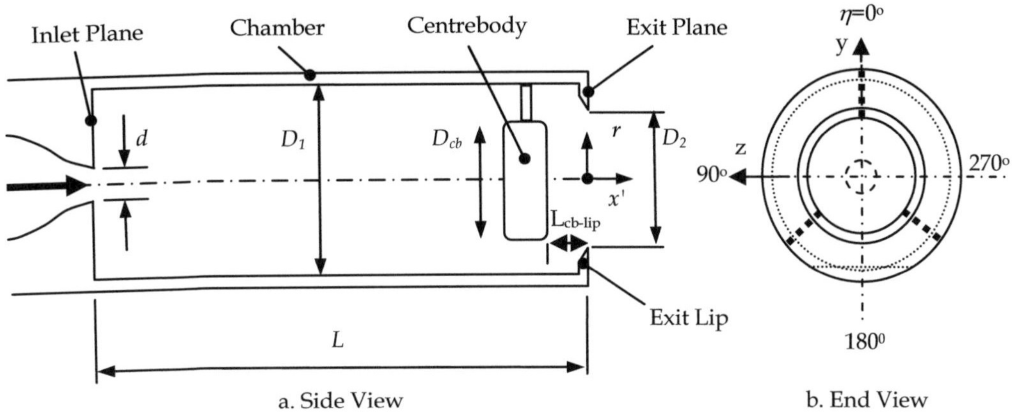

Fig. 3. Coordinate system used in the PIV experiments. d=15.79mm, D_1=80mm, D_2=64mm, D_{cb}=60mm and L=216, t_{cb}=16.8mm, t_{lip}=4mm, $L_{cb\text{-}lip}$=21.6mm and rod diameter of centrebody support is 5mm.

The TSI seeder produces a particle distribution with a modal-mean diameter of 0.6μm, while the nozzle particle generator uses olive oil to generate particle seeding (density, ρ=874kg/m^3 dynamic viscosity, μ=0.026kg/(m.s)) with a nominal particle diameter of approximately 1μm ±0.5μm (Kahler, Sammler and Kompenhans, 2002). Liquid droplets from the nozzle particle generator were used to seed the co-flow around the FPJ chamber and were distributed by means of a ring-type distributor located at the base of the flow conditioner. A cylindrical shroud of about 4.4 times the diameter of the FPJ nozzle was positioned such that the top edge was aligned with the exit plane of the FPJ nozzle. This is used to confine the co-flow seeding within the region of interest. The whole rig is positioned under an exhaust hood which produces a co-flow velocity of about 0.1 m/s based on PIV measurements. To reduce room draughts, all the doors and windows of the experimental laboratory (with a length, breadth and height of 6m, 6m and 5m respectively) were closed.

2.2.3 LDA system

The LDA system is a Dantec two-component LDA system in the burst and back-scatter mode using a Coherent Innova 70 5-W continuous wave Argon-ion laser. In this chapter, only the axial component of velocity is reported here. The 514.5nm (Green) beam line was used. To remove directional ambiguity, one of the split beam line was frequency-shifted by 40MHz. The LDA optical head had a beam separation of 64mm and a focal length of 310mm, resulting in a probe volume with a waist diameter of 0.17mm and a length of 1.65mm. The system was mounted on a Dantec 57G15 three-axis traverse with a position accuracy of ±0.05mm for all axes. Transit-time weighting method of correction was applied to all the velocity data here because a burst mode data sampling was used.

For phase-averaging experiments, a 3-mm diameter open-ended metal tube bevelled at 45° with its bevelled end facing upstream and protruding 10mm into the chamber was mounted half-way between the centrebody and the exit lip. The other end of the tube was connected to a pressure transducer with a 2mV/Pa sensitivity via a 300mm length of PVC tube. The pressure signal was low-pass filtered at 10 Hz by a 6-pole Butterworth filter and the subsequent signal passed to an oscilloscope which generated a TTL trigger pulse from each falling edge of the filtered pressure pulse. The TTL signal was stored by a Dantec 57N20 enhanced Burst Spectrum Analyser as a false velocity reading that was used a a reference marker for the phase-averaging algorithm. The phase-averaging algorithm divided each 360°-precession cycle into 36 segments based on the reference marker. For each radial measurement location, the axial velocity data in each segment was averaged over each 10°-segment. This technique assumes that the precession mode, precession frequency and phase speed do not change. Variations in these parameters will reduce the measured phase-averaged velocities and will increase the measured *r.m.s.* velocity fluctuations.

2.2.4 PIV experimental arrangement

The PIV system consisted of a light generation system, light delivery system, shaping optics, a pulse delay generator, a PIV camera and suitable lenses.

2.2.4.1 Light generation system

Light was generated by a Quantel Brilliant Twins pulsed Nd:YAG laser system rated at 380mJ per pulse at a frequency-doubled wavelength of 532nm. The Twins has two separate laser systems which produce a fundamental harmonic wavelength of 1064nm. A reference beam and a secondary beam were initially parallel with the secondary beam offset from the reference. The secondary beam was combined down-beam by a series of mirrors and lenses to share a common geometric axis with the primary beam. The combined beams were passed through a second harmonic generator which frequency-doubled the light to generate a (green) wavelength of 532nm. These beams were used in the PIV experiments. Each beam pulse had a manufacturer-specified pulse duration of about 6ns at FWMH (full width at maximum half-height). The short pulse width provided excellent temporal resolution for the current flow. The pulse width duration was also verified by observing on an oscilloscope the electronic signals produced by a high-speed light detector positioned in front of the scattered particles. After optimising the lasers for equal beam strength, the actual power measured was typically 180mJ per pulse per laser cavity for a flashlamp-to-Q-switch delay of 300μs.

2.2.4.2 Light delivery and shaping optics

The laser beam was delivered, via a series of mirrors, to the shaping optics. These comprised a diverging cylindrical lens (focal length 200mm) followed by a cylindrical focusing lens with a focal length of 260mm. Optimising this arrangement of lenses produced a suitable light sheet with a non-uniform thickness that varied between 1mm and 2mm in the region of interest. This thickness was chosen because of the significant out-of-plane motion in the emerging precessing jet. The variation in light sheet thickness was found to have little impact on the results since the calculations for particle out-of-plane motion are based on the minimum light sheet thickness (1mm) and the probe resolutions of the interrogation windows are all larger than the thickest part of the lightsheet (2mm).

2.2.4.3 Pulse delay generator

The laser system comprises two independently-controlled laser oscillators, each of which fire at a nominal rate of 10Hz. The lasers were controlled by a Stanford Research Systems

DG-535 digital pulse delay generator, with a pulse-to-pulse jitter of less than ±5ns. The pulse-to-pulse jitter (or *r.m.s.* fluctuation) between laser pulses was measured to be about ±0.7μs, slightly more than the manufacturer rated jitter of ±0.5μs. This value was used to estimate the precision of the PIV measurements. The pulse delay generator produced a signal whenever an internal timing trigger initiated a timing cycle. The flashlamp-to-Q-switch delay for both lasers was typically set to 300μs to provide a laser energy output of about 180mJ per oscillator. The output signal from the internal timer was sent to the camera which, following an external trigger, had an internal delay of 20μs before exposing the first frame. The exposure time was determined by the transfer pulse width, which was set to 304μs. The first laser pulse was synchronised to illuminate the region of interest during this period. Following this, a transfer pulse delay of 1μs occured before the second exposure was activated for a fixed time period of up to 30ms. The second laser pulse fired at a time Δt after the first pulse to be recorded by frame two. Since a 532nm±10nm narrow bandpass filter was not available, minimum room lighting was used to reduce the effect of background light to negligible levels and background light was later subtracted from each image pair.

2.2.4.4 PIV camera and lenses

The PIV camera was a Kodak Megaplus ES 1.0 (maximum frame-rate of 30Hz), which employed a charged-coupled device (CCD) with an array of 1008 by 1018 pixels (width and height) respectively. The quantum efficiency of the camera is approximately 36% for light at a wavelength of 532nm. Each pixel is approximately 9μm square and has a fill factor of 60%. Each pixel contains three main regions: light sensitive region 1, light sensitive region 2 and a dark region. It was purpose designed to allow it to collect two images in rapid succession. In the "triggered double-exposure mode" or the "frame-straddling mode", light sensitive region 1 is activated to collect photons for an extremely short period. This period, known as the transfer pulse delay, can range from 1 to 999μs. After the transfer pulse delay, a transfer pulse width event occurs before light sensitive region 2 is activated. This time can range from 1 to 5μs. During the transfer pulse width event, integrated light signals falling on region 1 are sequentially shifted into the dark region, which is then further transferred to the random access memory (RAM) of the camera capture card. Light sensitive region 2 is activated for a fixed time of up to 30ms before the data are transferred into the dark region and ultimately into the RAM. Data in RAM are later transferred to permanent storage in local hard disk drives.

The camera was connected to an AF Zoom-Nikkor 70-300mm f/4-5.6D ED lens set at an *f*-number ($f_\#$) of 5.6. The $f_\#$ is the ratio of the focal length, f, to the aperture diameter, D_a. This $f_\#$ value ensured that the image distortion around the edges of the image was minimised, while allowing adequate light to the image array via the lens. Raffel, Willert and Kompenhans (1998) noted that this kind of systematic perspective distortion due to the lens arrangement is generally neglected in most experiments and the only way to quantify this error for highly three-dimensional flows is to measure all three components of velocity. This was not possible using two-component PIV employed in the present investigation.

The depth-of-field of the lens system was checked to ensure that particles within the light-sheet were adequately imaged and focused. The depth-of-field of a lens is the distance along the optical axis over which an image can be clearly focused. For PIV, it is calculated based on the following set of equations from Raffel, Willert and Kompenhans (1998):

$$M = \frac{z_o}{Z_o} \quad (2\text{-}1)$$

$$f_{\#} = \frac{f}{D_a} \quad (2\text{-}2)$$

$$d_{diff} = 2.44 f_{\#}(M+1)\lambda \quad (2\text{-}3)$$

$$\delta_z = \frac{2 f_{\#} d_{diff}(M+1)}{M^2} \quad (2\text{-}4)$$

where,

M = magnification factor, z_o = distance between image plane and lens [m]; Z_o = distance between object plane and lens [m]; $f_{\#}$ = f-stop; f = focal length of lens [m]; D_a = aperture diameter [m]; d_{diff}= diffraction limited minimum object diameter [m]; λ= wavelength of light used [m], (532nm for frequency doubled Nd:YAG); δ_z= depth-of-field [m].

To estimate the particle image diameter, the following equation is used neglecting effects of lens aberrations:

$$d_\tau = \sqrt{\left(M d_p\right)^2 + d_{diff}^2} \quad (2\text{-}5)$$

where, d_τ= particle image diameter [m] and d_p= particle diameter [m].

The depth-of-field was calculated for each experimental setup, assuming a mean particle diameter of about 1μm, $f_{\#}$ = 5.6, laser wavelength of 532nm, and a nominal image pixel size of 9μm. In general, the depth-of-field exceeds 5.75mm for a magnification (px/mm) of 15. Since all the experiments have light sheet thicknesses of less than 3mm and a magnification of not more than 15px/mm, the depth-of-field used is appropriate and all the particles moving into or out of the light sheet were thus in focus. For a typical magnification of 10.6px/mm, the field-of-view is about 95mm wide by 95mm high.

2.2.5 Phase-and-precession-direction-resolved PIV

Various methods of triggering a data collection system based on external reference conditions have been used by researchers to study time-dependent flows. As noted above, it is necessary to develop a data-collection system that resolves both precession direction and phase. This is achieved with a triggering system using a pair of hot-wire probes to detect the phase and direction of precession. The components to be synchronised are the PIV lasers, the camera system and the triggering system. A block diagram showing the interaction between each system is presented in Fig. 4a while the timing diagram for the trigger system is presented in Fig. 4b.

2.2.5.1 Overall system timing

An ATMEL microprocessor interfaced the laser and camera system with the external hot-wire sensor-trigger system. An important criterion for stable and reliable operation of the pulsed Nd:YAG laser is that the 10-Hz flashlamp frequency must not vary by more than ±0.5Hz. The laser manufacturer advises that power output decreases dramatically when the flashlamp frequency is outside of the nominal flashlamp frequency (Quantel, 1994).

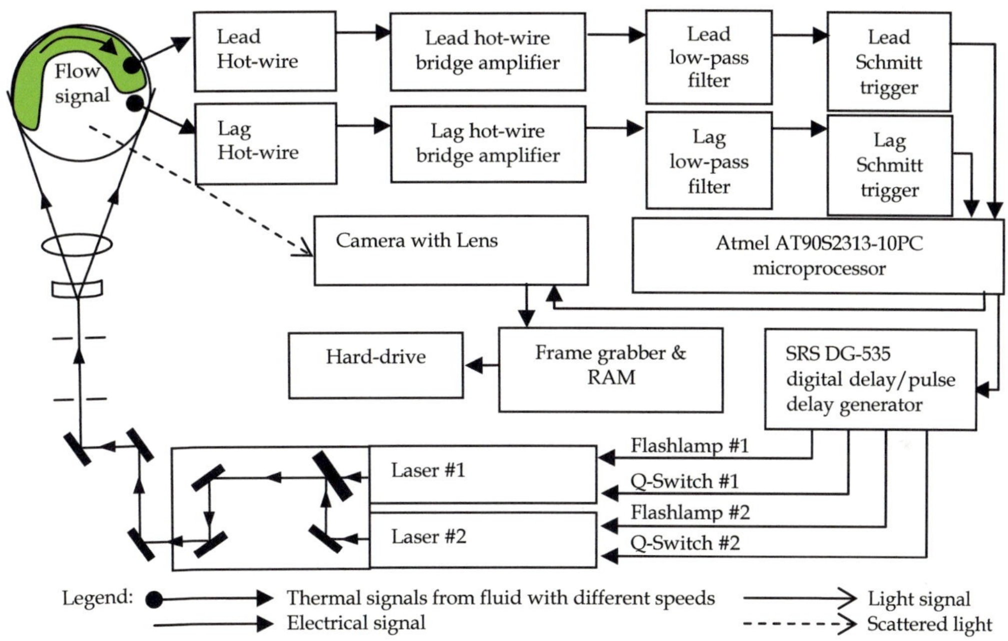

Fig. 4.a. Block diagram of the relationship between the triggers, laser and camera systems.

The following establishes some criteria for the timing system of the microprocessor.

$$f_{fl\_min} < f_{fl\_nom} < f_{fl\_max}\ ;\ 9.5\text{Hz} < f_{fl\_nom} < 10.5\text{Hz, and } \delta_{fl\_nom} = 1/\ f_{fl\_nom}.$$

therefore, δ_{fl_max}=105ms > δ_{fl_nom} > δ_{fl_min}=95ms.

where, f_{fl_nom}=Nominal flashlamp frequency [Hz]; f_{fl_min}=Minimum flashlamp frequency [Hz]; f_{fl_max}=Maximum flashlamp frequency [Hz]; δ_{fl_nom}=Nominal flashlamp time interval [sec]; δ_{fl_min}=Minimum flashlamp time interval [sec], and δ_{fl_max}=Maximum flashlamp time interval [sec].

The next step is to choose an appropriate time window that selects a particular band of precession frequency. In the present case, the precession frequency is nominally 5Hz and a frequency range between 3 and 6Hz was chosen. The 3 and 6Hz cut-off frequency range was chosen to match the -3dB frequency interval for a 5Hz precession frequency.

If the frequency range were to be shifted to a higher precession frequency envelope, i.e., from 3 and 6Hz to 4 and 7Hz, fluid structures having a higher precession frequency will be recorded. Although slightly different sizes of structures may be observed, the overall flow topology of the higher precession frequency flow is not expected to be markedly different from the lower frequency flow topology. Data collection time may be increased for the 4 to

7Hz range since the probability of detecting an event decreases as the range is shifted from the -3dB range (located between 3 and 6Hz). This argument similarly applies if the frequency range was shifted to a lower one, such as a range between 2 and 5Hz.

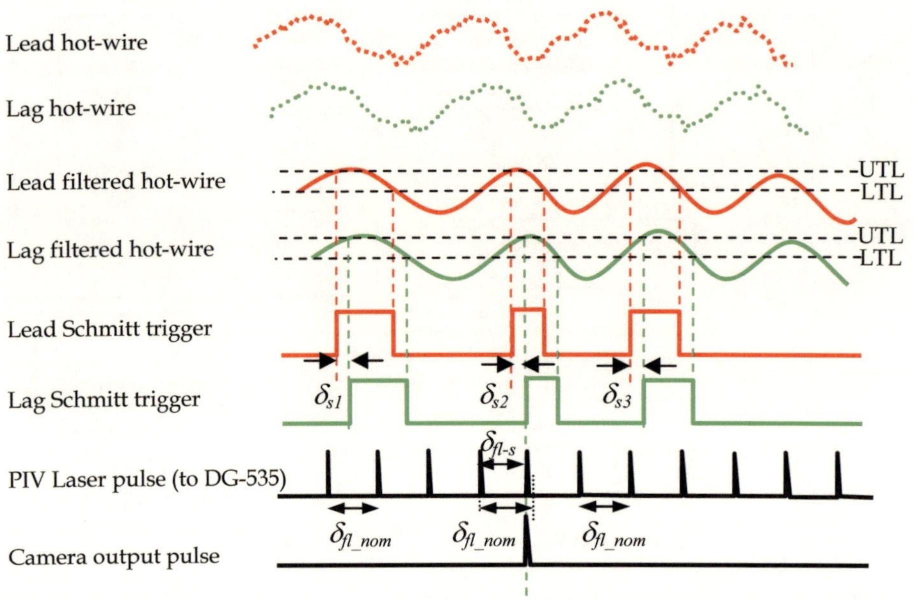

Legend:
UTL: Upper trigger level
LTL: Lower trigger level
Assume δ_{s1} and δ_{s3} are not within the *valid separation time* between Schmitt triggers and δ_{s2} is within the valid separation time between Schmitt triggers

Fig. 4.b. General timing diagram for the phase and precession-direction-resolved PIV experiment.

The main parameters that determined the time window calculations are: linear (or circumferential) separation distance between the two hot-wire probes, radial distance from the hot-wire probe sensing element to the geometric centre of the FPJ, and a frequency bandwidth that characterises the precessing jet (in this case between 3 and 6Hz). The filtering criteria accompanying these parameters are: the jet does not 'flip' across the exit face of the nozzle during one cycle of jet precession, rather the jet takes a circular path about the axis of the chamber, the output from the two-probe arrangement is adjusted to give similar voltage responses to similar velocity fluctuations, and the probes are separated sufficiently to not to interfere with each other.

Following the selection of the frequency bandwidth, a time window for valid inter-trigger times can be set in the microprocessor. The inter-trigger time was the time between triggers that was accepted by the microprocessor as a valid coincidence window for which another comparison can be made with the arrival of a valid laser-trigger. When a valid signal trigger time is sensed after the lag hot-wire probe, an internal comparison is made immediately by the microprocessor to see if a valid laser trigger window has arrived. If there is a match, a laser trigger and a camera signal are sent simultaneously. If there is no match, the laser trigger will still send a signal to the lasers 100ms after the last known laser trigger, and PIV laser pulses are still produced without the activation of the PIV camera. This was required to maintain stability in the laser system.

2.2.5.2 Atmel microcontroller

The microcontroller employed for this work was an Atmel A90S2313-10PC integrated circuit with a clock frequency of 10MHz. Thus the time resolution for this system can be as small as 0.1μs for each clock cycle. Programming the trigger acceptance window was carried out by transferring the microcontroller to an STK500 programmer's board and downloading the instruction set to the microcontroller's non-volatile memory.

2.2.5.3 Trigger system

Adjustable Schmitt triggers were used to convert the hot-wire signals to TTL-compatible phase pulses with minimal phase lag. Also to minimise the phase lag in the triggering time, short wire-runs were made between the trigger system and the microprocessor. The Schmitt trigger is made from a CMOS 4050 buffer (Lancaster, 1997) with two triggering voltage levels, an upper trigger level (UTL) and a lower trigger level (LTL). Whenever a signal voltage level rises above the upper trigger voltage level, the Schmitt trigger immediately outputs a step change to a high voltage value (+V). If the signal voltage level falls over time, but remains above the lower trigger voltage level, the output continues to remain high. However, when the signal level decreases below the lower trigger voltage level, a step change from high to low (~0V) signal is output from the CMOS 4050 buffer. The region between the trigger levels is known as a "dead band" in which the Schmitt trigger ignores any other voltage signals.

The width of the "dead band" is adjusted with the use of various resistance values (Lancaster, 1997). The use of such a trigger is advantageous to the phase detection of the precessing jet flow because it is possible to trigger the flow on the rising edge of the unfiltered hot-wire signal just before the centre of the jet sweeps past. By using a low-pass filter with a phase lag equal to the time from the trigger to the peak signal region of the jet, it was possible to capture an image of the peak flow coinciding with the physical location of the lag hot-wire probe.

A limitation in the use of the Schmitt trigger is its inability to adjust to variable signal voltage levels. This introduces a "phase jitter band" which varies with fluctuating signal amplitudes. This is estimated to be about ±7° for a repeatable flow (such as a mechanical precessing jet flow) and ±30° for the present FPJ flow. Furthermore, to obtain a small phase jitter band, the upper trigger level had to be set high, so that only a limited large amplitude flow precession would activate the trigger system. This increases the experimental time considerably by reducing the sampling rate (to approximately 1 image for every 300 flow cycles).

2.2.5.4 Verifying the operation of the direction sensors

A simple check was made to verify the direction selectivity of the sensor system with a mechanically precessing jet (MPJ) (Schneider *et al.*, 1997a and 1997b). For brevity, details of the experiment are not shown here; see Wong (2004) for further details. However, the physical rotation of the MPJ nozzle allows confirmation that the lead hot-wire probe is always exposed to the advancing jet before the lag hot-wire probe. This was found to result in a constant stream of camera signals being sent to the PIV system for data collection, as shown in Fig. 5. To confirm the directional selectivity of the system, the rotation direction of the MPJ was then reversed. This resulted in no triggering of the camera, demonstrating that the system is effective at discriminating precession direction. These experiments also showed that the mean phase lag between the lag hot-wire and the core of the jet is approximately 23.8° with an error band (or jitter) of ±7° as shown in Fig. 5. This phase jitter is caused by several factors, namely, the 3Hz frequency window chosen, the varying exit angle of the jet, and the selection of the UTL of the Schmitt trigger (that relates to the selection of higher magnitudes of velocities before the data collection system is allowed to trigger).

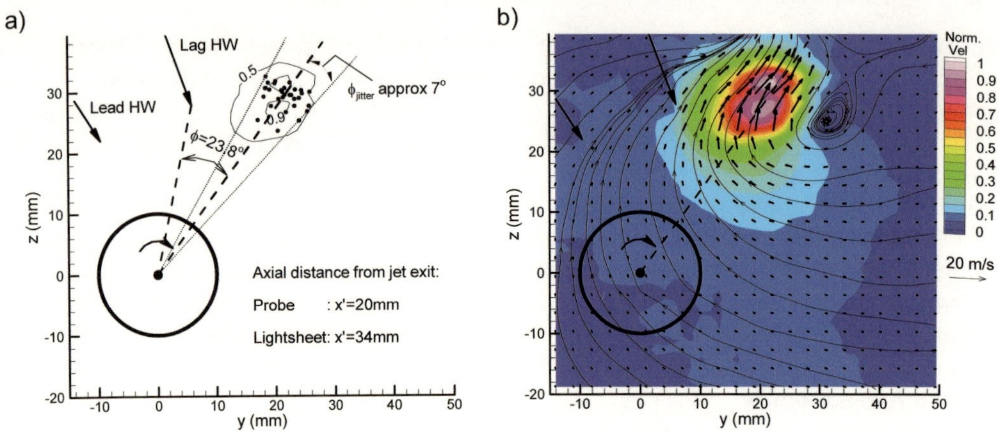

Fig. 5. Verifying the selectivity and phase of the direction sensors with a mechanical precessing jet (d_e=3mm) exiting at θ=45° from the spinning axis rotation in a clockwise fashion at f_p=9Hz. The nozzle is centred at y=0 mm and z=0 mm. **a)** Phase-lag (ϕ) of the jet relative to the lag hot-wire probe, where ϕ_{jitter} refers to phase-jitter in the measurements; **b)** Pseudo-streamlines are shown with velocity vectors while coloured contours represent in-plane velocity magnitudes $(v^2+w^2)^{1/2}$ normalised by $(v^2+w^2)^{1/2}{}_{max}$=18m/s. Measurements are for a light sheet at x'/d_e =11.3 and probe at x'/d_e =6.7 (and r/d_e=11) for an ensemble of 50 image pairs.

2.2.5.5 Optimising the location of the trigger probes in the FPJ flow

To measure the FPJ flow reliably, a suitable location is needed for the hot-wire probes. Nathan (1988) demonstrated that the introduction of a foreign object (such as a long rod) within the FPJ nozzle cavity will unnecessarily perturb the precessing flow mechanism. In order to avoid this, the probes were positioned outside, but as close as practical to the exit plane of the nozzle to reduce jitter, minimize the interference of the probes and probe holder, and maximize the signal strength of the flow. The lag hot-wire probe was positioned parallel to the nozzle axis with its centre at r/D_2=0.61 in the y-z plane for a total of 4 downstream locations, namely x'/D_2=0.12, 0.25, 0.50 and 0.75. An ensemble average of 50 image pairs are presented in Fig. 6, in which the phase-averaged velocities $(v^2+w^2)^{1/2}$ are non-dimensionalised by the maximum in-plane velocity for each measurement plane and shown as contours. The difference between each contour level is 0.2. The location of the light sheet was fixed at x'/D_2=0.05. For each experiment, the hot-wire parameters were re-adjusted to suit the new flow conditions for each hot-wire probe location.

The criterion for the best location was chosen by a combination of visual inspection of each successive image pair to decide on the location that results in the least spatial jitter and also by ensemble-averaging the results. In Fig. 6, the locations for x'/D_2=0.12, 0.25 and 0.75 appear to produce similar results in the total magnitude of in-plane velocities. Locating the probe at x'/D_2=0.50 appears to average data over two precession directions, as shown by the larger 'footprint' at the $0.2(v^2+w^2)^{1/2}$ contour. This location coincides with the zone of flow convergence (location of time-averaged, but not instantaneous, flow convergence downstream from the exit plane). Thus, any probe location which senses the magnitude of fluctuating velocities of the emerging precessing jet, other than at a location near to the zone of flow convergence, is a valid choice for the present experiment. The probes were finally located at x'/D_2=0.12 and r/D_2=0.61. This location, which is also approximately 130° to 150° from the peak in-plane velocity magnitude, resulted in the least influence of the probe and probe holder on the overall flow and was not in the plane of the light-sheet. Experiments conducted in the x'-r plane of the local jet centreline show that the instantaneous jet emergence angle (θ=53°) is relatively constant within x'=40mm (or x'/D_2=0.625) (Wong, Nathan & Kelso, 2008). Based on the $0.5(u^2+v^2)^{1/2}{}_{max}$ contours, the range of emergence angles was estimated to be approximately ±15° about the local jet centreline. This range of angles is two times the value of the minimum phase resolution (±30°) in the azimuthal direction (y-z plane) (Wong, Nathan & Kelso, 2008).

2.2.5.6 Selecting the separation distance of the probes

The separation distance between the probes also plays a key role in determining the success of the technique. The lag hot-wire probe was positioned at a radial distance of approximately r/D_2=0.61 and a downstream location of x'/D_2=0.12, as determined from Section 2.2.5.5. To solve for an appropriate probe separation distance, s_c, the following empirical criteria should be satisfied: the highest precession frequency of interest (i.e. 6Hz) should be at least 100 times lower than the smallest clock cycle of the Atmel controller, and the circumferential distance, s_c, travelled by the precessing flow at a fixed radius, r_s, from the nozzle axis should not be more than 10% of the circumference (s_f) described by the radius r_s. That is, $s_c < 0.1\, s_f$.

We can derive the time it takes for the flow to travel between the lead and the lag hot-wire probe (i.e. the separation time, T_{sep}) given the precession frequency of interest (i.e. 3 or 6 Hz), from the following equation:

$$T_{sep} = \frac{s_c}{2\pi r_s f_p} \qquad (2\text{-}6)$$

Where, s_c = probe separation distance [m], r_s = radial distance between the lag hot-wire probe and the nozzle axis [m], and f_p = precession frequency of interest [Hz].

To calculate the separation time between the lead and lag hot-wire probe for the ATMEL controller time, we use the following conversion formula:

$$T_{atmel} = \frac{T_{sep} \times 10 \times 10^6}{1024} \tag{2-7}$$

The smallest clock resolution (i.e. 1 integer clock cycle) of the ATMEL microcontroller was approximately 0.1ms. Based on the criterion mentioned earlier and for accuracy, a minimum clock time of approximately T_{atmel}=10ms (or 97 clock cycles) for f_p=6Hz was used. Subsequently, T_{atmel}=20ms (or 195 clock cycles) was selected for f_p=3 Hz. The circumference, s_f, travelled by the flow around a circular path for r_s=0.6D_2 is calculated from:

$$s_f = 2\pi r_s \tag{2-8}$$

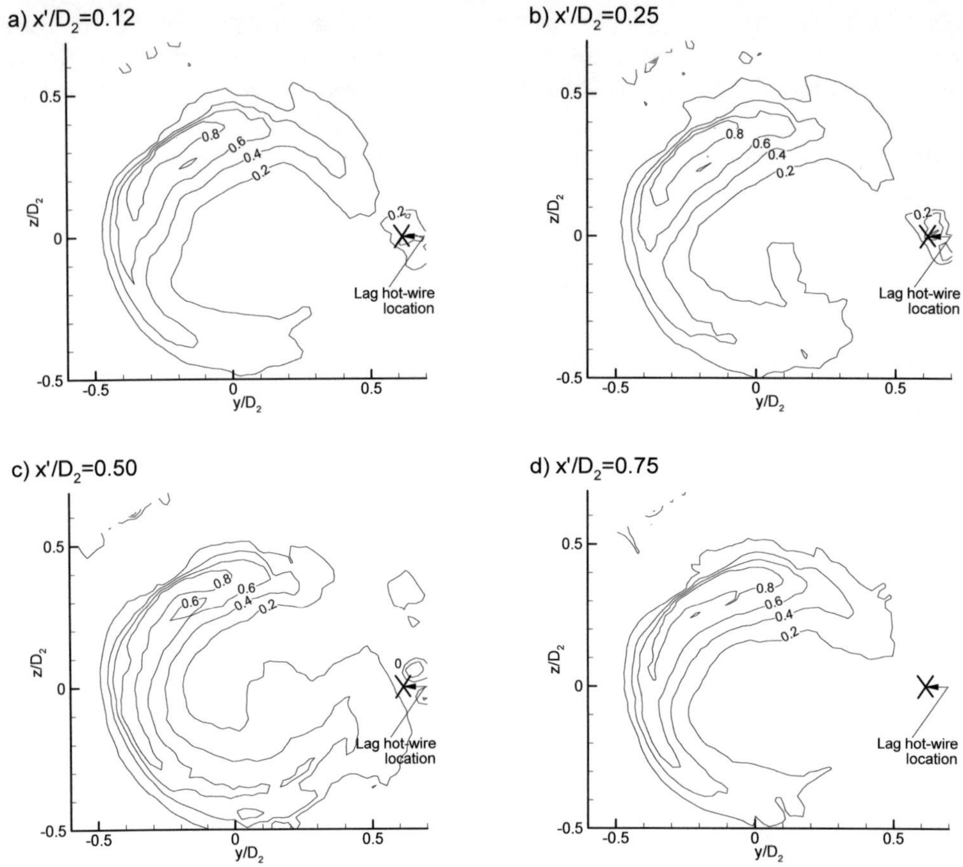

Fig. 6. In-plane velocity magnitude contours, $(v^2+w^2)^{1/2}$, for 4 spatial positions that were tested for the best placement of probes. Velocities are non-dimensionalised by maximum in-plane velocity for respective locations. Precession of the flow from an observer looking upstream is in a clockwise direction. Re_d=59K, St_d=0.00117.

This gives us a circumference of approximately 245mm. If we allow the flow at that radius (r_s) to travel at most 10% of that circumference every cycle, then an appropriate separation distance would need to be less than 24.5mm (that is, an angular separation between the wires should not exceed 36°). Alternative probe separation distances of s_c=5, 10, 15 and 20mm were also considered. However, for this case, a probe separation of 15mm was chosen as best satisfying the two criteria mentioned earlier.

2.2.5.7 Final setup of the system

The overall PIV system was combined with the phase-and-precession-direction stationary hot-wire sensors. In addition, a Tektronix TDS 210 Real Time Oscilloscope was used to monitor important signal outputs while a PC30D multi-channel A/D converter was used to collect various voltage signals from the Schmitt trigger inputs/outputs, the camera outputs and the laser outputs. The hot-wires were fully annealed prior to the experiments and the hot-wire system was designed for a short warm-up time and turned on at least 2 hours prior to use. If the hot-wire system was not allowed to stabilize, the data collected would be contaminated by electronic drift and lead to a false activation of the Schmitt triggers, and hence introduce errors in the detection of directional flow changes. Both hot-wire probes were held by probe holders and positioned at x'/D_2=0.16 and r/D_2=0.61. They were initially spaced approximately 1mm apart so that they sensed the same magnitude and frequency of flow. Both hot-wire channels (using an overheat ratio of 1.2) were adjusted to respond equally to a 1KHz square wave test. The signals from each hot-wire channel were passed to 4-pole Butterworth (maximally flat) Krohn-Hite filters (Model 3322) and low-pass filtered at the appropriate frequency (30Hz in the present experiments). Both filters were previously adjusted to give the same roll-off and filter frequency response for the same velocity fluctuations. Air was supplied to the FPJ at the experimental condition required and the output of each wire was adjusted to match and to fit within a 0 to +5 volts window. After calibration, the lag hot-wire probe was positioned at a radial distance of approximately r/D_2=0.61, a downstream location of x'/D_2=0.12 and a distance of 15mm from the lead hot-wire probe. Note that time-averaged PIV experiments employed the same experimental arrangement except with the phase-triggering de-activated and the hot-wire system removed.

2.2.6 PIV image pair data processing

Before each experiment, a calibration ruler was positioned within the plane of the laser sheet and its image recorded. This allowed the magnification (in px/mm) of each image to be recorded. An iterative process was then carried out to optimise the experiment so that each run had a velocity data yield of greater than 95%. Firstly, an estimate of the flow velocities was made and a few images taken to assess the relative movement of the correlation peak for a specific interrogation window and a tolerable dynamic velocity range (DVR). The DVR is defined by Adrian (1997) as the ratio of the largest resolvable velocity to the smallest resolvable velocity by the PIV system. An optimum out-of-plane motion of ¼ the lightsheet thickness suggested by Keane and Adrian (1990) was chosen for particles moving perpendicular to the lightsheet. The out-of-plane velocity for the FPJ flow was assumed to be of similar magnitude to the in-plane velocities. The software used for all the image processing, PivView 1.7, was assumed to have a sub-pixel resolution of 0.1 pixel. An acceptable level of cross-correlation signal was achieved by setting the time delay, Δt (typically 10μs), so that an out-of-plane particle motion of ¼ the lightsheet thickness was achieved. Thus, a maximum particle movement of between 3 and 4 pixels was achieved for an interrogation window (IW) of 16x16 pixels with no overlap. A resampling technique after

Hart (2000) reduced the number of spurious vectors due to no overlap used. A global histogram method was used to interactively select the region of interest and to exclude potential outliers. The software stored the 3 highest correlation peaks per correlation calculation. If the highest peak was an outlier, the next highest peak was selected and so on. When a spurious vector was marked, it was replaced by bi-linear interpolation using information from surrounding vectors. When several immediate neighbours were also outliers, then a Gaussian-weighted interpolation scheme was used. The measurement volume was typically 1.5 x 1.5 x 2mm^3 for a scaling of 10.6px/mm, Such small pixel displacements may result in a poor dynamic velocity range in these experiments. Nevertheless, broad features of the flow can still be obtained. A total of 350 image pairs (recorded over 35 seconds) were recorded for the time-average PIV experiments and due to resource limitations 50 image pairs (taking up to 2 hours to collect) were recorded for each phase angle of the phase-average PIV experiment. With only 50 samples collected in the phase-averaged PIV experiment, only the mean data are presented in this chapter.

2.2.7 LDA and PIV experimental uncertainties

Experimental uncertainty calculations follow the method proposed by Martin, Pugliese and Leishman (2000) and Kline and McClintock (1953). For LDA, the calibration uncertainties associated with measurement of probe length, measurement of beam separation and data acquisition uncertainties are approximately ±3.4% of the measured velocity for 95% confidence interval (c.i.). For PIV, the uncertainties associated with camera calibration and jitter in the laser-timing system are approximately ±7% of the measured velocity at 95% c.i.. Other sources of PIV errors (or *r.m.s.* uncertainties) as suggested in Raffel et al (1998) attributable to particle image diameter, particle image displacement, ratio of particle image size and pixel size, particle image shift, image quantisation, background noise of the image, displacement gradients and out-of-plane motions are estimated to be ±1.3% of the measured velocity.

Another major source of error for both techniques is due to phase uncertainties of the jet motion in the azimuthal and axial planes. These were estimated to be ±30° and ±15° respectively for the directionally-resolved phase-averaged PIV experiments, and potentially ±60° and ±15° respectively for the non-directionally resolved phase-averaged LDA experiments. Larger phase uncertainties in both planes contribute to a reduction of the phase-averaged magnitudes compared to velocities obtained during an instantaneous event.

2.2.8 Surface flow visualisation

The purpose of surface flow visualisation is to make visible the patterns of skin-friction lines (Maltby, 1962) and critical points on the surface of (wind-tunnel) models, so that the flow around the surface can be interpreted and understood (Perry & Chong, 1987). This method relies on the response of the visualisation media to skin friction stresses generated by the air-stream flowing over the object's surface, and on the media film being thin enough to be unaffected by pressure gradients (Tobak & Peake, 1979). It should be noted that the response time of this technique is much too long to allow it to represent the instantaneous state of the flow. Instead, the results reflect a pattern caused by the flow's interaction with the surface over a period of time commensurate with the drying time of the media (Hunt et al, 1978). The axial vorticity fields obtained using the phase-precession-direction resolved PIV revealed the presence of a number of vortical features near to the FPJ exit plane at x'/D_2=0.11. (Wong, Nathan & Kelso, 2008). The presence of these structures was further studied by conducting time-averaged surface flow visualisation on the downstream surface of the centrebody using an alcohol-powder mixture. While these experiments are qualitative, they provide important insight into some of the features measured from the PIV experiments, which do not measure the surface flow field.

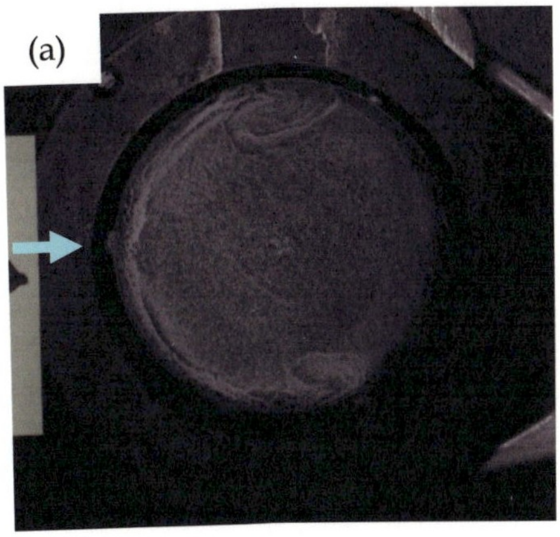

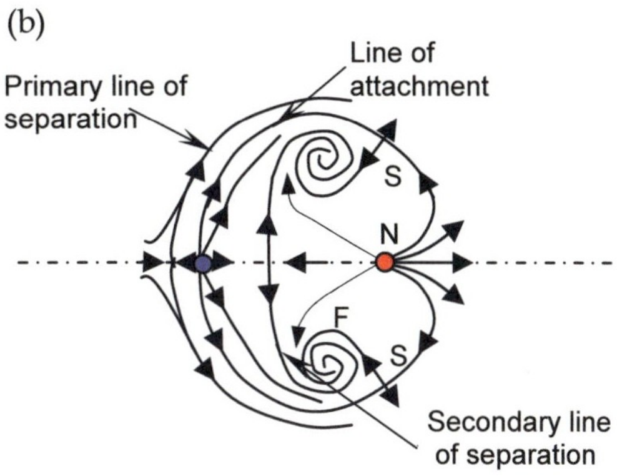

Fig. 7. a) Surface flow visualisation on FPJ nozzle at 10m/s free-stream inlet jet velocity. b) Interpretation of flow topology (Adapted from Wong *et al.* 2008 and reprinted with permission from Cambridge University Press). Steady deflected flow emerging from the left is deflected to the right and emerges at θ~45°. Re_d~10K, St_d=0. `F' is focus, `S' is saddle and `N' is node.

A steady (non-precessing) jet that deflects at a large exit angle to the nozzle axis was obtained by removing the chamber-section of the FPJ nozzle (chamber-lip-centrebody arrangement) from the supply pipe and introducing an alternative source air at an eccentric location within the inlet plane. The air was supplied by diverting a portion of the flow from a wind tunnel using a suitable turning vane. The deflected air-stream emerging from the exit lip was checked with a cotton tuft attached to a metal rod. As expected, with this arrangement no precession of the jet was detected, but instead the emerging jet was deflected steadily at approximately 45° relative to the nozzle axis.

Zdravkovich *et al.* (1998) studied coin-like cylinders using a talcum powder-paraffin film mixture; and Potts and Crowther (2000) used kerosene and fluorescent dye powder to study inclined disc-wings. In the present experiments, the use of white talcum powder and methylated spirit (96% ethanol @ 20°C: μ=5.23×10^{-7}kg/(m.s), ρ=802kg/m^3) was found to provide a suitable mixture for surface flow visualization on the centrebody surface. The centrebody and nozzle exit were first painted matt black to provide good contrast from the white powder. The mixture was stirred thoroughly before being applied as a thin and uniform film over the region of interest which was initially positioned horizontally. Excess mixture was drained off by slightly tilting the model until the desired film thickness was achieved. An optimum film is achieved when the mixture just 'wets' the sample's surface. This was an iterative process as it was difficult to quantify the optimum film thickness due to the rapid evaporation of the methylated spirit. The FPJ nozzle was positioned vertically at the working section with the deflector facing upstream, such that the flow emerging from the nozzle exit plane was deflected by about 45° as indicated by a cotton tuft held by a thin rod positioned in the centre of the exit lip, just above the centrebody.

The most obvious flow features, such as the flow reattachment location and the foci, could be seen to form immediately on the matt black surface of the centrebody (Fig. 7a). More features were revealed as the methylated spirit evaporated over time, leaving only the white talcum powder behind. An interpretation of surface flow topology is shown in Fig. 7b. The influence of the centrebody supports were also considered in the experiments, however, it was found that the orientation of the supports does not greatly affect the distinctive flow patterns observed on the upper surface of the centrebody. As such, a study on the centrebody supports is not further discussed here.

3. Final discussion of key results and issues

3.1 Comparison of time-averaged PIV and LDA results

This section discusses the time-average results for the Chamber-Lip-Centrebody (Ch-L-CB) case for both the 1-D LDA and PIV results. For the LDA experiments, only the jet was seeded and local seeding was applied to seed the ambient fluid during each measurement. This explained the higher validated LDA bursts (nominally 1000) in the locations $-0.7<r/R_2<+0.7$ during the experiments compared to between 10 and 1000 bursts/second outside these locations. Since this configuration was found to generate the precessing jet mode most of the time, any attempts to seed the region near to $r/R_2=0$ was problematic since the presence of the seeding device may have an unwanted influence over the emerging flow. The poor seeding explains a decrease in the LDA burst rate beyond the region $-0.7<r/R_2<+0.7$ in these measurements. However, this did not unduly affect the present results in Fig. 8a and b, (except, for the orifice case) since an acceptable number of validated LDA burst samples (generally greater than 4000) and data rate (at least 10 times higher than the local precession frequency) were achieved. Refer to Wong (2004) for more details. The results of the PIV experiments are also provided in Fig. 8c and Fig. 8d and specific details pertaining to both these experiments are also found in Table 1 and Table 2.

Generally, the axial velocity trends in the PIV and LDA data in the region $-0.8<r/R_2<+0.8$ typically vary by $u/u_i=0.02$. However, in the regions beyond this, the LDA data are expected to over-estimate the true velocity values due to insufficient global seeding. This means that unseeded ambient fluid entrained into the FPJ nozzle will not be recorded, while seeding from the bulk jet will be sampled most of the time. This introduces a velocity bias towards a higher velocity flow, especially in the region beyond $r/R_2=\pm0.8$. Thus, the flow structure beyond the exit lip cannot be elucidated clearly. This is evident in the contrasting results obtained by means of PIV, which employed uniform global seeding instead of localised seeding. The PIV results in Fig. 8c (Mean) and Fig. 8d (RMS) beyond $r/R_2=\pm1.0$ reflect a truer velocity profile as opposed to the LDA results in that region. As explained before, this is due to poor localised seeding beyond the exit lip radius, R_2.

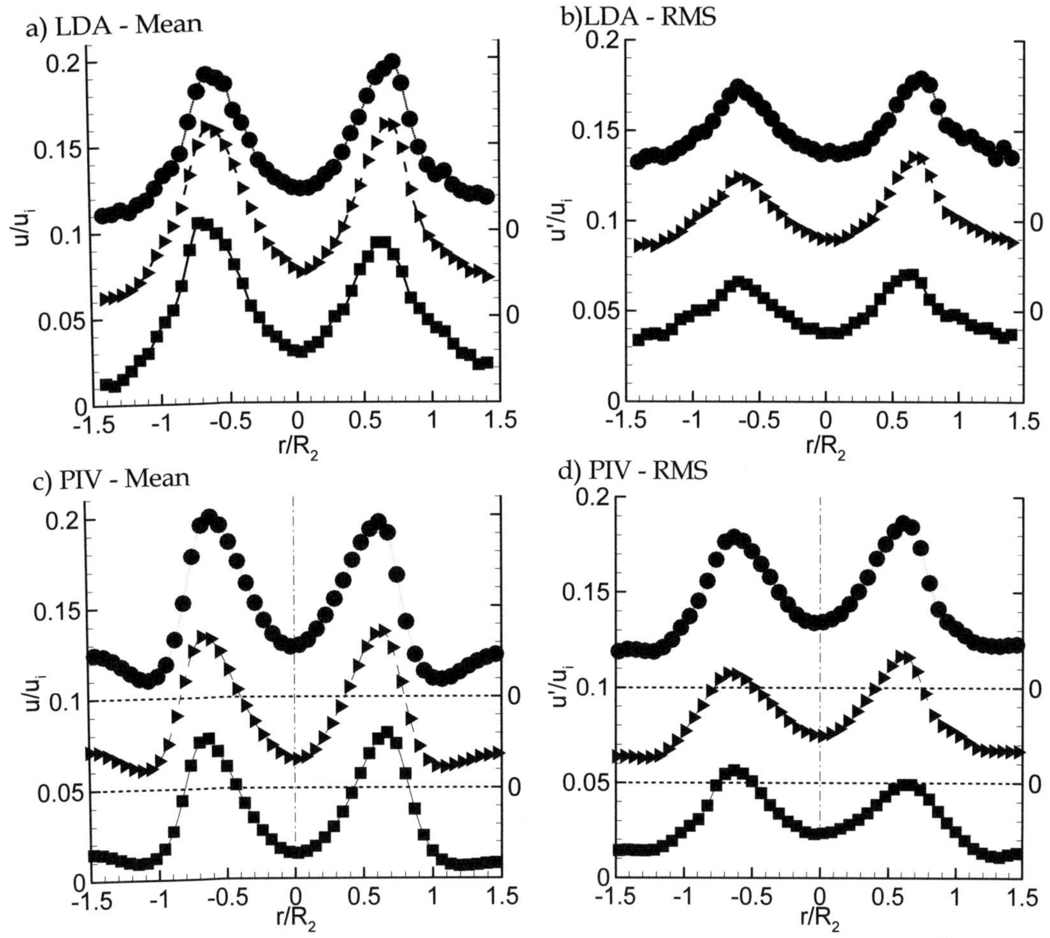

Fig. 8. Comparison between time-averaged LDA and PIV axial velocities at x'/D_2=0.16. The origins of the upwards-shifted ordinates are indicated on the right axis for contraction (▶) and orifice (●) inlets respectively. The left axis refers to the pipe case (■). All velocity values are normalised with the respective bulk inlet velocity u_i, for each inlet case. See Table 1 and Table 2, for LDA and PIV conditions respectively.

3.2 Comparison of conditional-directionally averaged PIV and conditional non-directionally averaged LDA results

Fig. 9a shows the radial distributions of phase-averaged axial velocity at x'/D_2=0.16, obtained using both LDA and PIV techniques. These were measured for Re_d=84,500, St_d=0.0015 and u_i=78.7m/s for the LDA, and Re_d=59,000, St_d=0.0017 and u_i=55m/s for the PIV technique respectively. The closed squares represent the transit-time weighted mean LDA results while the open triangles represent the mean PIV results (obtained at η=60°). Upper and lower confidence intervals (C.I.) at 95% confidence for the PIV measurements are represented as dotted lines and solid lines respectively.

Type of Inlet	Bulk Velocity, u_i m/s	Reynolds Number Based on u_i measured at x/d=1	Average number of LDA bursts per location	Probe volume: axial, radial, tangential (mm)	Step size (mm)
Pipe	103	111,000	13150		2
Contraction	140	150,000	14436	0.17×1.65×0.17	2
Orifice	96	103,000	15465		2

Table 1. Summary of experimental conditions used for the FPJ nozzles measured using LDA.

Type of Inlet	Bulk Velocity, u_i m/s	Reynolds Number based on u_i and d measured at x/d=1	N	Probe dimensions: axial, radial, tangential (*mm*)	Step size: axial, radial, mm	Δt (μs)	px/mm
Pipe	34.5	37,100	700	3.39×3.39×2	1.74	10	9.45
Contraction	29.7	31,900	1049	3.59×3.59×2	1.82	12	8.91
Orifice	26.4	28,400	1047	3.68×3.68×2	1.9	12	8.69

Table 2. Summary of experimental inlet conditions, in the absence of the nozzle chamber, used for the FPJ nozzles measured in time-averaged PIV experiments.

The phase-averaged LDA results broadly agree well, in terms of profile shapes and velocity magnitudes, with the PIV results within the range $-0.5 \le r / D_2 \le 0.1$. The LDA velocity measurements are generally larger than the PIV velocity measurements in the ranges $-0.48 < r/D_2 < -0.5$ and $r/D_2 > +0.1$. This can be explained by the differences in seeding methods. In the LDA experiments, the co-flow was only seeded from one local azimuthal location, i.e. it was seeded asymmetrically, so that the two measurements agree well for the side that was well seeded. However, the LDA will over-estimate the velocity for the region in which the ambient fluid was poorly seeded, so was subject to a velocity bias. In contrast, the seeding in the PIV experiment was uniform in the azimuthal direction.

In general, as shown in Fig. 9a, the axial velocity trend using both techniques rises from u/u_i=0 near to r/D_2= - 0.5 and peaks at approximately u/u_i=0.10, after which it decreases to a minimum near to r/D_2=+0.1 before rising again to a second maximum of u/u_i~0.085 at r/D_2=+0.30 for the PIV measurements and u/u_i~0.04 at r/D_2=+0.38 for the LDA measurements. There is an apparent shift of the second peak for the LDA results and also a decrease in the LDA axial velocities between $0.1<r/D_2<0.4$, despite transit-time weighted corrections to account for velocity bias. These differences may be attributed to a better phase and directional accuracy for the PIV measurements than for the LDA measurements.

Fig. 9b and Fig. 9c present the radial distributions of the phase-averaged PIV and LDA results for downstream locations at x'/D_2=0.48 and 0.79 respectively. Again, both techniques show similar axial velocity trends and have comparable velocity magnitudes, differing only by u/u_i=0.02 and 0.035 respectively. The larger magnitudes in the LDA results relative to the PIV measurements on the right-hand side of the jet are attributed to LDA velocity bias towards the higher velocities, as discussed above, noting also that the effectiveness of the external seeding of the measurement region decreases with distance downstream from the FPJ exit plane. Again, these differences may also be attributed to a better phase and directional accuracy for the PIV measurements than for the LDA measurements.

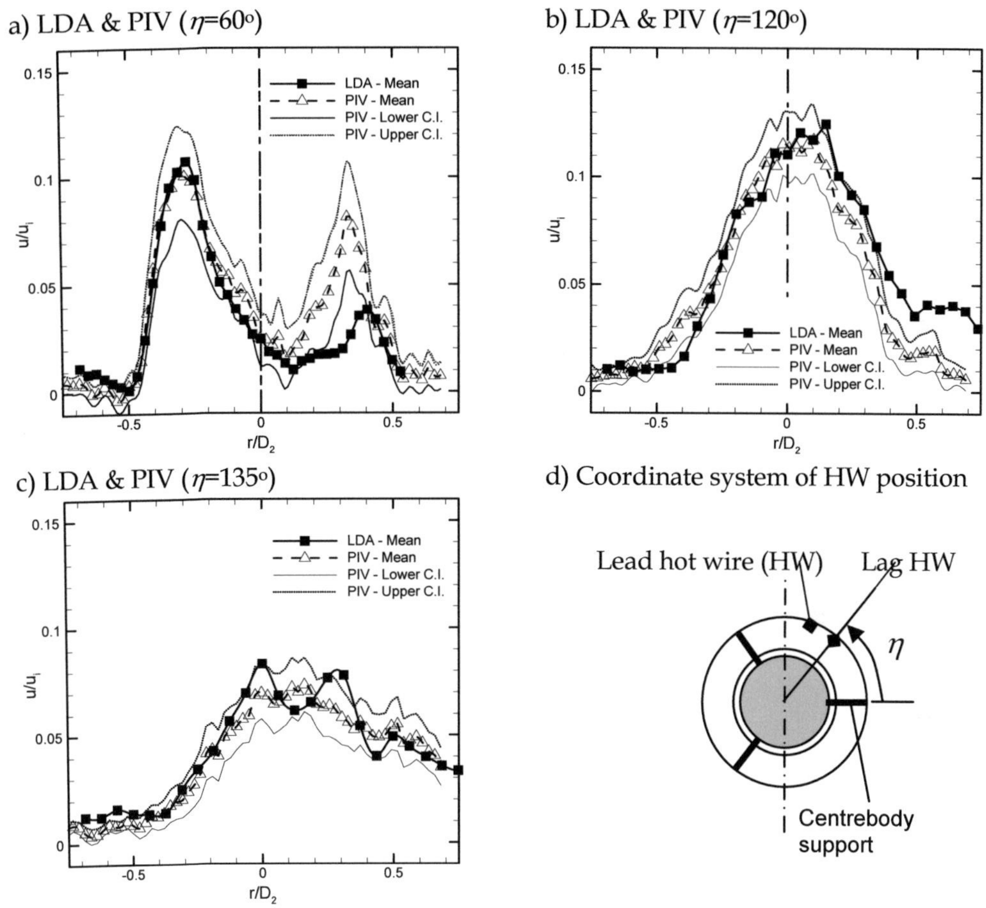

Fig. 9. Comparison between phase-averaged LDA and PIV axial velocities at x'/D_2=0.16, 0.48 and 0.79. Shown also are lower and upper 95% confidence interval (C.I.) for PIV results. Axial velocities are non-dimensionalised by u_i=78.7m/s for LDA and u_i=55.0 m/s for PIV. LDA:Re_d=84,500 and St_d=0.0015. PIV:Re_d=59,000 and St_d=0.0017.

3.3 Construction of flow topology

Fig. 10 illustrates the development of the 0.50 and 0.9 $(v^2+w^2)^{1/2}{}_{x',max}$ contours for various downstream sections from the exit lip taken from phase-averaged PIV measurements in the y-z plane. The jet is initially kidney-shaped (Fig. 10a), but as it converges to x'/D_2=0.58 (Fig.

10 f), it becomes an almost round jet. The approximate trajectory of the jet centreline in the transverse cross-section initially deflects to the left of the advancing side of the jet, but this deflection is most noticeable after the convergence location at approximately x'/D_2= 0.58.

Wong, Nathan and Kelso (2008) summarised the important features identified in the phase-averaged longitudinal (x'-r plane) PIV results. These were, a saddle above a reversed flow region near to the exit plane in the central region of the nozzle, a region of separation on the surface of the exit lip whereby the head of the focus points inwards, the mean extent of the flow convergence region, a region of separation on the surface of the exit lip where the head of the focus points outwards, and a conjectured streamline pattern describing the structure of an Edge 2 and an Edge 3 vortices. Some of these features are shown in Fig. 11a.

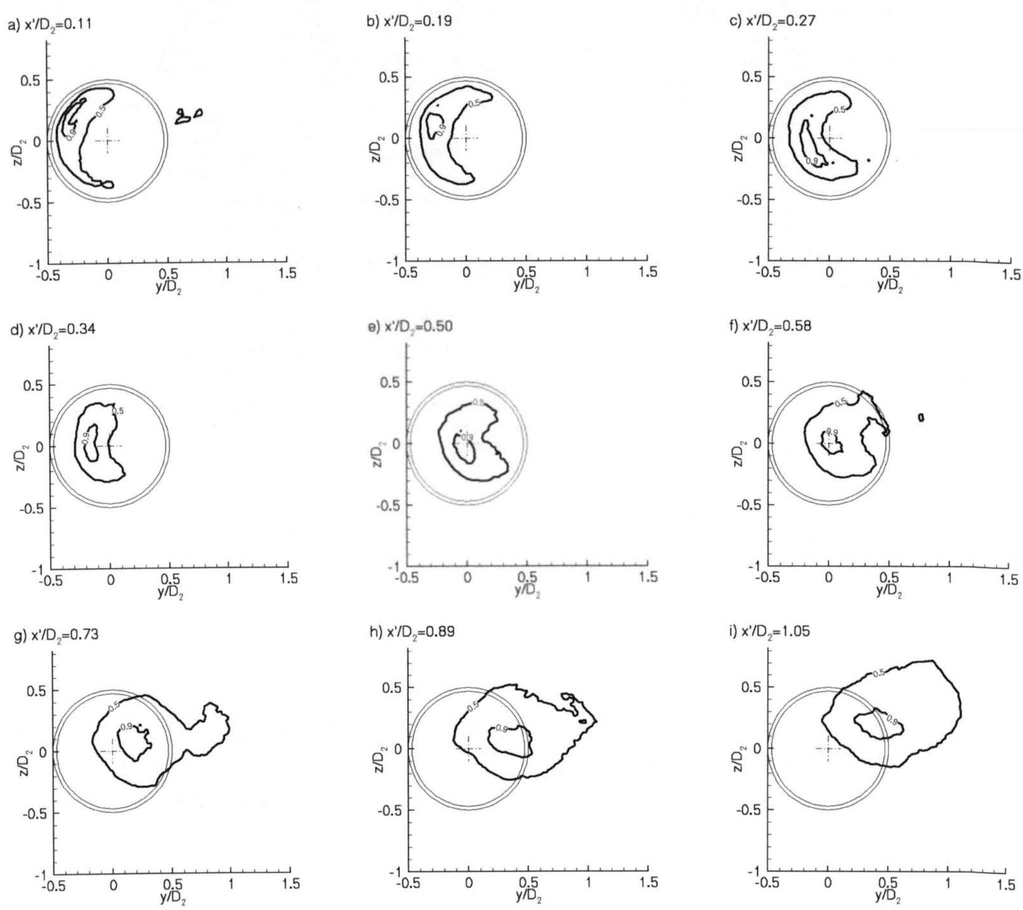

Fig. 10. Cross-sectional development of the jet downstream showing 0.5 and 0.9 $(v^2+w^2)^{1/2}{}_{x',max}$ contour levels. Note that by x'/D_2=0.58, the jet is almost round. D_2=64mm. Re_d=59K, St_d=0.0017.

The pair of vortices emerging from the middle of the centrebody represents the centrebody (CB) vortex pair which originate from the foci located to either side of the centrebody surface. These vortices move closer to each other with downstream distance and finally annihilate one another by $x'/D_2 \sim 0.5$. The thin black dashed line represents the vortex that is thought to originate from the edge of the centrebody (that is, the Edge 1 vortex). The 'legs' of this vortex form a pair of longitudinal vortices downstream of the centrebody (the Edge 1 vortex pair). These appear as positive and negative vorticity regions as seen by a downstream observer. This vortex pair departs the centrebody at an inclined angle, cutting the lower measurement planes at a large angle. Thus, the vortex pair appears as two kidney-shaped patterns near to the leading edge of the high velocity jet stream in the vorticity results of the transverse PIV measurements as given in Fig. 11a. Wong, Nathan and Kelso (2008) showed that the negative leg of this Edge 1 vortex follows the trajectory of the jet fairly closely. The positive Edge 1 vortex is initially located in the vicinity of y/D_2=+0.1 and z/D_2=+0.4, but at a distance x'/D_2=0.58 it moves towards y/D_2=+0.6 and z/D_2=+0.15 with increasing radius from the nozzle axis while departing in a direction opposite to the jet precession.

The thin blue solid line in Fig. 11c represents the core of a vortex that sits slightly above the surface of the exit lip and then lifts away from the exit at the rear side of the exiting jet (opposite the side from which the jet emerges). The lift-off occurs at an azimuthal angle of 130° to 150° from the center of the emerging jet. The legs of this lifted vortex form the 'Edge 2 vortex pair'. The vorticity associated with this vortex is observed most clearly in the longitudinal phase-averaged PIV experiments provided in Fig. 11b. The emerging jet separates at the exit lip to form a saddle-focus pattern at r/D_2=0.6 at x'/D_2=0.05 as illustrated in Fig. 11b. This pattern appears to be generated by the separation of the external flow entrained along the outside surface of the FPJ nozzle, and the separation of the exiting FPJ flow at the edge of the lip. The focus seen in Fig. 11b has a direction of rotation (clockwise) that is consistent with the vorticity that would be shed from the outer surface of the FPJ.

By inspection of Fig. 11a, Fig. 11b and the conjectured topology of Fig. 11c, the Edge 1 vortex and the Edge 2 vortex are of the same sign of circulation. If the emerging jet were to depart along the geometric axis (i.e. without deflection), then the Edge 2 vortex should remain in a closed 'circular' loop (or ring) sitting above the exit lip. However, the asymmetry of the emerging jet causes one end of this vortex loop to be entrained into the emerging jet flow between approximately 130° and 150° ahead of the advancing jet (Wong, Nathan & Kelso, 2008). This range of angles is similar to the separation angles of the surface flows in jets in crossflow (Fric & Roshko, 1994). Thus the edge vortex forms a horseshoe-shaped loop with either side of the loop being entrained into the advancing and receding sides of the jet. The lift-up of the Edge 2 vortex 'pair' from the surface of the exit lip appears initially to influence the trajectory of the Edge 1 vortex. The trajectories of the two 'legs' of the Edge 2 vortex follow the same trend as the Edge 1 vortex, in terms of the y-z trajectory with downstream distance, albeit with a slight delay in phase. The presence of the Edge 2 vortex pair below the core of the jet was detected in earlier phase-averaged LDA studies (Wong *et al.*, 2003) and in the $0.5(v^2+w^2)^{1/2}{}_{max}$ velocity contours of the phase-and-precession-direction-resolved PIV studies (see Fig. 10g).

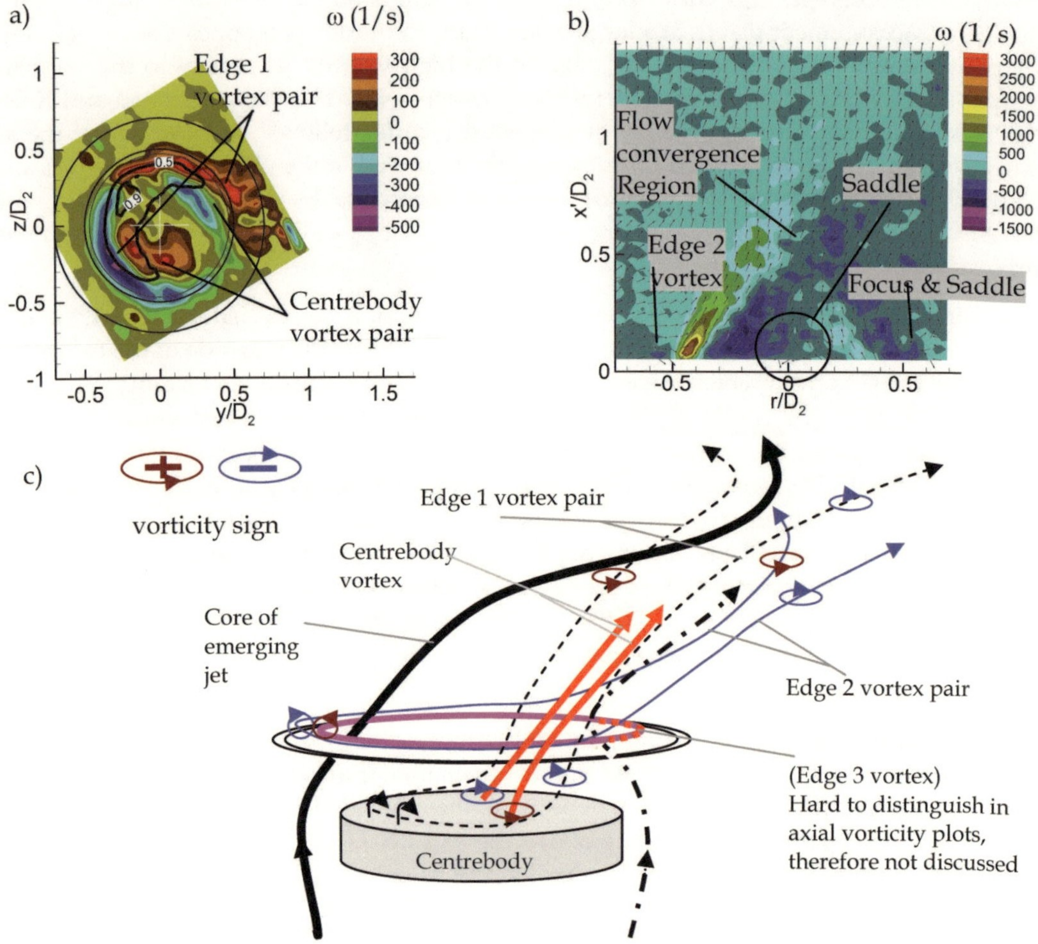

Fig. 11. Typical vorticity fields in the a) *y-z* plane @ x'/D_2=0.19 and b) *x'-r* plane @ η=60°. c) Qualitative interpretation of experimental data. (Adapted from Wong *et al.* 2008 and reprinted with permission from Cambridge University Press).

4. Conclusion

The study illustrates that a pre-requisite to the detailed application of laser diagnostic tools such as PIV and LDA is an adequate understanding of the qualitative features of the flow of interest. This is necessary for the design and careful application of appropriate triggering conditional statistics. When dealing with turbulent flow coupled with time-variant three-dimensional fluidic oscillations, it is important to match the flow conditions with the technical limits of the flow instrumentation. For the investigation of the fluidic precessing jet flow, the Schmitt filter was found to be effective in allowing two carefully-placed hot wire probes to provide a trigger to discriminate for both phase and direction of the oscillation, both of which vary randomly. This trigger allowed resolution of the azimuthal component of the flow velocity in the PIV data, which were smoothed out in the LDA measurements and also triggered on phase only, and not on direction. For the measurement of unconditioned statistics, such as the mean and *r.m.s.* statistical quantities, good agreement was found here between the PIV and LDA techniques for those regions of the flow in which both techniques were well seeded. As expected, some disagreements were found in those regions of the flow in which the seeding of the ambient air was poor, due to the non-uniform seeding technique, where the strong entrainment of unseeded ambient fluid led to a systematic velocity bias for the LDA case.

5. Acknowledgment

The authors acknowledge the Australian Research Council, Schools of Mechanical and Chemical Engineering (University of Adelaide) and Division of Mechanical Engineering (University of Wales, Cardiff) for various in-kind assistance, equipment and facility support. CYW also acknowledges the International Postgraduate Research Scholarship that enabled this research to be conducted. The authors are also grateful to the various peer-reviewers who have reviewed this manuscript.

6. References

Adrian R.J. (1997) Dynamic ranges of velocity and spatial resolution of particle image velocimetry. *Meas. Sci. Technol.*, 8, pp.1393-1398.

Durst F., Melling A. & Whitelaw J.H. (1981) *Principles and practice of laser-Doppler anemometry.* Academic press, London.

Fernandes E.C. & Heitor M.V. (1998) On the extension of laser-Doppler anemometer to the analysis of oscillating flames. In, *Proc. 9th Int. Sym. On Appl. Laser Techniques to Fluid Mechanics,* Lisbon, Portugal, 1, pp.3.3.1-3.3.9, 13-16 July.

Fick W., Griffiths A.J. & O'Doherty T. (1997) Visualisation of the precessing vortex core in an unconfined swirling flow. *Optical Diagnostics in Engineering,* 2(1), pp.19-31.

Fric T.F. & Roshko A. (1994) Vortical structure in the wake of a transverse jet. *J. Fluid Mech.,* 279, pp.1-47.

Guo B.Y. (2000) *CFD simulation of flow instability in axisymmetric sudden expansions.* PhD Thesis, Dept. of Chem. Eng., The University of Sydney, Australia.

Guo B.Y., Langrish T.A.G. & Fletcher D.F. (2001). Numerical simulation of unsteady flow in axisymmetric sudden expansions. Trans. ASME, *J. Fluids Eng.*, 123, pp.574-587, September.

Hart D.P. (2000) PIV error correction. *Exp Fluids*, 29, pp.13-22.

Hill S.J. (1992). *Characterisation of the flow fields produced by the enhanced mixing nozzle: A development towards high temperature and multi-phase applications.* Internal report, Dept. Mech. Eng., University of Adelaide.

Hill S.J., Nathan G.J. & Luxton R.E. (1992). Precessing and axial flows following a sudden expansion in an axisymmetric nozzle. In, *Proc. 11th Australiasian Fluid Mechanics Conference,* University of Tasmania. Hobart, Australia, pp.1113-1116, 14-18 December.

Hill S.J., Nathan G.J. & Luxton R.E. (1995). Precession in axisymmetric confined jets. In, Bilger R.W., editor, *Proc. 12th Australasian Fluid Mechanics Conference*, Sydney, pp.135-138, 10-15 December.

Hunt J.C.R., Abell C.J., Peterka J.A. & Woo H. (1978). Kinematical studies of the flows around free or surface mounted obstacles; applying topology to flow visualization. *J. Fluid Mech.*, 86, pp.179-200.

Kähler C.J., Sammler B. & Kompenhans J. (2002) Generation and control of tracer particles for optical flow investigations in air. *Exp. Fluids*, 33, pp.736-742.

Keane R.D. & Adrian R.J. (1990). Optimization of particle image velocimeters. Part I: Double pulsed systems. *Meas. Sci. Technol.*, 1, pp.1202-1215.

Kline S.J. & McClintock F.A. (1953) Describing uncertainties in single-sample experiments. Mechanical Engineering, 75, pp. 3-8, Jan.

Lancaster D. (1997). *CMOS Cookbook, Second Edition.* Newnes, United States of America.

Luxton R.E. & Nathan G.J. (1988). *Mixing of fluids*, Australian Patent Office (Patent Application No. 16235/88, International Patent Application No. PCT/AU88/00114).

Maltby R.L. (1962). *Flow visualization in wind tunnels using indicators.* AGARDOGRAPH, No.70.

Manias C.G. & Nathan G.J. (1994). Low NO_x clinker production. *World Cement*, 25(5), pp.54-56, May.

Martin P.B., Pugliese G.J. & Leishman J.G. (2000) Laser Doppler velocimetry uncertainty analysis for rotor blade tip vortex measurements. *AIAA 2000-0263*, 15pp.

Mi J. & Nathan G.J. (2000). Precession Strouhal number of a self-excited precessing jet. In, Ping Cheng, Ed., *Proc. Symposium on energy engineering in the 21st Century (SEE2000)*, Begell House, Hong Kong, 4, pp.1609-1614.

Mi, J., & Nathan, G.J. (2005) "Statistical analysis of the velocity field in a mechanical precessing jet flow", *Physics of Fluids*, 17, (1), 015102, 1-17.

Nathan, G.J. (1988) *The enhanced mixing burner.* PhD thesis Dept of Mech Eng, Uni of Adelaide, Australia.

Nathan G.J., Hill S.J. & Luxton R.E. (1998) An axisymmetric 'fluidic' nozzle to generate jet precession. *J. Fluid Mech.*, 370, pp.347-380.

Nathan G.J. & Luxton R.E. (1992) The flow field within an axi-symmetric nozzle utilising a large abrupt expansion. In, Zhuang F.G., editor, *Recent advances in experimental fluid mechanics*, International Academic Publishers, pp.527-532.

Nathan G.J., Mi, J., Alwahabi, Z.T. Newbold, G.J.R. & Nobes, D.S. (2006) Impacts of a jet's exit flow pattern on mixing and combustion performance, *Prog. Energy Combust. Sci.*, 32, (5-6), 496-538.

Newbold G.J.R. (1997) *Mixing and combustion in precessing jet flows*, Ph.D. Thesis, Dept. Mech. Eng., The University of Adelaide, Australia.

Nobes D.N. (1997). *The generation of large-scale structures by jet precession*, Ph.D. Thesis, Dept. Mech. Eng., The University of Adelaide, Australia.

Quantel (1994). *Quantel twins Q-switched double pulse Nd:YAG laser. Instruction manual issue no.1.* Quantel, France, May.

Ouwa, Y., Watanabe, M. & Asawo, H. 1981 Flow visualisation of a two dimensional water jet in a rectangular channel. *Japan. J. Appl. Phys.* 20, 243-247.

Parham, J.J. (2000). *Control and Optimisation of Mixing and Combustion from a Precessing Jet Nozzle*, Ph.D. Thesis, Dept. Mech. Eng., The University of Adelaide, Australia.

Perry A.E. & Chong M.S. (1987). A description of eddying motions and flow patterns using critical-point concepts. *Ann. Rev. Fluid Mech.*, 19, pp.125-155.

Potts J.R. & Crowther W.J. (2000). The flow over a rotating disc-wing. In, *Proc. Royal Aeronautical Society Aerodynamics Research Conference*, London, UK, Apr.

Raffel M., Willert C. & Kompenhans J. (1998) *Particle image velocimetry – a practical guide.* Springer, Germany.

Schneider G.M., Froud D., Syred N., Nathan G.J. & Luxton R.E. (1997a) Velocity measurements in a precessing jet flow using a three dimensional LDA system. *Exp. Fluids*, 23, pp.89-98.

Schneider G.M., Hooper J.D., Musgrove A.R., Nathan G.J. & Luxton R.E. (1997b) Velocity and Reynolds stresses in a precessing jet flow. *Exp. Fluids*, 22, pp.489-495.

Tobak M. & Peake D.J. (1979). Topology of two-dimensional and three-dimensional separated flows. In, *Proc. AIAA 12th Fluid and plasma dynamics conference*, Williamsburg, Virginia, USA, 79-1480, 23-25 July.

Videgar R. (1997). Gryo-therm technology solves burner problems. *World Cement Case Studies*, Nov.

Wong C.Y., Lanspeary P.V., Nathan G.J., Kelso R.M. & O'Doherty T. (2003). Phase-averaged velocity in a fluidic precessing jet nozzle and in its near external field. *J. Exp. Therm. Fluid Sci.*, 27, pp.515-524.

Wong, C.Y. (2004) *The flow within and in the near external field of a fluidic precessing jet nozzle.* Ph.D Thesis, Department of Mechanical Engineering, The University of Adelaide, Australia.

Wong, C.Y., Nathan, G.J. & O'Doherty, T. (2004). The effect of initial conditions on the exit flow from a fluidic precessing jet nozzle. *Exp. Fluids*, 36, pp.70-81.

Wong C.Y., Nathan G.J. & Kelso R.M. (2008) The naturally oscillating flow emerging from a fluidic precessing jet nozzle. *J. Fluid Mech.*, 606, pp.153-188.

Zdravkovich M.M., Flaherty A.J., Pahle M.G. & Skelhorne I.A. (1998). Some aerodynamic aspects of coin-like cylinders. *J. Fluid Mech.*, 360, pp.73-84.

Section 3

Micro-PIV Applications

10

μ-PIV for the Analysis of Flow Fields near a Propagating Air-Liquid Interface

Eiichiro Yamaguchi, Bradford J. Smith and Donald P. Gaver III
Department of Biomedical Engineering
Tulane University
USA

1. Introduction

1.1 Background

Understanding the micro-scale multi-phase fluid dynamics in the respiratory system during the airway re-opening process has been considered as one of the key aspects necessary to improve the survival rate of clinical treatments for various respiratory diseases.

In a healthy lung, the inside surface of respiratory airways is coated with a thin film of lipids and proteins called lung surfactants (LS) which dynamically alter the surface tension and stabilize airways and reduce the work of breathing. Certain clinical events, such as gastric content aspiration, pneumonia, near-drowning, toxic gas inhalation, or chest/lung trauma will trigger lung surfactant "inactivation". This inactivation is caused by the presence of LS inhibitors, including a variety of water-soluble and surface-active serum protein substances, such as albumin, fibrinogen, and IgG that are normally absent from the airway. These can leak through the capillary membrane from the damaged cells, causing LS to lose its ability to lower surface tension necessary for the lung function (Zasadzinski et al. 2010). Mechanical ventilation is inarguably a necessary life-sustaining form of medical intervention at this stage. However, since the leaked proteins inactivate the lung surfactants, the air-liquid interface exerts a wider range of excessive, irregular mechanical stresses and strains on the delicate and highly-sensitive tissues that make up the airways and alveoli of the lung, resulting ventilator-induced lung injury (VILI). This progressive failure will eventually cause more complex pathological conditions such as Acute Lung Injury (ALI) or Acute Respiratory Distress Syndrome (ARDS). Likewise surfactant deficiency, a hallmark of Respiratory Distress Syndrome of premature infants, leads to large interfacial stress that can damage airways and alveoli during an infant's first breaths which clears amniotic fluid from the airways is it introduces air to the deep lung and alveoli.

Multiple protective techniques to minimize airway epithelial damage have been developed including low-volume ventilation, positive-end-expiratory-pressure, and surfactant replacement therapy (Clements and Avery 1998). Although these techniques have contributed greatly to reducing instances of VILI, further improvements still must be made to better treat patients with fewer resulting complications (Gaver III et al. 2006). There are about 200,000 ARDS cases per year in U.S. with mortality rate of 40-46% (Mclntyre et al. 2000; Ware and Matthay 2000).

Gaver III et al. (1990; 1996) modeled the airway re-opening process by propagating a semi-infinite bubble into a narrow fluid-occluded channel with walls coated by pulmonary epithelial cells. Bilek et al. (2003) and Key et al. (2004) demonstrated using *in vitro* studies that the rate of cell damage increases with decreasing bubble propagation speed. With the aid of computational simulations, further analyses of the experimental results have indicated that the pressure gradient, not shear stress, is primarily responsible for cellular damage. The pressure gradient is the only stress component to increase with decreasing bubble velocity in the target flow conditions, indicating that excessive pressure gradients damage the cellular lining of the airway. That study has confirmed that the widely accepted stress prediction model of the collapsed airway reopening (Fig. 1) can be applied to rigid tube models.

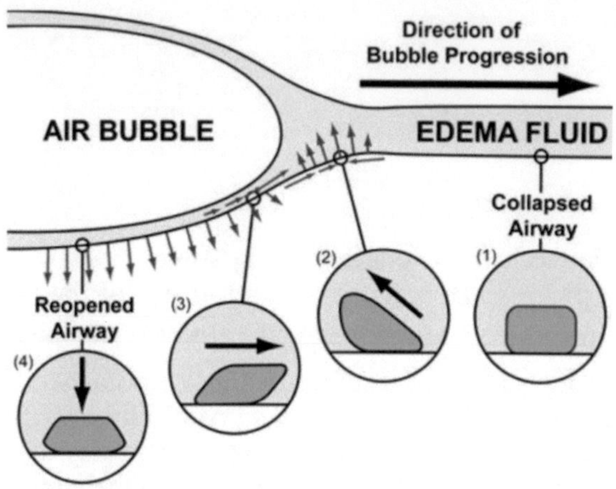

Fig. 1. Schematic of the stress distribution prediction from a model of flexible airway reopening (Gaver III et al. 1996). The cells far downstream are unstressed (1). As the bubble approaches, the cell is pulled up and toward the bubble (2). As the bubble passes, the cell is pushed away from the bubble (3). After the bubble has passed, the cell is pushed outward (4).

Follow-up studies by numerous groups have investigated this system and were found to be consistent with the theoretical predictions (Jacob and Gaver III 2005; Huh et al. 2007; Yalcin et al. 2007). Furthermore, Jacob and Gaver III (2005) included the influence of cell topography into the computational model, and found that the pressure gradient acting on the epithelial cells was even more pronounced than the original computations had demonstrated. These results have led to an interest in more detailed information of local fluid dynamics near the tip of the penetrating semi-infinite bubble. Smith & Gaver (2008) computationally investigated the flow fields surrounding a pulsating finger of air as it propagated along a rigid tube filled with a viscous Newtonian fluid. That study provides highly dynamic and intriguing spatial and temporal characteristics of the flow dynamics surrounding the bubble tip, and further elucidates the established relationship between pressure gradient and bubble velocity.

Since inactivated or insufficient amount of lung surfactants are one of important factors for controlling the cellar damage during the airway reopening, diffusive molecular transport of LS near the air-liquid interface including mechanism of competitive adsorption between lung surfactant and the inhibitors are well-studied areas (Refer to the review article of the surfactant inactivation by Zasadzinski et al. (2010) for detail.). It is predicted that for the airway reopening case, collapsing and re-spreading process of LS layer at the surface of the progressing semi-infinite bubble tip is dominated by convective transport driven by hydrodynamics movement of the liquid phase (Fig. 2). The complex convective transport dynamics occurring during pulsatile bubble propagation are further explored using finite time Lyapunov exponents (Smith et al. 2011). During the bubble forward and retrograde bubble propagation, converging and diverging stagnation regions exist on the bubble surface due to the flow circulation in the liquid phase. It causes non-uniform non-equilibrium surfactant concentration distribution along the interface. The experimental data of pulsating bubble surfactometer (Fig. 2-i) indicates that the distribution will cause significant difference of surface tension along the interface resulting non-uniform normal stress and tangential Marangoni stress distributions across the interface. Since the shape of the bubble tip interface and thickness of liquid deposited in the wake of propagating bubble can be defined by capillary number, $Ca = \mu U / \gamma$, where μ is the dynamic viscosity of the fluid, γ is the interfacial tension, and U is the mean flow velocity (Fairbrother and Stubbs 1935), wide deviation of the interface shapes will be expected from range of surface tension during the pulsatile motions (Fig. 2-ii). The non-equilibrium stresses and interface shapes during the retracting motion will reduce both the pressure drop and deleterious mechanical stresses on the epithelial cells (Ghadialli and Gaver, 2003; Ratulowski and Chang, 1990; Yap and Gaver, 1998). Theoretical studies by Zimmer et al. (2005) suggest that pulsatile motion

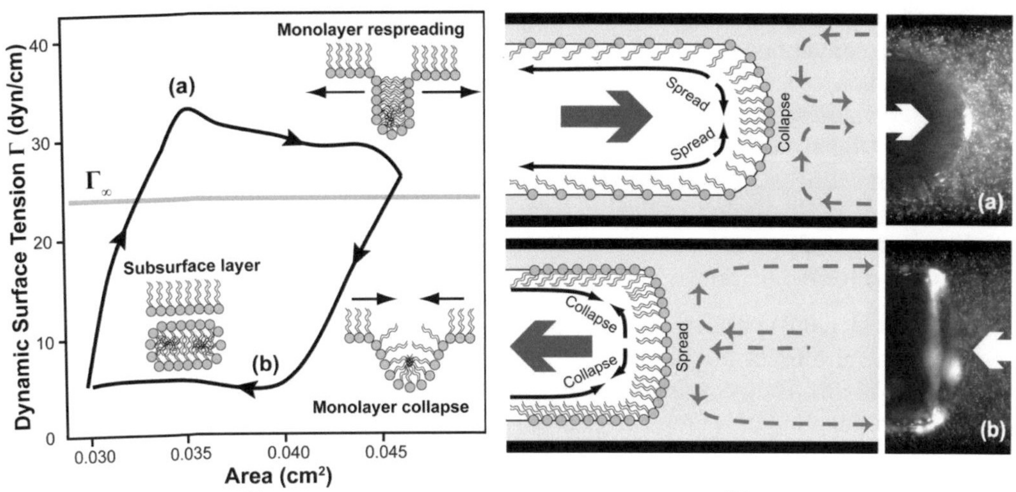

(i) (ii)

Fig. 2. (i) Hysteresis loop of dynamic surface tension from the pulsating bubble surfactometer, and graphical schemes of LS layer formation during (a) spreading and (b) collapsing of the surface (Krueger and Gaver III 2000). (ii) Prediction schemes of LS collapse and spread patterns during (a) forward progressing and (b) retracting mean motion of bubble tip. Bright spots on the fluorescent particle images on the right panels show particle flocculation indicating converging/diverging stagnation points in corresponding mean flow motions (Yamaguchi et al. 2009). The experimental observations and computer simulations indicate that pulsatile bubble motion generates a non-uniform non-equilibrium LS concentration across the interface, and the resulting Marangoni stress distribution alters the bubble interface shape accordingly.

may concentrate surfactant in high resistance regions near the progressing bubble. The bulk pressure drop measurement of a pulsating progressing bubble with LS in rigid cylindrical capillary (Pillert and Gaver III 2009) and *in-vitro* experiment of cell damage during the simulated airway reopening (Glindmeyer et al. 2011) confirmed this theory, and shown that pulsatile motion significantly reduced cell damage in comparison to linear motion airway reopening.

Due to the technical difficulties associated with whole field fluid dynamics measurements of such an unsteady interface, no experimental data are available to confirm the micro-scale relationship between the convection patterns in the sub-layer and LS distributions with local pressure and shear stress distributions. These factors also limit the investigation of LS inactivation, since identifying the difference between the inactivation and lack of LS in terms of the interface shape and the convection pattern will provide significant insight to understand the competitive adsorption mechanics.

1.2 Flow visualization of progressing semi-infinite bubble tip

Particle image velocimetry (PIV) is a well established and highly versatile technology for investigating the hydrodynamic properties of macro- and micro-scale flows (Adrian 2005). In micro-particle image velocimetry (μ-PIV) experiments, the entire test section volume is illuminated using a cone of light emanating from the recording lens, rather than the light sheet used in conventional PIV. The narrow focal depth of the objective lens results in a pseudo 2-D plane of focused particles for the correlation procedure (Santiago et al. 1998). A proper set of fluorescent seeding particles and an optical high-pass filter are employed to separate the signal from background noise. These innovations reduce the difficulties involved in creating a light sheet at the micron scale and in fabricating a microfluidic apparatus with multi-angled optical access windows that would otherwise be necessary with conventional PIV. As such, this technique is suitable to obtain instantaneous whole-field velocity information on the micro-scale.

Yamaguchi et al. (2009) performed μ-PIV investigations of flow fields surrounding a steadily progressing semi-infinite bubble tip in a smoothed wall straight glass capillary having a diameter of 312 μm. They employed a computer-controlled linear actuator system to control the steady propagation of a long finger of air, and let the bubble tip pass repeatedly under the fixed microscope observation window at controlled timing. The repeated data acquisition was necessary because the contributions of Brownian motion to an instantaneous velocity field are relatively significant and cannot be ignored due to the micron-size of the seeding particles, combined with their small displacements near the capillary tube wall. The location of the air-liquid interface in the fluorescent particle image was statistically estimated from accumulated instantaneous vector fields by identifying the significant rise of velocity fluctuations across the interface. The ensemble averaging method successfully reduced related errors for the steady low-Reynolds number flow measurements. The obtained ensemble averaged flow fields agree very well with the previously obtained computational simulations. Later the group developed a translating microscope stage system (Fig. 3) to vastly improve the capability and efficiency of image acquisition under complex flow patterns (Smith et al. 2010). The translating stage is a computer controlled sliding microscope stage that is programmed to give a counter motion opposite to the mean flow input. Therefore, in theory, the progressing bubble tip stays in the microscope observation window regardless of the direction of bubble propagation. The combination of linear actuators and the translating stage allowed the tracking of progressing bubble tip over 10-12 cm continuously, and made it possible to obtain the flow fields under realistic airway reopening patterns, such as simulated pulsatile ventilation.

Even though it is still possible to statistically estimate the geographical location and shape of the air-liquid interface, the method does not provide sufficient accuracy to examine the bubble tip shapes under a complex flow input where the interface may experience large deformations imposed by a quick alteration of local surface tension resulted from collapsing and re-spreading of the lung surfactant on the interface. Separately obtained ensemble averaged shadowgraph images under the same flow condition could help to improve the estimation of the bubble shape deformations in relatively stable case. Smith, et al. (2011) utilize shadowgraph and µ-PIV data to examine molecular transport phenomena near the interface of the pulsating bubbles by combining the experimental data and the Finite-Time Lyapunov Exponent (FTLE), and the Boundary Element Method (BEM). They used a strobe flash lamp as the backlight illumination source, and repeat the same data acquisition process they did for µ-PIV with the addition of a custom-made microscale position indicator. Since the shadowgraph image does not contain any spatial or dynamic information to connect to corresponding PIV data, the absolute position of the bubble tip was the only information used to estimate the mean velocity input at a given instant. The different refractive index of air and liquid phases imposes a clearly visible interfacial shadow on the image. The instantaneous interfacial shapes were then averaged in discrete intervals determined by the position indicator. The ensemble-averaged bubble shapes were then superimposed on the previously obtained flow fields to give a complete description of the flow field near the interface. Although the separate shadowgraph and PIV acquisition provided satisfactory results for this simple case, the ambiguity on the estimation of mean velocity from the position indicator, difficulty of reproducing multiple identical experiments, and complex post-processing scheme will pose clear limitations for the detailed flow analysis of more unstable and realistic flow conditions.

In order to overcome these disadvantages, the simultaneous use of PIV and a pulsed shadowgraph technique to detect the precise position of air-liquid interface in slug flows for relatively large size capillaries (about 2-3 cm of diameter) was performed by Norueria et al. (2003). That group utilized a Nd:YAG laser sheet illumination technique originally used for PIV and a sheet of LED arrays for uniform backlight illumination for the pulsed shadowgraph. A single camera was used to acquire PIV and shadowgraph data simultaneously. The image was then separated during the post-processing to give an accurate instantaneous interface shape for the flow field analysis. Although it has provided sufficient accuracy and great simplicity for the relatively large scale slug flow measurement, the single camera and post-process separation are not suitable for the current microscale multiphase application. The reliability of the post process separation is limited by the low signal-to-noise (S/N) ratio at the center of the tube and near the air-liquid interface. Here, the combination of the volumetric illumination technique (µ-PIV) in conjunction with a non-uniform channel depth provides the greatest amount of noise along with strong random refraction patterns of the fluorescent emission. To compensate for this type of problem, Meinhart et al. (2000) proposed a phase-averaged velocity field algorithm to overcome these low signal-to-noise ratio situations; however, this technique is difficult to implement in a multi-phase flow scenario. Additionally, Mielink and Saetran (2006) proposed a selective seeding method to reduce out-of-focus noise. Unfortunately, the 3-D nature of the flows investigated in our study make this technique untenable.

It is, therefore, necessary for the current application to optically separate the simultaneously acquired μ-PIV and shadowgraph images, and then record these images on two separate identical monochrome CCDs. The separation can be done by combining the proper choice of illumination sources, fluorescent materials, and epi-fluorescent filters. The images from two CCDs are then combined in post-processing to give accurate air-liquid interface information for the vector interrogation of the fluorescent images. Even though the implementation of the two cameras increases the complexity of the system development especially for optical adjustments and calibrations, it is expected to; i) be more affordable than using a three-CCD PIV capable camera, ii) have cleaner image separation by the eliminating signal interferences, and iii) have simpler post-processing.

2. Experiments

The development and descriptions of two-camera two-laser μ-PIV/Shadowgraph simultaneous data acquisition system will be described in this section. The includes the experimental setup and a brief explanation of the previously developed flow controlling system and the translating stage. Since the actual observation of the complex micro-multi-phase flows that include measurement of fluid flow fields in the neighborhood of propagating semi-infinite bubble having unsteady motions under influence of the lung surfactant can only be accomplished by combining and controlling the custom-made apparatus, the flow generating system, and the translating stage for continuous data acquisition, it is important to have dedicated subsections for each component.

2.1 Assembly of flow apparatus and the translating stage system

Continuous data acquisition of the progressing semi-infinite bubble is required for the current applications. It would not be practical to have a combination of a wide view lens with a super hi-resolution CCD in order to maintain enough spatial resolution for such a moving interface. Instead, the implementation of an interface tracking device and a moderate resolution CCD to keep the bubble tip within the fixed microscope observation window is a more practical and versatile solution to the target application. The computer-controlled translating microscope stage has been developed for this purpose. In summary, the translating stage keeps the relative position of the bubble tip within a fixed microscope observation window by mechanically sliding the entire microscope observation stage to cancel the given flow input. (a tread mill would be the closest analogy.) Precision 3-axis adjustable mechanisms for the flow apparatus allow alignment under the microscope for a flexible channel and the objective lens choices. The stage motion is controlled by attaching a computer-programmable linear actuator. A detailed technical description of this device is found in Smith et al. (2010) (Fig. 3-i provides a computer rendering of the translating stage.).

A uniform cylindrical channel with a smooth wall is the appropriate choice for the current experimental setup to the simulate airway reopening process, since most of the previous theoretical/experimental studies were associated with the similar geometrical configuration as discussed in previous section. A capillary tube made of fused silica coated with a polyimide protective layer (Flexible Fused Silica Capillary Tubing, Polymicro Technologies, AZ) was selected because of good optical transparency and flexibility. The capillary tube was trimmed to approximately 30cm in length and the middle section (about 10 cm) of the external protective layer was carefully removed to increase the optical transparency of the observation region. It should be noted that the removal of the protecting layer will drastically decrease resistance to any kind of stress inputs, and the testing section will become extremely fragile. However, it was necessary to overcome the low transparency of the test liquid (as discussed below in the sample preparation section) in order to maintain the required fluorescent seeding density. Therefore, the removal of the protective coating would not be required for certain low seeding density applications. The capillary generally provides good transparency under Nd:YAG laser with the protective layer (Natrajan et al. 2007).

Two identical capillary tubes were closely aligned in parallel, and glued onto a 12.5x5 cm microscope slide at both ends. In order to make the capillary section straight, a custom-made holding clamp was used to provide a small amount of tension while adjusting the capillary positions on the slide glass. A custom fabricated water jacket was attached on the top of the glass slide and carefully sealed, allowing temperature control of a water reservoir at 37C during the experiments. The finished apparatus has at least 10cm of continuous observation section for two identical capillary tubes in parallel. The setting of the parallel channels was necessary to capture the bubble tip motion (the upper channel) and the stage movement (the lower quiescent channel) simultaneously (Fig. 3-iii). Even though the motion of the translating stage system is controlled by a precision linear motor, it was impossible to eliminate many sources of mechanical vibrations during the experiment. Therefore the acquisition of precise instantaneous stage movement was necessary to accurately recover actual subtracted mean velocity from the pulsatile bubble motion. To do so, the μ-PIV was used to simultaneously measure the solid body motion of a 'tracking channel' attached to the stage in order to identify the stage velocity. The 3:7 of Glycerol:Water solution with 0.02 vol% of fluorescent particles (see the sample preparation for the seeding particle choice) was mixed to achieve a neutral buoyancy, and the tracking channel was filled with the liquid and sealed on both side to prevent any external input during the experiment. The experimental channel was connected to 1/16" PEEK tubing through sets of micro-tubing attachments (Upchurch, CA) to connect the flow generating unit. Finally, the apparatus was set on the translating stage to perform continuous data acquisition of a progressing bubble over 10-12 cm of capillary under various temperatures in single experiment.

2.2 Flow generating system and the pulsatile flow input setting

An electromagnetic precision actuator system (Electromagnetic direct linear motor P01-23x80/30x90 and E200-AT, LinMot Incorporated, Switzerland) was employed to control the bubble movement. The linear motor system consists of a magnetically driven actuator with a computer-based feedback and control system. The actuator is composed of a fixed stator with a linearly-aligned electromagnet and a position sensor for feedback, paired with a magnet-filled slider. It is driven by a servo controller that is run using supplied programming and monitoring software (LinMot-Talk, LinMot Inc.). The actuator has a speed range of 1 mm/sec to 20 cm/sec with less than 0.1 % position error at a 10 cm stroke. The translating stage and appropriate sizes of micro-syringes were attached directly to the slider to provide a wide range of flexibility and control.

For the current experiment, the pulsatile flow input for simulated airway reopening is defined as the linear addition of a constant mean velocity and a sinusoidal oscillation (Fig. 3). The equation is given as

$$Ca(t) = Ca_M + Ca_\Omega \sin(\Omega t), \tag{1}$$

where Ca_M, and Ca_Ω are the mean and oscillatory capillary numbers (dimensionless velocity expression) where $Ca(t) = \mu U(t)/\gamma$, with μ is fluid viscosity, γ is the surface tension, $U(t)$ is the cross-sectional average bubble velocity, $U(t) = Q(t)/\pi R^2$, where $Q(t)$ is the time-dependent flow rate. Therefore integration over the experimental duration gives the instantaneous actuator position for the flow input of the pulsatile motion;

$$Z(t)_{total} = U_M t + A\cos(\Omega t), \tag{2}$$

where V_M is the mean velocity, and A is amplitude of the oscillation.

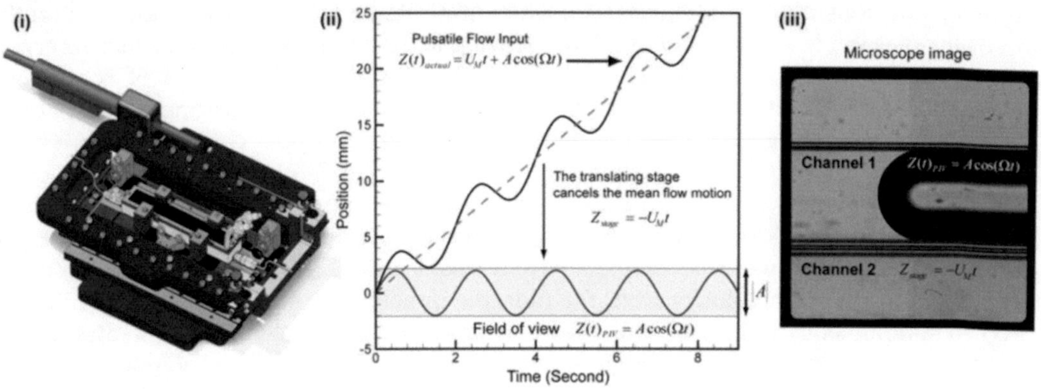

Fig. 3. (i) Scheme of the translating stage system (Smith et al. 2010), (ii) description of the flow generating system input/output, and (iii) sample microscope image of the testing section. The flow generating system is connected to channel 1, while channel 2 in Fig. 3-iii is used as the stage velocity indicator. The translating stage slides the entire microscope stage to cancel the mean bubble motion. The oscillating motion stays in the fixed microscope observation window. It significantly increases data acquisition ability and efficiency of measuring complex time dependent flows.

In order to examine the significant interfacial deformation and alterations in the flow pattern, parameters must be carefully selected to allow bubble tip retraction during the oscillation cycle without applying unrealistic frequencies and amplitudes. For the current experiment, mean bubble velocity U_M=5.5 mm/sec, amplitude A=2.0 mm, and frequency Ω=2 Hz provide a balance of the continuous acquisition efficiency over 10 seconds and the accuracy of the flow driving system. Since the target capillary tube diameter was 552 μm (see sample preparation section), a 25 μl syringe (Gastight Syringe 1707, Hamilton Company, NV) and 1 μl syringe (Microliter Syringe 7000.5, Hamilton Company, NV) were chosen to drive the mean and oscillatory component respectively. Therefore the input parameters for actuator operation were determined to be $U_{M\text{-}actuator1}$=2.97 mm/sec, $A_{actuator2}$=26.97 mm, and $\Omega_{actuator2}$ =2 Hz. The parameters fit within the range of acceptable operational error of less than 0.1 %. The same linear actuator attached was also attached to the translating stage, and was programmed to give $U_{M\text{-}actuator3}$=-5.5 mm/sec during the data acquisition. Therefore the fixed microscope window captures the bubble tip throughout the experiment. As mentioned above, a secondary capillary was used as the instantaneous stage velocity indicator, so the recovery of absolute input velocity and cancellation of mechanical noise from the translating stage can easily be obtained from this information.

2.3 System configuration and optical setup

The μ-PIV/Shadowgraph simultaneous data acquisition system developed for the current application is depicted in Fig. 4. An inverted microscope (Nikon Eclipse TE2000-U, Nikon Corporation, Japan) with a 10x objective lens (NA=0.30 Plan Flour, Nikon Corporation, Japan) and two identical 2048x2048 pixel CCD cameras (12 bit, 4MP, TSI POWERVIEW Plus, TSI Incorporated, MN) provides the observation area of 1523x1523 μm with a theoretical pixel resolution of 0.744 μm/pixel. Two cameras are attached to the side optical port by using double port adapter (Y-QT, Nikon Corporation, Japan).

For the μ-PIV, the volumetric illumination was provided by a dual pulse Nd:YAG laser (λ=532 nm, Power=15mJ/pulse, Pulse duration=4 ns, New Wave Laser Pulse Solo Mini, New Wave Research, CA). The beam was directed via fiber optics to an optical port on the microscope. It was then refracted by the epi-fluorescent prism/filter cube and guided through the objective lens to volumetrically illuminate the whole test section under the field of view. The sample liquid inside of the capillary tube was seeded with d_p=1 μm fluorescent particles (refer to sample preparation section for details) that have excitation/emission peaks at 535/575 nm. Therefore only returning emission from the particles passed the dichroic filer (λ>550 nm) in the first cube. Finally, it was navigated to CCD Camera A through the second filter set (epi-fluorescent prism/filter cube 2 in Fig. 4.) to provide a pure fluorescent particle image of the test section.

For the shadowgraph, an LED pulsed red laser (λ=660 nm, Power=2nJ/pulse, Pulse duration=15ns, MPL-III-660, Opto Engine LLC, UT) was selected for source of the backlight illumination. The laser is directed via fiber optics to the top of the flow apparatus where the collimator (NA=0.25, f=36.01mm, F810FC-780, ThorLabs, NJ) is attached directly above the observation window providing approximately 1.0 cm diameter spot of uniform illumination. The projected shadowgraph image signal shares the same optical path with the μ-PIV signal. It passes the first and second dichroic filters, and was navigated to CCD Camera B to record the shadowgraph image of the air-water interface.

The epi-fluorescent prism/filter cube 2 was selected to minimize cross-talk especially on CCD Camera A due to the signal-to-noise ratio sensitivity of vector interrogation process. Since images on CCD Camera B were used only to detect the bubble interface shape, and unmistakable sharp contrast at the air-liquid interface, clarity of the shadowgraph data was often dominated by setting of the illumination strength. A dichroic prism having a combination of a low-pass filter at 625 nm (625DCLP, Chroma Technology Corp, VT) and a cleaning band pass filter of 595nm/±40nm (D595/40x, Chroma Technology Corp, VT) provided excellent separation of the fluorescent particle image for CCD Camera A.

Timing of the cameras and lasers were controlled by a multi-channel laser pulse synchronizer (Model 610035, TSI Inc.) which also sends a queue signal for the linear actuator

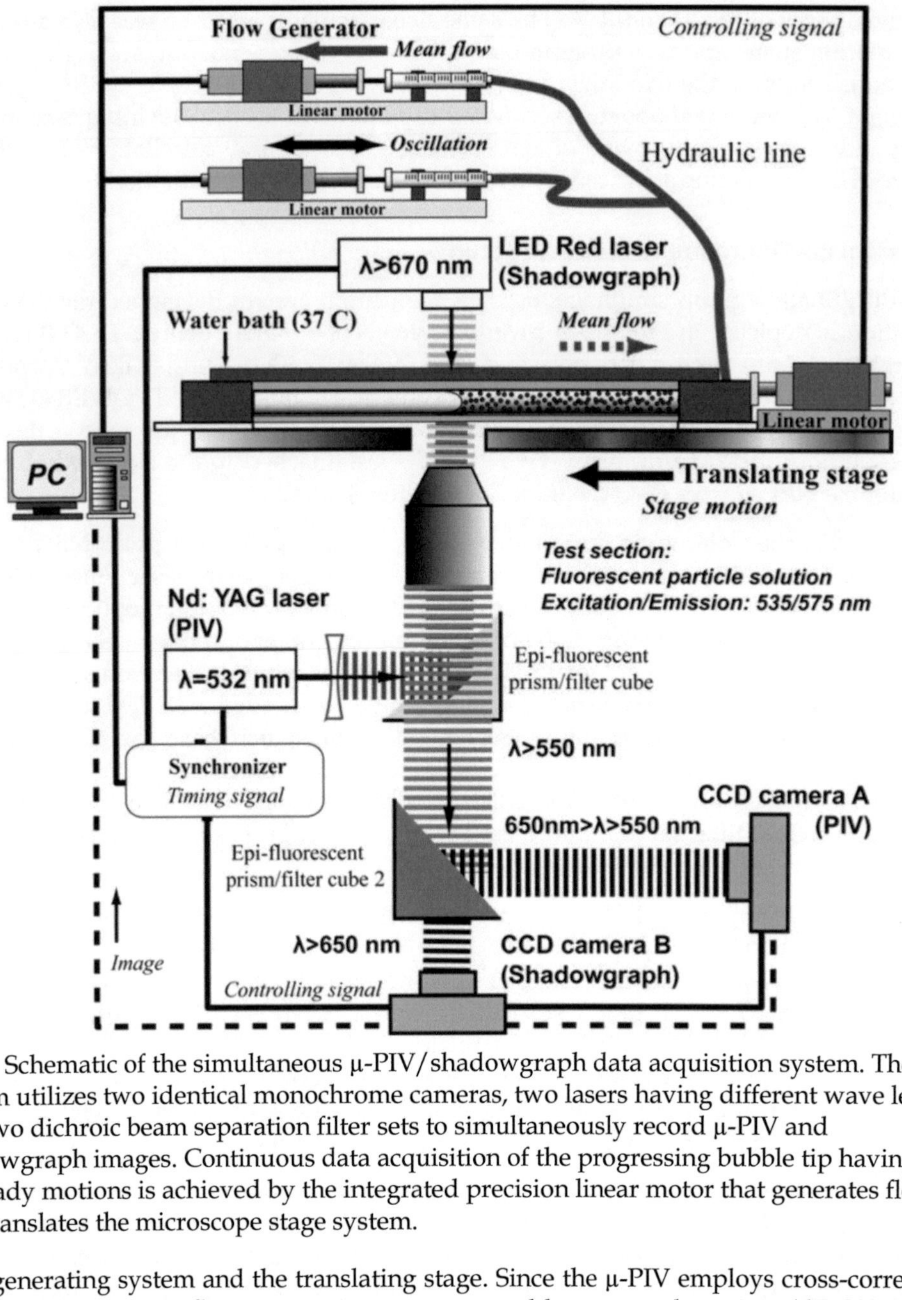

Fig. 4. Schematic of the simultaneous μ-PIV/shadowgraph data acquisition system. The system utilizes two identical monochrome cameras, two lasers having different wave length, and two dichroic beam separation filter sets to simultaneously record μ-PIV and shadowgraph images. Continuous data acquisition of the progressing bubble tip having unsteady motions is achieved by the integrated precision linear motor that generates flow and translates the microscope stage system.

flow generating system and the translating stage. Since the μ-PIV employs cross-correlation analysis, two separate fluorescence images separated by a very short time (dT=200-700 ns) were captured at every data acquisition time-point. On the other hand, the shadowgraph requires only one frame and one exposure to obtain the data at the same time-point. The timing and control signals for the cameras and lasers were, therefore, adjusted to capture one shadowgraph image at the timing of the first frame exposure of the μ-PIV. The system management, data acquisition control, and image display are controlled by Insight 3G (TSI Inc., MN).

2.4 Preparation of sample testing liquid and preliminary stability test

Nile Red was selected as fluorescent material for the current μ-PIV application due to excitation/emission peaks combination at λ=535/575 nm, since the emission spectrum was ideal for the combination with the LED red laser (λ=675 nm) for the shadowgraph. Then fluorescent particle diameter was determined as d_p=1 μm (Nile Red FluoSpheres, Invitrogen Corporation, CA) based on previously discussed pixel resolution of the observation window.

The vector interrogation window size (32x32 pixels with 50 % overlap) was determined by combination of a pixel resolution, and capillary diameter. For the current research objectives, it was preferred to have at least 40 vectors across the diameter of the capillary in order to allow detailed examination near the air-liquid interface. Since the capillary diameter D=552 μm has been set by balancing the slowest reliable velocity of the linear motors and the maximum experimental duration of the chosen syringe size, *552/(0.744*40)=18.55* pixels/vector provide sufficient resolution. Particle seeding density was determined by trial and error using measurement of Poiseulle flow with the same capillary tube and the same pixel resolution after initial estimation of density by using the methodologies described in Olsen and Adrian (2000) and Meinhart et al. (1999). Visibility is lowest at the center of the capillary because the transparency of the liquid is limited by the concentration of lung surfactant and other proteins. Therefore, the optimal seeding concentration was determined by limiting vector loss during the interrogation at the centerline to less than 1.0 % and the axial vector fluctuation of the time ensemble averaged velocity to within 2.0 % in the $r < 0.9R$ region, where r is radial coordinate and R is radius of capillary. The seeding density 0.02 vol% with d_p=1.0 μm particle for the current setup was determined to meet there criteria.

Dulbecco's Phosphate Buffered Saline 1x (DPBS) (Invitrogen, CA) was used as a base buffer solution. This is the buffer solution adjusted to have the same osmolarity and ion concentration as that of human serum. It also contains the necessary ions, such as sodium chloride, sodium phosphate, and potassium phosphate, for the lung surfactant proteins to be functional. Infasurf (calfactant) (35 mg/ml concentration, ONY Inc, NY) was chosen for the current experiment as LS. It is a lavage of natural calf lung surfactant with no tissues which is used clinically for RDS treatment. The mixture of DPBS and 0.02 vol% of the fluorescent particle was set as the standard 'DPBS solution' for the series of experiments. 0.01 mg/ml of Infasurf was added to make 'Infasurf solution' for the current experiments. The concentration 0.01 mg/ml of Infasurf is very low in comparison with previously reported values (Glindmeyer et al. 2011). However it was a sufficient concentration to observe the surface deformation under influence of LS in the current application. It was also convenient to use the lowest concentration for the system evaluation in order to avoid the unwanted external noise factors, such as, colloidal stability and the wall contamination issues. The Infasurf solution had to be kept at 37C before and during the experiment in order for LS to work properly. It was also necessary to use it within two to three days after opening original Infasurf bottle.

Colloidal stability of the fluorescent particle is critical to ensure the quality of μ-PIV data. For investigations of airway reopening phenomena, it is necessary to use specific electrolytes, organic compounds, and proteins in order to evaluate LS functions during the measurements. The functional properties of the fluorescent particle are highly sensitive to the balance of electrical charge between the particle surface and the buffer solution.

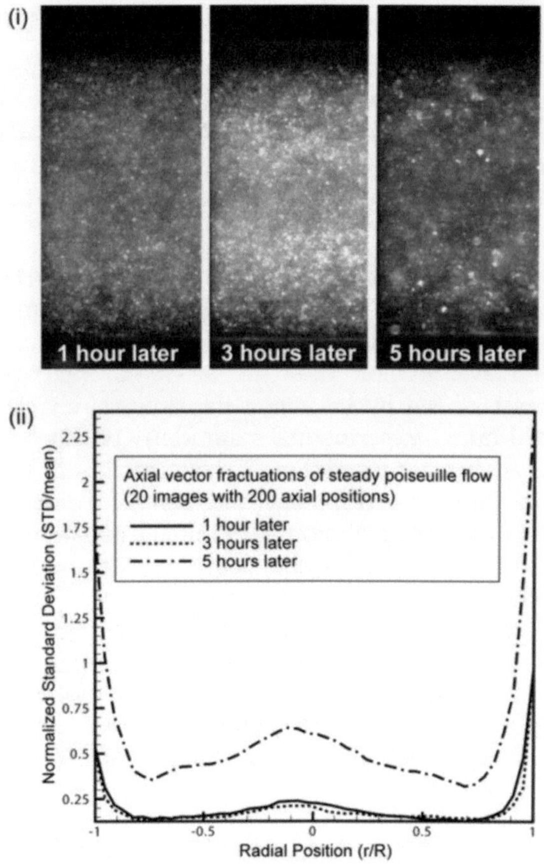

Fig. 5. Colloidal stability tests of 1.0 mg/ml Infasurf solution with 5.0 mg/ml albumin. (i) μ-PIV images of fluorescent particle flocculation at1, 3, and 5 hours. (ii) Velocity fluctuations along the channel radial direction over 100 axial positions during steady Poiseuille flow (U=5.5 mm/sec) in the cylindrical channel. After 5 hours, the particle aggregations are visible in the fluorescent images and affect on the vector interrogation result.

Even though it is safe to assume that fluorescent particles in the standard PBS are stable, since the pH of DPBS is 7.1±0.1 and the surface of the particle is cationic (Invitrogen 2004), colloidal stability under influence of Infasurf and additional LS inhibitors, such as albumin, must be carefully evaluated prior to the design of the experimental process. In order to verify the colloidal stability of the particle solution, the accuracy of μ-PIV data with a solution of 1.0 mg/ml of Infasurf with 5.0 mg/ml albumin using steady Poiseuille flow was evaluated. The capillary was filled with the solution and kept at 37C. As soon as the system reached stable temperature, the linear motor system was set to give a constant forward mean velocity at U=5.5 mm/sec , and 20 μ-PIV images were recorded. The process was repeated every 30 minutes for 5 hours. The velocity fluctuations of the ensemble axial velocity component over 100 axial positions were computed for the reliability evaluation of the vector interrogation over the progressing particle flocculation (Fig. 5). The result shows that the solution was stable for the μ-PIV data acquisition at least 3 hours after it was mixed. However, after 5 hours particle flocculation was severe and clearly identifiable either by the visual conformation of the image or the quality of vector interrogation near the centerline. It should be noted that the experiment was design to 'boost' particle flocculation effect by adding excess amount of Infasurf and albumin. In fact, the effect of the flocculation could not be identified for the current 0.01 mg/ml Infasurf solution even after 50 hours.

2.5 Experimental procedure and data processing

The data acquisition frame rate was set to 7.5 images/sec (15 pulses/sec for the Nd:YAG laser, since the μ-PIV requires 2 frames to obtains one vector field.), and the duration of a single continuous trial was set to 60 images/run. This requires a minimum of 60/7.5=8 seconds of continuous data acquisition. Since the bubble mean velocity was defined to be 5.5 mm/sec and the translating stage provided about 10-12 cm of continuous observation window, this setting gave enough safety margin for pre- and post- adjustment time for the flow generating system.

The pulsatile frequency was Ω=2 Hz, therefore each experimental run could capture 16 cycles with 3.75 images/cycle. Repeating data acquisition and cycle averaging were employed to increase the temporal resolution (the number of image capturing points per cycle) and accuracy of ensemble averages (total number of images at the same instantaneous flow condition) to fully analyze the effect of LS on flow patterns. The stability of the testing solution shown in the previous section ensures repeatability of experimental trials. By initiating each trial with a different cycle phase, the temporal resolution was increased by accumulating the images from each repeating trial. Cycle averaging over the same experimental run was also applicable to the current application to increase the temporal resolution. Pulsating bubble surfactometer experiments have shown that cycle-averaged surface tension and the hysteresis loops of Infasurf solution were saturated after 3-4 cycles and became independent of the number of cycles (Krueger and Gaver III 2000). Therefore, the data acquisition was started after the solution passed at least ten cycles. Since the frame timing returned to the same point in the cycle every 15 image acquisition (every 4 seconds, or 4 cycles) a single continuous acquisition with cycle averaging provided 15 images/cycle with 4 images for each time-point in the cycle. In the current experimental setup, the oscillation amplitude was A=2 mm, which is wider than the microscope's field of view. Since at least 550 μm of the liquid phase ahead of the bubble tip was necessary to obtain the instantaneous downstream velocity for each image, in order to compute ensemble average, it was necessary to maintain the bubble at least 1000 μm from end of the image. Therefore at least two acquisitions focusing on different part of the cycle were necessary to capture the entire cycle with sufficient liquid phase visible in the field of view.

The simultaneous μ-PIV/shadowgraph data acquisitions for the previously described pulsatile flow condition were preformed approximately 20 times (total 1200 images) for the DPBS and the Infasurf solutions. Roughly 30% of the images were selected for analysis because they had a sufficient length of liquid phase to compute the downstream mean velocity. There images were then interrogated and filtered by using the bubble tip and channel wall geometrical information from the corresponding simultaneous shadowgraph images. The interrogation employs a recursive Nyquist grid with a FFT correlation engine and a Gaussian peak algorithm with a 64x64 pixels first interrogation window, and a 32x32 pixel second interrogation window. The resulting vector fields were validated by standard deviation, local magnitude difference, and velocity range filters. Finally, the vector fields were divided into 16 bins based on the instantaneous downstream velocity to compute the ensemble average of each point in the cycle.

3. Results and conclusions

3.1 Image quality and efficiency of the system

Examples of the simultaneously acquired instantaneous μ-PIV and shadowgraphy images are presented in Fig. 6. In this figure Camera A provides one of the two fluorescent particle

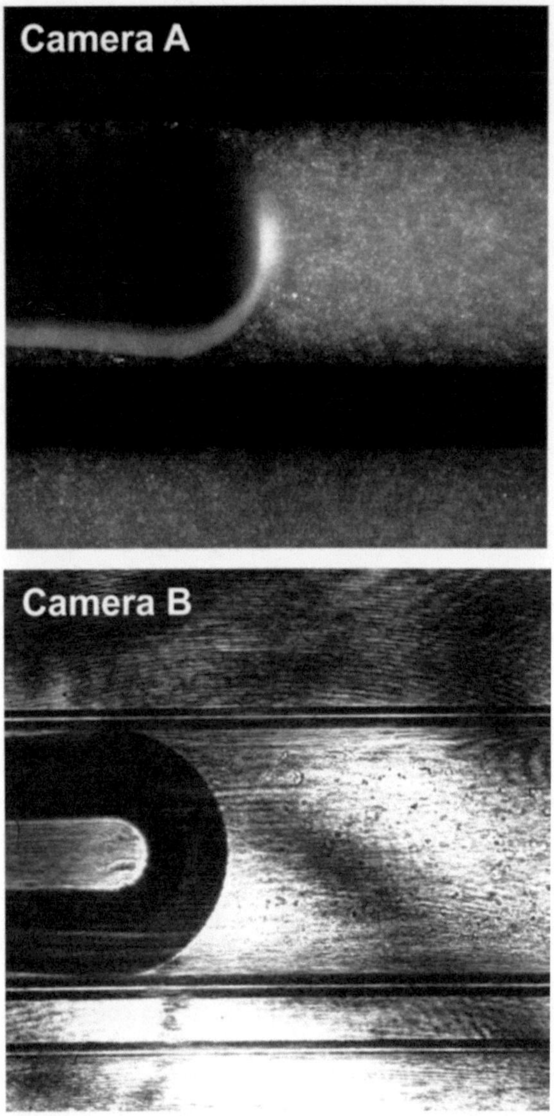

Fig. 6. Sample of simultaneously acquired μ-PIV (Camera A) and shadowgraph (Camera B) images under pulsatile flow input. Testing solution is the DPBS, and instantaneous downstream velocity is U=5.5 mm/sec.

images at an instantaneous downstream velocity U=5.5 mm/sec with the bubble progressing from left to right. U was calculated by taking the average velocity more than 1.5R downstream from the tip of the bubble. A band of fluorescent particles is visible in the lower part of image, which provides the stage velocity throughout the use of a second channel filled with the quiescent fluid. The figure demonstrates that the image separation through the dichroic filter is very good since there is virtually no cross-talk signal from the shadowgraph illumination in the particle image, and particles at the center of the channel are clearly distinguishable. As mentioned in the previous section the interfacial shape especially near the center of the channel is difficult to obtain from the fluorescent image which necessitates the use of a separate simultaneous technique to determine the interface shape.

Camera B in Fig. 6 presents the corresponding shadowgraph image. In this image, a large amount of random speckle pattern noise caused by coherent laser is visible. However, the degree of noise does not cause difficulty in determining the interfacial shape in the current case. The speckle could be eliminated by using diverse polarizations and wavelengths if necessary in future application. There is slight amount of cross-talk noise from the fluorescent particle emission due to the broad emission spectrum and less strict clean-up filter setting. This does not cause problems in the interface detection.

The interface geometry and channel wall information from Camera B were used to generate a grid mask for the vector interrogation of the Camera A image. The mask greatly increases the accuracy of the flow field measurement near the bubble tip especially during the reverse flow phase where significant interface deformations and instabilities exist. Two examples at the instantaneous downstream velocity U=-6.5 mm/sec (the bubble is progressing to the left) are shown in Fig.7. In both cases, the bubble shape estimation and the vector interrogation near the interface would be very inaccurate without the geometrical information from the corresponding shadowgraph image. Moreover, statistical analysis of the instability during the reverse flow phase is possible with this technology. This may provide information related to important physical characteristic of dynamic behavior of LS with and without competitive adsorption by using one-by-one corresponding interface shape information, flow patterns near the bubble tip and downstream velocity.

3.2 Data analysis of the bubble shape and the flow field surrounding semi-infinite bubble tip under influence of the pulmonary surfactant

The cycle averaged downstream velocities and bubble tip curvatures are displayed in Fig. 8. The data in Fig. 8-i clearly demonstrate excellent controllability of the flow generating system and the translating stage. The observed mean forward velocity is 5.5 mm/sec and the peak forward/backward velocities are about 18.5/-12.5 mm/sec. The reverse flow phase exists between $0.58 < t/T < 0.92$, where T=0.5 sec is period and t is time in the sinusoidal cycle. Both the DPBS and Infasurf solutions follow the input very closely with exception of slight phase delay on the Infasurf solution due to the large interface deformation. The corresponding ensemble-averaged interfacial curvatures of the bubble tip and fluctuations at each phase are shown in Fig. 8-ii. The interfacial curvature (κ) was calculated by first applying cubic smoothing splines on the interface and computing the arc length (s) around the centerline to obtain the radius of the oscillating interface at the centreline (Ro) (Thieman 2011). Details of the interfacial curvature and flow patterns (streamlines) at the reverse flow phase with and without the effect of Infasurf are shown in Fig. 9. These figures represent the ensemble-average of over 20 instantaneous images taken at point (b) in Fig. 8-i.

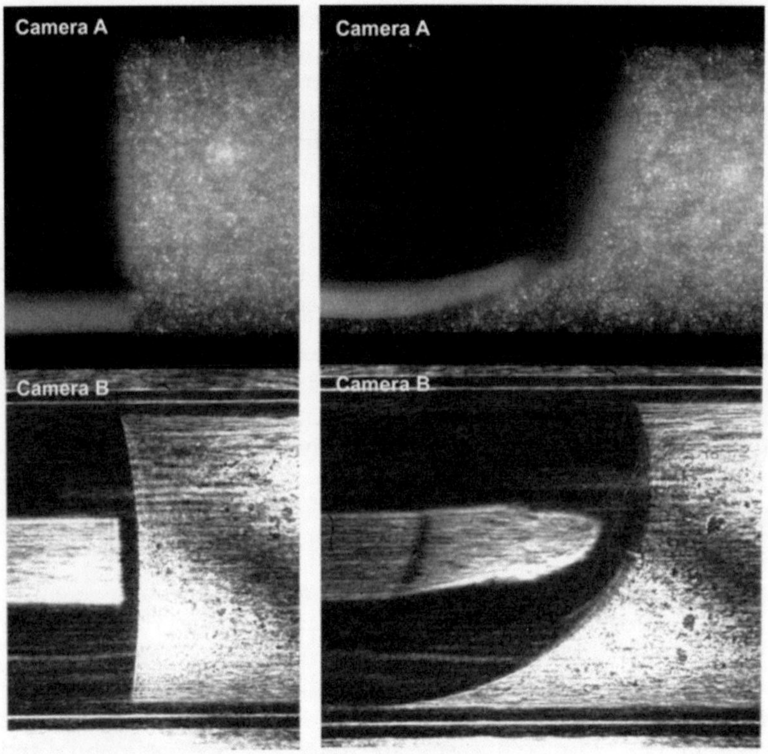

Fig. 7. Examples of the unsteady air-liquid interface during the reverse flow phase of downstream flow in pulsatile flow. The instantaneous downstream mean velocity was about U=-6.5 mm/sec for the both cases (position (b) in Fig. 8-i).

Significant interfacial deformation is only visible with the Infasurf solution during pulsatile flow. Almost negligible interface deformation is visible for the DPBS solution under the same pulsatile flow condition. The Infasurf solution with steady flow (U=5.5 mm/sec, gray straight line with circle in Fig. 8-ii) does not show any deformation. This experimental result is consistent with the hypothesis presented in Fig. 2 that predicts a significant interfacial deformation for the Infasurf solution during pulsatile flow.

During the reverse flow phase, converging stagnation region (ring in 3D) is located near the capillary walls (Fig. 9). Since the multi-layer formation and collapse of Infasurf dramatically lower the local surface tension relative to the expanding (diverging) region at the center (Fig. 2-i), the gradient of surface tension causes strong Marangoni stress tangential to the interface and rigidifies the interface. The very small curvature shift of the DPBS solution suggests insignificance of fluid dynamic effect on the interface shape. The curvature of the Infasurf solution during the forward flow phase is constantly lower than exists with steady flow due to the consistently lower surface tension caused by larger surface accumulation of Infasurf throughout the pulsatile motion. Glindmeyer et al. (2011) recently performed in-vitro experiments of LS effects on cell-laden rigid tube, and reported reduced cell damage during the reverse flow region. This finding coincides with the change of the interface and flow pattern near the bubble tip, and provides a physicochemical basis for the reduction of cell damage.

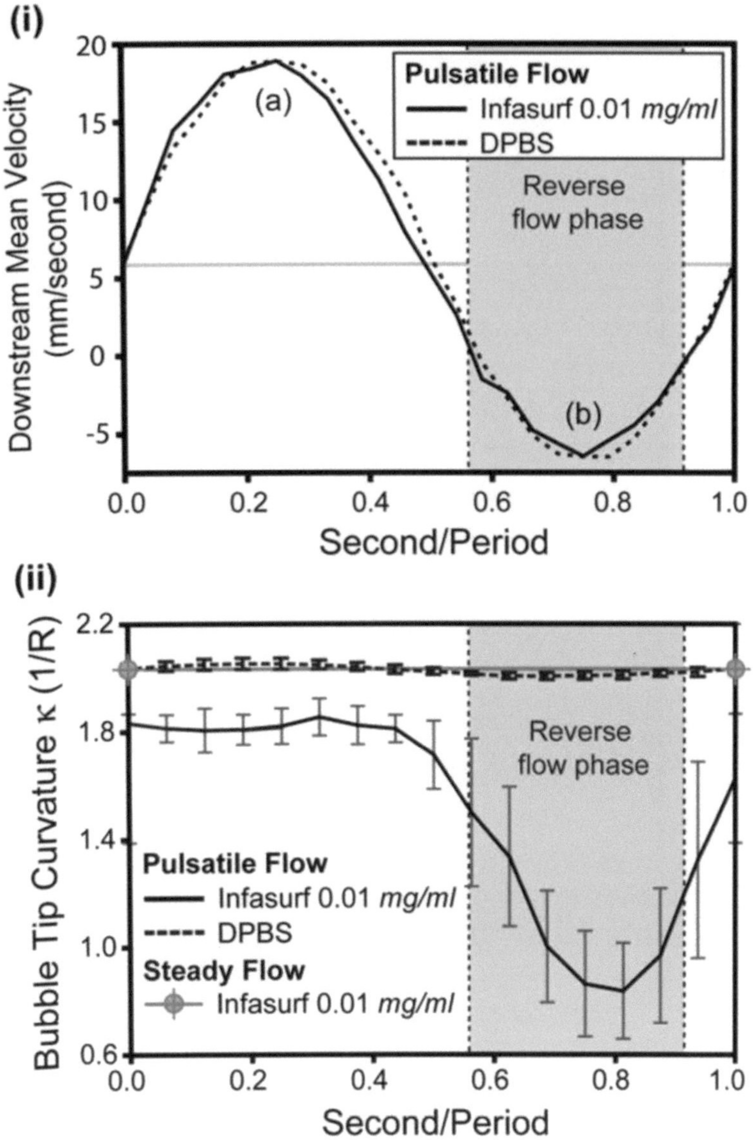

Fig. 8. (i) Measured downstream mean velocity in the cycle, and (ii) corresponding bubble tip curvature. Only the Infasurf solution display significant bubble interface deformation and fluctuation of curvature (see Fig. 7) during the reverse flow phase. The result is explained very well by the LS adsorption/desorption prediction model (Fig. 2), since the DPBS under the same flow condition and the Infasurf without sinusoidal flow input do not show similar deformations.

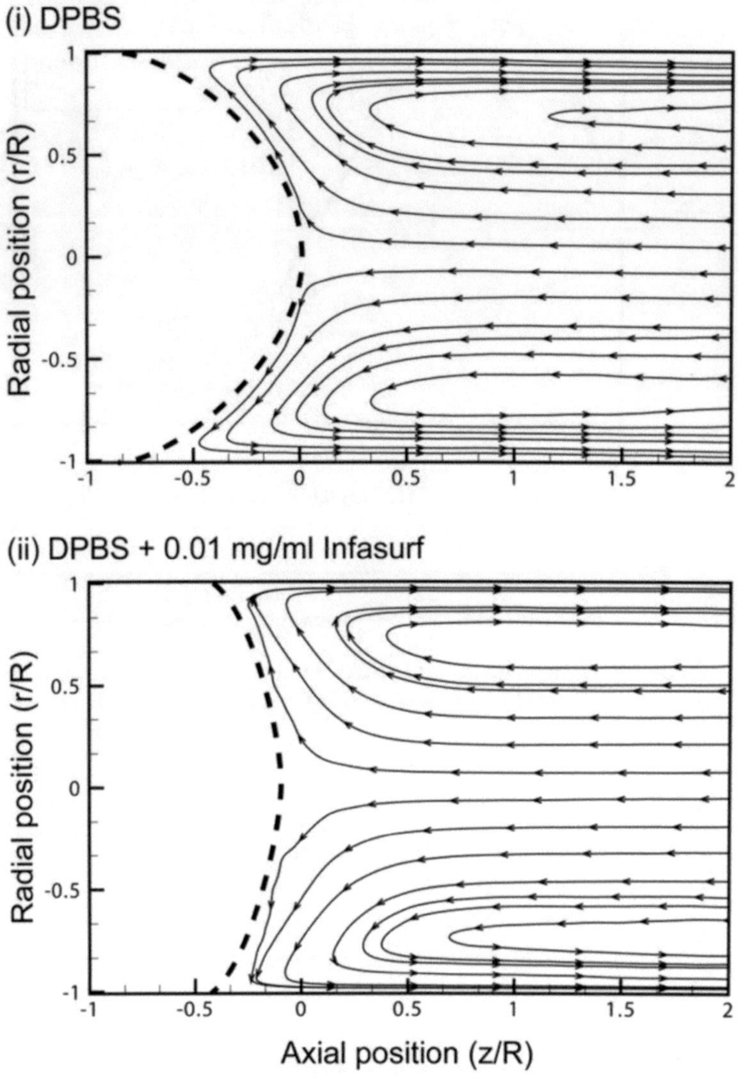

Fig. 9. Ensemble averaged streamlines and bubble shapes. Both figures have the same downstream mean velocity U=5.9 mm/sec (position (b) in Fig. 8-i). The bubble shape and flow pattern in (II) are significantly altered by accumulated LS around converging stagnation points (ring in 3D) near the wall. This difference has been observed only during the reverse flow region, and matches very well with reduced cellular damage that exists during reverse flow observed of in-vitro experiments (Glindmeyer et al. 2011).

An increasing curvature fluctuation (instability) is also noticeable during the reverse flow phase with the pulsatile flow and Infasurf (examples of unsteady interface are shown in Fig. 7). It can be hypothesized that the increasing fluctuation is a result of weakened wetting of the contaminated glass capillary wall since a glass surface is hydrophilic and has almost perfect wettability. Therefore if any kind of protein substances from the Infasurf solution adheres to the surface, it will certainly cause a loss of wettability. Since the trailing edge of tip of the semi-infinite bubble is a thin-film region, surface stability which is sensitive to surface energy balance is very low at the contaminated area causing dewetting regions. This is not a problem for a forwarding bubble, because the thin-film region is behind it. However, for the reverse flow the region is traversed by the returning bubble interface during a stage when it has the lowest local surface tension point very close to the wall. A re-wetting of the relatively hydrophobic region certainly causes stiff 'stick-slip' motions of the interface near the wall and results in widely diverged unevenly deformed instantaneous interface shapes during the reverse flow phase.

The simultaneous acquisition allows researchers to examine every single instantaneous velocity profile with a corresponding interfacial shape. Therefore vector fields of obviously skewed bubble tip shapes could be eliminated manually during the process of ensemble averaging, if they want to focus on LS effect and subsurface flow pattern. This will be increasingly important for the analysis of multi-phase fluid dynamics regarding the airway reopening problem, since obvious further investigations of the current application are to examine effects of higher Infasurf concentration to interface shape and interactions between LS inhibitors. It is possible to approach the contamination of the glass tube by developing a surface modification technique for the glass surface. For example, protein contaminations can be reduced by using a surface coating agent such as 2-[methoxy(polyethylenxy-propyl)trimethoxysilane (Tech-90, Gelest Inc., PA). This grafts poly(ethylene glycol) polymer chains on glass surface (Sui et al. 2006) and reduces nonspecific binding of proteins in the Infasurf solution. Using the simultaneous μ-PIV/shadowgraph acquisition system will also be a powerful tool to examine the effectiveness of various surface modification methods by evaluating the fluctuations during the reverse flow phase.

In summary, the simultaneous μ-PIV/shadowgraph acquisition system has been demonstrated as a powerful analysis tool for the complex multi-phase micro-biofluidic investigation, such as investigation of physicochemical interaction of lung surfactant during mechanically ventilated reopening of fluid-occluded respiratory airway. Complex flow controls, continuous data acquisition, and the simultaneous μ-PIV/shadowgraph acquisition with two separate cameras are all required to investigate this system, and the successful combination of the system opens the door to further investigations of this complex system and will allow the investigation of the role of LS inhibitors and the effect of surface properties to the flow fields.

4. Acknowledgment

This material is based upon work supported by the National Science Foundation under grant No. CBET-1033619, and the National Institutes of Health through grant R01-HL81266

5. References

Adrian, R. J. (2005). Twenty years of particle image velocimetry. *Experiments in Fluids*, Vol.39. pp.159-169.

Bilek, A. M., Dee, K. C. and Gaver III, D. P. (2003). Mechanisms of surface-tension-induced epithelial cell damage in a model of pulmonary airway reopening. *Journal of Applied Physiology*, Vol.94. pp.770-783.

Clements, J. A. and Avery, M. E. (1998). Lung surfactant and neonatal respiratory distress syndrome. *American Journal of Respiratory ad Critical Care Medicine*, Vol.157. pp.S55-S66.

Fairbrother, F. and Stubbs, A. E. (1935). Studies in electro-endosmosis VI. The bubble tube method of measurement. *Journal of Chemical Society*, Vol.1. pp.527-529.

Gaver III, D. P., Halpern, D., E., J. O. and Grotberg, J. B. (1996). The steady motion of a semi-infinite bubble through a flexible-walled channel. *Journal of Fluid Mechanics*, Vol.319. pp.25-65.

Gaver III, D. P., Jacob, A.-M., Bilek, A. M. and Dee, K. C. (2006). The significance of air-liquid interfacial stresses on low-volume ventilator-induced lung injury, *Ventilator-induced lung injury*, Dreyfuss, D., Saumon, G. and Hubmayr, R. D., New York, Taylor & Francis Group, Vol.215, pp.157-203.

Gaver III, D. P., Samsel, R. W. and Solway, J. (1990). Effects of surface tension and viscosity on airway reopening. *Journal of Applied Physiology*, Vol.69. pp.74-85.

Glindmeyer, H. W. I., Smith, B. J. and Gaver III, D. P. (2011). In Situ enhancement of pulmonary surfactant function using temporary flow reversal. *Journal of Applied Physiology*, Vol. (accepted).

Huh, D., Fujioka, H., Tung, Y.-C., Futai, N., Paine III, R. and Grotberg, J. B. (2007). Acoustically detectable cellular-level lung injury induced by fluid mechanical stress in microfluidic airway systems. *Proceedings of the National Academy of Science*, Vol.104, No.48. pp.18886-18891.

Invitrogen (2004), Propaties and Modifications, *Working with FluoSpheres fluorescent microspheres*: Vol.

Jacob, A.-M. and Gaver III, D. P. (2005). An investigation of the influence of cell topography on epithelial mechanical stresses during pulmonary airway reopening. *Physics of Fluids*, Vol.17. pp.031502.

Kay, S. S., Bilek, A. M., Dee, K. C. and Gaver III, D. P. (2004). Pressure gradient, not exposure duration, determines the extent of epithelial cell damage in a model of pulmonary reopening. *Journal of Applied Physiology*, Vol.97. pp.269-276.

Krueger, M. A. and Gaver III, D. P. (2000). A theoretical model of pulmonary surfactant multilayer collapse under oscillating area conditions. *Journal of Colloidal and Interface Science*, Vol.229. pp.353-364.

Mclntyre, R. C., Pulido, E. J., Bensard, D. D., Shames, B. D. and Abraham, E. (2000). Thirty years of clinical trials in acute respiratory distress syndrome. *Critical Care Medicine*, No.28. pp.9.

Meinhart, C. D., Wereley, S. T. and Santiago, J. G. (1999). PIV measurements of a microchannel flow. *Experiments in Fluids*, Vol.27. pp.414-419.

Meinhart, C. D., Wereley, S. T. and Santiago, J. G. (2000). A PIV algorithm for estimating time-averaged velocity fields. *Journal of Fluid Engineering*, Vol.122. pp.285-289.

Mielink, M. M. and Saetran, L. R. (2006). Selective seeding for micro-PIV. *Experiments in Fluids*, Vol.41. pp.155-159.

Natrajan, V. K., Yamaguchi, E. and Christensen, K. C. (2007). Statistical and structural similarities between micro- and macroscale wall turbulence. *Microfluid Nanofluid*, Vol.3. pp.89-100.

Nogueira, S., Sousa, R. G., Pinto, A. M. F. R., Riethmuller, M. L. and Campos, J. B. L. M. (2003). Simultaneous PIV and pulsed shadow technique in slug flow: a solution for optical problems. *Experiments in Fluids*, Vol.35. pp.598-609.

Olsen, M. G. and Adrian, R. J. (2000). Out-of-focus effects on particle image visibility and correlation in microscopic particle image velocimetry. *Experiments in Fluids*, Vol.Suppl. pp.S166-S174.

Pillert, J. E. and Gaver III, D. P. (2009). Physicochemical Effects Enhance Surfactant Transport in Pulsatile Motion of a Semi-Infinite Bubble. *Biophysical Journal*, Vol.96, No.1. pp.312-327.

Santiago, J. G., Wereley, S. T., Meinhart, C. D., Beebe, D. J. and Adrian, R. J. (1998). A Particle image velocimetry system for microfluidics. *Experiments in Fluids*, Vol.25. pp.316-319.

Smith, B. J. and Gaver III, D. P. (2008). The pulsatile propagation of a finger of air within a fluid-occluded cylindrical tube. *Journal of Fluid Mechanics*, Vol.601. pp.1-23.

Smith, B. J., Lukens, S., Yamaguchi, E. and Gaver III, D. P. (2011). Lagrangian transport properties of pulmonary interfacial flows. *Journal of Fluid Mechanics*, Vol.Accepted.

Smith, B. J., Yamaguchi, E. and Gaver III, D. P. (2010). A translating stage system for µ-PIV measurements surrounding the tip of a migrating semi-infinite bubble. *Measurement Sicence and Technology*, Vol.21, No.1. pp.15401-15413.

Sui, G., Wang, J., Lee, C.-C., Lu, W., Lee, S. P., Leyton, J. V., Wu, A. M. and Tseng, H.-R. (2006). Solution-phase surface modification in intect Poly(dimethylsiloxane) microfluidic channels. *Analytical Chamistry*, Vol.78, No.15. pp.5543-5551.

Thieman, J. W. (2011). The development of digital signal processing techniques to estimate stress fields in biological two-phase flows. The Department of Biomedical Engineering. New Orleans, Tulane University. Master of Science.

Ware, L. B. and Matthay, M. A. (2000). The acute respiratory distress syndrome. *The New England Journal of Medicine*, Vol.342, No.18. pp.1334-1349.

Yalcin, H. C., Perry, S. F. and Ghadiali, S. N. (2007). Influence of airway diameter and cell confluence on epithelial cell injury in an in vitro model of airway reopening. *Journal of Applied Physiology*, Vol.103. pp.1796-1807.

Yamaguchi, E., Smith, B. J. and Gaver III, D. P. (2009). µ-PIV measurements of the ensemble flow fields surrounding a migrating semi-infinite bubble. *Experiments in Fluids*, Vol.47, No.2. pp.309-320.

Zasadzinski, J. A., Stenger, P. C., Shieh, I. and Dhar, P. (2010). Overcoming rapid inactivation of lung surfactant: Analogies between competitive adsorption and colloid stability. *Biochimica et Biophysica Acta*, Vol.1798. pp.801-828.

Zimmer, M. E., Williams, H. A. R. and Gaver III, D. P. (2005). The pulsatile motion of a semi-infinite bubble in a channel: flow fields, and transport of an inactive surface-associated contaminant. *Journal of Fluid Mechanics*, Vol.537. pp.1-33.

11

Digital Micro PIV (μPIV) and Velocity Profiles *In Vitro* and *In Vivo*

Aristotle G. Koutsiaris
Bioinformatics Lab, School of Health Sciences,
Technological Educational Institute (TEI) of Larissa, Larissa,
Greece

1. Introduction

The term **Particle Image Velocimetry (PIV)**, in the technical world, is used to describe a powerful, automated flow visualization method which quantifies the instantaneous flow velocity field in 2 dimensions. PIV gives valuable information on how velocity field changes at a specific measurement plane, at regular time intervals, selected by the operator.

Correlation techniques were used extensively for the measurement of fluid velocity. In temporal correlation techniques, velocity is inversely proportional to the transit time for flow tracers to cross a fixed distance. In **spatial correlation techniques**, the velocity is directly proportional to the displacement of flow tracers in a fixed time interval. In this sense, many video techniques, high speed cinematography and PIV can be classified as spatial correlation techniques.

The term "**Digital**" refers to the fully digital implementation of the method, namely the use of a digital camera connected directly to a digital electronic computer in order to acquire flow images and then to estimate the correlation function. This may seem a common place today, but a lot of early implementations used to have special electrooptical and electromechanical hardware for the estimation of the correlation function (optical correlation) or film cameras with laser scanners (optical - digital correlation). Later, fully electronic PIV implementations appeared using analog together with digital hardware: analog CCD cameras connected to video recorders produced video tape images which were digitized by computers (Gardel et al., 2005). These implementations were limited by their analog components.

The term **"Micro"** refers to the length scale of the microfluidic environment ranging between 1μm and 1000 μm. The word "environment" is a general term for every possible microstructure inside which the fluid under study flows.

This chapter is a medium size review on **Digital micro PIV (μPIV)** and its applications on the velocity profiles *in vitro* and *in vivo*. Parts of sections 6 and 7 come from a mini review paper (Koutsiaris, 2010b).

2. Components of a Digital μPIV system

A complete PIV implementation is subdivided into four steps (Koutsiaris et al., 1999): 1) seeding of the flow, 2) illumination of the flow plane of interest with a light sheet, 3) image capture by a camera placed at right angles to the illuminated flow plane, and 4) data processing.

There are **three principal differences** between an ordinary macroscopic and a μPIV system. **First**, as it was mentioned in the introduction, it is the **length scale** of the fluidic circuit under study (hence the name "micro"). **Second**, as a consequence of the first difference, it is the design of the optical system, which has as a minimum requirement some kind of a **microscope** objective lens. **Third**, the thickness of the measurement plane is defined by the optical system using **volume illumination** and not by the thickness of an illuminating light sheet. Forming a light sheet with a thickness between 1 and 20 μm and aligning it properly in the microfluidic circuit is very difficult.

In this section, the main components of a Digital μPIV system are described: **1)** the microfluidic circuit with appropriate flow tracers, **2)** the optical system, **3)** the digital electronics system and **4)** the computer software. These components are presented in schematic form in figure 1.

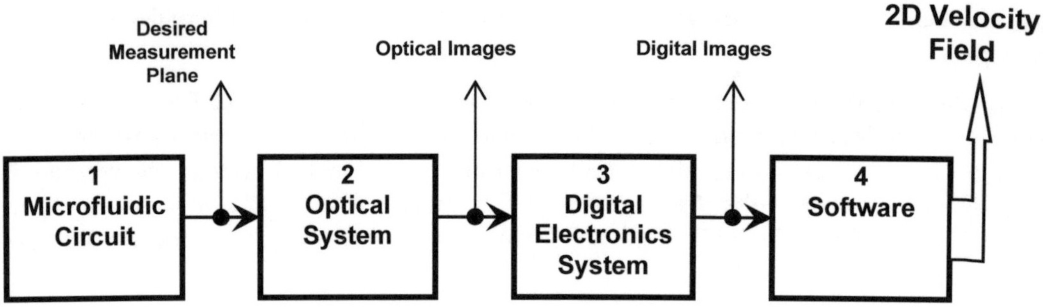

Fig. 1. The four main components of a Digital μPIV system.

2.1 Microfluidic circuit and flow tracers

The usual implementation of the microfluidic system comprises some kind of a microstructure-microchannel (microduct, microtube, capillary) connected to a syringe pump (figure 2). The microchannel can be made of various materials such as glass (Koutsiaris et al., 1999; Meinhart et al., 1999), biocompatible polymers and collagen gel inside silicone elastomer (Chrobak et al., 2006; Potter & Damiano, 2008). Biocompatible polymers can be classified to thermoplastic polymers such as **P**oly **M**ethyl **M**eth**A**crylate (**PMMA**, Timgren et al., 2008) and to elastomeric polymers such as **P**oly **D**i**M**ethi**S**iloxane (**PDMS**, Lima et al., 2008).

A lot of research was directed during the past 10 years to **biocompatible polymers**, due to their advantages over glass. PDMS in particular is permeable to gases such as oxygen and there is no need to use a refractive index matching liquid. The ultimate tensile stress and Young's elastic modulus are much closer to those of blood vessels in comparison to glass and sealing does not require any complex bonding technique. In addition, various microdevices such as micropumps can be integrated easily in a PDMS device. A more detailed discussion on PDMS advantages can be found in the paper of Lima et al. (2008).

The microfluidic system can be a part of a MicroElectroMechanical Systems (MEMS) device. Other typical microdevices are microheat sinks, micropumps, microturbines, microengines, micromixers and microsensors. More details on the design, fabrication and applications of MEMS devices can be found in review papers (Hassan, 2006) and in books edited by Gad-el-Hak (2005) and Breuer (2005) and written by Nguyen and Wereley (2002).

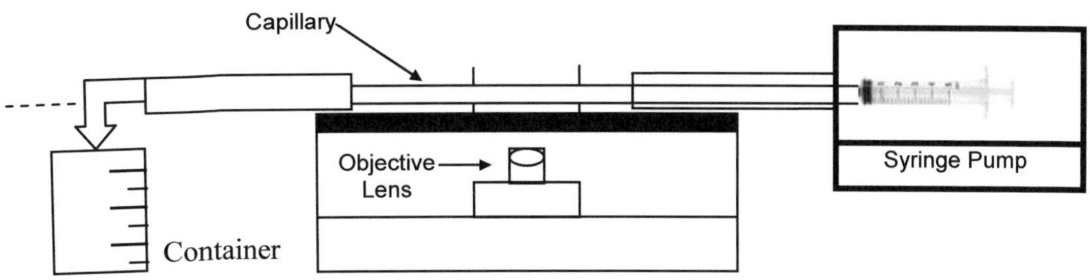

Fig. 2. A typical microfluidic circuit with a syringe pump, on an inverted microscope.

When the experimental conditions are right, the syringe pump is not needed since the necessary driving force can be controlled with the height difference (DH, figure 3). In this way a more economic set up is possible.

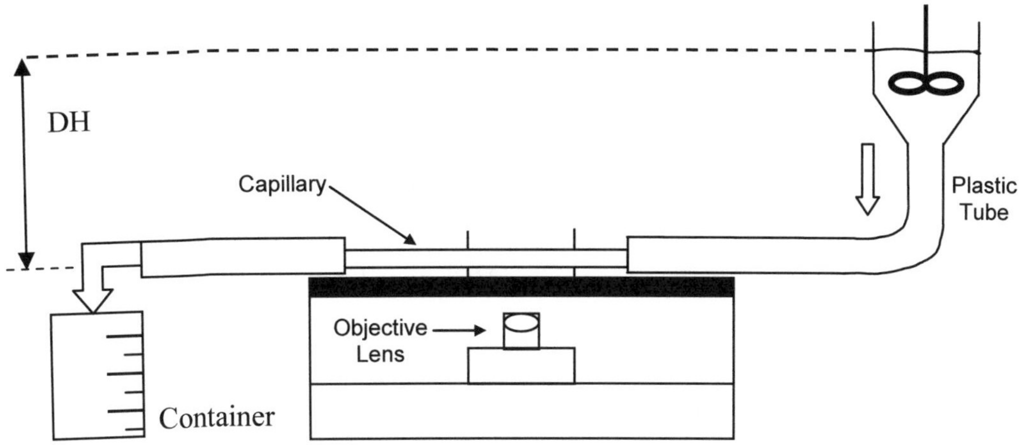

Fig. 3. A microfluidic circuit without any syringe pump (Koutsiaris et al., 1999). The driving force is provided by the height difference DH.

The basic characteristic of a successful flow tracer is that it should accurately follow the flow with an appropriate specific gravity matching (Stokes equation, Raffel et al., 2007). Flow tracers are usually made of fluorescent **P**oly**S**tyrene **L**atex (**PSL**) particles but the final choice depends on the fluid. When the particle diameter is less than ≈ 0.5 μm and characteristic velocities are less than ≈ 1mm/s, the Brownian motion should be taken into account (Werely and Meinhart 2010). More details on Brownian motion and the Saffman effect can be found in the textbook edited by Breuer (2005).

In addition, the particle diameter distribution should be monodispersive (diameter values distributed over a very limited range around the mean value). We shall see later that this makes easier the design and error evaluation.

2.2 Optical system

2.2.1 Basics

The basic optical characteristics carved on the metallic surface of every microscope objective lens are magnification (M) and **numerical aperture (NA)**. In simple terms, M shows how many times the perpendicular to the optical axis dimension of an object lying in the object plane is enlarged in the image plane on the other side of the lens. A schematic diagram of these planes is shown in figure 4.

Magnification is a characteristic easily perceived by the majority of people who often seem to neglect the second equally important optical characteristic of a microscope objective lens, the NA:

$$NA = \eta \sin\Theta \tag{1}$$

Where η is the refraction index of the material between the lens and the specimen and Θ is the half angle of the cone of light received by the objective lens (figure 4). It should be noted that η is by definition always greater or equal to 1 and is a function of the light wavelength λ ($\eta = f(\lambda)$).

The NA is a very important lens specification for µPIV applications because of the following reasons: **First,** it determines in a non-linear way the light gathering ability of the objective lens: the higher the NA the much higher the light gathering capacity. **Second,** it specifies, in an inversely proportional way, the resolving ability (RA) or simply "resolution" of the objective lens, i.e. the smallest discernible distance between two objects (points) in the object plane:

$$RA = 0.61\, \lambda_a / NA \tag{2}$$

The above formula was derived using the Rayleigh criterion and assuming paraxial conditions (optical rays close to the optical axis). λ_a is the wavelength of light in the air.

Third, the NA specifies in a non-linear way, the depth of field or depth of focus DoF (figure 4) of an objective lens: the higher the NA the much lower the DoF. A typical "simple" equation quantifying the DoF of an objective lens was proposed by Shillaber (1944, as cited in Delly, 1988):

$$DoF = \lambda_\alpha \frac{\sqrt{\eta^2 - NA^2}}{NA^2} \tag{3}$$

However, there seems to be a difference between the DoF seen by an observer through an eyepiece and the DoF_V when a video camera is attached on the microscope (Nakano et al., 2005):

$$DoF_V = \frac{\lambda_\alpha \cdot \eta}{NA^2} + \frac{\eta \cdot px}{M \cdot NA} \tag{4}$$

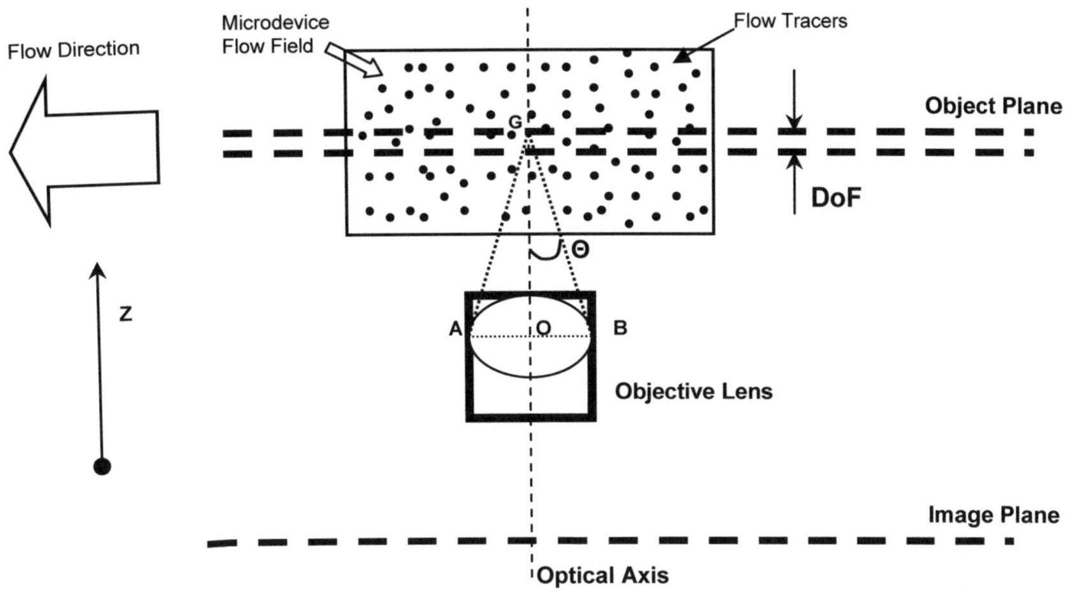

Fig. 4. The object and image planes of a microscope objective lens have a finite depth along the optical axis. The object plane depth is called **Depth of Focus (DoF)**. Θ is the half cone angle of the objective light rays focused on a **focal point G** inside the object plane. OG is the focal distance of the objective lens and AB is its diameter D.

Where px is the pixel size of the video camera. The difference in the DoF estimations given by the 2 formulas becomes important for NA > 0.5, as it is shown in Table 1.

Objective Lens (M/NA)	DoF (Shillaber) (μm)	DoF_V (Video) (μm)	Relative Difference $\frac{DoF - DoF_V}{DoF_V}$ (%)
10/0.25	8.5	9.3	- 9
20/0.5	1.9	2.3	- 17
40/0.6	1.2	1.6	- 25
40/0.75	0.6	1	- 35
60/1.4	Not defined	0.3	-

Table 1. The Depth of Field (DoF) and the video Depth of Field (DoF_V) depend on the objective lens. Numbers were estimated assuming: $\lambda_a = 0.55$ μm, $\eta = 1$, pixel size px = 16 μm.

Another optical characteristic of the microscope objective lens, used frequently in photographic cameras, is called f number (f#). The f# is closely related to the NA, but is defined in purely geometrical terms:

$$f\# = \frac{OG}{D} \tag{5}$$

Where OG is the focal distance of the objective lens and D is its diameter (figure 4). The mathematical relationship between NA and f# can be derived from equations 1 and 5:

$$f\# = \frac{1}{2}\sqrt{\left(\frac{\eta}{NA}\right)^2 - 1} \tag{6}$$

This is a strong non linear relationship which can be significantly simplified assuming paraxial conditions ($\tan\Theta \approx \sin\Theta \approx \Theta$, Bown et al., 2005):

$$f\#_P = \frac{\eta}{2NA} \tag{7}$$

The assumption of paraxial conditions in equation 7, leads to a relative error less than 12% when $f\# > 1$ or $(NA/\eta) < 0.45$.

2.2.2 Measurement Plane Width (MPW)

In classical PIV the measurement plane width (MPW) is defined by a laser light sheet because outside particles do not influence the result. However in μPIV, there is not any laser light sheet and most of the flow field is illuminated in a situation commonly referred to as "**volume illumination**" (figure 5). This means that an unknown amount of particles below and above the level of the object plane contributes to the position of the maximum of the correlation function and consequently to the velocity result (see Section 2.4). These particles are shown in the areas between the 2 solid lines and the 2 dashed lines of figure 5. The position of the 2 solid lines, depending on many factors related to the experimental set up (but primarily on the objective lens NA), can not be located accurately and therefore the MPW can not be determined in a rigorous way.

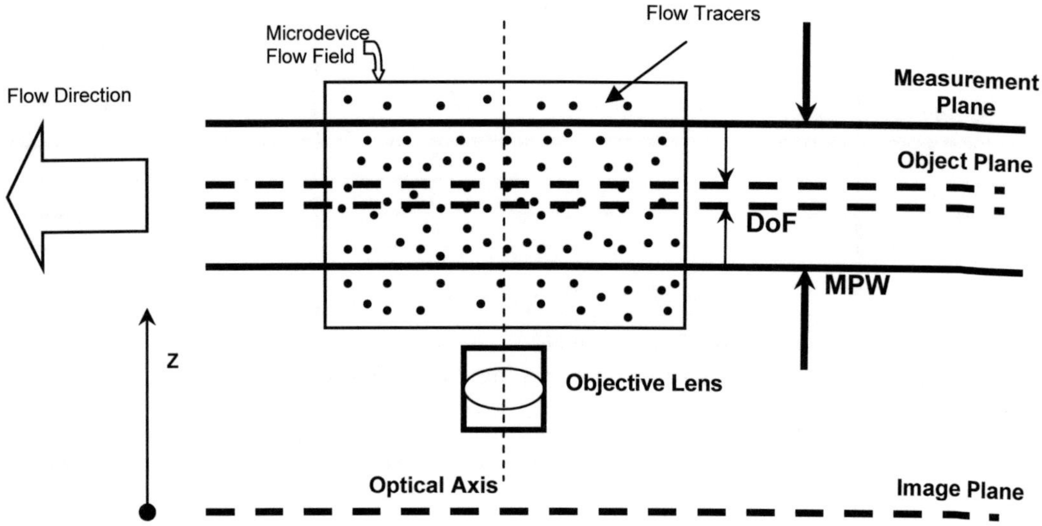

Fig. 5. The μPIV volume illumination principle: the depth of focus (DoF) is always much lower than the measurement plane width (MPW).

In an attempt to quantify the contribution of out of focus particles in the velocity result, Olsen & Adrian (2000) introduced the concept of the "correlation depth". The correlation depth (Zcor) was defined as the axial distance from the object plane in which a particle becomes sufficiently out of focus so that it no longer contributes significantly to the velocity estimation (Werely & Meinhart, 2005). The **measurement plane width MPW** (proposed by Werely et al., (2000), as cited in Olsen & Adrian, (2000)) can be defined as double the Zcor:

$$MPW = 2 \cdot Zcor = 2 \cdot \sqrt{\frac{1-\sqrt{\varepsilon}}{\sqrt{\varepsilon}}\left[f\#_p{}^2 d_p{}^2 + 5.95\frac{(M+1)^2 \lambda_a{}^2 f\#_p{}^4}{16 \cdot M^2}\right]} \qquad (8)$$

Where ε is the relative contribution of a particle located at a distance Zcor from the object plane, $f\#_P$ is the f number with paraxial approximation (equation 7), d_p is the particle diameter in the object plane, M is the objective lens magnification and λ_a is the wavelength of light in the air. According to Werely & Meinhart (2010), equation 8 agrees more closely to experiments than when using the exact f number relationship (equation 6).

A series of MPW values for various objective lenses and for a $d_p = 1\mu m$ are shown in Table 2 ($\varepsilon = 0.01$, $\lambda_a = 0.55\ \mu m$, $\eta = 1$, $px = 16\ \mu m$). The relative difference between the MPW and the DoF_V is **higher than 300%** for all kinds of objective lenses.

Objective Lens (M/NA)	DoF_V (Video) (μm)	f_p	MPW (μm)	Relative Difference $\frac{MPW - DoF_V}{DoF_V}$ (%)
10/0.25	9.3	2	37.4	302
20/0.50	2.3	1	10.4	352
40/0.60	1.6	0.83	7.5	367
40/0.75	1	0.66	5.35	435
60/1.4	0.3	0.36	2.40	700

Table 2. Video Depth of Filed (DoF_V) in comparison to the Measurement Plane Width (MPW).

In order to see the effect of the particle diameter d_p on the MPW given by equation 8, the reader can see table 1 in Bourdon et al. (2004) and in Werely & Meinhart (2010), even though there seems to exist a discrepancy in the values given by the 2 groups (in addition, Werely & Meinhart (2010) do not mention the light wavelength used in their calculations). For the effect of d_p on MPW using the exact definition of f number, the reader can see table 2.2 in Werely & Meinhart (2005) where it is shown that, for "dry" lenses, there is not a significant effect of particle diameter on MPW when $d_p < 0.5\ \mu m$.

2.2.3 Effective Diameter (d_E)

The recorded image of a tracer particle is the convolution result between the geometric image of the particle and the point response function of the imaging system, given diffraction limited optics. The diameter of the geometric image of a particle is Md_P. The particle diameter d_P needed for determining the geometric image dimension is usually measured using an optical microscope and an object micrometer (micrometric ruler). For particle diameters less than about 0.2 μm other techniques must be used such as atomic force microscopy and electron microscopy.

The diameter of the point response function d_S is given by the Airy function:

$$d_S = 2.44\,(M+1)\,f\#\,\lambda_a \tag{9}$$

Adrian and Yao (1985, as cited in Werely & Meinhart, 2005) found that the Airy function can be approximated accurately by a Gaussian function. Given spherical particles and diffraction limited optics the geometric image can also be approximated accurately by a Gaussian function and the total diameter d_E (effective diameter in microns) of the particle in the image plane can be approximated by:

$$d_E = \sqrt{(Md_P)^2 + d_S^{\,2}} \tag{10}$$

Then effective diameter can be converted to pixels by using the pixel size of the video camera. The contribution of the geometric and the point response components in the final d_E is shown in Table 3 for various particle diameters, in the case of a 10/0.25 objective lens (λ_a = 0.55 μm).

Particle Diameter d_P (μm)	Geometric Component $M\,d_P$ (μm)	Point Response Component d_S (μm)	Effective Diameter d_E (μm)	Relative Difference $\frac{d_S - d_E}{d_E}$ (%)
0.1	1	28.6	28.6	0
0.5	5		29.0	- 1.4
0.7	7		29.4	- 2.7
1.0	10		30.3	- 5.6
3.0	30		41.4	- 31.0
5.0	50		57.6	- 50.3

Table 3. The geometric and the point response component of the effective particle diameter d_E.

As it can be observed in the gray line of table 3, when the geometric component is **less than one third of** d_S, then it contributes by less than 5% in the final effective diameter d_E. In practice this means that for an objective lens 10/0.25, all particles with diameters $d_P < 1$μm will have an effective diameter approximately equal to d_S .

In Table 4, the point response diameter d_S is given for commonly used objective lenses and λ_a = 0.55μm. It should be noted that in the case of oil immersion objectives (60/1.4) the NA is equal to the refraction index of the medium and using equation 6 this gives f# = 0 (hence d_S = 0, which is not true). In this case, the f# should be estimated by its original definition (equation 5). In Table 4, for the case 60/1.4 it was assumed that f# = 0.2.

Objective Lens (M/NA)	10/0.25	20/0.50	40/0.60	40/0.75	60/1.4
F#	1.94	0.87	0.66	0.44	0.2
d_S (μm)	28.6	24.5	36.3	24.2	16.4

Table 4. Point response diameter d_S for various objective lenses.

The effective diameter d_E is very important because it contributes to the random and bias error components (Section 3) in the displacement and velocity estimation. It has been shown (Prasad et al., 1992) that, when the centroiding technique is chosen for locating the autocorrelation peak, the total displacement error is minimized by selecting $d_E \approx 2$ pixels. The same result was obtained when a Gaussian peak fit estimator was used (Westerweel 1997). Since the pixel size is usually fixed, depending on the CCD camera, the number of pixels corresponding to the effective diameter d_E can be arranged by selecting the appropriate optics and flow tracers. It should be noted however, that later authors preferred a design of $d_E \approx 4$ pixels (Buffone et al., 2005; Koutsiaris et al., 1999; Meinhart et al., 1999).

Finally, it should be noted that d_E depends also on the vertical distance in the z direction from the object plane (figure 5). So, for out of focus particles located outside the object plane, there is a z dependent third term added in equation 10. This is beyond the scope of this chapter (see Bown et al., 2005).

2.2.4 Fluorescence μPIV

Volume illumination can be implemented with 2 different experimental set ups: **fluorescence** and **brightfield** illumination. In the experimental set ups with fluorescence, **darkfield images** are produced by the camera and are sent to the PC for off line processing (figure 6). In a darkfield image the tracers appear bright in a dark background, as in the

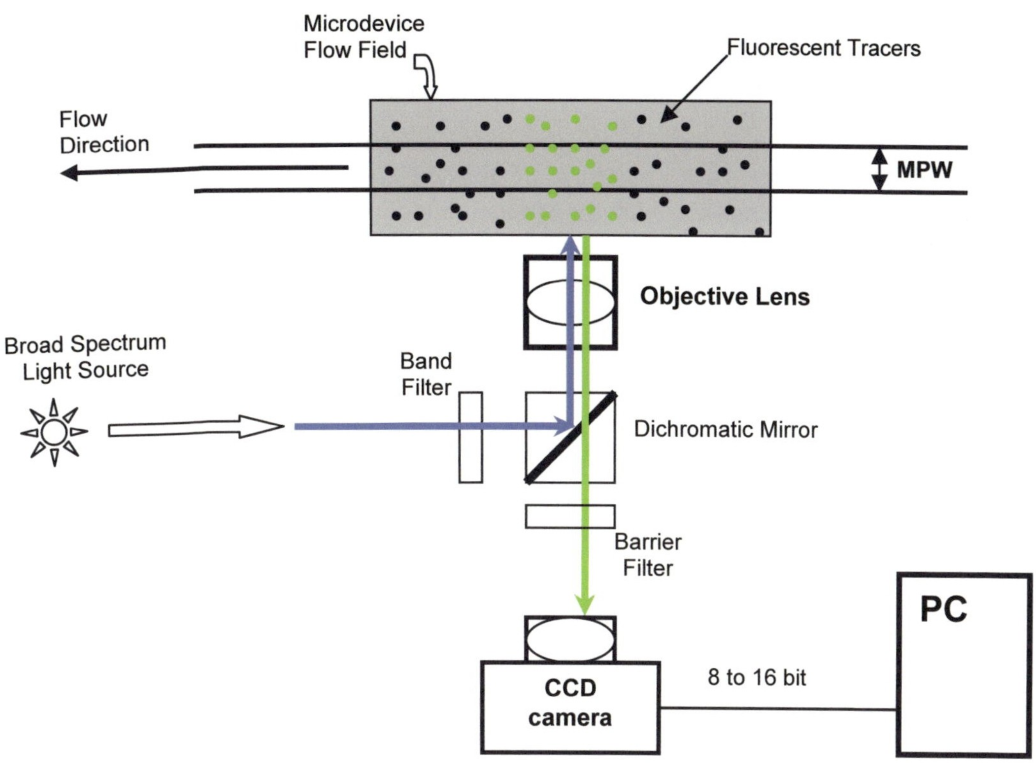

Fig. 6. Digital μPIV set up with fluorescence. MPW: Measurement Plane Width.

classical film negatives. When a monochromatic light source (laser) is used, the band pass filter is not necessary.

The first μPIV system using an upright **epi-fluorescent** microscope was reported by Santiago et al. (1998). The vast majority of fluorescent microscopes are epi-fluorescent (figure 6), meaning that image sensors detect the back scattered (reflected) fluorescence from the object plane and not the transmitted fluorescence. Santiago et al. (1998) used 300 nm in diameter, fluorescently labeled, PSL particles in the flow field between a microscope slide and a 170 μm thick coverslip. The liquid thickness between the slide and the coverslip was ≈ 5 μm and the spatial resolution was 6.9 x 6.9 x 1.5 μm. Because of the large time delay between successive images, they managed to measure only low velocities of the order of 35μm/s. One year later, Meinhart et al. (1999), using an inverted epi-fluorescent microscope, and a different digital camera illuminated by 2 Nd:YAG lasers, reduced the time delay between images by more than a factor of 100 and managed to measure velocities of the order of 10 mm/s.

2.2.5 Brightfield μPIV

In brightfield μPIV images, the tracers appear dark in a **bright background**, as the objects in the positive photos of everyday photographic albums. This happens because particles are not fluorescent and instead of emitting they actually absorb light.

The first μPIV system without fluorescence, using an inverted brightfield microscope (transmission mode) was reported by Koutsiaris et al. (1999). The vast majority of the brightfield microscopes work in **transmission mode**, meaning that video cameras detect the transmitted (forward scattered) light from the object plane (figure 7).

Koutsiaris et al. (1999) used borosilicate glass particles with a mean diameter of 10 μm to seed the flow of a suspension consisting of glycerol and water. The distinctive optical characteristic of this suspension was the similar refractive index with the glass capillary (figure 8). They demonstrated for the first time that μPIV is possible in cylindrical microtubes (150 to 300 μm internal diameter) by using the aforementioned suspension and by surrounding the capillary with a fluid having a similar refractive index with the capillary. The spatial resolution of their velocity fields was 26.2 x 335 x 10 μm, optimized for the velocity profile measurement inside the capillaries and velocities up to the order of 5 mm/s were measured. In addition, the application of the μPIV technique at the microvessels of a living animal was proposed for the first time and a discussion on possible obstacles of such an attempt was presented.

A brightfield μPIV set up can be considered as the dual counterpart of the fluorescence set up, but without requiring complex and expensive fluorescence hardware such as a special light source, filters and a dichromatic mirror. It seems that a brightfield set up has a higher in focus noise in comparison to a fluorescent set up but a smaller contribution to noise from out of focus particles. In addition, dark particles with diameters lower than about 0.5 μm can not be easily resolved by an optical microscope. It seems that these topics as well as the MPW and the d_E in the case of a brightfield set up have not been fully exploited by the research community yet.

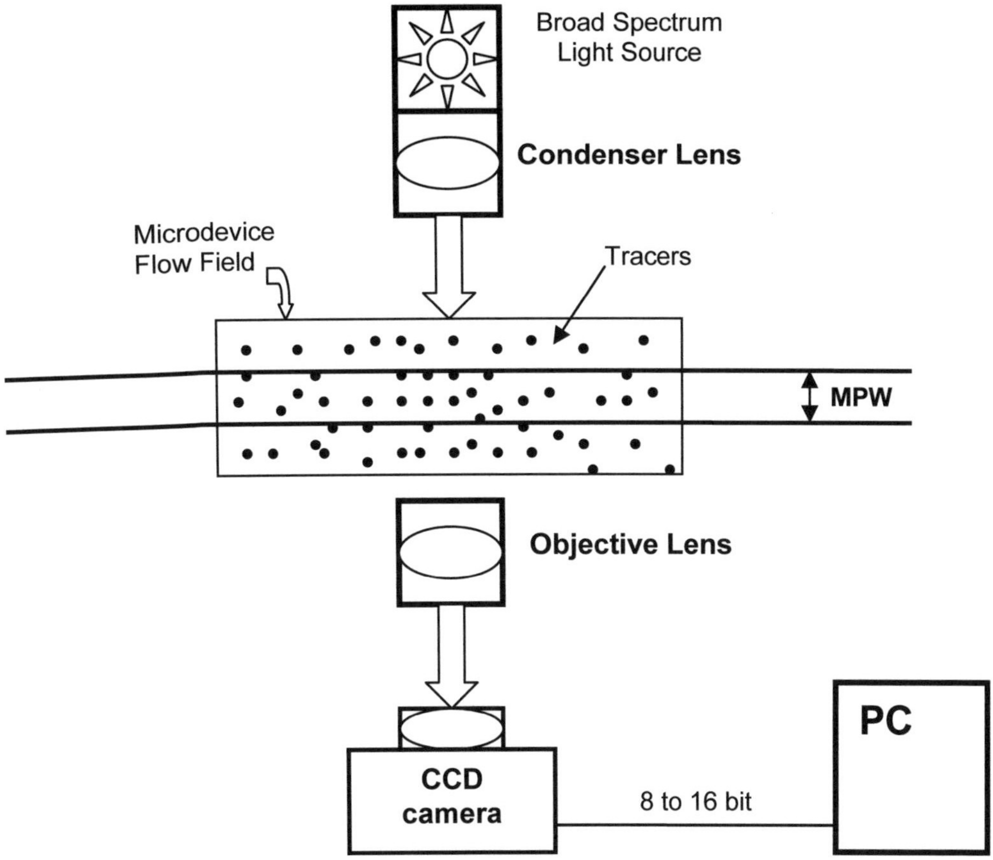

Fig. 7. Digital μPIV set up with brightfield illumination. MPW: Measurement Plane Width.

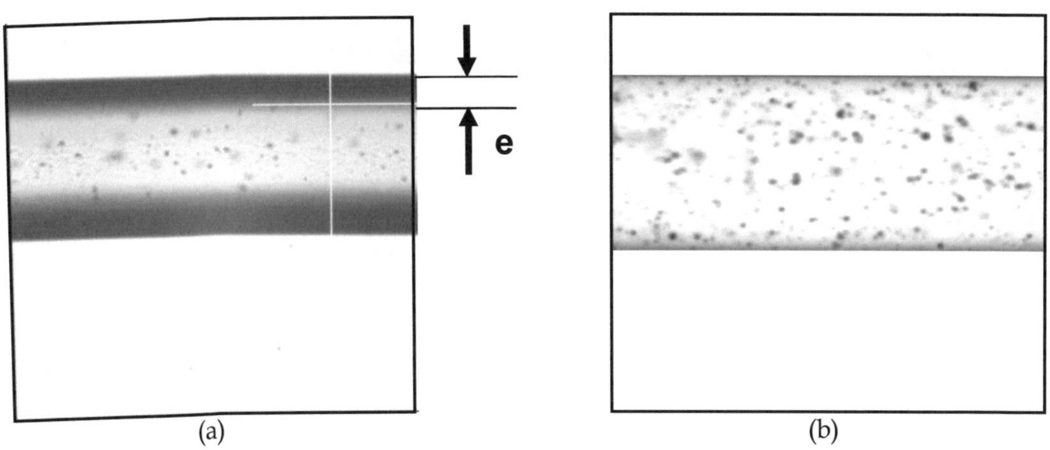

Fig. 8. Glass cylindrical capillaries with internal diameters of ≈ 275 μm. (a) Glass particles suspended in water: in the diametric plane, approximately 41% of the internal diameter is not visible (black band e ≈ 57 μm). (b) The black bands disappeared when the glass particles were suspended in a mixture of 75% glycerol and 25% water.

2.2.6 Confocal μPIV

The incorporation of confocal imaging in a μPIV experimental set up provides a powerful experimental tool introduced by Park et al. (2004). In a confocal optical system the object plane of interest is scanned point by point by a laser beam. Each scanned point of the measurement plane is the optical conjugate of a pinhole located in the image plane of the objective lens. This fact reduces the signal intensity passing through the pinhole, but also reduces dramatically the out of focus background light, producing a rigorously defined optical slice in the object plane with a thickness much smaller than that in ordinary optical systems.

The confocal μPIV has the following advantages over the classical μPIV: **first**, the difference between the depth of focus (DoF) and the measurement plane width (MPW) is significantly smaller due to the confocal superiority over classical optics in terms of axial resolution.

Second, the increased axial resolution implies sharp reduction of background light from out of focus images and higher signal to noise ratios for the correlation operation. **Third,** there are no different effective particle diameters produced from their different distances from the object plane.

Despite the numerous advantages of the confocal μPIV, there are some disadvantages stemming out from the operating principle of confocal imaging according to which a single point can only be imaged at one instance in time. This means that each image is produced by scanning a 2D plane point by point, which implies restrictions in the frame rate and increased cost due to complex hardware. Restrictions in the frame rate (hence in the maximum measurable velocity) can be overcome by using a superfast Nipkow disk (Lima et al., 2006; Park et al., 2004). However, this action further increases complexity and cost.

It seems that, confocal μPIV has not been applied without fluorescence. This would perhaps be feasible in relatively transparent media which do not attenuate light intensity significantly. In addition, image processing of confocal images could increase even more the image quality (signal to noise ratio).

Combining 2D velocity fields at different z positions (figure 5), confocal data can be used in steady flows to calculate the 3D velocity field by applying the continuity equation. Other methods for the quantification of the 3D velocity field are digital holography, stereo μPIV and particle image defocusing. In stereo μPIV (Bown et al., 2006; Lindken et al., 2006) all 3 velocity components in space can be extracted from a single 2D plane. More details on the 3D μPIV techniques can be found in the review papers of Lee & Kim (2009) and Werely & Meinhart (2010) and a comparison between stereo μPIV and the multiple 2D velocity field technique was presented in the paper of Bown et al. (2006).

2.3 Digital Electronics System (camera)

The digital electronics component of μPIV set up comprises a digital video camera, a frame grabber and a personal computer. The usual way is to put emphasis on the camera design, but its performance depends heavily on the frame grabber design so, they usually come together.

Most digital cameras are based on the **CCD (Charge-Coupled Device)** technology. Briefly, photons from the optical image of the measurement plane hit on the surface of the camera producing free electrons, which in their turn are trapped inside potential wells. Each potential well of the camera corresponds to a pixel. Therefore, the higher the light intensity on the camera sensor, the higher the amount of electrons trapped inside the potential wells, and the higher the gray value of the corresponding pixel. In fact it is the product of the light intensity and the exposure time which affects the final gray value of a pixel. The most important characteristics of a CCD camera are: **1)** the quality of the CCD sensor, **2)** the architecture of the CCD sensor, **3)** the sensitivity (lowest detectable light intensity), **4)** the frame rate capacity (frames per second, fps) and **5)** the gray level dynamic range where the camera operates linearly.

Monochrome CCD cameras are preferable to color CCD cameras (Breuer 2005), because colors usually increase the complexity and the cost of a camera at the expense of frame rate, sensitivity and dynamic range. In addition pixels should be square to avoid complexities related to the conversion from image domain (pixels) to real dimensions.

For μPIV systems, non interlaced high speed digital cameras are preferred. A camera is of the "**high speed**" type, when it can acquire more than 25 (EUROPE) or 30 (USA and JAPAN) full frames per second (fps). But in special μPIV applications, speeds up to 2000 fps may be necessary. A high speed camera is a prerequisite in μPIV applications because of the magnified measurement plane causing the virtual magnification of the real velocities.

In the interlaced format, first all the odd rows of the image frame are read out and then all the even rows. This introduces a series of complications in PIV applications. In order to avoid these complications, other architectures are recommended such as the frame transfer and the full frame interline transfer architecture. In the **frame transfer architecture** the camera sensor is divided into two different rectangular areas: 1) the active or light sensitive area used for image acquisition and 2) the inactive or masked storage area used for the temporal storage of each acquired image. In the **full frame interline transfer architecture** the masked area is composed of columns between the active columns of the sensor.

The original advantage of the frame transfer architecture was the 100% fill factor i.e. the absence of gaps between pixels. In contrast, full frame interline transfer cameras had always fill factors much less than 100%. Later, this disadvantage was almost eliminated by using micro lenses in front of the sensor.

The **sensitivity** of a Digital μPIV camera should be enough to acquire images of the tracer particles flowing inside the microstructure. This is not so trivial to estimate because it depends on many factors such as frame rate, exposure time and the wavelength depended factors of quantum efficiency (QE) and noise equivalent exposure (NEE).

In **CMOS (Complementary Metal Oxide Semiconductor)** cameras, photodiodes are used for detecting light instead of potential wells. A relatively new development is active pixel sensor (APS) technology which allows the integration of amplification on each pixel with an appropriate MOS-FET transistor. In a typical APS implementation **each pixel** consists of a photodiode and 3 transistors: a reset transistor to control integration time, a readout amplifier transistor (source follower) and a row select transistor.

Advantages of the **CMOS sensors in comparison to the CCD sensors** are the recording of high contrast images without blooming, higher pixel rates (less clock pulses per pixel read out) leading to higher frame rates at the same resolution, electronic shuttering integrated in each pixel (shutter transistor) and windowing (reading smaller windows of the sensor array at higher frame rates). More details on digital cameras can be found elsewhere (Holst, 1998; Raffel et al., 2007).

2.4 Software

According to the PIV method, successive recorded images of the flow field in the (x,y) measurement plane are divided into rectangular areas called "interrogation windows" (figure 9). The time interval Δt between the recorded images is defined by the frame rate of the digital camera.

The record of the successive images of the flow in the measurement plane can form "pairs": the first pair includes the first image and the second image after Δt, the second pair includes the second and third image, etc. Each image pair corresponds to a collection of interrogation window pairs.

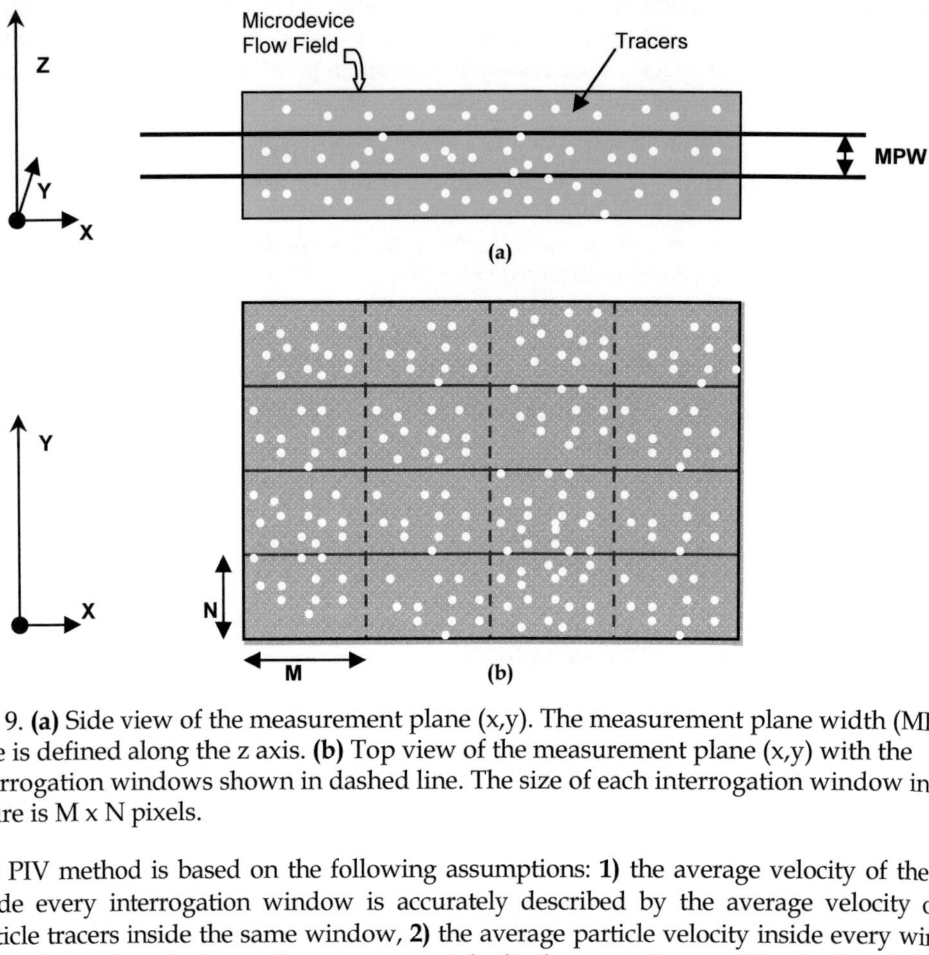

Fig. 9. **(a)** Side view of the measurement plane (x,y). The measurement plane width (MPW) here is defined along the z axis. **(b)** Top view of the measurement plane (x,y) with the interrogation windows shown in dashed line. The size of each interrogation window in this figure is M x N pixels.

The PIV method is based on the following assumptions: **1)** the average velocity of the fluid inside every interrogation window is accurately described by the average velocity of the particle tracers inside the same window, **2)** the average particle velocity inside every window can be found by calculating the average particle displacement **S** caused by the flow in and **3)** the average particle displacement S can be estimated using the correlation function on the 2 window particle images.

In more detail, let's take an interrogation window pair and assume that f(x,y) is the luminous intensity distribution of the first window and g(x,y) the distribution in the second. Then, the cross correlation function is given by the formula:

$$CCOR(s_x, s_y) = \iint f(x,y) g(x + s_x, y + s_y) dxdy \quad (11)$$

Where (s_x, s_y) is the variable vector **s** coordinates of the new CCOR window (figure 10). Assuming that the interrogation windows of the first and second image have the same dimensions M x N pixels (Fig. 9(b)), then the CCOR result window will have dimensions (M+M-1) x (N+N-1) = (2M-1) x (2N-1) pixels. For negative images, the CCOR function will have a maximum value (peak) at position (s_{xMAX}, s_{yMAX}) where the image of the particles in the first window coincides with the image of the particles in the second window. The CCOR peak is shown in white color in figure 10. The **vector distance S_M between the geometrical center of the CCOR window and the peak of the CCOR function** corresponds to the statistical mean displacement of the window particles.

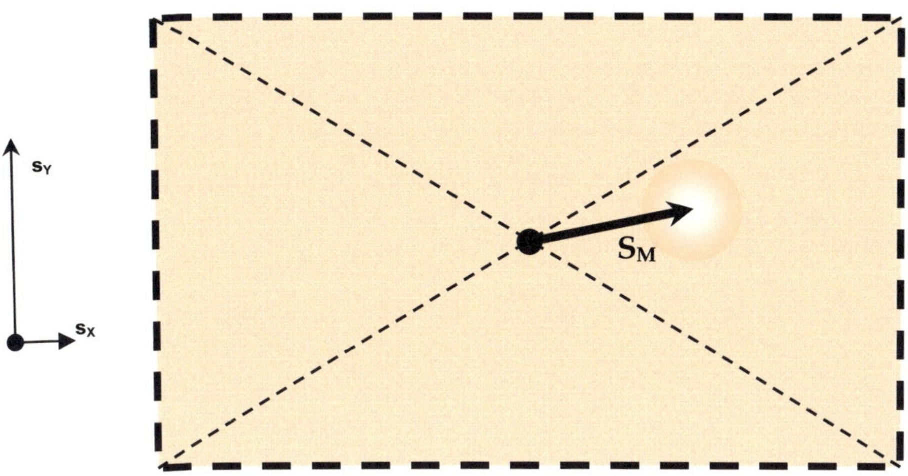

Fig. 10. The displacement vector **S_M**, starts from the geometrical center of the **CCOR window** and ends at the peak of the CCOR function shown in white. In the rest area of the window various values of the CCOR function are shown in dark color. Any other peak of the CCOR function giving a false estimate of the vector S is considered as "noise". The higher the difference of the "white" CCOR peak from the second highest peak, the higher the signal to noise ratio.

By analogy, when the original images are **positive** (Fig. 7 & 8), we are interested in the location of the **CCOR minimum** in order to estimate the vector distance S between the geometrical center of the CCOR window and the minimum of the CCOR function.

In a perfect situation of a white particle on a completely black background of the first window, the size of the CCOR window permits the detection of a displacement equal to M-1 pixels in the x direction and to N-1 pixels in the y direction. However, these conditions are never met in real experimental conditions with many "gray" particles from which some are lost between the first and second window. Therefore, the statistical correlation estimator CCOR has 3 components (Keane & Adrian 1992):

$$CCOR(s) = CCOR_D(s) + CCOR_C(s) + CCOR_F(s) \tag{12}$$

Where $CCOR_D$ is the displacement component (the peak of which we want), $CCOR_C$ is the mean intensity convolution component and $CCOR_F$ is the component of the convolution of the mean intensity of the first window with the fluctuating intensity of the second window. The components $CCOR_C$ and $CCOR_F$ are considered as "noise" creating a fluctuating background shown as a dark background in figure 10.

Since the computation of the cross correlation function in the space domain is time consuming, it is usually estimated in the frequency domain by using the **Fast Fourier Transform (FFT)**. Using the FFT algorithmic technique, the calculation of the CCOR is at least 10 times faster, depending on the information content of the interrogation windows. In some applications it can be 100 times faster. The shortcoming is that the CCOR window has now the same dimensions with the interrogation windows and consequently the dynamic range of the measured displacements is reduced.

Using a Gaussian interpolation in the determination of the cross correlation peak, **subpixel accuracy** can be achieved (Willert & Gharib, 1991) in the estimation of the $\mathbf{S_M}$ magnitude. Then a **conversion factor (CF)** measured in μm/pixel is needed to transform the pixel units of $\mathbf{S_M}$ to micrometers. Since the nominal magnification of a video microscopic system often deviates from reality, an object micrometer is the best way for the CF determination.

Knowing the time interval Δt between the two successive images from the frame rate of the camera, the statistical mean velocity **V** corresponding to each interrogation window pair can be estimated from the classical velocity definition:

$$\mathbf{V} = \frac{\mathbf{S_M}}{\Delta t} \tag{13}$$

Repeating the procedure described in the above paragraphs for all the interrogation window pairs belonging to an image pair, a computer software program can calculate the **whole velocity field**.

In the early PIV implementations many scientists preferred using the **autocorrelation (ACOR)** function on a single image with double exposure. This technique is not used frequently any more, mainly due to the fact that ACOR is a 5 component estimator (Adrian, 1991, as cited in Keane & Adrian, 1992). Two of the five components are displacement components at symmetrical positions with respect to the correlation window center. This has the following implications: **1)** directional ambiguity, **2)** increased hardware complexity to select the right displacement component and **3)** approximately double noise level in comparison to CCOR. In addition, the CCOR technique permits the second window to be larger than the first window in order to reduce the in plane loss of particles Fi. This is not possible with the ACOR function.

In a **bright field μPIV set up**, the images are positive and consequently the displacement vector S_M can be estimated by two different ways: **1)** images are inverted to their negative version and the same software algorithm described above is used and **2)** positive images are used but now there should be a difference in the algorithm: the minimum of the CCOR function is detected. In either case, the typical signal to noise ratio in a brightfield set up is low so, some kind of image processing is usually required in order to reduce the noise level or in other words to remove the background noise (Koutsiaris et al., 1999).

3. Error analysis (Uncertainty)

Using the velocity definition equation 13, there are two major error sources contributing to the total velocity measurement error: the error Δt_e in the estimation of the time interval between two images and the error ΔS_e of the displacement calculation. The time interval error is considered negligible hence we can focus on ΔS_e. Generally, ΔS_e can be considered equal to the square root of the sum of the squares of a random (RAN) and a mean bias (BS) component:

$$\Delta Se = \sqrt{RAN^2 + BS^2} \tag{14}$$

RAN or precision is considered proportional to the effective particle image diameter d_E and the streak length λ_s (Adrian, 1991) according to the formula:

$$RAN = c(d_E + \lambda_S) \tag{15}$$

Where c is a proportionality constant ranging between 1% and 10% depending on various experimental factors such as flow tracing, velocity gradients, non uniform illumination, CCD camera electronics, the cable and digitization noise.

Most of the factors contributing to the random component can also be sources of error for the mean bias component BS. In more detail, BS could have the following error components: **1)** flow tracing error, **2)** flow structure error, **3)** particle image generation error and **4)** interrogation procedure error.

Flow tracing error is determined by the quality of density matching between particles and fluid and the Brown motion.

Flow structure error is caused by fluid accelerations in Δt and from velocity spatial gradients inside the interrogation windows. Starting from fluid acceleration and assuming it causes a critical velocity change in time τ inside the interrogation window, then, if $\Delta t > \tau$, the result will be an integration of $\mathbf{V}(t)$ over the time difference Δt of successive images. For a proper PIV quantification of the **instantaneous velocity field**, Δt should be less than τ in all interrogation windows. The definition of the **time scale** $\boldsymbol{\tau}$ depends on the application and the wanted quantification quality (temporal resolution). It should be noted that in the case of linear fluid acceleration, Δt can be equal or greater than τ as long as an average estimation of the velocity is acceptable.

In analog terms with the above paragraph, assuming that a velocity gradient causes a critical velocity change in distance δ less than the size of the window, the result would be by definition an average velocity value over the surface area of the window. For a proper PIV quantification of the **spatial velocity field**, window size should be less than δ in all interrogation windows. The definition of the **space scale** $\boldsymbol{\delta}$ depends on the application and the wanted quantification quality (spatial resolution). It should be noted that in the case of a linear gradient along only a certain direction (x or y) the selection criteria of δ may be less strict.

Particle image generation error can be produced by many factors such as non-uniform illumination, out of focus particles, non-uniform scattering cross section, optical aberrations, electronic and digitization noise.

Interrogation procedure error can be produced by the size of the interrogation window, the displacement S_M estimation error and the interpolation technique for the location of the CCOR peak.

The above analysis was based on the first error quantification analysis on a μPIV application published by Koutsiaris et al. (1999). More details on PIV uncertainty can be found in other chapters of this textbook.

4. Performance of a μPIV system

From section 2.4 it should be obvious by now that the **spatial resolution** of the μPIV technique is defined by the size of the interrogation windows. So, which are the factors affecting window size? **First** of all, it is the size of the flow tracer particles. The smaller the particles are the smaller the interrogation windows and the higher the spatial resolution of the μPIV technique.

Second, it is the resolution capability of the optical system and the digital camera (electro-optical system). The higher the resolution of the electro-optical system is, the smaller the interrogation window. But the resolution of the optical system depends on its numerical aperture and the resolution of the digital camera on its pixel size. So, the higher the numerical aperture and the smaller the pixel size, the smaller the interrogation window.

Third, it is the image particle density. If window density Ni is defined as the mean number of particles per interrogation window, Fi the in plane loss of particles and Fo the out of plane loss of particles, both Fi and Fo ranging between 0 and 1, then **effective image density equals NiFiFo** (Keane & Adrian, 1992). Since PIV is a statistical method calculating the mean velocity over the entire interrogation window, the NiFiFo corresponds to a detection probability. The detection probability takes maximum values for NiFiFo > 7. For example, a NiFiFo = 5.6 corresponds to an approximate 92% detection probability.

Ni depends on the **particle concentration** (**Nc**) of the fluid, assuming that particles are dispersed homogeneously in the entire volume. It has been shown empirically that a volume per volume particle concentration between 1% and 2% gives an acceptable image density. When concentration is higher than say ≈ 5% agglomerates start to form and when it is lower than ≈ 0.5% effective image density becomes extremely low to give an acceptable detection probability and we have a situation called low image density (LID). However, as we are going to see later in the *in vitro* and *in vivo* applications of the technique, the researchers have found clever ways to overcome these limitations.

The values of the Fi and Fo depend on the **nature of the velocity field**. Fast flows cause large displacements and "aliasing" (Westerweel, 1997) reducing the Fi and the signal to noise ratio. In addition, spatial velocity gradients cause deformations influencing detection probability (Huang et al., 1993). As a starting rule of thumb the maximum particle displacement should be less than 25% the window dimension (for both directions).

The **dynamic velocity range (DVR)** of a μPIV system can be defined in a similar way with a macroscopic PIV system, i.e. the ratio of the maximum to the lowest velocity which can be measured (Adrian, 1997). A higher frame rate digital camera, can record images from faster flows increasing linearly the dynamic range of the system, but in μPIV applications care should be taken with exposure time. As the frame rate increases, the exposure time should decrease accordingly to avoid streaking but there is a down limit in the exposure time depending on the power of the illumination system and the sensitivity of the digital camera.

From the above paragraphs it is evident that the size of the interrogation windows can not be reduced indefinitely. There is a lower size limit depending on many factors. Similarly, the DVR can not increase indefinitely. More details on the spatial resolution and the DVR can be found in the work of Adrian (1997).

The overall performance of a μPIV system can be improved by using **appropriate software techniques** such as overlapping, averaging, single-pixel resolution, background removal, adaptive window shifting and image correction.

With the **overlapping** technique, new windows are defined which overlap up to a point with the old ones. For example, a 50% overlapping means that new windows occupying half the area of each old window, were introduced, doubling the number of interrogation windows along each direction (double minus one). One could argue that in this case the window size is now half, but we should keep in mind that the new windows produced by overlapping do not give new information. It is actually an interpolation technique.

The **correlation averaging** technique (or **ensemble correlation**) improves dramatically the signal to noise ratio, but it requires a steady laminar flow. The technique permits also the formation of a **background image** which can be used to remove the background noise by subtracting it from the image sample pairs. An extension of the ensemble correlation technique is the **single-pixel resolution ensemble correlation** (Westerweel et al., 2004) but it requires 1000 images or more in contrast to the 10-30 image pairs of the correlation averaging technique.

The PIV bias error can be reduced using the **adaptive window shifting** technique which requires iteration for optimum results. In adaptive window shifting, the second interrogation window of each pair is shifted proportionally to the expected average particle displacement (Forward Difference Interrogation). In an improvement of this technique called CDI (Central Difference Interrogation, Werely & Meinhart, 2001) both windows are shifted by half the expected average particle displacement: the first interrogation window is shifted backwards and the second window is shifted forwards. Using CDI the bias error can be neglected in comparison to the random error.

The random error can be reduced using **image correction** techniques. In image correction techniques (Huang et al., 1993) image patterns in the interrogation windows are deformed in a way depending on the spatial velocity gradients. In this way, the random error is reduced and the total error becomes approximately half.

More details on software techniques can be found in other textbooks (Breuer, 2005; Nguyen & Wereley, 2002; Raffel et al., 2007).

5. Research *IN VITRO*

As it is the usual practice, the first applications of the μPIV technique were *in vitro*. The fluorescence μPIV set up introduced by Santiago et al. (1998) produced negative images in accordance with the classical PIV systems. The μPIV set up introduced by Koutsiaris et al. (1999) produced positive images i.e. the background was brighter than the particles (figure 11 (a)) similar to the ordinary daylight photographs. This meant that images should be inverted before calculating the cross correlation or the software should be modified to detect the minimum of the correlation function. However, this set up was much simpler and cheaper than fluorescence set ups since it did not require the use of fluorescence apparatus (special particles, light sources, optics and low light CCD cameras).

With the use of the appropriate **image processing digital filters** one can have image inversion and simultaneous increase of the signal to noise ratio. For example, the Laplace filter (a 6 x 6 Kernel) shown in figure 11(c) enhances particle boundaries and removes slowly varying background shading. In addition, the same kernel helps in eliminating the out of focus particles, since the spatial frequencies corresponding to the in focus flow tracers can be maximized by adjusting the size of the filter (Koutsiaris et al., 1999). The determination of thresholds and optimum filter variable values could be the object of further research in digital image processing and digital filter design. The effects of the possible variable combinations on the μPIV flow field accuracy have not yet been examined thoroughly.

Cylindrical microtubes (Koutsiaris et al., 1999) simulate better the flow geometry of microvessels but their construction is demanding. In addition, they require a good optical matching otherwise image optical distortions must be corrected.

Rectangular microchannels are not good models for simulating animal microcirculation, but they possess a simple geometry with minimum optical distortions. The aspect ratio (AR) of a television screen is defined as the ratio of the longest (horizontal) over the shortest (vertical) side and is always greater than 1. Keeping the same definition in rectangular microchannels, the longest (horizontal) side of the cross section is usually referred to as "width" (W) and the shortest (vertical) as "height" (H). Research groups all over the world have studied flows inside microchannels with various ARs: 1 (W = 100μm, H = 100μm, Lima et al., 2006), 1.7 (W = 80μm, H = 48μm, Kuang et al., 2009), 4 (W = 800μm, H = 200μm, Lindken et al., 2006; W = 20mm, H = 5mm, Timgren et al., 2008), 6.7 (W = 300mm, H = 45mm, Lima et al., 2008) and 10 (W = 300μm, H = 30μm, Meinhart et al., 1999).

There are several analytical solutions for the steady Newtonian flow inside a rectangular channel such as those shown by Werely & Meinhart (2005), Lima et al. (2006) and Lindken et al. (2006). All these solutions contain infinite sum terms. In addition, Werely & Meinhart (2005) reported that their analytical solution failed to capture the trends of the measured profile near the wall and that in order to amend this they used a second order polynomial fit. Lima et al. (2008) reported results in the middle plane with analytical solution errors of less than 5%. However, it seems that these errors can be reduced to less than 1.6% (more than 3 times) using an alternative model equation discussed below (see figure 15 in section 7).

In 2001, Werely and Meinhart measured the flow of deionized water around human red blood cells placed in the gap between a microscope slide and a coverslip. The same year, Gomez et al. measured for the first time the flow of water based suspensions inside a microfluidic biochip designed for impedance spectroscopy.

(a) (b)

-1	-1	-1	-1	-1	-1
-1	-1	-1	-1	-1	-1
-1	-1	8	8	-1	-1
-1	-1	8	8	-1	-1
-1	-1	-1	-1	-1	-1
-1	-1	-1	-1	-1	-1

(c)

Fig. 11. (a) An original positive image from the diametric plane of a cylindrical glass tube with internal diameter of 262 μm. Numbers 1, 2 & 3 correspond to different positions downstream the flow. (b) The negative of the original image with improved signal to noise ratio, after the application of the kernel shown in (c). (From Koutsiaris et al., 1999)

Soon after the first μPIV measurements on Newtonian flows, researchers started the flow study of **red blood cell (RBC) suspensions**. Okuda et al. (2003) measured in round tubes with internal diameter of 100μm the flow of rabbit blood seeded with fluorescent particles; Bitch et al., (2005) measured in specially flattened tubes the flow of a RBC suspension with 60% hematocrit using RBCs as natural flow tracers (brightfield set up).

Lima et al. (2006) were the first to measure three-dimensional velocity profiles of human RBC suspensions inside a square glass microchannel (100 x 100 μm) using a special confocal experimental set up. One year later (Lima et al., 2007), using the same microchannel dimensions, found that low hematocrit (up to 17%) suspensions have parabolic averaged velocity profiles. In 2008 Lima et al. compared the velocity profile of physiologic saline with the profile of human blood with 20% heamatocrit inside a rectangular high aspect ratio (W = 300 μm, H = 45 μm) PDMS microchannel; microturbulences were encountered on averaged blood flow profiles. Next year (Lima et al., 2009), they combined a particle tracking velocimetry (PTV) system with a confocal microscope in order to quantify the RBCs trajectories in suspensions up to 20% in heamatocrit.

When the tube dimension is greater than 40-60 μm it is difficult for the light rays to penetrate blood at physiologic hematocrits both in transmission and reflectance mode. Even when a confocal set up is used, the microchannel dimension parallel to the optical axis should be less than ≈ 100 μm and the hematocrit less than ≈ 20%. In order to overcome these limitations, Kim and Lee (2006) tried a completely different approach using X-rays. The Fresnel diffraction pattern images can be used to calculate the blood velocity field at high haematocrits (20-80%) without tracer particles, in microchannels with dimensions higher than ≈ 500 μm. A disadvantage of the method is that sample thickness and heamatocrit affect image quality so more work is required to define the limits inside which valid measurements can be taken.

In 2011, Kaliviotis et al. used brightfield μPIV to measure the velocity field of 45% heamatocrit human blood samples inside an optical shearing system consisted of 2 circular parallel glass plates set apart by a gap h = 30 μm. The shear rate derived from the blood velocity field was used in their non-Newtonian (shear rate, time and aggregation depended) blood viscosity model.

Except for the biological flows, *in vitro* μPIV has a lot of other useful applications such as those regarding the **computer industry**. Meinhart and Zhang (2000) measured the evolution of the meniscus and the detailed flow fields inside a nozzle of an inkjet printhead. Buffone et al. (2005) investigated the thermocapillary Marangoni convection in the proximity of the liquid-vapor interface for an evaporating meniscus in horizontal capillary tubes (600, 900 and 1630 μm). Today, air cooling limitations are responsible for a barrier in the total power consumed by a processor chip (a power wall of approximately 100 Watts) and ultimately in the chip performance. However, this situation can change using microchannels for liquid flow cooling (either single phase cooling with water, or two phase cooling with refrigerant). This could be a revolution in the microelectronics industry since the maximum heat dissipation can increase approximately 10 times from 37 W/cm^2 (air cooling) to 314 W/cm^2 (Thome & Marcinichen, 2011). Typical components of a two phase microcircuit are a liquid micropump or vapor compressor, a microevaporator and a microcondenser.

Among **other applications**, Bown et al., (2005) demonstrated the application of μPIV to complex microchannel geometries and introduced a method of estimating the out of plane effects on the velocity measurements. Lindken et al. (2006) measured for the first time in the mixing region of a T-shaped micromixer. King et al. (2007) demonstrated the ability to perform μPIV measurements of aqueous plugs in two phase flow within circular tubing with a 762 μm internal diameter. Such two phase flows are encountered in lab-on-chip devices designed for various biological applications such as the Polymerase Chain Reaction (PCR). Similarly, Timgren et al. (2008) measured the velocity field inside silicon oil drops forming in a mixture of water and glycerol flowing in a rectangular channel.

Kuang et al. (2009) presented a **molecular tracer method** (Molecular Tracers Velocimetry, MTV) for measuring the velocity profile in cylindrical and rectangular microchannels with dimensions less than 100 μm. Their method was based on the Laser Induced Fluorescence Photobleaching Anemometry (LIFPA) technique.

More information on the **biological applications** of μPIV can be found in the review paper of Lindken et al. (2009) and more details on the numerical and experimental analysis of flows inside microdiffusers and valveless micropumps are presented in the review paper of Nabavi (2009).

6. Velocity profile measurements *IN VIVO*

The actual measurement of the blood velocity profile in microvessels is a known difficult task to accomplish. This is in part due to the many different scientific fields that need to cooperate and in part due to the expensive experimental set up. Therefore it is not surprising that apart from some preliminary efforts in the 70's (Schoenbein & Zweifach, 1975), the **first *in vivo*** quality velocity profile measurements were presented in 1986 by the group of professor Reneman (Tangelder et al., 1986). They measured velocity profiles in the arterioles of the rabbit mesentery using as flow tracers platelets labeled with a special fluorescence technique.

More than 15 years later, Sugii et al. (2002) and Nakano et al. (2003) measured the velocity profile in arterioles of the rat mesentery using a technique introduced some years earlier by Sugii et al. (2000) under the name **"High Resolution Particle Image Velocimetry" (HR-PIV)**. Regarding the software, this technique was a combination of the iterative cross-correlation method (Raffel et al., 2007) and the optical flow (or gradient or spatio-temporal derivative) method, providing excellent accuracy (as little as 0.01 pixels). Their hardware was a brightfield µPIV set up with the exceptional advantage of using natural blood flow thus avoiding any toxic effects from fluorescent materials and complex-expensive fluorescent equipment. The final velocity field spatial resolution in the object plane was 0.8 µm. The only drawback is that being a **fully automated** technique it is difficult to validate the results. Another fully automated technique was presented by Tsukada et al. (2000) based on circular correlation windows reduced down to the size of the erythrocytes.

In 2003, Hove et al. (2003) mapped the flow field inside the developing zebrafish heart 37 hours and 4.5 days post fertilization using RBCs as flow tracers. They found the presence of higher shear vertical flow than expected and evidence that **blood flow-induced forces act as a key epigenetic factor** in embryonic cardiogenesis. Later, Vennemann et al. (2006) improved the resolution of the flow field maps and boundaries, using fluorescent, long-circulating liposomes with a diameter of 400 nm, as flow tracers, in the embryonic avian heart. They noted the potential of µPIV to become a general tool in complex geometries in cardiovascular research. Both groups underlined the importance of examining the interplay between genetics and fluid shear forces in analyzing not only normal development but also the pathogenesis of embryonic cardiovascular defects.

In 2004, Long et al. (12 venules from male mice, 24 µm ≤ D ≤ 42.9 µm) and Damiano et al (9 light-dye treated venules from 3 mice, 24 µm ≤ D ≤ 42.9 µm) provided the most complete **manual** velocity profile measurements until now, with the best spatial resolution and the assumption of steady axisymmetric flow. They measured in the cremaster muscle of mice using fluorescent microspheres (0.47 ± 0.01 µm) as blood flow tracers.

In 2005, Nakano et al. (2005) were the first to measure the effect of an arteriolar bifurcation and confluence on the red blood cell velocity profile on the rat mesentery. More recently, Potter & Damiano, (2008) performed measurements in mice venules up to diameters of 101 µm but they mainly concentrated on the properties of the endothelium glycocalyx layer both *in vivo* and *in vitro*.

7. Velocity profile equations *IN VIVO*

7.1 Introduction

The accurate quantification of the velocity profile in a cylindrical axisymmetric flow is very important because it is the **first and unique step** required for the estimation of the wall shear rate (WSR) and volume flow (Q). In addition, the viscosity profile (and hence the apparent viscosity), the axial pressure gradient and the shear stress profile (and hence wall shear stress, WSS) can be estimated **under the assumptions** of a locally Newtonian fluid and the existence of plasma between red blood cells (RBCs) and vessel walls (Damiano et al., 2004).

Before the last third of the 20th century, the only equation available to the researchers studying laminar flows inside cylindrical tubes was the parabolic one. In the seventies (70s) it became evident that the flow of blood is quite different from simple Newtonian flows like that of water. Blood exhibits special shear thinning properties due to reasons which are still partly unexplained. For example the molecular and biochemical basis of the rouleaux (structures resembling coin piles) formation is still unknown. The shear thinning property means that blood viscosity diminishes (blood becomes thinner) as shear rate increases. For the case of blood, this property is quite evident taking into account that at high shear rates ($> 100\ s^{-1}$) its viscosity is many times lower than at shear rates $< 10\ s^{-1}$. This means that near the vessel axis where there are such low shear rates, blood is much more viscous, causing a characteristic "blunting" of the velocity profile (Bugliarello & Sevilla, 1970; Damiano et al., 2004; Gaehtgens et al., 1970; Long et al., 2004; Nakano et al., 2003; Schoenbein & Zweifach, 1975; Tangelder et al., 1986).

After 1970, the researchers proposed equations trying to describe the blunting of the velocity profile in the microvasculature with diameters (D) higher than ≈ 20 μm. The approximate diametric **down size limit of the 20 μm** (Cokelet, 1999) is imposed by the manifestation of the biphasic nature of blood in the smallest arterioles and venules and in the capillaries. In these microvessels RBCs flow separately, constituting a different liquid phase from plasma and therefore the flow medium can not be considered as a "continuum" and a velocity profile can not be defined in the ordinary sense. However, **averages of WSR, Q and WSS** can be estimated *in vivo* using empirical equations (Koutsiaris, 2005; Koutsiaris et al., 2007) requiring only axial RBC velocity measurements.

The available velocity profile equations today, could be divided in many ways, but in this section the criterion was whether they can be easily reduced to the classic parabolic equation (Group A) or not (Group B).

7.2 Available equations

Realistic assumptions of the blood flow in straight sections of microvessels with D > 20μm, several diameters downstream their entrance, are (Koutsiaris, 2009): 1) incompressible flow, 2) continuous medium, 3) viscous flow with Reynolds number less than one, 5) cylindrical vessel geometry with a radius R, 5) non-Newtonian medium with a time averaged velocity profile blunter than the parabolic with the same maximum velocity V_m, 6) the blood velocity is zero on the vessel wall (zero slip condition: $V(r) = 0$, at radial position $r = R$) and 7) axisymmetric velocity profile with maximum velocity V_m on the vessel axis.

The assumption of steady flow can be accepted for venules with D > 20μm (Koutsiaris et al., 2011), but not for arterioles. Recently, it was verified for the first time in humans, that the velocity pulse in arterioles is quite strong, even at the precapillary level (Koutsiaris et al., 2010c). Despite the strong pulse, the velocity profile can be measured at the same cardiac cycle phase taking advantage of the periodic nature of the flow. So, the flow can be considered as "steady" for the same phase.

All the above conditions are satisfied by the equations presented below except for the parabolic equation which can not satisfy the non-Newtonian condition, as it will be shown below.

7.2.1 Group A

This group comprises 3 velocity profile equations:

$$V_P(r) = V_m\left[1 - \left(\frac{r}{R}\right)^2\right] \tag{16}$$

$$V_{RS}(r) = V_m\left[1 - \left(\frac{r}{R}\right)^\kappa\right] \tag{17}$$

$$V_{KS}(r) = V_m\left[1 - \kappa_1\left(\frac{r}{R}\right)^2\right]\left[1 - \left(\frac{r}{R}\right)^{\kappa_2}\right] \tag{18}$$

Where $V_P(r)$, $V_{RS}(r)$, $V_{KS}(r)$ is the velocity at radial position r, for the parabolic (equation 16), the Roevros (equation 17, Roevros 1974) and the Koutsiaris (equation 18, Koutsiaris 2009) equation, respectively.

The parameters κ and κ_1, κ_2 **affect the velocity profile shape** of equations 17 and 18 respectively. For a velocity profile blunter than the parabolic one, with the same V_m, the following conditions must be satisfied: $\kappa > 2$ (equation 17), $0 < \kappa_1 < 1$, $\kappa_2 > 2$ and $(1-\kappa_1)\,\kappa_2 \geq 2$ (equation 18).

In **equation 17**, the higher the κ, the flatter the profile near the vessel axis and the higher the wall shear rate (Roevros 1974). Equation 17 reduces to parabolic when $\kappa = 2$. It should be noted that a modified version of equation 17 with 2 more parameters (a and b) was proposed in the 80s (Tangelder et al., 1986), but it does not satisfy the zero slip condition on the vessel wall which is true for any viscous flow.

In **equation 18**, the advantage is that the bluntness of the profile can be controlled near the axis and the wall: generally, as κ_1 approaches zero the profile becomes flatter near the axis and as κ_2 takes values higher than 2 the profile becomes flatter near the wall (Koutsiaris, 2009). Equation 18 reduces to the parabolic equation when $\kappa_1 = 0$, $\kappa_2 = 2$.

7.2.2 Group B

Damiano et al. (2004) made significant contributions in theory and experimental measurements and provided a way to estimate the viscosity profile as well as the effective viscosity *in vivo*. They proposed the following velocity profile equation which identically satisfies the momentum equation and boundary conditions (Long et al., 2004):

$$V(r) = V_m \frac{\int_{r/R}^{R} f(\sigma)d\sigma}{\int_{0}^{R} f(\sigma)d\sigma} \tag{19}$$

Where $f(\sigma)$ is a function of R and of two independent parameters c_1 and c_2 which can be found through non linear regression analysis that uses equation 19 to minimize the least-squares error (SSE) of the fit to the experimental velocity profile data sets (Long et al., 2004). With a suitable modification the equation can take into account the infinitesimal flow inside the microvascular glycocalyx layer. This is important for the transcapillary exchange and endothelium studies but from the volume flow estimation point of view it contributes little due to the very low velocities near the vessel wall. The fitting of this equation to experimental data from mouse venules can be seen in the relevant papers (Damiano et al., 2004; Long et al., 2004).

7.3 Fitting efficiency

At a preliminary evaluation of the **group A equations** (Koutsiaris et al., 2010a), fixed values for the parameters were selected: $\kappa = 9$ (equation 17) and $\kappa_1 = 0.58$, $\kappa_2 = 22$ (equation 18) assuming they all have the same axial velocity estimated from the nearest experimental point to the vessel axis (**axial fit**). Eight profiles from **mouse venules** were estimated **graphically** from the literature (Long et al., 2004). According to the results, the parabolic equation tends to underestimate blood velocity, reaching a maximum relative error of - 72% near the vessel wall. The Roevros equation tends to overestimate blood velocity reaching a maximum relative error of + 48% at a radial position between 70% and 80% of the vessel radius R. Equation 18 tends to approximate blood velocity with a relative error between - 8% and + 7%, for all radial positions, leading to an average volume flow error of less than 0.5%. An axial fit example of the 3 equations is shown in figure 12.

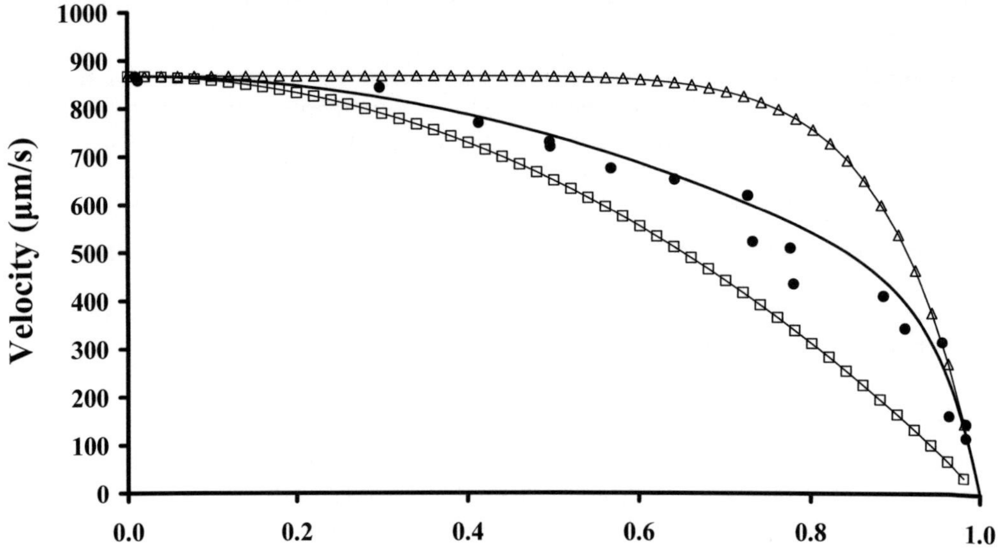

Fig. 12. Velocity profile equation 16, 17 and 18 (with the same axial velocity V_m) is shown in squares, triangles and solid black line respectively (Koutsiaris et al., 2010a). Velocity profile data from a 38.6 μm mouse venule are shown in black dots. Data were estimated graphically from Long et al. (2004).

Instead of using the same maximum velocity and fixed parametric values (axial fit), the classical approach would be to apply the **best fit**, i.e. finding the parametric values producing the minimum sum square error on each experimental velocity profile. Then statistics could be applied to the velocity relative errors of all the profiles, dividing the normalized radial position (r/R) at 10 different radial segments:

$$\frac{j-1}{10} \leq \frac{r}{R} \leq \frac{j}{10} \tag{20}$$

Where j is an integer index ($1 \leq j \leq 10$). Working in this way, Koutsiaris et al. (2011) compared statistically the best fits of equations 17 and 18 and the axial fit of equation 18 on the **original experimental data of 12 velocity profiles from mouse venules** between 21 and 39 μm in diameter (Long et al., 2004). As it is shown in figure 13, the most efficient fit is the best fit of equation 18 with average relative errors (RE_j) between -1% and +2% in all radial segments, but it requires the complete velocity profile data from each vessel. The axial fit of equation 18 gives results with acceptable error (RE_j less than 11%) and requires only one velocity measurement near the vessel axis (Koutsiaris, 2009; Koutsiaris et al., 2011).

Here, it should be noted that a lot of researchers working in the area of micro blood flow still use the parabolic equation. As it is proved by figures 12 and 13, according to the current experimental data, the parabolic equation can lead to serious errors, especially near the vessel wall.

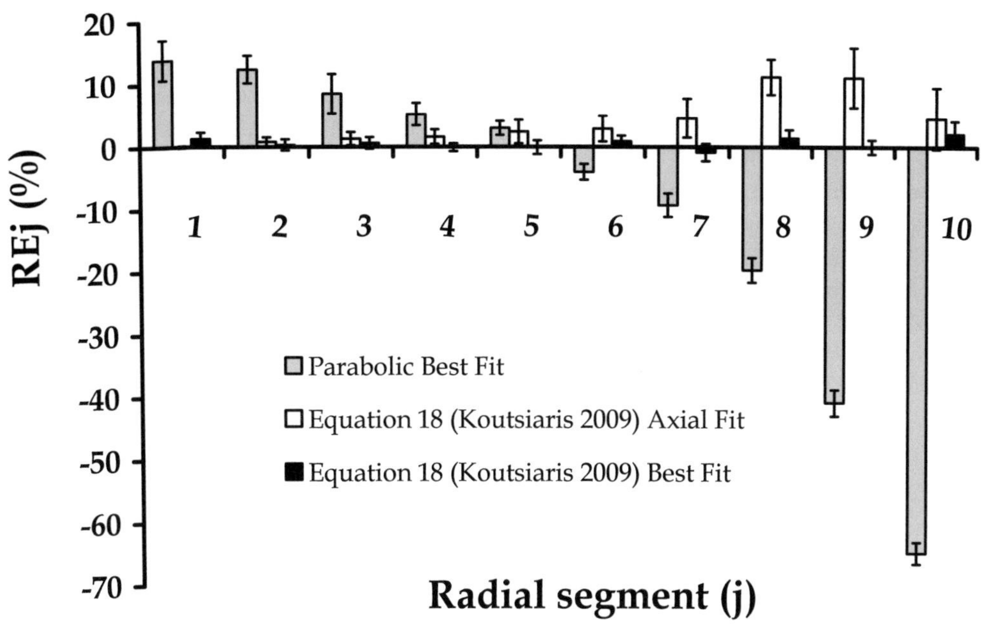

Fig. 13. Velocity **relative error (RE)** statistics on original experimental data from the mouse cremaster muscle (12 velocity profiles, Long et al., 2004). **Average RE of each segment j (RE_j)** is shown in columns and standard error of the mean is shown with black bars on each column. Gray, white and black columns represent the parabolic best fit, the equation 18 axial fit and the equation 18 best fit respectively. Data were taken from Koutsiaris et al. (2011).

In addition, it should be noted that the best fit of the Roevros equation was not examined here, because it has already been reported (Tangelder et al., 1986) that it tends to underestimate velocity in the center of the vessel and near the wall and to overestimate velocity in the range $0.5 \le (r/R) \le 0.8$.

As it was mentioned in the beginning of section 7.2, equations 16 to 19 could also be applied to the same phase of the **arteriolar blood flow**. The best fit of equation 18 to the **original velocity profile data** from a 25 μm rabbit arteriole at the diastolic phase (Tangelder et al., 1986) is shown in figure 14. The dispersion of the 151 experimental points around the black line in figure 14 (a) was caused by at least two factors: **1)** the inability of a volume illuminated system to define the measurement plane (out of focus effects) and **2)** small irregularities in the periodic arteriolar flow.

The first obstacle can now be overcome with modern confocal systems (section 2.2.6). An off-line way to overcome the out of focus effects is the use of the filtering criterion proposed by Damiano et al. (2004). The best fit of equation 18 on the 11 experimental points left after applying this criterion is shown in figure 14 (b).

Closing this section, it should be reported that equation 18 seems also to provide good fits, to velocity profile data coming from **Newtonian fluids inside rectangular microchannels**. It appears that rectangular cross sections with aspect ratios >> 1 (or << 1 depending on the definition; here the definition described in section 5 was accepted) distort the parabolic velocity profile of a Newtonian fluid producing blunt profiles similar to those of shear thinning fluids, such as blood, flowing inside cylindrical microchannels. The blunt profiles of the Newtonian fluids appear along the greatest side of the rectangular cross section.

An example of velocity profile along the greatest dimension (width W) of a rectangular cross section microchannel is shown in figure 15. The width was defined along the y direction, the height (H) was defined along the z direction and the length along the x direction (the direction of flow). The experimental velocity profile data shown in black dots, were measured in the central plane ($z = (H/2) = 22.5$ μm, Lima et al., 2008) and the width position y here was normalized with $(W/2) = 150$μm. The line represents the best fit approximation of equation 18. The parametric values of equation 18 giving the best fit were: $V_m = 0.451$ mm/s, $\kappa_1 = 0.109$ and $\kappa_2 = 12$. The κ_2 value was rounded down to the nearest even number, because a non-positive number can not be raised to a fractional power. Even with this approximating κ_2 value, the absolute relative error ($|RE|$) was less than 1.6% and the average $|RE|$ was 0.5%.

Perhaps, the parameters κ_1 and κ_2 could be expressed as functions of the aspect ratio (W/H) and the z position in order to use equation 18 for the velocity profile expression of a Newtonian fluid in every rectangular channel, but this requires more experimental data at various aspect ratios in order to be verified.

7.4 Application to humans

A fundamental remaining question is whether equation 18, which seems to approximate efficiently the blood velocity profile in the mouse and rabbit microcirculation, can be applied to the human microcirculation also. Taking into account the current state of technology, it would be rather difficult now, or in the near future, to measure the velocity profile of blood in human microvessels.

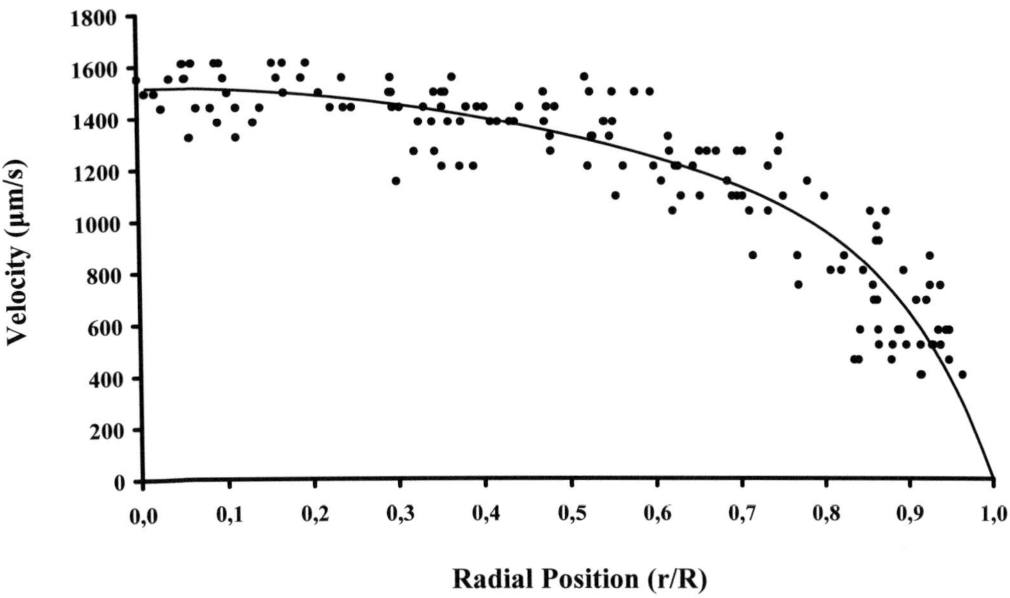

(a)

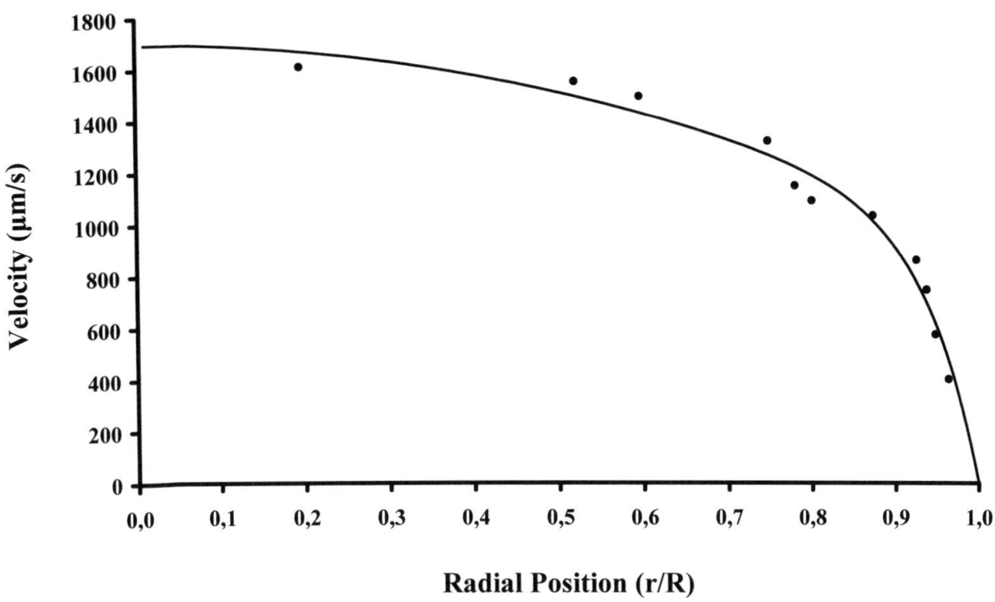

(b)

Fig. 14. Best fit of equation 18 to original velocity profile data from a 25 μm rabbit arteriole at the diastolic phase (Tangelder et al., 1986). **(a)** 151 experimental points. **(b)** 11 experimental points after filtering.

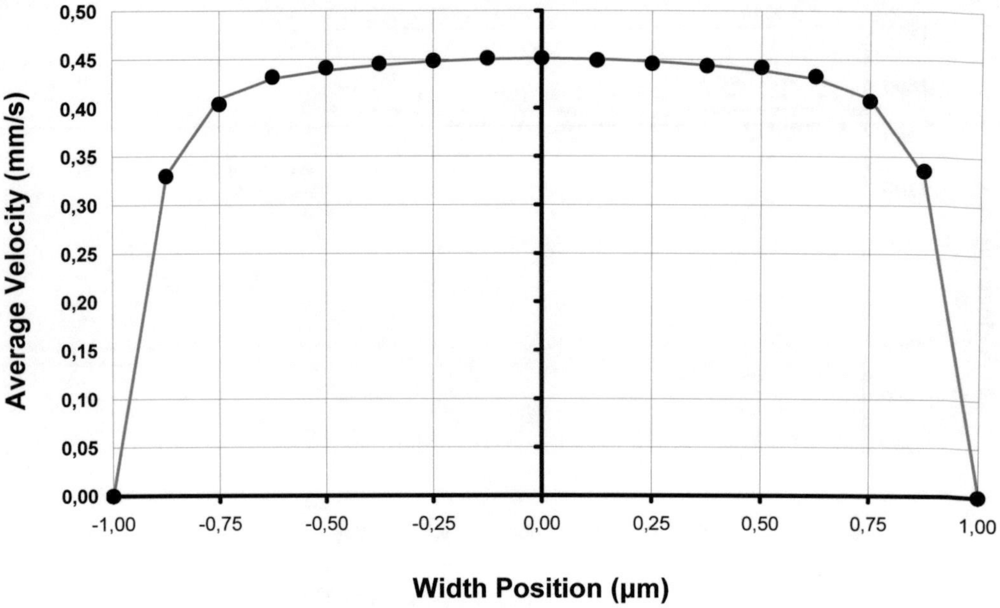

Fig. 15. Best fit of equation 18 (shown with a line) to velocity profile data (shown with black dots) from a rectangular microchannel (Lima et al., 2008, Fig. 5) with a Height (H) = 45 µm and a Width (W) = 300 µm. Here, the width position was normalized with (W/2) = 150µm. The absolute relative error of the fit was less than 1.6% at all experimental points.

An indirect way of finding the answer would be to measure the rheological differences between mouse and human blood and more specifically the viscosity differences at low shear rates. In case these differences prove to be high enough, presumably the profiles would be different and a change of the equation fixed parametric values would be required.

It is already known (Windberger et al., 2003; Windberger & Baskurt, 2007) that in humans and other athletic species like horses, **whole blood viscosity (WBV)** is higher than in mice. However, this WBV difference becomes important at very low shear rates (< 10 s^{-1}) occurring close to the vessel axis. Using the fixed parametric form of equation 18, the shear rate of 10 s^{-1} corresponds to a radial position r = 0.18 R, or to a surface area of only 3.3% of the total cross sectional area of the vessel. Therefore, it would be logical to assume that the fixed parametric form of equation 18 could be applied to the human microvessels as well. However, one could argue that this region near the vessel axis was responsible for the profile blunting in the first place.

A more detailed viscometric experiment comparing mouse and human blood samples would involve WBV measurements at many different shear rates < 10 s^{-1}, at physiologic temperatures (human blood at 36.6 ^{o}C and mouse blood at 38 ^{o}C). Even so, there would be some remaining issues such as the use of anticoagulants in the blood samples and the selection of the appropriate hematocrit level, since according to the Fahraeus effect, the average microcirculatory hematocrit is lower than the systemic hematocrit (Hs). As it was proved by Lipowsky et al. (1980), for diameters of approximately 20 µm, the average hematocrit *in vivo* would be approximately 0.28Hs in the venous side. A suggested set of experiments would comprise WBV measurements at hematocrits of 0.28Hs, 0.38Hs and 0.50Hs.

8. Conclusion

In this chapter a medium size review was given on digital micro particle image velocimetry (μPIV). The four main components of a Digital μPIV system were described in subsections of section 2: 2.1) microfluidics, 2.2) optics, 2.3) electronics and 2.4) software.

In the optics subsection, important insight was given into the concepts of volume illumination, depth of focus (DoF), f number (f#), measurement plane width (MPW) and effective diameter (d_E). In addition, the distinction between fluorescent and brightfield μPIV systems was clarified.

The latest development in μPIV optics is the "confocal" experimental set up which has the advantage of defining the width of the measurement plane (optical slice thickness). It seems that this will be the long term future of the digital μPIV set ups. However, at the moment, this is an expensive solution and there is always the drawback of the single point scanning, meaning that the image of the measurement plane is not strictly acquired at the same time. It is the view of the writer that low cost brightfield systems have not been examined adequately together with the use of image processing techniques.

In the electronics subsection some hinds were given on digital camera species and architecture. In the software subsection the basic algorithm was described together with the basic assumptions and differences between fluorescent and brightfield set ups.

In addition, a brief but incisive description was given of the error (uncertainty, section 3) and performance (section 4) of a μPIV system, two subjects that often are not given the proper attention.

The promising future of the Digital μPIV technique stems from the wide range of applications in micromechanics and microelectronics industry and research. Separate sections (sections 5 and 6) were devoted to the *in vitro* and *in vivo* experiments with a historical research overview.

Special emphasis was given on the microflows of biological fluids and especially in the velocity profiles of blood *in vivo*. A separate section (section 7) was dedicated on the velocity profile equations *in vivo* and their evaluation based on original experimental data.

All current experimental evidence suggests that the classical parabolic velocity profile is inappropriate for describing blood flow in the mammalian microvasculature. As it was shown in section 7, there are now **new equations** and **fitting techniques** for the expression of the average velocity profile of small mammals *in vivo* with acceptable error (figures 12, 13 and 14). In conclusion these new equations are at the moment the best choice for the description of blood flow in the human microcirculation. It should also be noted that equation 18 seems to describe well the Newtonian flow inside rectangular microchannels.

9. Acknowledgments

The author of this chapter would like to thank Professor Edward Damiano from Boston University, USA, Professor Geert Tangelder from VU University Medical Center, Amsterdam, The Netherlands and Professor Rui Lima from Polytechnic Institute of Braganca, Portugal, for kindly sending the necessary original experimental data.

10. References

Adrian, R.J. (1991). Particle-imaging techniques for experimental fluid mechanics. *Annu. Rev. Fluid Mech.*, Vol.23, pp. 261-304, ISSN 0066-4189.

Adrian, R.J. (1997). Dynamic ranges of velocity and spatial resolution of particle image velocimetry. *Meas. Sci. Technol.*, Vol.8, pp. 1393-1398, ISSN 0957-0233.

Bitsch, L.; Olesen, L.H.; Westergaard, C.H.; Bruus, H.; Klank, H. & Kutter, J.P. (2005). Micro particle-image velocimetry of bead suspensions and blood flows. *Exp. Fluids*, Vol.39, pp. 505-511, ISSN 0723-4864.

Bourdon, C.J.; Olsen, M.G. & Gorby, A.D. (2004). Validation of an analytical solution for depth of correlation in microscopic particle image velocimetry. *Meas. Sci. Technol.*, Vol.15, pp. 318-327, ISSN 0957-0233.

Bown, M.R.; MacInnes, J.M. & Allen, W.K. (2005). Micro-PIV simulation and measurement in complex microchannel geometries. *Meas. Sci. Technol.*, Vol.16, pp. 619-626, ISSN 0957-0233.

Bown, M.R.; MacInnes, J.M. & Allen, W.K. (2006). Tree-component micro-PIV using the continuity equation and a comparison of the performance with that of stereoscopic measurements. *Exp. Fluids*, Vol.42, pp. 197-205, ISSN 0723-4864.

Breuer, K. (Ed.). (2005). *Microscale Diagnostic Techniques*, Springer, ISBN 3-540-23099-8, Heidelberg.

Buffone, C.; Sefiane, K. & Christy, J.R.E. (2005). Experimental investigation of self-induced thermocapillary convection for an evaporating meniscus in capillary tubes using micro-particle image velocimetry. *Physics of Fluids*, Vol.17, No 5, pp. 052104/1-18, ISSN 1070-6631.

Bugliarello, G. & Sevilla, J. (1970). Velocity distribution and other characteristics of steady and pulsatile blood flow in fine glass tubes. *Biorheology*, Vol.7, pp. 85-107, ISSN 0006-355X.

Chrobak, K.M.; Potter D.R. & Tien, J. (2006). Formation of perfused, functional microvascular tubes in vitro. *Microvasc. Res.*, Vol.71, pp. 185-196, ISSN 0026-2862.

Cokelet, G.R. (1999). Viscometric *in vitro* and *in vivo* blood viscosity relationships: how are they related? *Biorheology*, Vol.36, pp. 343-358, ISSN 0006-355X.

Damiano, E.R.; Long, D.S. & Smith, M.L. (2004). Estimation of viscosity profiles using velocimetry data from parallel flows of linearly viscous fluids: application to microvessel haemodynamics. *J. Fluid Mech.*, Vol.512, pp. 1-19, ISSN 0022-1120.

Delly, J.G. (1988). *Photography through the microscope*, Eastman Kodak Company, ISBN 0-87985-362-X, New York, USA.

Gad-el-Hak, M. (Ed.). (2005). *The MEMS Handbook* (2nd Edition, 3 Volume Set), CRC press (Taylor & Francis Group), ISBN 0-84932-106-9, Florida, USA.

Gaehtgens, P.; Meiselman, H.J. & Wayland, H. (1970). Velocity profiles of human blood at normal and reduced hematocrit in glass tubes up to 130 μ Diameter. *Microvascular Research*, Vol.2, pp. 13-23, ISSN 0026-2862.

Gardel, M.L.; Valentine, M.T. & Weitz, D.A. (2005). Microrheology, In: *Microscale Diagnostic Techniques*, Breuer K.S. (Ed.), pp. 1-49, Springer, ISBN 3-540-23099-8, Heidelberg.

Gomez, R.; Bashir, R.; Sarakaya, A.; Ladisch, M.R.; Sturgis, J.; Robinson, J.P.; Geng, T.; Bhunia, A.K.; Apple, H.L. & Werely, S.T. (2001). Microfluidic biochip for impedance spectroscopy of biological species. *Biomed. Microdevices*, Vol.3, No 3, pp. 201-209, ISSN 1387-2176.

Hassan, I. (2006). Thermal-Fluid MEMS Devices: A Decade of Progress and Challenges Ahead. *J. Heat Trans-T ASME*, Vol.128, No.11, pp. 1221-1233, ISSN 0022-1481.

Holst, G.C. (1998). *CCD arrays cameras and displays*, SPIE Press, ISBN 0-8194-2853-1, Bellingham, Washington, USA.

Hove, J.R.; Köster, R.W.; Forouhar, A.S.; Acevedo-Bolton, G.; Fraser, S.E. & Gharib, M. (2003). Intracardiac fluid forces are an essential epigenetic factor for embryonic cardiogenesis. *Nature*, Vol.421, pp. 172-177, ISSN 0028-0836.

Huang, H.T.; Fiedler, H.F. & Wang, J.J. (1993). Limitation and improvement of PIV, part II. Particle image distortion, a novel technique. *Exp.Fluids*, Vol.15, pp. 263–273, ISSN 0723-4864.

Kaliviotis, E.; Dusting, J. & Balabani, S. (2011). Spatial variation of blood viscosity: Modelling using shear fields measured by a μPIV based technique. *Medical Engineering & Physics*, Vol.33, pp. 824-831, ISSN 1350-4533.

Kean, R.D. & Adrian, R.J. (1992). Theory of cross-correlation analysis of PIV images. *Applied Scientific Research*, Vol.49, pp. 191-215, ISSN 1386-6184.

Kim, G.B. & Lee, S.J. (2006). X-ray PIV measurements of blood flows without tracer particles. *Exp. Fluids*, Vol.41, pp. 195-200, ISSN 0723-4864.

King, C.; Walsh, E, & Grimes, R. (2007). PIV measurements of flow within plugs in a microchannel. *Microfluid Nanofluid*, Vol.3, pp. 463-472, ISSN 1613-4982.

Koutsiaris, A.G.; Mathioulakis, D.S. & Tsangaris, S. (1999). Microscope PIV for velocity field measurement of particle suspensions flowing inside glass capillaries. *Meas. Sci. Technol.*, Vol.10, pp. 1037-1046, ISSN 0957-0233.

Koutsiaris, A.G. (2005). Volume flow estimation in the precapillary mesenteric microvasculature *in-vivo* and the principle of constant pressure gradient. *Biorheology*, Vol.42, No 6, pp. 479-491, ISSN 0006-355X.

Koutsiaris, A.G.; Tachmitzi, S.V.; Batis; N., Kotoula; M.G., Karabatsas, C.H.; Tsironi, E. & Chatzoulis, D.Z. (2007). Volume flow and wall shear stress quantification in the human conjunctival capillaries and post-capillary venules *in-vivo*. *Biorheology*, Vol.44, No 5/6, pp. 375-386, ISSN 0006-355X.

Koutsiaris, A.G. (2009). A velocity profile equation for blood flow in small arterioles and venules of small mammals *in vivo* and an evaluation based on literature data. *Clinical Hemorheology and Microcirculation*, Vol.43, No.4, pp. 321-334, ISSN 1386-0291.

Koutsiaris, A.G.; Tachmitzi, S.V.; Kotoula, M.G. & Tsironi, E. (2010a). Old and new velocity profile equations of the blood flow *in vivo*: a preliminary evaluation. *Series on Biomechanics*, Vol.25, No 1-2, pp. 111-116, ISSN 1313-2458.

Koutsiaris, A.G. (2010b). Velocity profile equations for microvessel blood flow in mammals. *Bulletin of the Portuguese Society of Hemorheology and Microcirculation*, Vol.25, No 1, pp. 5-1-, ISSN 0872-4938.

Koutsiaris, A.G.; Tachmitzi, S.V.; Papavasileiou, P.; Batis, N.; Kotoula, M.G. & Tsironi, E. (2010c). Blood velocity pulse quantification in the human conjunctival pre-capillary arterioles. *Microvascular Research*, Vol.80, pp. 202-208, ISSN 0026-2862.

Koutsiaris, A.G.; Tachmitzi, S.V. & Giannoukas, A.D. (2011). How good are the fittings of the velocity profiles in vivo? 3rd *Micro and Nano Flows Conference*, Thessaloniki, Greece, 22-24 August, ISBN 978-1-902316-98-7, In press.

Kuang, C.; Zhao, W.; Yang, F. & Wang, G. (2009). Measuring flow velocity distribution in microchannels using molecular tracers. *Microfluid Nanofluid*, Vol.7, pp.509-517, ISSN 1613-4982.

Lee, S.J. & Kim, S. (2009). Advanced particle-based velocimetry techniques for Microscale flows. *Microfluid Nanofluid*, Vol.6, pp. 577-588, ISSN 1613-4982.

Lima, R.; Wada, S.; Tsubota, K. & Yamaguchi T. (2006). Confocal micro-PIV measurements of three-dimensional profiles of cell suspension flow in a square microchannel. *Meas. Sci. Technol.*, Vol.17, pp. 797-808, ISSN 0957-0233.

Lima, R.; Wada, S.; Takeda, M.; Tsubota, K. & Yamaguchi T. (2007). In vitro confocal micro-PIV measurements of blood flow in a square microchannel: The effect of the heamatocrit on instantaneous velocity profiles. *Journal of Biomechanics*, Vol.40, pp. 2752-2757, ISSN 0021-9290.

Lima, R.; Wada, S.; Tanaka, S.; Takeda, M.; Ishikawa, T.; Tsubota, K.; Imai, Y. & Yamaguchi T. (2008). In vitro blood flow in a rectangular PDMS microchannel: experimental observations using a confocal micro-PIV system. *Biomed Microdevices*, Vol.10, pp. 153-167, ISSN 1387-2176.

Lima, R.; Ishikawa, T.; Imai, Y.; Takeda, M.; Wada, S. & Yamaguchi T. (2009). Measurement of individual red blood cell motions under high hematocrit conditions using a confocal micro-PTV system. *Annals of Biomedical Engineering*, Vol.37, pp. 1546-1559, ISSN 0090-6964.

Lindken, R.; Westerweel, J. & Wieneke, B. (2006). Stereoscopic micro particle image velocimetry. *Exp Fluids*, Vol.41, pp. 161-171, ISSN 0723-4864.

Lindken, R.; Rossi, M.; Große, S. & Westerweel, J. (2009). Micro-particle image velocimetry (PIV): recent developments, applications and guidelines. *Lab on a Chip – Miniaturisation for Chemistry and Biology*, Vol.9, No 17, pp. 2551-2567, ISSN 1473-0197.

Lipowsky, H.H.; Usami, S. & Chien, S. (1980). *In vivo* measurements of "Apparent Viscosity" and microvessel hematocrit in the mesentery of the cat. *Microvascular Research*, Vol.19, pp. 297-319, ISSN 0026-2862.

Long, D.S.; Smith, M.L.; Pries, A.R.; Ley, K. & Damiano, E.R. (2004). Microviscometry reveals reduced blood viscosity and altered shear rate and shear stress profiles in microvessels after hemodilution. *PNAS*, Vol.101(27), pp. 10060-10065, ISSN 0027-8424.

Meinhart, C.D.; Werely, S.T. & Santiago, J.G. (1999). Measurements of a microchannel flow. *Exp Fluids*, Vol.27, pp. 414-419, ISSN 0723-4864.

Meinhart, C.D. & Zhang, H. (2000). The flow structure inside a microfabricated inkjet printhead. *Journal of Microelectromechanical Systems*, Vol.9, No 1, pp. 67-75, ISSN 1057-7157.

Nabavi, M. (2009). Steady and unsteady flow analysis in microdiffusers and micropumps: a critical review. *Microfluid Nanofluid*, Vol.7, pp. 599-619, ISSN 1613-4982.

Nakano, A.; Sugii, Y.; Minamiyama, M. & Niimi, H. (2003). Measurement of red cell velocity in microvessels using particle image velocimetry (PIV). *Clinical Hemorheology and Microcirculation*, Vol.29, No., pp. 445-455, ISSN 1386-0291.

Nakano, A.; Sugii, Y.; Minamiyama, M.; Seki, J. & Niimi, H. (2005). Velocity profiles of pulsatile blood flow in arterioles with bifurcation and confluence in rat mesentery, *JSME International Journal Series C*, Vol.48, No. 4, pp. 444-452, ISSN 0914-8825.

Nguyen, N.T. & Wereley, S.T. (2002). *Fundamentals and Applications of Microfluidics*, Artech House, ISBN 1-58053-343-4.

Okuda, R.; Sugii, Y. & Okamoto, K. (2003). Velocity measurement of blood flow in a microtube using micro PIV system. Proceedings of PSFVIP-4, F4084, June 3-5, 2003, Chamonix, France.

Olsen, M.G. & Adrian, R.J. (2000). Out of focus effects on particle image visibility and correlation in microscopic particle image velocimetry. *Exp. Fluids*, Suppl, pp. S166-S174, ISSN 0723-4864.

Park, J.S.; Choi, C.K. & Kim, K.D. (2004). Optically sliced micro-PIV using confocal laser scanning microscopy (CSLM). *Exp. Fluids*, Vol.37, pp. 105-119, ISSN 0723-4864.

Potter, D.R. & Damiano, E.R. (2008). The hydrodynamically relevant endothelial cell glycocalyx observed *in vivo* is absent *in vitro*. *Circ. Res.*, Vol.102, pp. 770-776, ISSN 0009-7330.

Prasad, A.K.; Adrian, R.J.; Landreth, C.C. & Offutt, P.W. (1992). Effect of resolution on the speed and accuracy of particle image velocimetry interrogation. *Exp. Fluids*, Vol.13, pp. 105-116, ISSN 0723-4864.

Raffel, M.; Wereley, S.T.; Willert, C.E. & Kompenhans, J. (2007). *Particle Image Velocimetry*, Springer, ISBN 3-540-72307-3, Heidelberg.

Roevros, J.M.J.G. (1974). Analogue processing of C.W.-Doppler flowmeter signals to determine average frequency shift momentaneously without the use of a wave analyser, In: *Cardiovascular Applications of Ultrasound*, Reneman, R.S. (Ed.), pp. 43-54, ISBN 978 044 4106 31 5, Amsterdam-London.

Santiago, J.G.; Werely, S.T.; Meinhart, C.D.; Beebe, D.J. & Adrian, R.J. (1998). A particle image velocimetry system for micro fluidics. *Exp Fluids*, Vol.25, pp. 316-319, ISSN 0723-4864.

Schmid-Schoenbein, G.W. & Zweifach, B.W. (1975). RBC velocity profiles in arterioles and venules of the rabbit omentum. *Microvascular Research*, Vol.10, pp. 153-164, ISSN 0026-2862.

Sugii, Y.; Shigeru, N.; Okuno, T. & Okamoto, K. (2000). A highly accurate iterative PIV technique using a gradient method. *Meas. Sci. Technol.*, Vol.11, pp. 1666-1673, ISSN 0957-0233.

Sugii, Y.; Shigeru, N. & Okamoto, K. (2002). *In vivo* PIV measurement of red blood cell velocity field in microvessels considering mesentery motion. *Physiol. Meas.*, Vol.23, pp. 403-416, ISSN 0967-3334.

Tangelder, G.J.; Slaaf, D.W.; Muijtjens, A.M.M.; Arts, T.; Egbrink, M.G.A. & Reneman, R.S. (1986). Velocity profiles of blood platelets and red blood cells flowing in arterioles of the rabbit mesentery. *Circ. Res.*, Vol.59, pp. 505-514, ISSN 0009-7330.

Thome, J.R. & Marcinichen, J.B. (2011). On-chip micro-evaporation: experimental evaluation of liquid pumping and vapour compression cooling systems. *3rd Micro and Nano Flows Conference*, Thessaloniki, Greece, 22-24 August, ISBN 978-1-902316-98-7, In press.

Timgren, A.; Tragardh, G. & Tragardh, C. (2008). Application of the PIV technique to measurements around and inside a forming drop in a liquid-liquid system. *Exp Fluids*, Vol.44, pp. 565-575, ISSN 0723-4864.

Tsukada, K.; Minamitani, H.; Sekizuka, E. & Oshio C. (2000). Image correlation method for measuring blood flow velocity in microcirculation: correlation "window" simulation and *in vivo* image analysis. *Physiol. Meas.*, Vol.21, pp. 459-471, ISSN 0967-3334.

Vennemann, P.; Kiger, K.T.; Lindken, R.; Groenendijk, B.C.W.; Stekelenburg, S.; Hagen, T.L.M.; Ursem, N.T.C.; Poelmann, R.E.; Westerweel, J. & Hierck, B.P. (2006). In vivo micro particle image velocimetry measurements of blood-plasma in the embryonic avian heart. *Journal of Biomechanics*, Vol.39, pp. 1191-1200, ISSN 0021-9290.

Werely, S.T. & Meinhart, C.D. (2001). Adaptive second-order accurate particle image velocimetry. *Exp Fluids*, Vol.31, pp. 258-268, ISSN 0723-4864.

Werely, S.T. & Meinhart, C.D. (2005). Micron-Resolution Particle Image Velocimetry, In: *Microscale Diagnostic Techniques*, Breuer K.S. (Ed.), pp. 51-112, Springer, ISBN 3-540-23099-8, Heidelberg.

Werely, S.T. & Meinhart, C.D. (2010). Recent Advances in Micro-Particle Image Velocimetry. *Annu. Rev. Fluid Mech.*, Vol.42, pp. 557-576, ISSN 0066-4189.

Westerweel, J. (1997). Fundamentals of digital particle image velocimetry. *Meas. Sci. Technol.*, Vol.8, pp. 1379-1392, ISSN 0957-0233.

Westerweel, J.; Geelhoed, P.F. & Lindken, R. (2004). Single-pixel resolution ensemble correlation for micro-PIV applications. *Exp. Fluids*, Vol.37, pp. 375-384, ISSN 0723-4864.

Willert, C.E. & Gharib, M. (1991). Digital particle image velocimetry. *Exp. Fluids*, Vol.10, pp. 181-193, ISSN 0723-4864.

Windberger, U.; Bartholovitsch, A.; Plasenzotti, R.; Korak, K.J. & Heinze, G. (2003). Whole blood viscosity, plasma viscosity and erythrocyte aggregation in nine mammalian species: reference values and comparison of data. *Experimental Physiology*, Vol.88, pp. 431-440, ISSN 0958-0670.

Windberger, U. & Baskurt, O.K. (2007). Comparative Haemorheology, In: *Handbook of Hemorheology and Hemodynamics*, Baskurt, O.K.; Hardeman, M.R.; Rampling, M. & Meiselman, H.J. (Eds), pp. 267-283, IOS Press, ISBN 978-1-58603-771-0, The Netherlands.

Section 4

PIV and PTV

12

Measurements of Particle Velocities and Trajectories for Internal Waves Propagating in a Density-Stratified Two-Layer Fluid on a Slope

Motohiko Umeyama, Tetsuya Shintani,
Kim-Cuong Nguyen and Shogo Matsuki
Department of Civil & Environmental Engineering, Tokyo Metropolitan University
Japan

1. Introduction

The existence of internal waves was first recognized by a measurement of water temperature by Helland-Hansen and Nansen (1926). Later, the vertical structure of internal waves has been detected by observation of temperature, salinity, or ocean current, while the propagation of internal waves has been identified by images from radar or acoustic Doppler and echo sounder. Shand (1953) found internal wave fronts appearing on aerial photographs. From observation of offshore temperature variations, LaFond (1962) found that time-dependent isotherms are flattened for a shallower thermocline and peaked for a deeper thermocline on the wave crests. Apel et al. (1975, 1976, 1985) reported a series of research results for the internal waves observed in pictures from satellites, space shuttles, and aircraft. In addition to these indirect photographs, they also used various instruments such as the expendable bathythermograph (XBT), acoustic echo sounding, and ship radar.

In contrast, numerous researchers have carried out laboratory experiments to examine the shoaling and breaking of internal waves on various topographic features. Thorpe (1968) studied the breaking and runup of internal waves in a two-layered system, and found that internal waves steepen at the front as the lower layer becomes shallow, but the crests break backwards unlike surface breakers. On the upper slope, the wave (bolus) behaves like the front of a gravity current, and the dense fluid returning down the slope from previous waves flows over the top of the overcoming waves. Using a hydrogen-bubble wiring system, Kao et al. (1985) measured the particle velocity profile during the passage of internal waves. Wallace and Wilkinson (1988) found that overturning is initiated by the interaction of an incident wave with the backflow from the preceding boluses. Helfrich (1992) performed experiments to observe the interaction of an internal wave of depression with a sloping bottom, and recognized the importance of the backflow that produces significant mixing. He found a rapid offshore flow of lower-layer water while the front face of the incident wave moved up the slope.

There have been significant developments in measuring fluid velocities by using laboratory equipment, but the use of tools such as the electric–magnetic current meter and laser Doppler anemometer has not solved kinetic problems for internal waves. Experimental studies for fluid velocities have generally relied on point measurement techniques. However, recent advances in technology provide two and three–dimensional features of the velocity field for several flow problems. Visualization techniques have played an essential role in fluid flows during the previous decade, because they yield both qualitative and quantitative insights in geophysical or environmental fluid mechanics. Now, imaging techniques enable high–resolution images of several kinds of unsteady flows. The latest developments in particle image velocimetry (PIV) and particle tracking velocimetry (PTV) have led to the visualization of velocity fields and particle paths. Both techniques are analysis methods for image pairs taken in a seeded flow field with known temporal separation. Water velocities due to the propagation of internal waves have been measured by Michallet and Ivey (1999), and Walker et al. (2003) using PIV, and by Grue et al. (1999, 2000) using PTV.

The mechanism of internal waves in a two-layer system comprising homogeneous fluids of slightly different densities has been studied using visualization techniques in the Hydraulic Laboratory at Tokyo Metropolitan University. In the research of Umeyama (2002), a digital video camera was first used to illustrate the internal waves propagating in a fluid of finite depth over a flat bed. After analysing continuous pictures recorded on a DV tape, the density interface was determined from a set of luminance values using image-processing software. The temporal and spatial variations of the density interface were compared with the analytical results based on the third-order finite amplitude approximation. Umeyama and Shintani (2004) installed a Plexiglas plate in the same wave tank to observe the runup and breaking of long internal waves over an artificial slope. The profile of internal waves and the mixing between upper and lower layers, were visualized by adding a blue dye (Anilin Blue) in lower salt water. Later, Umeyama and Shintani (2006) performed more precise laboratory tests, by considering additional aspects such as transformation, attenuation, set-down, and setup during the shoaling and breaking events. The method of characteristics, energy dissipation model with radiation stress, and momentum balance equation were used to confirm the experimental results. To study the transformation processes using instantaneous and mean velocity fields and nonlinear properties using interfacial displacements over a uniform slope, Shimizu et al. (2005) developed a PIV system that consisted of a Nd:YAG pulsed laser and a CCD camera. The analysis, however, highlighted the limitations in the measuring range. This PIV system of the laser sheet could not measure a large area while maintaining a fine resolution. The PIV measurements were repeated three times to cover one wavelength and the total depth in adjoining areas for each case. In contrast, Umeyama (2008) and Umeyama and Shinomiya (2009) developed a new PIV system that utilized halogen lamps and three high-definition digital video cameras in which the maximum resolution was 2,016×1,134 pixels and the images were recorded on a hard disk. Each video camera, operating simultaneously side by side, covered a larger area with a frame rate 16:9. Since the water surface was almost flat during all experiments, the halogen lamps were set in a line along the wave tank at its top, and the light sheet of 3 mm width was emitted from the upper side of the wave tank. Umeyama and Matsuki (2011) recently measured the similar physical quantities with two frequency-doubled Nd:YAG lasers of 50 mW energy at 532 nm as the illumination sources. In addition to the common use of the PIV technique, the knowledge has been extended to visualize water particle paths and mass transport variations due to the propagation of internal waves in the tank.

From a Eulerian prospective, the motion of an incompressible fluid is distinctive if the velocity vectors occupy an instantaneous velocity field. As PIV is used to represent a regular array of velocity vectors, it is convenient to define the Eulerian velocity based upon the average particle motion in the possible space. In contrast, PTV traces the individual particle path from a sequence of images in a system. From a Lagrangian viewpoint, PTV is better suited than PIV for handling unsteady flow. Quantitative results for Lagrangian fluid motion can be obtained through computerized analyses of the particle images in the modern PTV technique. Generally, PIV has determined the Eulerian velocity field from a sequence of images, while PTV has estimated the Lagrangian velocity by tracing individual particle paths. Umeyama (2011) employed a PTV system with single-exposure images to track particle displacements for surface waves with or without a steady current. In addition to the basic use of PTV, an alternative measurement technique was proposed to describe particle trajectories in a Eulerian scheme through PIV analysis. Later, Umeyama et al. (2011) made similar measurement and analysis for the water particle velocity and trajectory using a PIV system with two Nd:YAG lasers of 50 mW. In this experiment, DIAION and micro bubbles were chosen as tracer. More recently, Umeyama (2012) measured the water particle velocity and trajectory in a pure wave motion using a new PIV system with an 8 W Nd:YAG laser. These previous validation results are presented by comparing the measurements of the PIV system to those of the electromagnetic current (EC) meter.

This article investigates the spatial and temporal variations of the density interface, wave height, celerity and setup due to shoaling and breaking of internal waves, using an image processing technique, and the velocity and trajectory of water particles in an internal wave motion, using particle image velocimetry (PIV). The writers illustrate 2D instantaneous displacements of density interface from the observed data, and analyse them to obtain the wave height, celerity, and setup. These results are compared with the calculated values by the method of characteristics, the simple shoaling model with energy dissipation, and the momentum balance equation based on a radiation stress concept. The vector fields and vertical distributions of velocities are presented at several phases in one wave cycle. The PIV technique's ability to measure both temporal and spatial variations of the velocity is proven after a series of attempts. This technique is applied to the prediction of particle trajectory in a Eulerian scheme. The measured particle path is compared to the positions found theoretically by the method of characteristics.

2. Experiments and data analysis

2.1 Wave tank and wavemakers

Experiments on internal waves were carried out in the 6.0-m-long, 0.15-m-wide, and 0.35-m-deep wave tank. It was constructed of 12 Plexiglas panels, 10 stainless flanges, and a stainless bottom. Each glass panel was 91.0 cm long, 27.0 cm high and 1.0 cm thick. A wave generator was placed at one end. Fig. 1 shows a sketch of the apparatus for the wave tank and its photograph.

In the earlier stages of experiments, Umeyama (2002) and Umeyama & Shintani (2004, 2006) used a horizontal flap-type wavemaker. Recently, experimental studies were undertaken to test several types of wave-generating paddles, and a different kind of wavemaker was set at one end of the wave tank. The sketch of the new wavemaker is presented in Fig.2. This slide-type wavemaker consists of a D-shaped paddle and a linear actuator that moves smoothly and programmatically within a given stroke length. As the D-shaped paddle limits the mixing of the upper and lower waters, the intermediate density layer does not grow quickly during each test. Umeyama (2008), Umeyama & Shinomiya (2009), and Umeyama & Matsuki (2011) examined internal waves generated by the slide-type wavemaker to obtain insight into the nonlinear properties and kinematics.

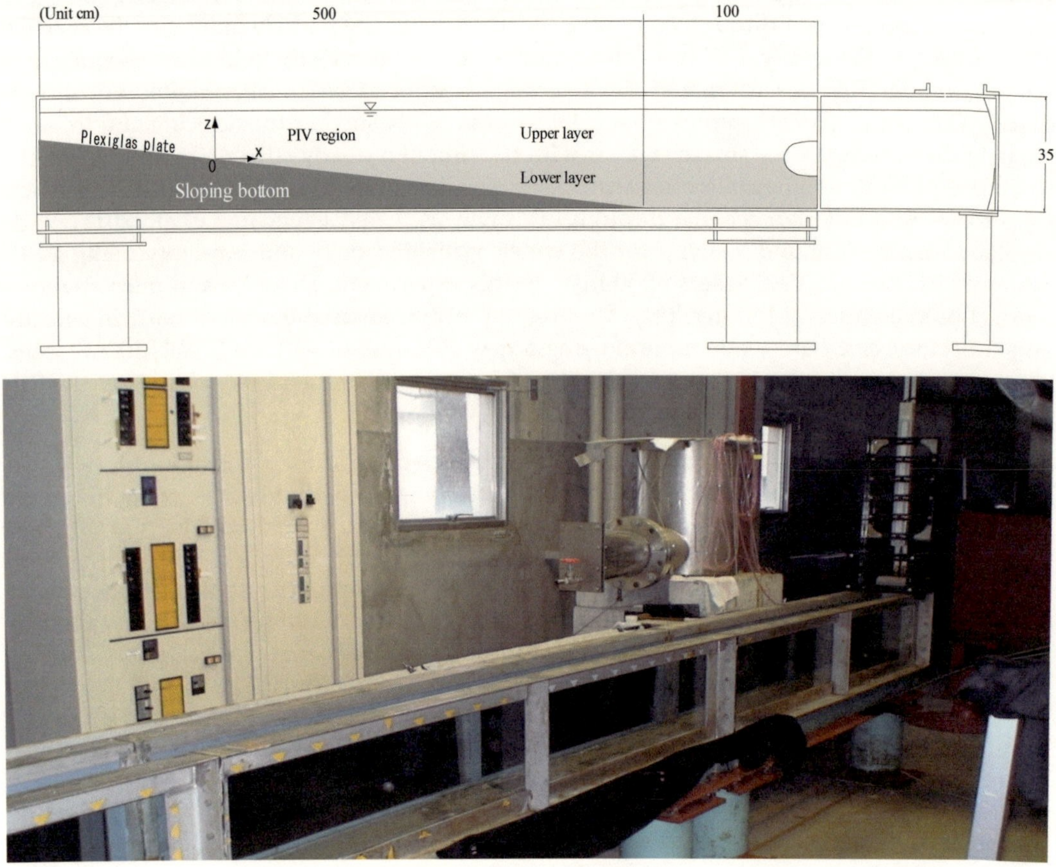

Fig. 1. A schematic diagram of experimental arrangements and a photograph of wave tank

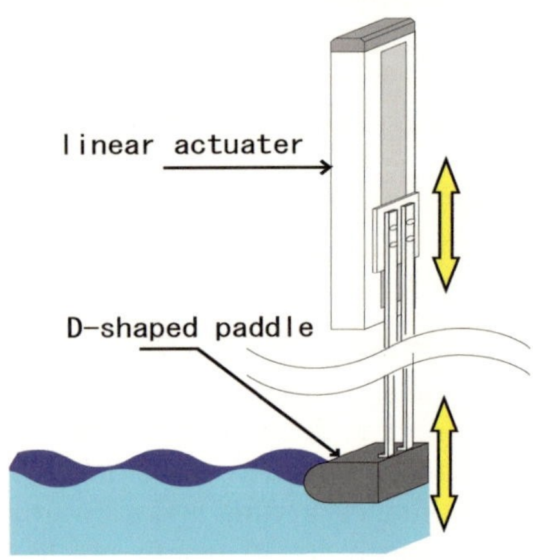

Fig. 2. A slide-type wavemaker with a D-shaped paddle

In the present study, a 1-cm-thick Plexiglas plate, which served as the plane seabed, was fabricated between 100 and 600 m from the wavemaker. A density-stratified fluid consisting of fresh water and salt water was prepared for the experiments, and the density of the salt water was 1,030 mg/cm^3. The total water depth from the bottom of the wave tank was kept at 30 cm. The density of the salt water was assumed to be constant initially, but subsequent disturbances produced deviations from the original density value.

2.2 Flow visualization

Using an image processing technique, Umeyama & Shintani (2004) illustrated the temporal and spatial variations of the density interface and the mixture of upper and lower layer waters in the density-stratified two-layer fluid on a sloping bed. The numerically predicted density variation due to mixing was compared with a set of luminance data analyzed by the image processing technique. Later, Umeyama & Shintani (2006) and Umeyama (2008) extended the aspects of internal wave mechanics during shoaling and breaking events. The objective of these studies was to observe internal waves approaching an upper slope in a wave tank using a video recording system, convert the measured data to the wave profile, celerity, height and setup, and compare the experimental values with theoretical ones. For the flow visualization and video recording technique in detail, refer to Umeyama (2008). The colour image file from the video recorder was transferred to an 8-bit (256 grades) grayscale image file. Using the method of image segmentation, individual pixels in a grayscale image were marked as object or background pixels. The density interface was estimated from the spatial distribution of the threshold. Finally, the temporal displacements of the density interface were obtained at several points, and the wave height and mean level were estimated by averaging these displacements about 10 wave periods. This duration may be assumed stationary at the location where data are gathered, because artificial internal waves are considered approximately stationary only for a few minutes. Beyond the duration, their properties are expected to change in the wave tank.

An internal wave gauge was also used to measure the vertical displacement of density interface at a location 100 cm from the wave maker. A salinity sensor at the tip of the internal wave gauge could run after the position of the prescribed salt density layer. Under each set of experimental conditions, experimental data were collected continuously at a frequency of 47 Hz.

2.3 PIV technique

The instantaneous water particle velocities induced by internal waves were measured using PIV. The basic principle of PIV is evaluating the instantaneous velocities through recording the position of images of small tracers, suspended in the fluid, at successive instants in time. In practice, when two successive images of tracers illuminated in a thin and intense light sheet are acquired, the velocity is calculated from the known time difference and measured displacement.

In this study, the water particle velocity was measured using a single-exposure image PIV system. The instantaneous vector field was trapped in the 91-cm-long and 27-cm-high glass panel using a frequency-doubled Nd:YAG laser of 8-W energy at 532 nm. A 2-mm-thick light sheet was emitted from the upper side: this light sheet had a very uniform intensity and covered the total area in the glass panel. The system included two high-definition digital video cameras (SONY HXR-NX5J) with a maximum resolution of 1920×1080 pixels. The video camera was arranged linearly 1.28 m from the sidewall of the wave tank. The camera image area was centered in the light sheet; consequently, the corresponding viewing area up to 100.0 cm×28.0 cm was chosen using the optical arrangement. According to Austin and Halikas (1976), the index of refraction of clear water is 1.335, while that of salt water with a density of 1,030 mg/cm³ is 1.341. As the index of refraction varies by less than 0.5% (the maximum difference in angle of refraction is approximately 0.1 degree), the difference in geometric distortion was not corrected. DIAION (DK-FINE HP20SS) was used to capture the high-contrast images with the particle tracer in both layers. It consisted of ion-exchange resin with the homogeneous matrix structure inside the particle. This kind of matrix gave micropores formed by the polymeric networks, so that water could pass through these pores. Before each experiment, DIAION was mixed in the salt water. Analysis of the displacement of images in each interrogation window by means of the cross-correlation method leads to an estimated average displacement of particles. The resolution is directly related to the size of the interrogation window. The displacement vector computed at any location is the spatially averaged transitional motion of particles. Vector fields could be obtained with the PIV system processing a pair of images, using an interrogation window of 64×64 pixels in a candidate region of 128×128 pixels. Since the internal wave topography does not change significantly over 0.1 s, the resultant displacement of topographic features for two images spaced in Δt =0.1 s was chosen for a direct calculation of the velocity vectors.

2.4 Eulerian/Lagrangian method

A cross-correlation method was performed to calculate the water particle displacement and local velocity by processing a pair of image frames. Although the representation of the velocity vector field in a Eulerian system is a typical example of the PIV method, the result can be applied to a particle tracking process in a Lagrangian system. Umeyama & Matsuki (2011) used the velocity given at the spatially discrete nodal point to estimate the imaginary velocity and location of a particle.

The following explains the particle motion within a tracking time step Δt along an arbitrary trajectory across a general mesh of quadrilateral cells (Fig.3). The algorithm has two steps: (1) the velocity value of a Lagrangian point A where t is obtained by interpolating the neighboring velocity values (u_1 at P_1, u_2 at P_2, u_3 at P_3, and u_4 at P_4, where P_1, P_2, P_3, and P_4 are corners of the mesh), and (2) the particle associated with the Lagrangian point at t is traced to a hypothetical location (A′) at $t+\Delta t$. Thus, these Lagrangian velocities u_L at t and u_L' at $t+\Delta t$ are

$$u=\frac{\frac{u_1}{l_1}+\frac{u_2}{l_2}+\frac{u_3}{l_3}+\frac{u_4}{l_4}}{\frac{1}{l_1}+\frac{1}{l_2}+\frac{1}{l_3}+\frac{1}{l_4}} \text{ at } t=t\text{, and } u'=\frac{\frac{u_1'}{l_1'}+\frac{u_2'}{l_2'}+\frac{u_3'}{l_3'}+\frac{u_4'}{l_{14}'}}{\frac{1}{l_1'}+\frac{1}{l_2'}+\frac{1}{l_3'}+\frac{1}{l_4'}} \text{ at } t=t+\Delta t$$

where l_1, l_2, l_3, l_4, l_1', l_2', l_3', and l_4' are distances to the imaginary location of a particle from the Eulerian grid points.

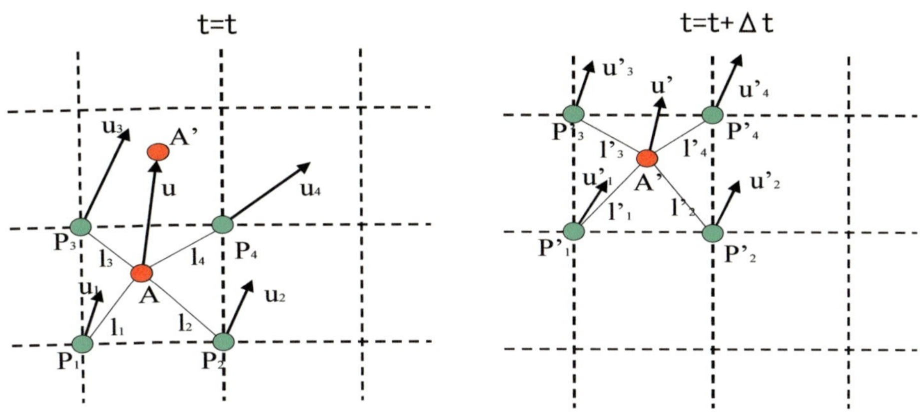

Fig. 3. Lagrangian particle trajectory on Eulerian grids

2.5 Experimental conditions

The experimental conditions are listed in Table 1 that includes the present results in the Hydraulic Laboratory of Tokyo Metropolitan University.

Case	Slope	Density	h_I:h_{II}	Wave height*	Wave period
1	3/50	1,030 mg/cm^3	15 cm:15 cm	2.6 cm	5.2 s
2				2.7 cm	7.2 s
3	3/50	1,030 mg/cm^3	10 cm:20 cm	2.8 cm	5.2 s
4				2.4 cm	7.2 s

* Wave height was measured using wave gauge at 100 cm from the wavemaker.

Table 1. Experimental cases

3. Theory

3.1 Stokes internal waves

First, we consider Stokes internal waves propagating along the interface between two homogeneous incompressible and inviscid fluids of different density in a constant depth. The origin of the axes is located in the undisturbed interface. The density and depth of the upper layer are ρ_I and h_I, respectively, and those of the lower layer are ρ_{II} and h_{II}, respectively. The vertical displacements of the free surface and the density interface are $\eta_I(x,t)$ and $\eta_{II}(x,t)$, respectively. Let $\phi_I(x,z,t)$, and $\phi_{II}(x,z,t)$ denote the velocity potentials in the upper and lower layers, so that the Laplace equations are

$$\phi_{Ixx} + \phi_{Izz} = 0 \tag{1}$$

$$\phi_{IIxx} + \phi_{IIzz} = 0 \tag{2}$$

where x =horizontal coordinate; and z =vertical coordinate. The kinematical and dynamical boundary conditions at the free surface are

$$g(\eta_I + h_I) + \phi_{It} + \frac{1}{2}(\phi_{Ix}^2 + \phi_{Iz}^2) = 0 \text{ on } z = h_I + \eta_I \tag{3}$$

$$\eta_{It} + \eta_{Ix}\phi_{Ix} - \phi_{Iz} = 0 \text{ on } z = h_I + \eta_I \tag{4}$$

where g= gravity acceleration; and t= time. The boundary conditions at the density interface are

$$\eta_{IIt} + \eta_{IIx}\phi_{Ix} - \phi_{Iz} = 0 \text{ on } z = \eta_{II} \tag{5}$$

$$\eta_{IIt} + \eta_{IIx}\phi_{IIx} - \phi_{IIz} = 0 \text{ on } z = \eta_{II} \tag{6}$$

$$\rho_I\{g\eta_{II} + \phi_{It} + \frac{1}{2}(\phi_{Ix}^2 + \phi_{Iz}^2)\} = \rho_{II}\{g\eta_{II} + \phi_{IIt} + \frac{1}{2}(\phi_{IIx}^2 + \phi_{IIz}^2)\} \text{ on } z = \eta_{II} \tag{7}$$

The bottom boundary condition is

$$\phi_{IIz} = 0 \text{ on } z = -h_{II} \tag{8}$$

In the finite-amplitude wave theory, the perturbation method is used to solve the above basic equations and boundary conditions. These solutions have been obtained to the second order by Umeyama (1998), and to the third order by Umeyama (2000).

When the displacement of fluid interface is given by a linear profile such as $\eta_{II}(x,t) = a\cos(kx - \sigma t)$, the horizontal and vertical velocity components for the water particle can be obtained from the velocity potentials where $u_I = -\partial\phi_I / \partial x$, $w_I = -\partial\phi_I / \partial z$, $u_{II} = -\partial\phi_{II} / \partial x$, and $w_{II} = -\partial\phi_{II} / \partial z$. Therefore, the velocity components in both layers are expressed as

$$u_I = \frac{ak}{\sigma}(\alpha\cosh kz - \frac{\sigma^2}{k}\sinh kz)\cos(kx - \sigma t) \tag{9}$$

$$w_I = -\frac{ak}{\sigma}(\alpha\sinh kz - \frac{\sigma^2}{k}\cosh kz)\sin(kx - \sigma t) \tag{10}$$

$$u_{II} = -\frac{a\sigma}{\sinh kh_{II}}\cosh k(z + h_{II})\cos(kx - \sigma t) \tag{11}$$

$$w_{II} = \frac{a\sigma}{\sinh kh_{II}}\sinh k(z + h_{II})\sin(kx - \sigma t) \tag{12}$$

where a =amplitude of internal waves; k =wave number; and α =constant that is given by

$$\alpha = \frac{\rho_{II} - \rho_I}{\rho_I}g - \frac{\rho_{II}}{\rho_I}\frac{\sigma^2}{k}\coth kh_{II}$$

The dispersion relation may be written as

$$\sigma^4(\coth kh_I \coth kh_{II} + \frac{\rho_I}{\rho_{II}}) - \sigma^2(\coth kh_I + \coth kh_{II})gk + \frac{\rho_{II} - \rho_I}{\rho_I}(gk)^2 = 0 \quad (13)$$

Solving Eq.(13) for $C = \sigma / k$ and assuming $\rho_{II} \approx \rho_I$, it becomes

$$C = \varepsilon_g \sqrt{\frac{1}{k(\coth kh_I + \coth kh_{II})}} \quad (14)$$

where $\varepsilon_g = \sqrt{\frac{\rho_{II} - \rho_I}{\rho_I} g}$.

Thus the group velocity for internal waves is given by

$$C_g = \frac{C}{2}\left(1 + \frac{\dfrac{kh_I}{\sinh^2 kh_I} + \dfrac{kh_{II}}{\sinh^2 kh_{II}}}{\coth kh_I + \coth kh_{II}}\right) \quad (15)$$

3.2 The method of characteristics for long internal waves on a slope

Umeyama and Shintani (2004) investigated the runup of internal waves on a plane impermeable slope. A reliable solution for the displacement of the density interface and the horizontal velocity of the internal waves was derived by means of the method of characteristics. Let $u_{II}(x,z,t)$ denotes the horizontal velocity for long internal waves so that the governing equations are

$$2(1 + \frac{\eta_{II} + h_{II}}{h_I})C_t + u_{IIx}C + 2(1 + \frac{\eta_{II} + h_{II}}{h_I})C_x = 0 \quad (16)$$

$$u_{IIt} + u_{II}u_{IIx} + 2(1 + \frac{\eta_{II} + h_{II}}{h_I})^2 CC_x = g\frac{\rho_{II} - \rho_I}{\rho_{II}} h_{IIx} \quad (17)$$

It is convenient to define the lower-layer thickness for a uniform slope as

$$h_{II} = h_{II}(x) = m\frac{\rho_I}{g(\rho_{II} - \rho_I)}x$$

where m =constant. By adding and subtracting Eqs.(16) and (17), and assuming $\eta + h_{II} << h_I$, the results can be written in the familiar form:

$$[\frac{\partial}{\partial t} + (u_{II} \pm C)\frac{\partial}{\partial x}](u_{II} \pm 2C - mt) = 0 \quad (18)$$

Use of the method of characteristics will make it possible to describe η_{II} and u_{II} such as

$$\eta_{II} = -\frac{A^2}{2}\frac{h_I + sx}{h_I sx}[J_1(X)\cos T + \{J_0(X) - \frac{J_1(X)}{X}\}\sin T]^2 + A\{J_0(X)\sin T + J_1(X)\cos T\} \quad (19)$$

$$u_{II} = A\varepsilon_g \sqrt{\frac{h_I + sx}{h_I sx}} [J_1(X)\cos T + \{J_0(X) - \frac{J_1(X)}{X}\}\sin T] \tag{20}$$

where A =constant; and J_p =the Bessel function of order p. The dependent variables are given by

$$X = \frac{2\sigma}{s\varepsilon_g}\sqrt{\frac{h_I sx}{h_I + sx}}, \text{ and } T = -\sigma t$$

For convenience, the dependent variable X will not be shown hereafter.

A displacement of the interface will cause an associated surface displacement. In the upper layer, the linearized momentum equation may be simply given by

$$\frac{\partial u_I}{\partial t} = -g\frac{\partial \eta_I}{\partial x} \tag{21}$$

The continuity equation can be determined in the same manner:

$$u_I = \frac{C(\eta_I - \eta_{II})}{h_I + (\eta_I - \eta_{II})} \approx -\frac{C}{h_I}\eta_{II} \tag{22}$$

Combining Eqs.(21) and (22), the following differential results:

$$\begin{aligned}\eta_I = &-\frac{A^2\sigma C}{g}\int \frac{h_I + sx}{h_I^2 sx}[J_1 \sin T - \{J_0 - \frac{J_1}{X}\}\cos T][J_1 \cos T + \{J_0 - \frac{J_1}{X}\}\sin T]dt \\ &- \frac{A\sigma C}{gh_I}\int \{J_0 \cos T - J_1 \sin T\}dt\end{aligned} \tag{23}$$

3.3 Radiation stress, setup, and attenuation for internal waves

During the passage of surface waves, there are mean transport of water upward the shoreline and depression of the mean water level from the still water level. Longuet-Higgins and Stewart (1964) introduced the radiation stress concept to prove these mechanisms, and Umeyama (2006) applied it to internal waves. The principal component of the radiation stress for internal waves can be defined as

$$S_{xx} = \overline{\int_{\eta_{II}}^{\eta_I + h_I} (p_I + \rho_I u_I^2)dz} - \int_0^{h_I} p_{I0}dz + \overline{\int_{-h_{II}}^{\eta_{II}} (p_{II} + \rho_{II} u_{II}^2)dz} - \int_{-h_{II}}^{0} p_{II0}dz \tag{24}$$

where S_{xx} =radiation stress for internal waves; p_{I0} =hydrostatic pressure in the upper layer; p_{II0} =hydrostatic pressure in the lower layer; and the over-bar denotes averaging in time over a wave period. Assuming that the upper limit of integration may be replaced by $z = h_I$, the displacement of the density interface is small relative to the wavelength, and the mean mass flux of vertical momentum across a horizontal plane balances with the weight of the water above it, Eq.(24) reduces to

$$S_{xx} = \int_0^{h_I} \rho_I \overline{(u_I^2 - w_I^2)} dz + \int_{-h_{II}}^{0} \rho_{II} \overline{(u_{II}^2 - w_{II}^2)} dz \qquad (25)$$

Substituting Eqs.(9)-(12) into Eq.(25), the radiation stress of internal waves becomes

$$S_{xx} = \frac{a^2}{2}\{\rho_I(\frac{k^2 h_I}{\sigma^2}\alpha^2 - \sigma^2 h_I) + \frac{\rho_{II}\sigma^2 h_{II}}{\sinh^2 kh_{II}}\} \qquad (26)$$

The change in radiation stress leads to a change in the mean level of the density interface when internal waves encounter a sloping beach. The equilibrium between radiation stress change and average slope of the density interface yields

$$\frac{dS_{xx}}{dx} - \{\rho_I - \rho_{II}\}g(\overline{\eta} + h_{II})\frac{d\overline{\eta}}{dx} = 0 \qquad (27)$$

Generally, the attenuation of internal waves in the continental slope is a complicated hydrodynamic process. For the purpose of the present study it may be helpful to use a model in which the divergence of the energy flux is balanced by the dissipation. Therefore, the equation of energy conservation can be expressed as (Umeyama & Shintani 2006)

$$\frac{dF}{dx} = \frac{d(EC_g)}{dx} = -f_w \frac{\rho_I + \rho_{II}}{3\pi}(\frac{\sigma a}{\sinh kh_{II}})^3 - \beta\frac{(\rho_I + \rho_{II})g}{4\pi\sqrt{2}}\sigma a^2 \qquad (28)$$

where F =energy flux per unit of width; E =depth-integrated, time-averaged wave energy per unit area; f_w =friction coefficient; and β =energy dissipation coefficient.

3.4 Theoretical water particle trajectory

In the case of internal waves on a slop, we denote $(x_I(t), z_I(t))$ and $(x_{II}(t), z_{II}(t))$ the instantaneous water particle positions at time t in the upper and lower layers, respectively. The corresponding horizontal and vertical velocities become $u_I = \partial x_I / \partial t$, $w_I = \partial z_I / \partial t$, $u_{II} = \partial x_{II} / \partial t$, and $w_{II} = \partial z_{II} / \partial t$. When the mean position of a water particle is given at $(\overline{x_I}, \overline{z_I})$ for the upper layer or at $(\overline{x_{II}}, \overline{z_{II}})$ for the lower layer, the instantaneous water particle position is denoted as $x_I(t) = \overline{x_I} + \varsigma_I(t)$ and $z_I(t) = \overline{z_I} + \xi_I(t)$, or $x_{II}(t) = \overline{x_{II}} + \varsigma_{II}(t)$ and $z_{II}(t) = \overline{z_{II}} + \xi_{II}(t)$, where ς_I, ξ_I and ς_{II}, ξ_{II} are the horizontal and vertical displacements in the upper layer and lower layer, respectively. Thus, expansion of velocities in the Taylor series and integration of those velocities with respect to time yield

$$\begin{aligned}\varsigma_I = &\frac{A^2 C}{2\sigma h_I}\frac{h_I + sx}{h_I sx}\{-J_1^2(\frac{T}{2} + \frac{\sin 2T}{4}) + J_1 J_{01}\frac{\cos 2T}{2} + J_{01}^2(-\frac{T}{2} + \frac{\sin 2T}{4})\} \\ &- \frac{AC}{\sigma h_I}(J_0 \cos T - J_1 \sin T)\end{aligned} \qquad (29)$$

$$\begin{aligned}
\varsigma_{II} &= -\frac{A\varepsilon_g}{\sigma}\sqrt{\frac{h_I+sx}{h_I sx}}(J_1 \sin T + J_{01}\cos T) \\
&+\frac{A^2\varepsilon_g^2}{\sigma^2}(\frac{h_I+sx}{h_I sx})[\frac{1}{2sx^2}\frac{h_I sx}{h_I+sx}(J_1^2\frac{\cos 2T}{4}+J_1 J_{01}\frac{\sin 2T}{2}-J_{01}^2\frac{\cos 2T}{4}) \\
&-J_1 J_{01}\frac{dX}{dx}\frac{\cos 2T}{4}+J_0\{J_1\frac{dX}{dx}-\frac{1}{X}(\frac{1}{2sx^2}\frac{h_1 sx}{h_1+sx}J_1-J_{01}\frac{dX}{dx})\}(-\frac{T}{2}+\frac{\sin 2T}{4}) \\
&-J_{01}^2\frac{dX}{dx}(\frac{T}{2}+\frac{\sin 2T}{4})-J_{01}\{J_1\frac{dX}{dx}-\frac{1}{X}(\frac{1}{2sx^2}\frac{h_1 sx}{h_1+sx}J_{01}\frac{dX}{dx})\frac{\cos 2T}{4\sigma}\}]
\end{aligned} \quad (30)$$

where $J_{01} = J_0 - J_1 / X$.

4. Results

4.1 Interfacial displacement and celerity distribution

Fig.4 shows the spatial displacements of the density interface for two different experiments, i.e. (a) Case 1 and (b) Case 2. The interval between two panels is $T/4$ for each case. The abscissa is the distance from the intersection between the sloping bed and stationary level of the density interface, and the ordinate is the elevation above that level. The circular symbols show the density interface by using the image processing technique. The solid curve shows the analytical solution based on Eq.(19). For Case 1, the absolute value of the crest level is larger than that of the trough level, and this tendency is prominent in the upper-slope region. Inspecting the four figures reveals that the test data agree with the theoretical displacement in the horizontal range of $15 < x < 50$ cm. To obtain closer agreement in the rest of range, several appropriate assumptions such as breaking, bottom friction, reflection and return flow may be proposed to account for the observed wave profiles. For Case 2, the experimental waveform is peaked near the crest but declined near the trough except in the upper-slope region, when comparing it to the theoretical profile. The method of characteristics appears to adequately predict the details of the runup profile on the sloping bottom. Umeyama and Shintani (2006) examined the similar measured distributions with the $k-\varepsilon$ model and the method of characteristics, and found that the former is quantitatively superior to the latter in the upper-slope region.

Fig.5 depicts the corresponding variations of celerity for Cases 1 and 2. The celerity distributions obtained from the linear wave theory (Eq.14) and the method of characteristics by Umeyama & Shintani (2004) are presented to compare with the experimental data. The measured celerity is smaller than these predicted ones except in the upper-slope region where most measured values exceed the predicted ones by the linear wave theory for Case 1 and those by the method of characteristics for Case 2. In the present cases, i.e. the linear bottom slope of 3:50, the solution by the linear wave theory shows a fair agreement with the measured celerity when compared with the solution by the method of characteristics. According to Umeyama and Shintani (2006), the former underestimates the measurements but the latter overestimates them by about 0-30% in some experimental cases for the bottom slope of 1:28. These quantitative discrepancies could be due to the difference of boundary conditions of the experiment and the theory. The theoretical solution was derived for a topography consisting of a plane sloping beach, although the experimental data were obtained over a different topography consisting of a constant depth between 0 and 100 cm from the wavemaker and a slope between 100 and 600 cm from it.

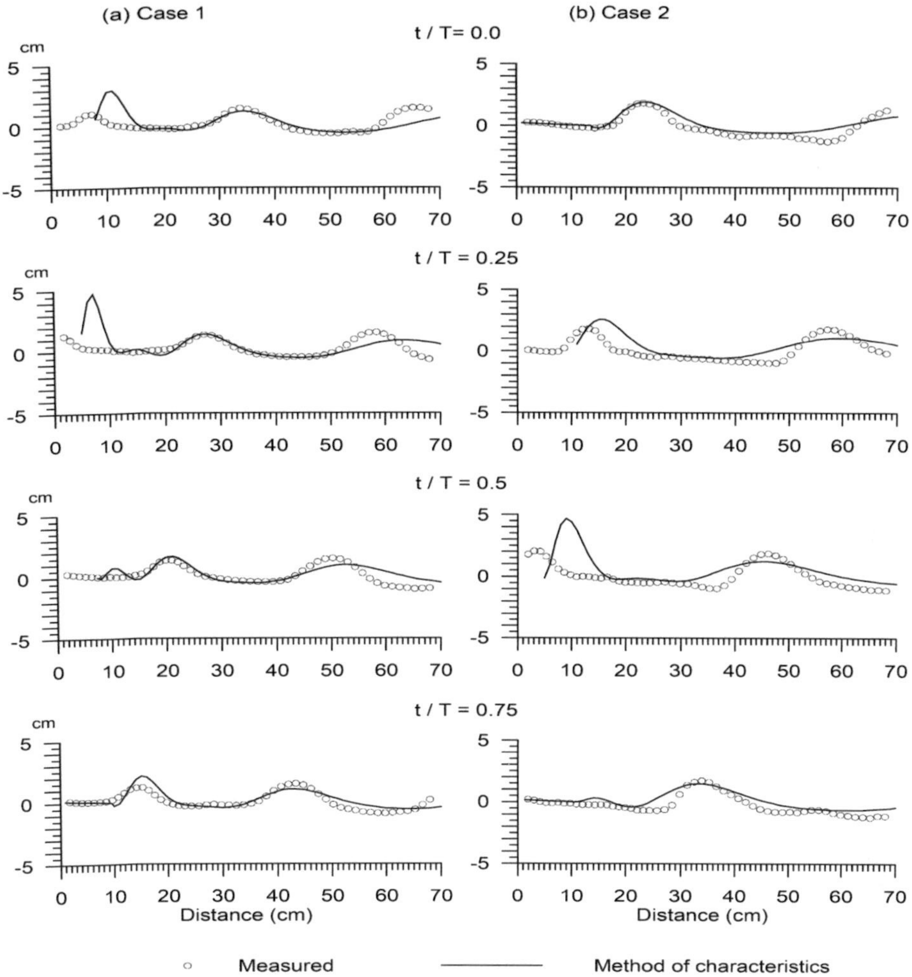

Fig. 4. Spatial displacements of density interface of $h_I : h_{II}$ =15 cm:15 cm

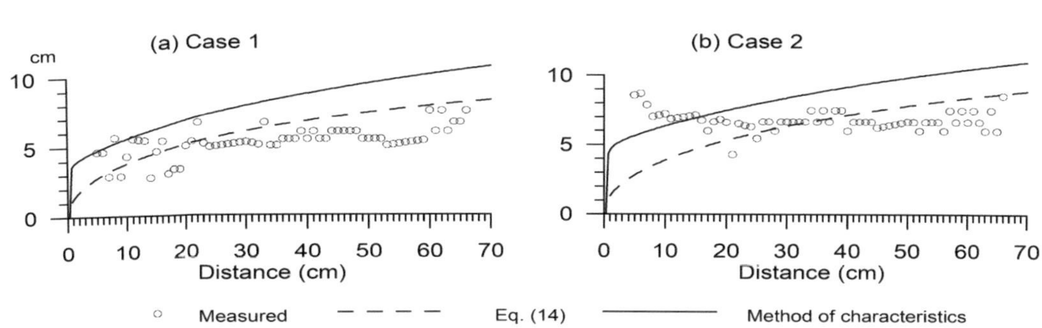

Fig. 5. Spatial profiles of celerity

4.2 Attenuation and setup for internal waves

The laboratory data in this section were obtained for the upper and lower thickness ratio of $h_I : h_{II}$ = 15 cm:15 cm. Fig.6 represents a series of temporal displacements of the density interface for Case 1 at six horizontal locations along the wave tank. All waveforms differ considerably from a sinusoidal profile, but the general characteristics at any given location are quite similar. The interfacial displacements show strong nonlinearity. These internal waves exhibit a higher rise at the crest and a depression near the trough in the upper-slope region. When waves are moving closer to the origin, a prominent feature of the cnoidal-type fluctuations continues from one wave to the next. The wave height abruptly decreases, while its mean density interface slowly increases.

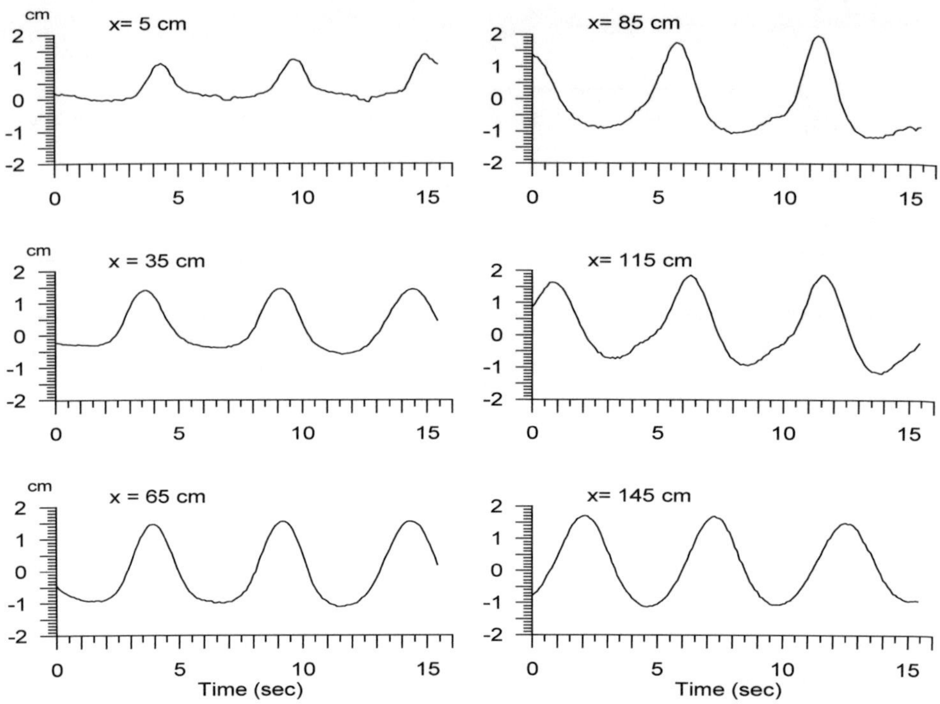

Fig. 6. Temporal displacement of density interface for Case 1

Fig.7 shows comparisons of predicted and observed variations of the local wave height and the mean density interface for Cases 1 and 2. Illustrated in the solid curve are the wave transformation predicted using Eq.(28) and the wave setup or set-down using Eq.(27). A portion of the incident wave motion is converted to the forward translation of the water mass by the breaking process. This results in forming a bolus that runs further up the face of the slope. The maximum runup elevation depends on the wave steepness and the bottom angle. Although the reflection coefficient should be determined in advance of the calculations of attenuation and setup, the reflection was neglected. In order to calculate the wave-height distribution, the value of β in Eq.(28) was determined using a trial-and-error procedure. In the present computation, it was equal to 0.020 for Case 1 and 0.025 for Case 2.

For shoaling waves on a given bottom slope, the wave height decreases gradually with decreasing water depth but does not vanish at the origin. Good agreement is found between the measured and calculated values of wave height in the recorded region. The discrepancy between these measured and calculated distributions is probably due to the neglect of the reflection or the overestimation of shear stress parameters. The radiation stress for partial standing waves differs from that for incident waves; it is presented by the sum of the incident and reflected waves. For surface waves, Longuet-Higgins and Stewart (1964) found that the mean surface level for a standing wave train slightly increases at the antinodes and correspondingly decreases at the nodes, having twice the frequency of an incident wave train. A somewhat similar relationship was found in the experimental result by Umeyama (2008) for partial standing internal waves. Partial standing internal waves develop a longitudinal mean interfacial oscillation that is a half period out of phase. Although it is difficult to detect these oscillations from the observation data, generally, the theoretical mean interfacial displacement agrees with the measured setup.

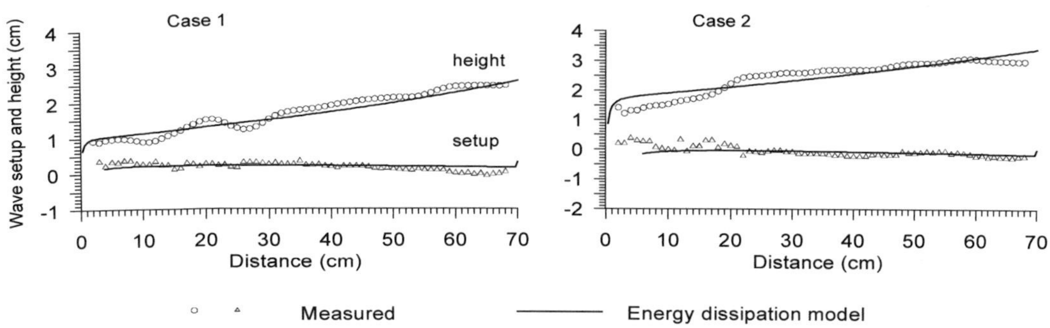

Fig. 7. Attenuation and setup of internal waves on a slope

4.3 Instantaneous velocity fields of internal waves

Fig.8 depicts the experimental velocity fields of internal waves for the layer thickness ratio of $h_I : h_{II}$ =15 cm:15 cm by a PIV measurement. Each vector image consists of two pictures taken simultaneously using two high-definition digital video cameras. Internal waves propagate from right to left. Fig.8 (a) indicates four velocity fields at an interval of $T/4$ for Case 1, in which the internal waves were generated with a period of T = 5.2 s, and the measured wave height was H =2.6 cm at a location 100 cm from the wavemaker. In these experimental images, an array of asymmetric vortices forms along the wave tank and their scale decreases as waves progress into the shallow-water region. The mean velocity in the lower layer appears to be relatively larger than that in the upper layer. The thinner clockwise vortex alternates with the thicker counterclockwise vortex, and the array of vortices sakes vertically from the mean density interface (z =0). Although the distortion to the pair of vortices occurs in the upper slope, the velocities still show quasi-elliptical trends. Closer the density interface, the magnitude of the velocity vector reaches its maximum. The center of vortex located within a zone between the crest and trough of internal waves. Generally, the center of ellipse starts to depart upward from the density interface before the wave crest arrives, while it deviates downward slightly from the mean interface level before the wave trough arrives. The flow under the density interface converges in the front of the wave crest and diverges behind it. This contributes to creating trajectory systems in the upper and lower layers, without crossing the density interface. Fig.8 (b) depicts four velocity fields for Case 2. The clockwise and counterclockwise vortices are in an orderly line when compared with those in Fig.8 (a). This fact suggests that an increase of wave period leads to

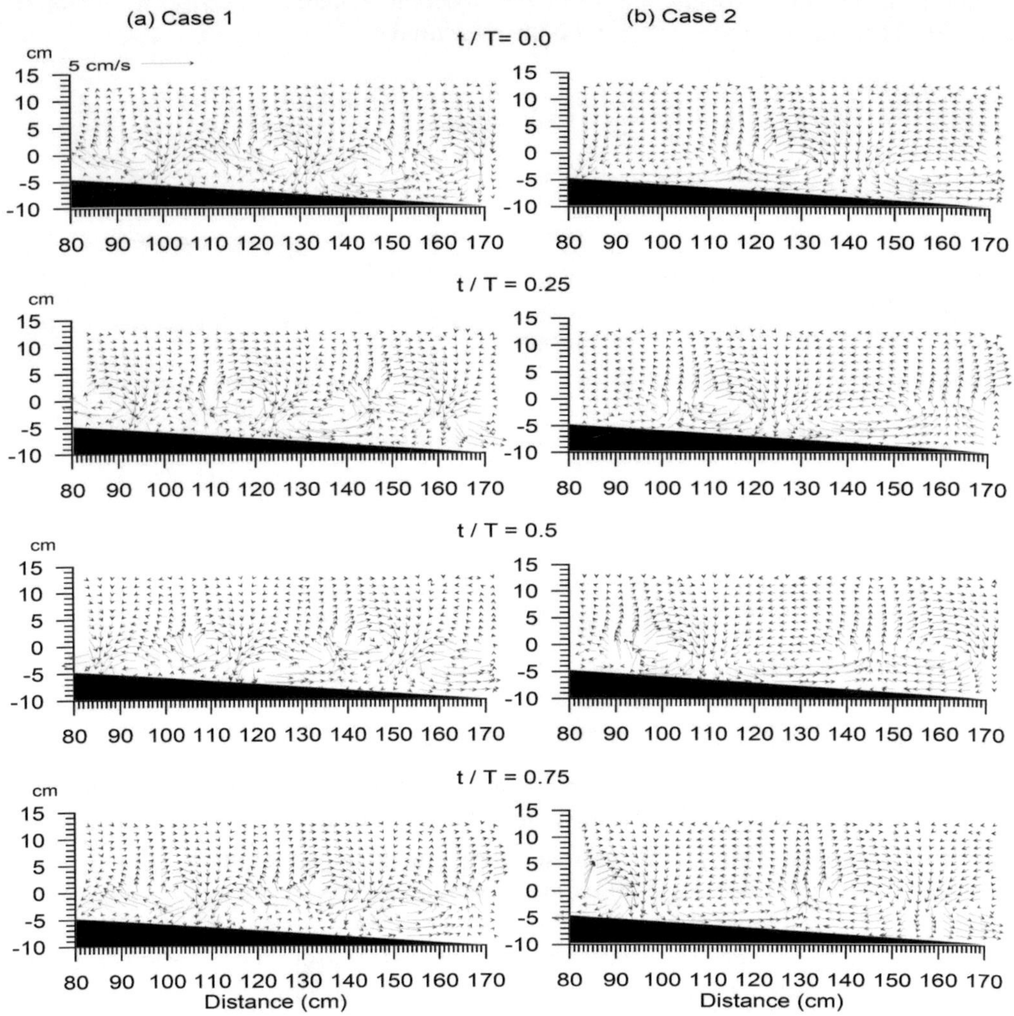

Fig. 8. Instantaneous velocity fields of internal waves ($h_I : h_{II}$ = 15 cm:15 cm)

an increase in the stability of vortices that gradually decrease the size with distance up the slope. From a pair of counterrotating vortices, one can expect the nonlinearity of internal wave, although the difference of two vertical positions for the counterrotating vortices is relatively small. The flow is no more symmetric with respect to the vertical line at the node in these cases.

Fig.9 shows a similar comparison for the generated internal waves in the density-stratified water where the upper and lower thickness ratio is $h_I : h_{II}$ = 10 cm:20 cm (Cases 3 and 4).

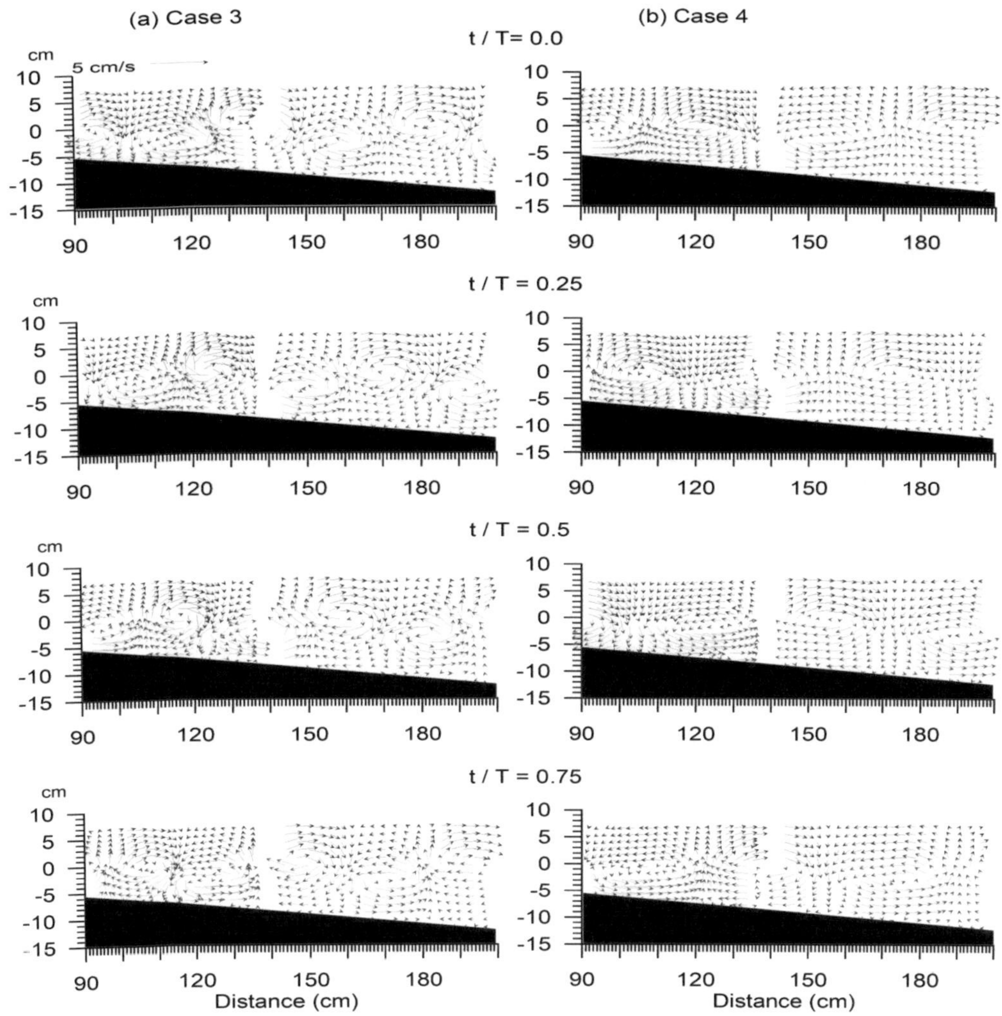

Fig. 9. Instantaneous velocity fields of internal waves ($h_I : h_{II}$ = 10 cm:20 cm)

The experimental data points are missing due to the flume flange in the range of 135 cm< x <145 cm. A pair of counterrotating vortices in the experiment shows still a satisfactory pattern. Umeyama & Matsuki (2011) found that the vortex pair turns inconspicuous by changing the thickness rate from $h_I : h_{II}$ =15 cm:15 cm to $h_I : h_{II}$ =5 cm:25 cm in a fluid of finite depth over a flat bed. The present improvement may be attributed to the replacement of the PIV system from two frequency doubled Nd:YAG lasers of 50-mW energy to a frequency-doubled Nd:YAG laser of 8-W energy. It could be confirmed from a series of experiments that the water particle movement in clockwise vortices is stretched in the horizontal direction while the anticlockwise vortices become less elliptical in the longitudinal direction over the slope.

4.4 Vertical distributions of velocity components

In this section, the measured profiles of the horizontal and vertical components of velocity at a cross-section are examined. An example of these profiles is shown in Fig.10 for Case 2.

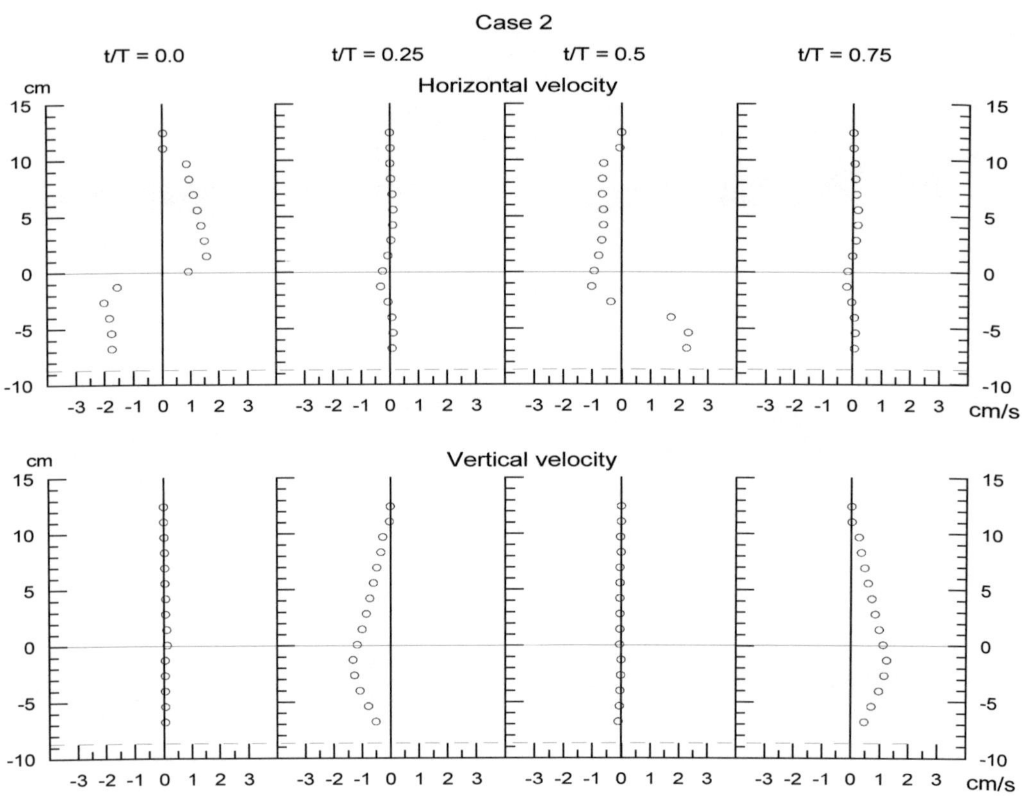

Fig. 10. Vertical distributions of velocity components at x =145 cm for Case 2

The PIV data at x =145 cm are exhibited individually for four different phase values: i.e., t/T =0.0, 0.25, 0.5, and 0.75. The horizontal solid line indicates the density interface, while the dash line indicates the bottom at this cross-section. The wave direction is from right to left in the horizontal velocity panel. The time-marching exhibitions were performed to examine the application of the PIV method to the velocity variations in a total depth. Under the crest of an internal wave at t/T =0.0, the vertical velocity is nearly zero over the depth. The horizontal velocity in the wave direction appears in the total lower layer, and reaches the maximum slightly below the wave trough level ($z \approx -2.0$ cm). In the upper layer, the horizontal velocity is opposed to the wave direction. It increases from the density interface until the wave crest level $z \approx -1.0$ cm, and slowly decreases to the surface. Under the node of an internal wave at t/T =0.25, the upward velocity starts to increase from the bottom,

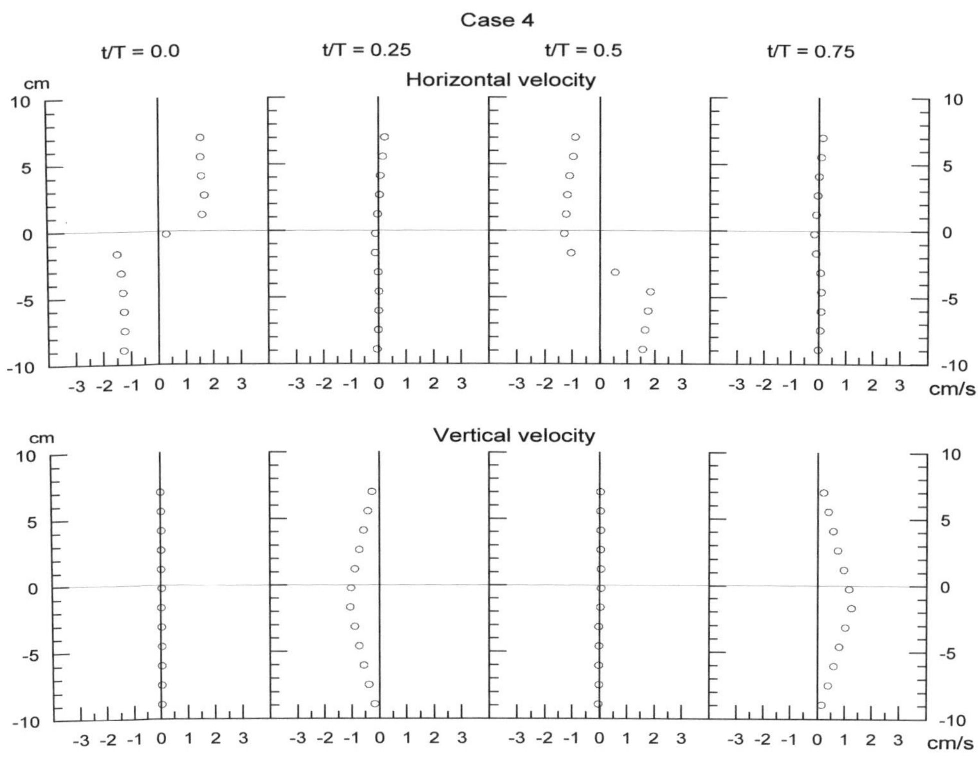

Fig. 11. Vertical distributions of velocity components at x =170 cm for Case 4

becomes the peak at an elevation slightly below the density interface, and decreases to the surface. The horizontal velocity is zero except in the region where there is a slight rise of velocity in the wave direction near the density interface. Under the trough at t/T =0.5, the direction of horizontal velocity changes in the lower layer. The velocity peak occurs at $z \approx -1.0$ cm in the wave direction and $z \approx -5.5$ cm in the direction opposing to the wave. The velocity decreases slowly toward the surface in the upper layer but it keeps a relatively large value at the bottom. Under the other node at t/T =0.75, the downward velocity dominants the phase. The vertical velocity profile is reverse to that at t/T =0.25. The horizontal velocity in the wave direction can be only found in the region from -1.0 cm $\leq z \leq 0.0$ cm.

Fig.11 depicts similar distributions of the horizontal and vertical components of velocity for Case 4, in which the experiment was performed for the layer thickness ratio of $h_I : h_{II}$ =10 cm:20 cm. Plotted are two velocity components at x =170 cm. Under the crest and trough, the horizontal velocities are maxima, while the vertical velocities are zero, and under the nodes, the opposite is true. At t/T =0.0 and 0.5, the horizontal velocity decreases slowly to the surface in the upper layer or the bottom in the lower layer but does not vanish near these boundaries. In contrast, the horizontal velocity is zero over the depth at t/T =0.25 and 0.75 when the velocity peak in the vertical direction appears slightly below the mean density interface.

4.5 Water particle trajectories

Fig.12 displays a plot of the particle orbit geometries for $h_I : h_{II}$ = 15 cm:15 cm and $h_I : h_{II}$ = 10 cm:20 cm. Shown in the diagram are attained for four wave conditions: i.e. (a) Case 1; (b) Case 2; (c) Case 3; and (d) Case 4. The data were taken approximately at the center of the frame in the digital video camera. The circular symbol indicates the instantaneous position of a water particle based on the PIV technique in which each Lagrangian point was determined using the present algorithm (see Fig.3). All trajectories are not elliptical and not closed for each case. Near the density interface in the lower layer, the horizontal and vertical displacements of the water particle are large relative to those near the mid-depth and surface region. The particle tends to take less distance to complete one wave cycle in the wave direction for larger wave period. Generally, the vertical excursion of the particle becomes smaller than its horizontal excursion with distance up in the water column. At these elevations, the particle moves backward from the wave direction to maintain a balance between the following and opposing fluxes in the wave tank. In contrast, the particle marches forward in a large nonclosed loop at the density interface, implying that each particle has a periodic motion per wave cycle but yields a maximum forward drift. The horizontal particle displacement in the direction opposing to the wave propagation increases with distance down in the water column for Cases 3 and 4. Fig.13 shows the theoretical trajectory calculated for Cases 1, 2, 3, and 4 by equations (19), (29) and (30) whose higher terms lead to an asymmetry of the particle orbit. Note that ideal trajectory data cannot be obtained without removing extraneous effects such as reflection from the wall and higher harmonics generated by the wavemaker, although the size of the wave tank restricts these effects. Inspecting the present comparisons for internal waves, however, the experimental path is qualitatively in good agreement with the theoretical trajectory just above and below the density interface.

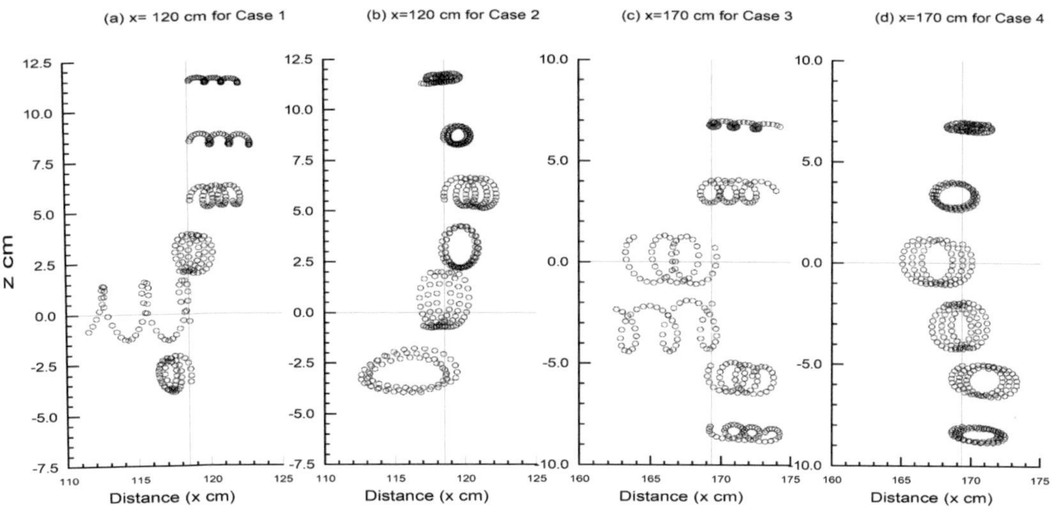

Fig. 12. Water particle trajectories by PIV/PTV method

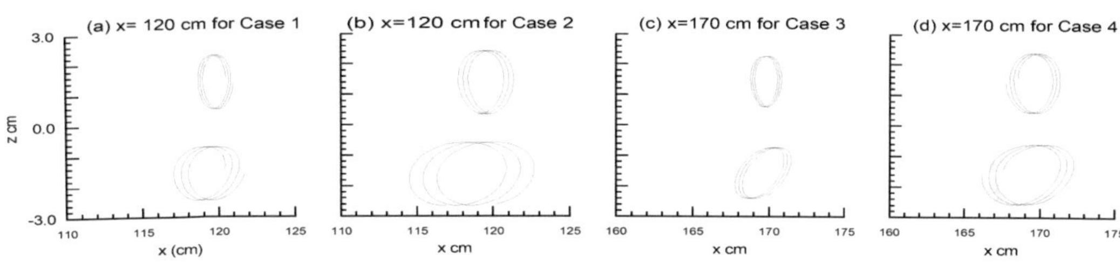

Fig. 13. Calculated particle paths by the Method of characteristics

5. Conclusion

In the present study, an image processing technique was used to quantify the spatial and temporal displacements of the density interface, and a PIV system was examined to illustrate the velocity fields in layers of fresh and salt water over a slowly varying water depth. The wave generation was made by a slide-type wavemaker. Wave celerity, attenuation and setup of internal waves were investigated theoretically and experimentally. The method of characteristics was adapted to express the internal waves at the interface of two fluids of slightly different densities on a sloped topography. A comparison of theoretical and measured variations of internal waves indicated that the agreement with these values improves when the target area is located in the region slightly apart from the origin. Over the range of interest, the spatial variations of internal waves were well predicted by the nonlinear long-wave equation. In addition, the wave celerity was obtained by analyzing the experimental data and compared with the result by the method of characteristics. To predict the local wave-height attenuation and the setup of density interface, we used the energy dissipation model, the energy conservation equation, and the momentum balance equation based on the radiation stress concept for internal waves by Umeyama and Shintani (2006). Note that the spatial variations of wave height and setup for all experimental conditions were well reproduced by the theoretical solution.

Using a PIV system, we measured the water particle velocity to obtain successive velocity fields over one wave cycle. The instantaneous velocity vector map clearly illustrated a vortex pair when the thickness ratio was relatively large in a fluid of two density layers. The vertical distributions of velocities analyzed from the PIV data were also presented at different phases. In addition to the common use of the PIV technique, we extended the knowledge to the visualization of the mass transport due to the propagation of internal waves. The algorithm for PIV was employed to compute Lagrangian velocity and track water particle displacements in Eulerian grids. The particle trajectories in a cross-section were simulated using a solution based on the definition of the Lagrangian approach to the method of characteristics. The agreement between the computed and measured results near the density interface was reasonable, apart from the restriction resulting from the apparatus. Thus, the proposed approach of the PIV can be applied to the Lagrangian description of the trajectory of a water particle when internal waves propagate above a sloping bottom.

6. Acknowledgment

The authors wish to thank Shinya Watanabe, Yasuhiro Takei and Ryota Kobayashi for preparing all experimental data.

7. References

Apel, J. R., Byrne, H. M., Proni, J. R., and Charnell, R. L. (1975). Observations of oceanic internal and surface waves from the earth resources technology satellite. *J. Geophys. Res.*, Vol.80, No.6, pp.865–881, ISSN: 0148–0227.

Apel, J. R., Byrne, H. M., Proni, J. R., and Sellers, R. L. (1976). A study of oceanic internal waves using satellite imagery and ship data. *Remote Sens. Environ.*, Vol.5, pp.125–135, ISSN: 0034-4257.

Apel, J. R., Holbrook, J. R., Liu, A. K., and Tsi, J. (1985). The Sulu Sea internal soliton experiment. J. Phys. Oceanogr., Vol.15, No.12, pp.1625–1651, ISSN: 0022-3670.

Austin, R.W. and Halikas, G. (1976). "The index of refraction of seawater." Scripps Institution of Oceanography, Ref. No. 76-1.

Grue, F., Jensen, A., Rusas, P.-O. and Sveen, J.K. (1999). Properties of large-amplitude internal waves. *J. Fluid Mech.*, Vol.380, pp.257-278, ISSN: 0022-1120.

Grue, F., Jensen, A., Rusas, P.-O. and Sveen, J.K. (2000). Breaking and broadening of internal solitary waves. *J. Fluid Mech.*, Vol.413, pp.181-217, ISSN: 0022-1120.

Helfrich, K. R. (1992). Internal solitary wave breaking and run-up on a uniform slope. *J. Fluid Mech.*, Vol.243, pp.133–154, ISSN: 0022-1120.

Helland-Hansen, B., and Nansen, F. (1926). The eastern North Atlantic. *Geofys. Publ.*, Vol.4, No.2.

Kao, T. W., Pan, F.-S., and Renouard, D. (1985). Internal solitons on the pycnocline: Generation, propagation, and shoaling and breaking over a slope. *J. Fluid Mech.*, Vol.169, pp.19-53, ISSN: 0022-1120.

LaFond, C. (1962). Internal waves, Part 1, In: *The sea*, M. N. Hill, ed., Interscience, New York, 1, pp.731–751.

Longuet-Higgins, M. S., and Stewart, R. W. (1964). Radiation stress in water waves: A physical discussion, with applications. *Deep-Sea Res.*, Vol.11, pp.529–562, ISSN: 0967-0637.

Michallet, H., and Ivey G. N. (1999). Experiments on mixing due to internal solitary waves breaking on uniform slopes. *J. Geophys. Res.*, 104(C6), pp.13467-13477, ISSN: 0148–0227.

Shand, J. A. (1953). Internal waves on Georgia Strait. *Trans.*, AGU, Vol.34, No.6, pp.849–856.

Shimizu, R., Shintani, T. and Umeyama, M. (2005). Instantaneous and Lagragian velocity fields of internal waves on a slope by PIV measurement and numerical simulation. Annu. *J. Coastal Eng.*, Vol.52, pp.1-5.

Thorpe, S. A. (1968). On the shape of progressive internal waves. *Phil. Trans. the Royal Soc. A*, Vol.263, pp.563–614, ISSN: 1471-2962.

Umeyama, M. (2002). Experimental and theoretical analyses of internal waves of finite amplitude. *J. Waterway, Port, Coastal, and Ocean Eng.*, ASCE, Vol.128, No.3, pp.133-141, ISSN:0733-950X.

Umeyama, M. (2008). PIV techniques for velocity fields of internal waves over a slowly varying bottom topography. *J. Waterway, Port, Coastal, and Ocean Eng.*, ASCE, Vol.134, No.5, pp.286-298, ISSN:0733-950X.

Umeyama, M. (2011). Coupled PIV and PTV measurements of particle velocities and trajectories for surface waves following a steady current. *J. Waterway, Port, Coastal, and Ocean Eng.*, ASCE, Vol.137, No.2, pp.85-94, ISSN:0733-950X.

Umeyama, M. (2012). Eulerian/Lagrangian analysis for particle velocities and trajectories in a pure wave motion using particle image velocimetry. *Phil. Trans. the Royal Soc. A*, 370(1964), pp.1687-1702, ISSN: 1471-2962.

Umeyama, M. and Matsuki, S. (2011). Measurements of velocity and trajectory of water particle for internal waves in two density layers. *Geophysical Res. Lett.*, AGU, Vol.38, No.3, L03612.

Umeyama, M. and Shinomiya, H. (2009) Particle image velocimetry measurements for Stokes progressive internal waves. *Geophysical Res. Lett.*, AGU, Vol.36, No.6, L06603.

Umeyama, M. and Shintani, T. (2004). Visualization analysis of runup and mixing of internal waves on an upper slope. *J. Waterway, Port, Coastal, and Ocean Eng.*, ASCE, Vol.130, No.2, pp.89-97, ISSN:0733-950X.

Umeyama, M. and Shintani, T. (2006). Transformation, attenuation, setup and undertow of internal waves on a gentle slope. *J. Waterway, Port, Coastal, and Ocean Eng.*, Vol.132, No.6, pp.477-486, ISSN:0733-950X.

Umeyama, M., Shintani, T. and Watanabe, S. (2011). Measurements of particle velocities and trajectories in a wave-current motion using PIV and PTV. *Proceedings of 32nd Inter. Conf. of Coastal Eng.*, ASCE, ISSN: 2156-1028, Shanghai, July, 2010.

Walker, S. A., Martin, A. J., Easson, W. J., and Evans, W. A. B. (2003). Comparison of laboratory and theoretical solitary wave kinematics. *J. Waterway, Port, Coastal, and Ocean Eng.*, ASCE, Vol.129, No.5, pp.210–218, ISSN:0733-950X.

Wallace, B. C., and Wilkinson, D. L. (1988). Run-up of internal waves on a gentle slope in a two-layered system. *J. Fluid Mech.*, Vol.191, pp.419–442, ISSN: 0022-1120.

13

Applying of PIV/PTV Methods for Physical Modeling of the Turbulent Buoyant Jets in a Stratified Fluid

Valery Bondur[1], Yurii Grebenyuk[1], Ekaterina Ezhova[2], Alexander Kandaurov[2], Daniil Sergeev[2] and Yuliya Troitskaya[2]
[1]*AEROCOSMOS Institute for Scientific Research of Aerospace Monitoring*
[2]*Institute of Applied Physics Russian Academy of Sciences*
Russia

1. Introduction

Modern optical methods of flow visualization: Particle Tracking Velocimetry (PTV) and Particle Image Velocimetry (PIV) (see (Adrian, 1981)) are widely used in experimental investigations of air and liquid flows for scientific and industrial purposes. They are common in exploration of such processes as heat and mass transfer in power plants, flows in aircraft and shipbuilding, and medico-biological applications. Nowadays this technique is also employed in the laboratory modeling of geophysical flows (Sergeev & Troitskaya, 2011) including air-sea interaction (Reul et al., 1999; Veron et al., 2007); flows in water column and their interaction with bottom topography (Umeyama, 2008; Zhang et al., 2007); vortex flows (Beckers et al., 2002; Heist et al., 2003).

The present work is devoted to the investigation of the new class of flows by PIV/PTV-methods - oscillating buoyant jets (fountains) in stratified fluid. (Bondur & Grebenyuk, 2001; Bondur 2004, 2011; Bondur et al. 2006, 2009). Among the applications of these flows are heat and moisture exchange in the atmosphere and cloud formation (Turner, 1966). The investigation of buoyant jets in stratified fluid is extremely important for the development of new technologies of sewage disposal by coastal cities (Koh & Brooks, 1975). The end body of such systems is a submerged collector. A typical diffuser of a modern collector is a pipe with lots of outlets. Sewage water (which has almost the density of fresh water after sewage disposal plants) is discharged into ambient salt water to produce buoyant jets (Bondur, 2011; Bondur et al, 2006, 2009). One of the main questions when constructing diffusers of these collectors is a regime of jet flows when they are trapped by pycnocline and don't reach the surface.

Sewage disposal to the ocean is an example of the man's impact, influencing mass transfer, hydrodynamics, hydrobiology and coastal ecosystem state as a whole (Bondur, 2011). The complex investigation of these processes includes mathematic modeling of jets dynamics (Bondur, 2011; Bondur et al, 2006, 2009; Bondur & Grebenyuk, 2001; Koh & Brooks, 1975; Ozmidov 1986), contact methods (Bondur 2006; Bondur & Tsidilina, 2006; Gibson et al., 2006 a), and airborne and spaceborne remote sensing (Bondur, 2004, 2006, 2011; Bondur

&Grebenyuk, 2001). The investigation of physical mechanisms, responsible for surface manifestations of sewage flows, is a challenging problem. In the papers (Bondur, 2004, 2006, 2011; Bondur &Grebenyuk 2001) mechanisms, resulting from surface deformation by rising vortices or internal waves are suggested, and in works (Bondur et al., 2005; Gibson et al., 2006 b, 2007a,b) surface manifestations are explained by the complex interaction between turbulence, internal waves, tides and bottom topography.

The intensive generation of internal waves by buoyant jets from submerged wastewater outfalls in stratified fluid was investigated by contact methods in experiments on the laboratory scale modeling of these systems (Bondur et al, 2009, 2010 a, 2010 b; Troitskaya et al., 2008). However, in order to explore the possibility of internal waves manifestation and mechanisms of their generation by buoyant jets, the additional experiments were needed which allowed high-precision measurements of velocity fields of flows on the surface and in the water column. Contact sensors are not applicable for such systems since they cause essential perturbations in the investigated flow. Thus, in new experimental series non-invasive methods were employed, based on the flow visualization (modified PIV/PTV-methods), that allowed the effective solution to this problem.

The Chapter is organised as follows. The laboratory scale modeling of buoyant jets and the investigation of surface flows induced by these jets with the application of PIV-technique is described in Section 2. Section 3 is devoted to the testing of PIV-methods for the measurement of velocity fields in buoyant jets in stratified fluid. The last Section deals with the application of this method for the laboratory scale modeling. The parameters of oscillating jets are compared with the characteristics of internal waves. Special methods of video processing are employed to investigate the structure of jet perturbation modes.

2. Experimental studies on measurements of the surface flows induced by submerged buoyant jets in the thermostratified tank with the modified PTV-method

The main purpose of investigations, described in this part, is to assess the possibility of internal waves manifestation caused by a submerged sewer system on the sea surface. To solve this complicated problem it is necessary to obtain the properties of inhomogeneous flow fields created on the sea surface.

The experimental study of these processes was carried out on the basis of laboratory scale modeling at the Large Thermally Stratified Tank (LTST) of the Institute of Applied Physics, Russian Academy of Sciences (IAP RAS). The experiments included measurements in the water, and the main attention was drawn to the measurements of the surface flow, where the PTV- method was used.

2.1 Experimental setup for the scale laboratory modeling of the surface flows induced by the typical submerged sewer system in the LTST

The principal scheme of experiments is shown in Fig1. The LTST dimensions are as follows: 20 m in length, 4 m in width, and 2 m in depth. The temperature (density) stratification in the LTST is generated through liquid heating and cooling with heat exchangers installed along the tank walls (Arabadzhi et al., 1999; Bondur et al., 2009). It results in the formation of the inhomogeneous vertical distribution of the temperature (density) in the tank.

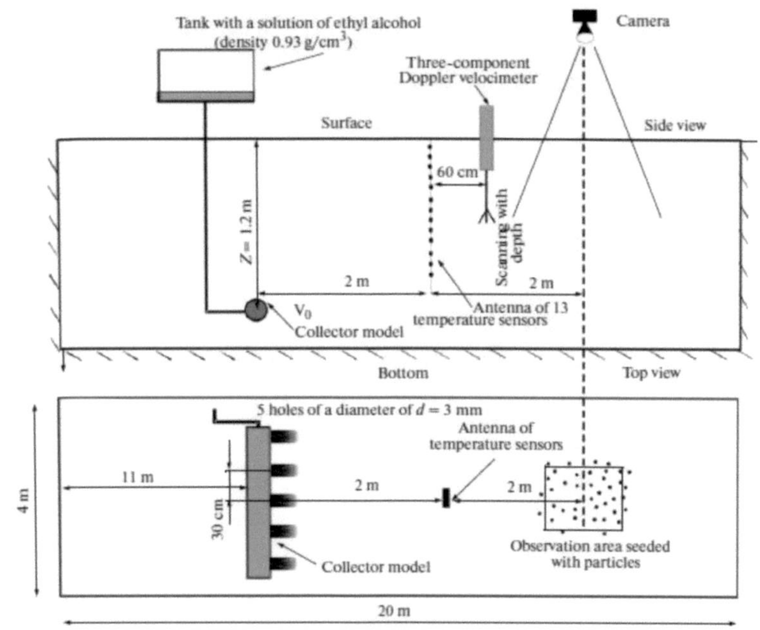

(a)

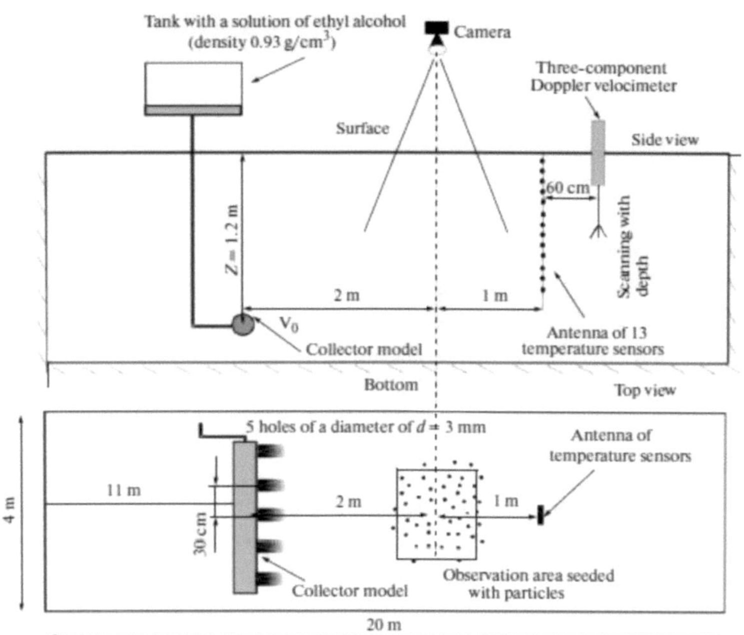

(b)

Fig. 1. General scheme of LTST IAP RAS experiments with the location of temperature and velocity sensors before the area of surface PTV measurements series *S1* - (*a*), and behind *S2* - (*b*)

The experiments were carried out using distribution of temperature with shallow thermocline depth of 13–15 cm on the average (profile 1 in Fig.2), with the total water depth of 130 cm.

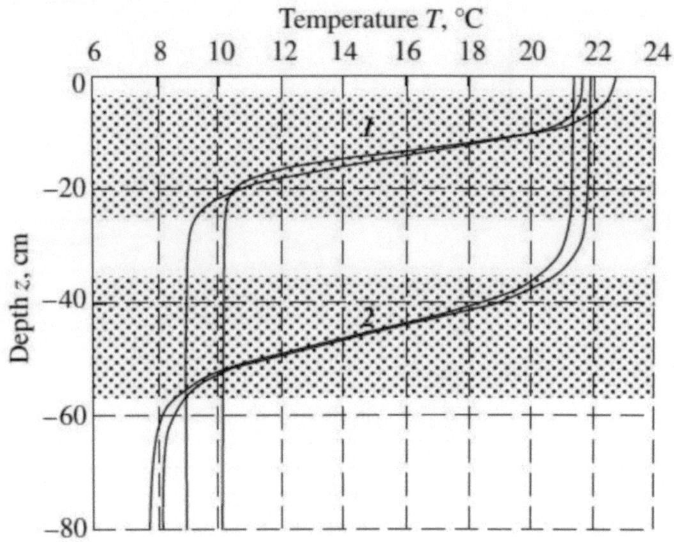

Fig. 2. Operating profiles of temperature stratification in the LTST. 1) shallow thermocline 2) deep thermocline. Two curves for each case indicates borders of stratification changing during experiments.

The scaled model of the diffuser of the sewer system was a metallic blanked-off tube at one end with the length of 1.3 m and the diameter d of 1.2 cm; the model includes 5 holes with the diameter of 3 mm located at a distance of 30 cm from each other at the same level. The tube is oriented horizontally across and in the middle of the tank (Fig. 1). The holes axes are oriented horizontally. The model is connected by a hose with a tank containing the solution of ethyl alcohol, the density of which during the experiments was kept constant (0.93 g/cm^3). Thus initial difference in density between jet and ambient fluid $\Delta\rho_0$ was 0.07 g/cm^3. The outflow rate V_0 from the diffuser model was: 40, 70, 100, and 150 cm/s (controlled by measuring the volumetric time rate). The variation of the flow rate through the change in the tank-solution level during the experiment does not exceed 10% (the average value was 5%).

The parameters of the induced jet flow and stratification in the LTST, enable us to simulate the typical conditions of the coastal area parameters of the submerged disposal system (Koh & Brooks, 1975) with respect to the Ri, Str numbers (Bondur et al., 2009):

$$Ri = \frac{g\Delta\rho_0 d}{\rho_0 V_0^2}$$ - global Richardson number,

$$Str = \frac{N_0^2 d\rho_0}{g\Delta\rho_0}$$ - parameter of ambient stratification,

where ρ_0 - mean density of ambient fluid, N_0 - maximum buoyancy frequency of the ambient stratification, g - gravitational acceleration, and a geometric similarity on the scale of 1:27. In this case, the Reynolds number at the output was around 3000, which ensures the developed turbulence mode of buoyant jets in the laboratory experiment.

2.2 Measuring technique

Preliminary estimates indicated that the amplitudes of surface flows induced by internal waves would be only few mm/s. For these conditions, the modified Particle Tracking Velocimetry (PTV) method was used to measure the velocities of the surface flows induced by internal waves.

Thus, the limited area of basin surface under observation was seeded by particles of black polyethylene (density 0.98 g/cm^3) with a characteristic size of about 1.5 mm for creating a contrast to the white background of the bottom (see the experimental setups in Figs. 1 a, b).

The motion of particles was recorded by a CCD camera from above (25 frames/sec), and the resulting time series of images were then processed on the computer. It turned out to be impossible to combine the particle observation region with the sensor allocation area (the distance between sensors and the observation area center is a minimum) in the conditions of this experiment, because the sensor images hindered the correct identification of particles on the images. That is why, two series of experiments were carried out with the sensors and the observation area allocated differently relative to the diffuser model.

The series *S1* (see Fig. 1 *a*) included the installation of 13 temperature sensors vertically at a distance of 200 cm from the diffuser model for measuring proprieties of internal waves; a scanning 3D ultrasound doppler anemometer (ADV) (at a distance of 260 cm); and the center of the observation area (at a distance of 400 cm), which is a 60×48 cm^2 rectangle was oriented by its longer side along the direction of the jet flow.

During the series *S2* (see Fig. 1 *b*), the surface observation area was located at a distance of 200 cm from the model. The area had dimensions 100×80 cm^2 and was oriented perpendicular to the flow motion. Further, at a distance of 300 cm from the collector model, temperature sensors were located and, at a distance of 360 cm, an ADV was located.

Two important problems concerning the measurements of surface flows in the LTST appeared during these experiments. The first one was that, when the temperature stratification with a shallow thermocline is created (Fig. 2), large-scale flows as a system of cyclonic and anticyclonic vortexes occurred on the surface layer. The measurements performed by the PTV method indicated that the scale of an individual vortex was 1.5–2.5 m, the maximum velocities of flows in it reached 1.3 cm/s, and the average velocities were about 4 mm/s. These values were calculated approximately, because the size of the observed area in both series of experiments did not allow one to cover the whole area at least one vortex. It can be assumed that vortexes appeared by inhomogeneous horizontal heating of the fine upper layer of thermocline. The presence of the large-scale flows made a low frequency trend in the dependence of the velocity on time, which was eliminated by the low frequency filtering of the results (see Subsection 2.3).

The second factor making it difficult to perform velocity measurements of surface flows is the presence of a Surface active substance (SAS) or dust film, which could not be completely removed.

We measured the parameters of the SAS film using the technique proposed in (Ermakov & Kijashko, 2006).

2.3 Modified algorithms of the PTV-method for studying weak surface flows

A classical PTV-method measures the velocity of each particle with respect to its motion on frames separated by a time interval and then reconstructs the velocity field by the velocity values at the points of the particle location. In these experiments, due to the presence of a background flow (see Subsection 2.2), the number of seeding particles in the observation area quickly decreased with time. For this reason we were able to determine only the averaged (over the observation area) value of the flow velocity. We found the trajectories of particles passing through this area and calculated their velocity at each moment. Then we calculated the velocity value averaged over all particles that were found in the observation area at the given time moment. The uncertainty of the PTV measurements was defined by the accuracy of the center of particle determination. It was about 10 % (based on results of special test experiments). To exclude the background low-frequency trends in the particle velocity, caused by large-scale flows in the LTST, we performed the low-frequency filtering of distributions of signals at frequencies below 0.02 Hz. The examples of such time dependences of the averaged surface flow velocity are shown in Fig. 3. These data were compared with oscillations of isotherms obtained from the temperature measurements (see Bondur et al., 2010 a).

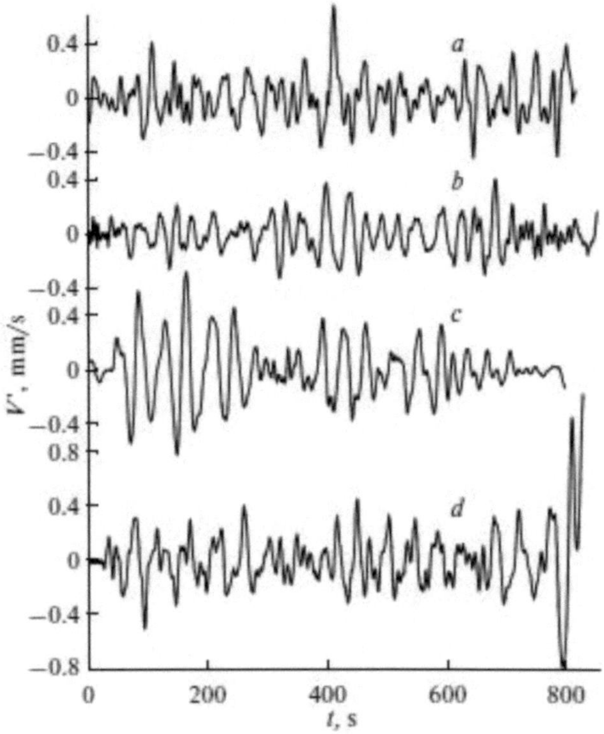

Fig. 3. Amplitudes of disturbances at the surface measured by the PTV method for rates of outflow from the diffuser model: *a*, 40 cm/s, *b*, 70 cm/s, *c*, 100 cm/s, and *d*, 150 cm/s.

2.4 Results - Comparing with theory forecasts

Basing on measured time realizations, we calculated the rms values of the surface velocity $\langle v'^2 \rangle^{1/2}$. The dependence of $\langle v'^2 \rangle^{1/2}$ on the outflow rate from the diffuser model is shown in Fig. 4. It can be seen from here that $\langle v'^2 \rangle^{1/2}$ varies in the range between 0.1 and 0.3 cm/s.

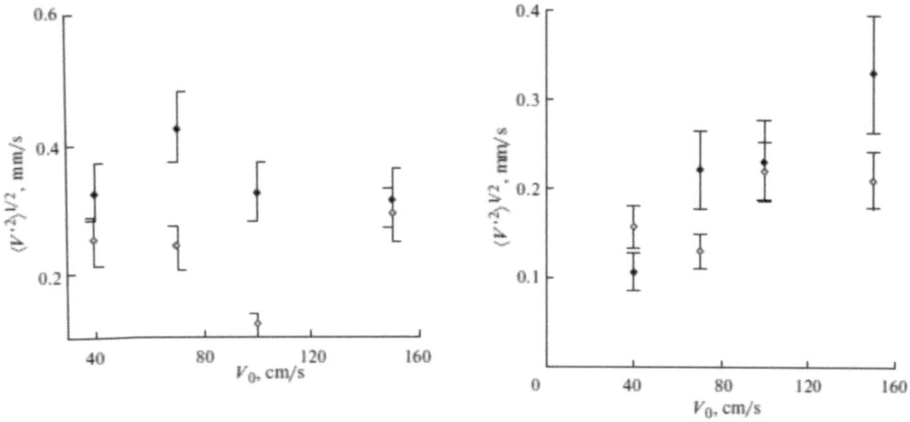

Fig. 4. Amplitudes of surface flow velocities (series *S1* left and series *S2* right): the black diamonds indicate theoretical estimates while accounting for the influence of the film and the light diamonds indicate the measurement results obtained by the PTV- method. Error bars corresponding to the uncertainty of the PTV measurements 10%, and uncertainty of theoretical estimations (determined by the accuracy of SAS films properties measurements 20 %).

The experimental data were compared with theoretical estimations of the value of the velocity on the surface according the theory of internal wave propagation proposed in (Bondur et al., 2010 a). Estimations were obtained taking into account the influence of the SAS films. The estimate for the error in theoretical values was determined by the measurement error (constituted 20%) (Ermakov & Kijashko, 2006). It is well seen from the figures that the results are in good agreement.

Thus, the experiments revealed that in the presence of internal waves induced by jets during outflow from the submerged sewer system model, surface flows with spatial periods controlled by the properties these waves appeared. The wavelength was from 30 to 160 cm. The standard deviations of velocities in the presence of a SAS film at the water surface varied in the range from 0.1 to 0.3 mm/s, corresponding to the amplitudes from 0.15 to 0.45 mm/s. For pure water the surface would constitute 1 to 2 mm/s. It should be pointed out that these experiments were performed for the conditions of scaled modeling.

In view of this, using the coefficients of scale modeling, one can estimate the flows created near the submerged sewer system in nature conditions: the parameters of internal waves (induced by jets), and the prosperities of their surface manifestations, as well as make conclusions on the possibility of remote diagnostics of these waves. In our laboratory experiment the coefficient of scale modeling with respect to the velocity value constitutes 1:3 and the coefficient of geometric similarity was 1:27. In view of this, we revealed the fact that this experiment simulated stratification with a pycnocline thickness of about 4 m, an internal wave with the length from 8 to 43 m, celerity from 5 to 10 cm/s, and the surface flow velocity from 0.3 to 0.6 cm/s. For such flows it yielded an estimate (see (Bondur et al., 2010 b)) for the contrast in the field of short waves of 0.12–0.3, which can be confidently detected by modern remote sensing methods (Bondur, 2004, 2006, 2011; Bondur, Grebenuyk, 2001).

3. The experimental study of the interaction of buoyant turbulent jet with pycnocline stratification with the PIV-method

A theory describing the relation of characteristics of surface manifestations with the operational parameters of sewer systems should be made. For this propose it is necessary to investigate the dynamics of the buoyant jets in the pycnocline region, where the trapping of the jets by stratification occurred. The PIV-method is widely used for studying the velocity fields of the jet flows. However, there are several problems of carrying out measurements by PIV-methods in LTST (see Section 4). That is why preliminary test experiments of applying PIV-methods were provided in a small reservoir with saline stratification.

3.1 Experimental setup in saline stratification

To study the properties of the buoyant turbulent jets and ambient stratified liquid interaction, we performed preliminary test experiments in a small plexiglas basin with saline stratification. The scheme of the experiment is shown in Fig.5.

Here, a salt stratification of the pycnocline type is created. The distributions of density and buoyancy frequency are shown in Fig. 6.

In this experiment, the diffuser model had only a single vent with a diameter of 1.2 mm that allowed fresh water to flow out with the rate of 50 cm/s and form buoyant jet in ambient salt water.

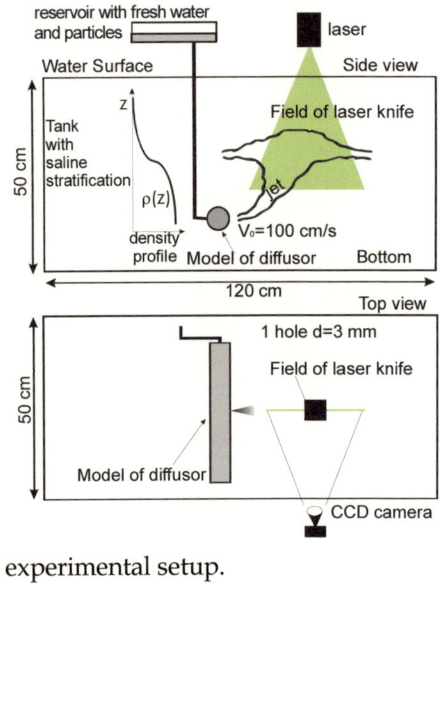

Fig. 5. Principal scheme of experimental setup.

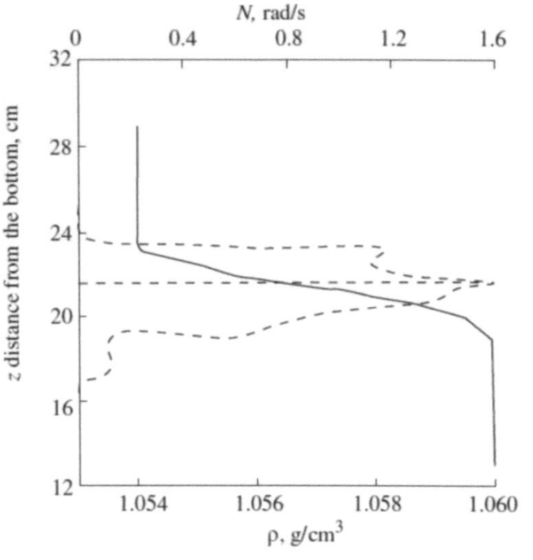

Fig. 6. Distributions of density (solid line) and buoyancy frequency $N = \sqrt{\frac{g}{\rho}\frac{d\rho}{dz}}$ (dashed line) in the small reservoir.

The PIV method was used to study the jet flow. Polyamide particles 50 μm were added to the reservoir with freshwater. We put particles only in the jet, because we want to see and measure the form and boundary of the jet precisely (oscillations of the top of the jet). The motion of particles in the jet was visualized by a vertical laser sheet along the jet axis. The source - CW NdYag laser (532 nm wavelength, 0.5 Wt power). The lateral view was recorded on a CCD-camera (example of the buoyant plume is shown in Fig. 7) with the rate of 25 frames per second and exposure time of 5 ms. The displacement was less than 1 pxl during time exposure.

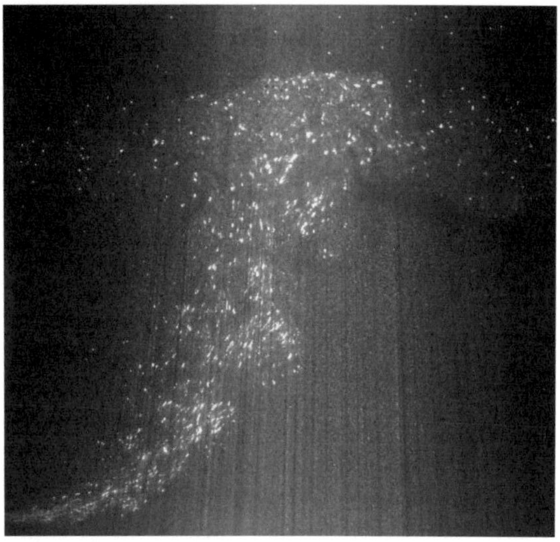

Fig. 7. Example of the buoyant plume.

3.2 PIV-processing

The main attention was drawn to the area of the front of the fountain. The processing of the resulting frame sequences by PIV algorithms made it possible to obtain the velocity field in the laser-sheet cross section at consecutive time points with a step of 0.25 s by the way of cross correlation processing successive pairs of frames. The interrogation window size was 32* 32 pix, with 50 % overlapping. The Gaussian approximation of the correlation function was used to avoid the effect of peak locking. The measurements uncertainty by PIV-method was about 3 %. It was obtained from the processing of synthetic images with determined displacement. Fig. 8 shows examples of measured instantaneous velocity fields.

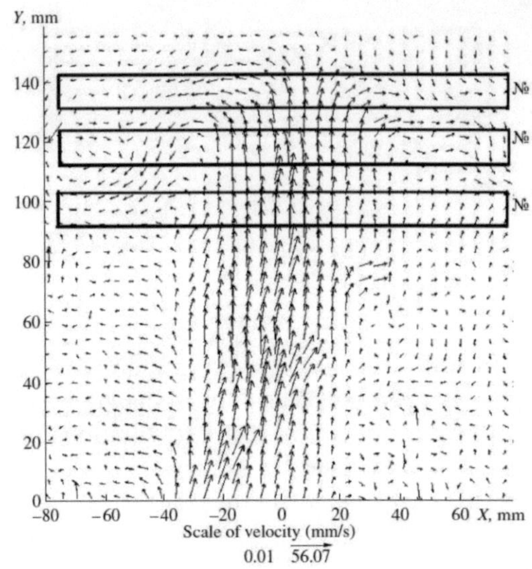

Fig. 8. Velocity field of buoyant turbulent jet.

It can be clearly seen that the jet is trapped by the stratification and propagates at the neutral buoyancy level, which is located on the lower boundary of the pycnocline. The video recording also indicated that the upper boundary of the jet oscillates in the vertical plane; the oscillation spectrum of the front clearly expressed peak at a frequency of 0.1 Hz (Fig. 9).

3.3 Main results of the experiment

Fig. 10 shows the instantaneous velocity profiles in different cross sections of the jet. To smooth the turbulent fluctuations that arise on these profiles, averaging by the coordinate along the jet axis over three domains shown in Fig. 8 was used. This includes the calculation of the mean profile of velocity on the basis of three adjacent profiles located 4.8 mm away from each another. It can be seen from the Fig. 10 that the counterflow exists in the region of pycnocline.

A stability analysis for the resulting profiles of flow velocities performed by the method of normal modes has revealed that, for the jet with the counterflow region, the condition of absolute instability by the Briggs criterion (Briggs, 1964) for axisymmetric jet oscillations is satisfied. The stability of nonparallel currents is normally analyzed in the following way: the current is divided into parts, each of which is taken to be quasi-parallel and treated by the method of normal modes. It testifies to the fact that the globally instable mode (Monkewitz et al., 1993) is actuated. The estimates for oscillation frequencies of the globally instable mode are well consistent quantitatively with the measured spectrum of jet oscillations (see. Fig 9).

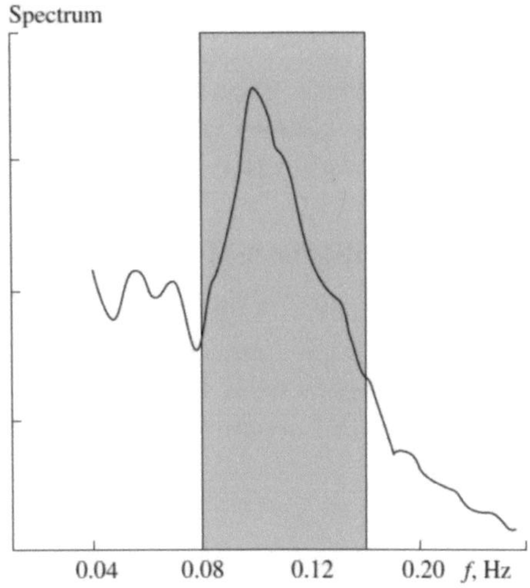

Fig. 9. Spectrum of oscillations of the upper boundary of the jet, the gray field indicates boundaries of theoretically predicted frequency of oscillations basing on the stability analysis (semi-logarithmic scale)

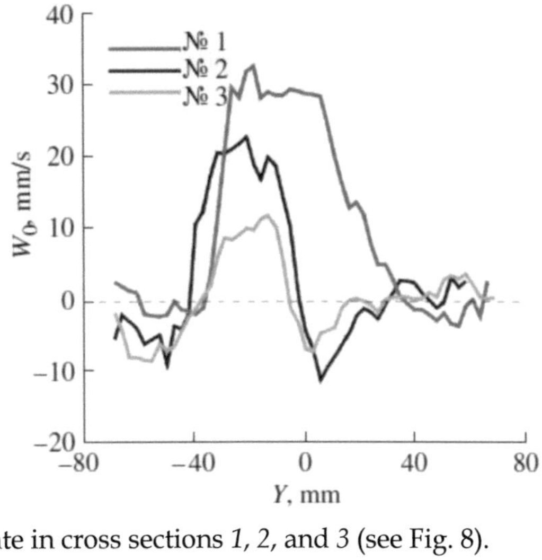

Fig. 10. Profiles of jet rate in cross sections *1*, *2*, and *3* (see Fig. 8).

Thus, the following mechanism of internal wave generation by buoyant plume is proposed. (Huerre et al., 1990). Self induced oscillations of the globally unstable mode appear during the interaction of a buoyant plume with the pycnocline. Internal waves are intensely generated if the frequency of these oscillations turns out to be lower than the maximal buoyancy frequency in the pycnocline.

4. Investigation of the mechanisms of internal waves generation by buoyant jets within the laboratory modeling of submerged sewage in a stratified ocean with the PIV-methods

Basing on the results of successful applying of the PIV-methods in small reservoir with saline stratification a series of similar experiments in the LTST for the conditions of scale modeling of the typical sewer system (see Section 2) were carried out for approving the hypothesis offered in section 3. In these experiments for the first time simultaneous measurements of the jet characteristics (source of oscillations) with PIV-method and the properties of internal waves by contact methods were performed.

4.1 Experiments in LTST with the application of the PIV-technique

The simplified general experimental scheme is shown in Fig. 11. This scheme is similar to the one we used for the experimental investigation of surface flows (see Section 2). In this experimental series the temperature stratification was created in LTST with the thermocline center located at the depth of 43-45 cm. Full water depth in the Tank was 160 cm (profile 2 in Fig. 2).

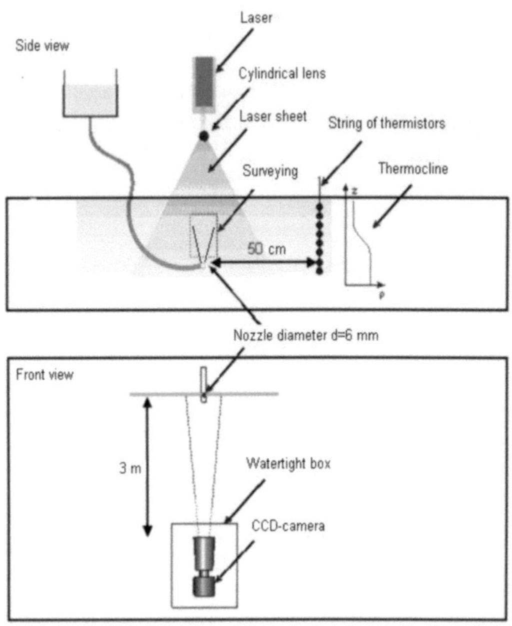

Fig. 11. Experimental setup: visualization of a turbulent jet in LTST

Opposite to the previous experimental series in LTST (see Section 2), a jet was discharged vertically from a П-shaped round pipe of 6 mm diameter (see Fig. 11), i.e. only one nozzle was used. Previously we worked with several buoyant jets (alcohol solution) discharged horizontally from a diffuser model at different rates.

The laboratory modeling of the jets from disposal systems showed that jets do not merge until they reach pycnocline. The visualization of such flows and the application of the PIV-technique in LTST were a very complicated task. Horizontally discharged buoyant jet comes to the pycnocline nearly vertically at some distance from the nozzle (see the jet photo from the experiment in Fig. 7, and calculations result in Fig. 12). In order to perform successful jet-thermocline interaction survey of a high space resolution we have to know this distance with a very high accuracy just to place the camera and the laser system correctly. It should be noted, that this distance depends strongly on a jet flow rate resulting in difficulties with equipment positioning

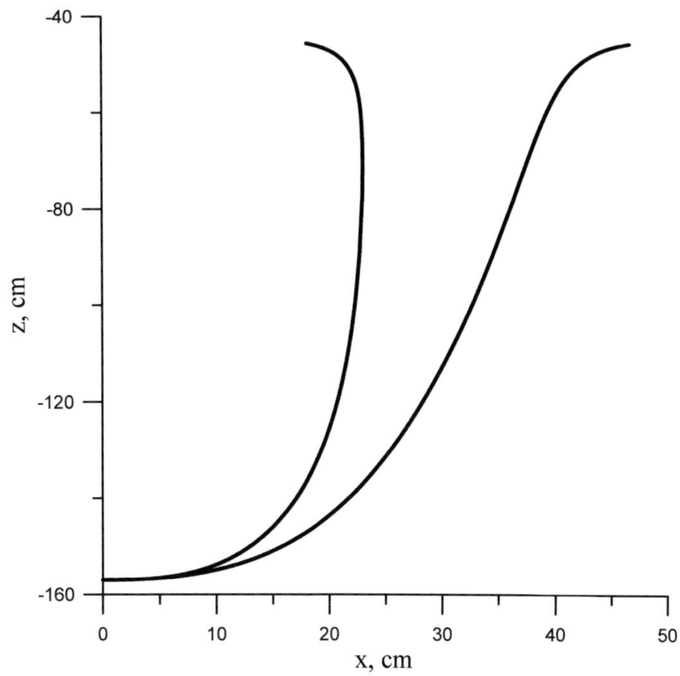

Fig. 12. Jet trajectory (boundaries) calculated for typical stratification in LTST basing on the integral model (thermocline center is at 44 cm depth). Side view.

Alcohol solutions also cause problems for visualization (chemical interaction with polyamide particles). Thus, in the new experimental series we used one jet of neutral buoyancy, discharged vertically at several rates. Jet exit velocity and a distance from the nozzle to the thermocline were chosen to provide jet parameters in the thermocline coincident with those in the previous series on the laboratory scale modeling of sewage disposal systems (see Section 2). Jet parameters in the thermocline for the series with buoyant liquid were determined from the direct numerical solutions to the system of equations for integral parameters of a turbulent jet in ideal incompressible liquid (Fan 1968).

New series consisted of 10 experiments: 2 for each flow rate. Jet parameters in the thermocline measured experimentally are shown in Fig.13.

Velocity fields in a jet were studied by the PIV-method. For jet visualization 20 μm polyamide particles were added to jet liquid. Only jet was seeded, for precise measurements of boundary oscillations. Particles were put to a reservoir 5 minutes after the beginning of each experiment. The laser sheet was produced by the same laser we used in the previous experimental series. Since LTST walls are not transparent, the digital camera (frame rate 25 frames/sec, time exposure 10 ms) was placed into a specially designed waterproof box (we performed underwater survey). The time exposure was 2 ms. Displacement of the particles less 1 pxl during one frame. The camera was submerged to the thermocline level at the distance 3 m from the nozzle (a maximum possible distance). The digital data from the camera were processed out with "Vortex" program which was also used previously to calculate the velocity fields, and a specially developed "SMPD" algorithm for simultaneous work with the survey data and calculated velocity fields.

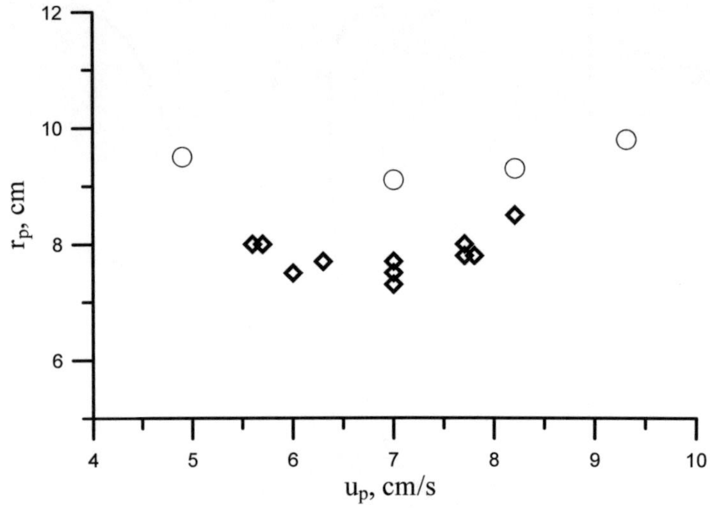

Fig. 13. Experimental jet parameters (maximum longitudinal velocity and radius): O – calculated in the framework of the integral model for discharge rates of buoyant jets 40, 70, 100 and 150 cm/s; ◊ - parameters, measured in the experiments with one jet.

Temperature oscillations in the LTST during jet discharge were measured as in previous experiments by contact methods (a string of 13 thermistors was employed). The string was placed at 50 cm distance from the nozzle. Using the data from thermistors we calculated isotherms.

The underwater survey in each experiment lasted 20 min (during this time particles were added to the reservoir from time to time) and temperature oscillations were measured during the whole time of the experiment – 1 h.

4.2 Processing technique for video and velocity fields

We studied jet oscillations in the thermocline using the SMPD algorithm. It allows calculating the average intensity of a frame in the user-defined rectangular area, thus, intensity dependence on time can be easily derived. If a rectangular area is chosen in the jet oscillations region, the intensity dependence detects these oscillations due to the light particles in the jet. We also employed the next considerations when chose the area. One could expect 2 types of unstable jet modes to develop: the first was axisymmetric and the second was spiral or helical. An axisymmetric mode results in the jet top oscillations in the vertical plane, and a spiral one in the laser sheet plane looks like jet oscillations from left to right. In order to detect both modes, one rectangular side was chosen along the jet axis and the other side separated the area with particles from the rest, simply black area (see Fig. 14). Upper and bottom rectangular sides indicated the maximum and minimum jet top positions. Jet top oscillations were calculated for each experiment (an example is shown in Fig. 15a).

It should be noted, however, that particles concentration in a jet changed in time causing changes in average intensity. This resulted in the average intensity trend. At the moments of particles injection in the reservoir one could observe sudden changes of intensity. Besides, slow intensity decrease due to decreasing particles concentration could also lead to incorrect ratio of power spectral peaks, and oscillations corresponding to the film beginning or particles injection moments would be the most powerful.

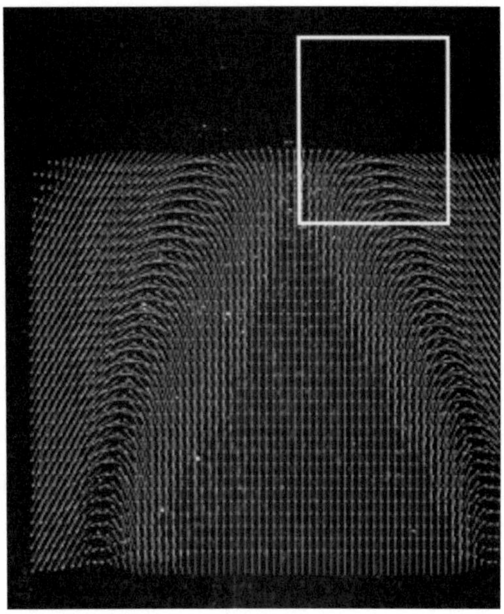

Fig. 14. An example of rectangular area for investigation of jet top oscillations. Mean velocity field is calculated on the PIV-measurements base.

Thus, the calculated intensity had to be corrected. For each experiment we found the intensity trend due to varying particles concentration in a jet. This was performed by choosing a maximum possible rectangular area in a jet fully occupied by particles and calculating its average intensity dependency on time (Fig.15 b). The bigger this area is, the smaller high-frequency pulsations of intensity are. Jet top oscillations can be represented by a formula

$$y(t)=I(t)*f(t),$$

where $f(t)$ - is jet oscillations function, as it would be for the constant particles density; $I(t)$- function, corresponding to the average intensity trend (or particles concentration trend).

Then a desired function is $f(t)=y(t)/I(t)$ (see example in Fig. 15 c). When processing data out, high-frequency oscillations were filtered from $I(t)$.

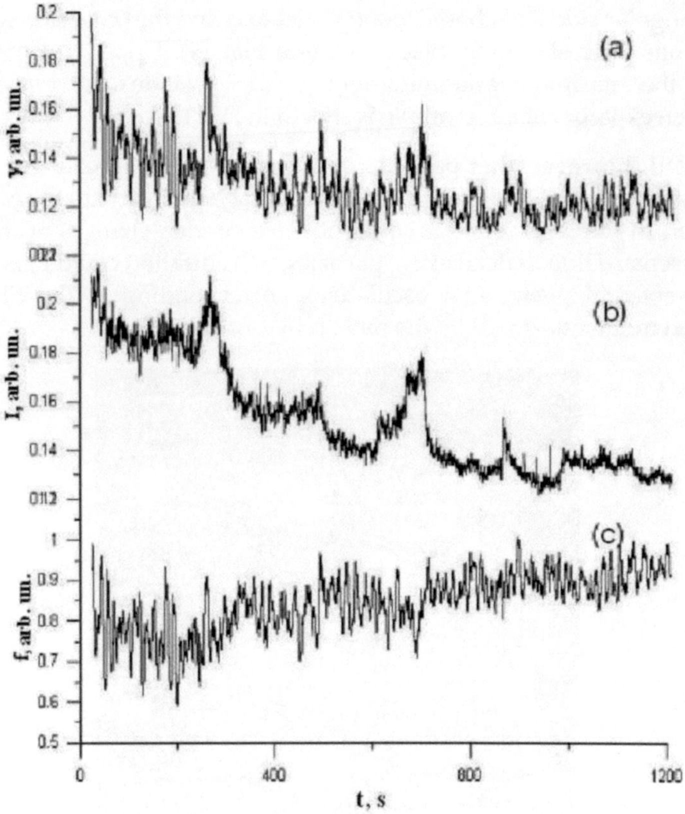

Fig. 15. Average intensity oscillations of the digital image (the gradations of grey color - 12 bit) (a), intensity change due to varying particles concentration (b); jet top oscillations (c). All taken from rectangular area in Fig 14

4.3 Experimental results. The comparison between internal waves and jet top oscillations spectra

During the experiments the turbulent jet, as in previous series, spread at the level of neutral buoyancy, forming a horizontal shear flow under the thermocline, and generated intensive internal waves. We calculated internal waves spectra for all experiments and revealed pronounced peaks in the frequency interval between f_{min}=0.02 Hz и f_{max}=0.05 Hz, with maximum buoyancy frequency being 0.07 Hz.

Jet top oscillations spectra (spectra of functions *f(t)*) were found for all experiments and compared to spectra of isotherms *T*=16°C, close to the thermocline center (from the 1h realization we cut 20 min corresponding to survey time). The examples of such spectra are shown in Fig. 16 for experiment with discharge rate 150 cm/s. There exists a pronounced peak at the frequency close to 0.7N_0 in the internal waves spectrum. It can be seen from the figure, that jet oscillations at frequency 0.7N_0 generate internal waves most effectively. Theoretical analysis performed basing on the work [Bondur et al, 2010a] for source parameters taken from our experiments confirms the most effective generation of internal waves at this particular frequency.

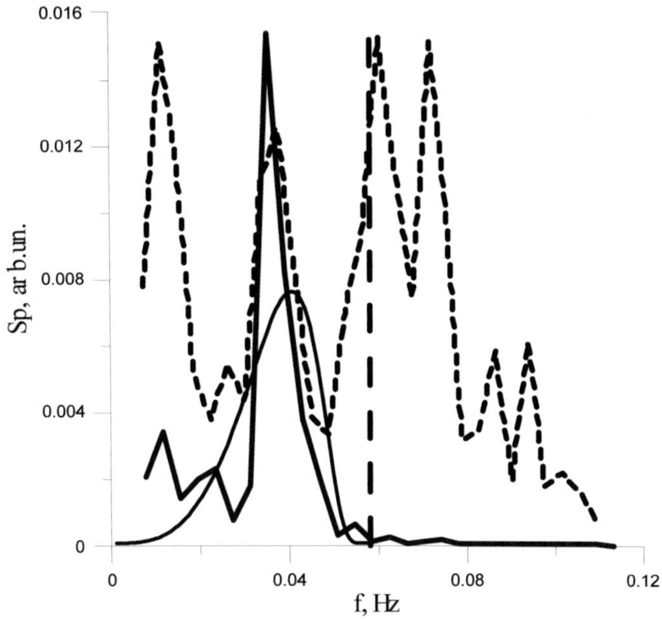

Fig. 16 Jet top oscillations spectrum (hatch line) and spectrum of isotherm (solid line), corresponding to the thermocline center for the experiment with discharge rate 150 cm/s. The straight line marks the maximum buoyancy frequency. The smooth curve is a frequency dependence of the internal waves excitation coefficient.

4.4 Jet mode structure investigation with the application of the PIV-method

In this Subsection we determine an unstable jet mode type generating internal waves. For this purpose we used a modified method, developed in (Yoda et al., 1992), where the jet mode structure was investigated basing on the experimental data for a turbulent round jet at a distance $x / d >> 1$ (x is a distance from the nozzle, d is a nozzle diameter). The method is based on cross-correlation processing of the digital survey data. First, in each frame from the film jet boundaries were determined – areas of the same intensity defined by the authors. Then for several jet cross sections they obtained jet boundaries dependencies on time (2 for every cross section) and calculated their cross-correlations. Basing on these data, a prevailing jet mode was determined.

Let's illustrate this method application on the example. Let the axisymmetric jet mode dominate. Jet boundaries dependencies for this case are shown in Fig. 17. If one of them is reflected with respect to the jet axis, the curves *a* and *b* coincide. Their cross correlation function is periodic and has a maximum in zero. If a spiral mode prevails (Fig. 17), the function has a minimum at t=0.

We modified this method: jet boundaries were determined using longitudinal jet velocity profiles, calculated by means of the PIV-method. In order to reduce high-frequency fluctuations, we averaged velocity fields by time and coordinate along the jet axis. The averaging time and length were chosen small as compared to characteristic time and spatial scales. Characteristic time corresponding to the frequency $0.7N_0$ was 25 s. The averaging time was t_{avg} = 2 c. The wavelength of the unstable jet mode had to be of jet diameter order, which was 15-16 cm in the thermocline. Thus, 5 neighbour jet velocity profiles were averaged, located at a distance 1 cm from each other and the average length was, consequently, l_{avg} = 4 cm (see Fig. 18).

For each experimental frame sequence consisting of 30000 frames, 5 rectangular areas were chosen, as a rule, 2 of them were in thermocline, 2 - above it and 1 – below (see Fig. 18). For each rectangular area the SMPD algorithm was employed to make a file containing averaged by t_{avg} and l_{avg} longitudinal velocity profiles for successive time moments at interval t_{avg}. Using these data we obtained the maximum velocity and jet boundaries dependencies on time for each rectangular area. Jet boundaries were determined by e times velocity decrease from the maximal meaning.

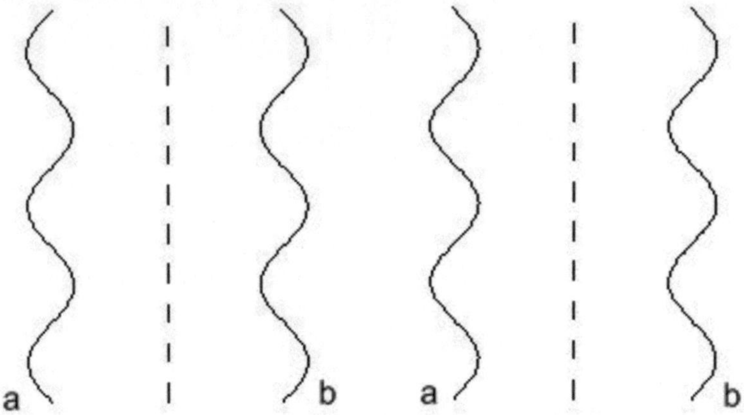

Fig. 17. Jet boundary dependencies on time for some jet cross-section: left – axisymmetric mode, right – spiral mode.

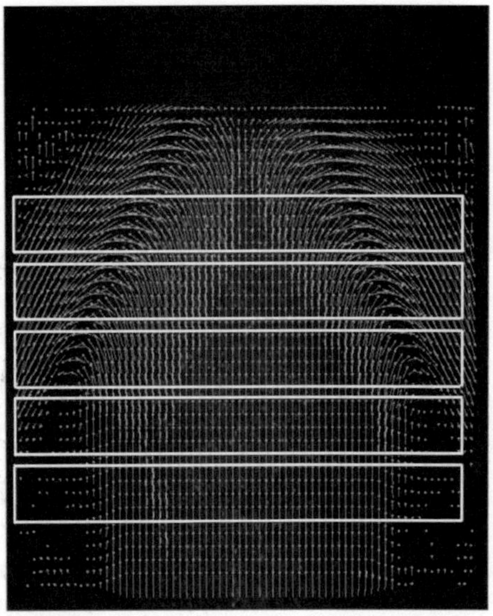

Fig. 18. An example of rectangular areas for velocity profiles averaging

In order to determine a mode, generating internal waves, every jet boundaries function was filtered with cut-off frequencies 0.02 Hz and 0.05 Hz. For each rectangular area we calculated jet boundaries cross-correlation functions, see Fig. 19 as an example. It can be seen from this figure, that functions has maxima at t = 0, consequently, the axisymmetric mode prevails at generation frequency of internal waves.

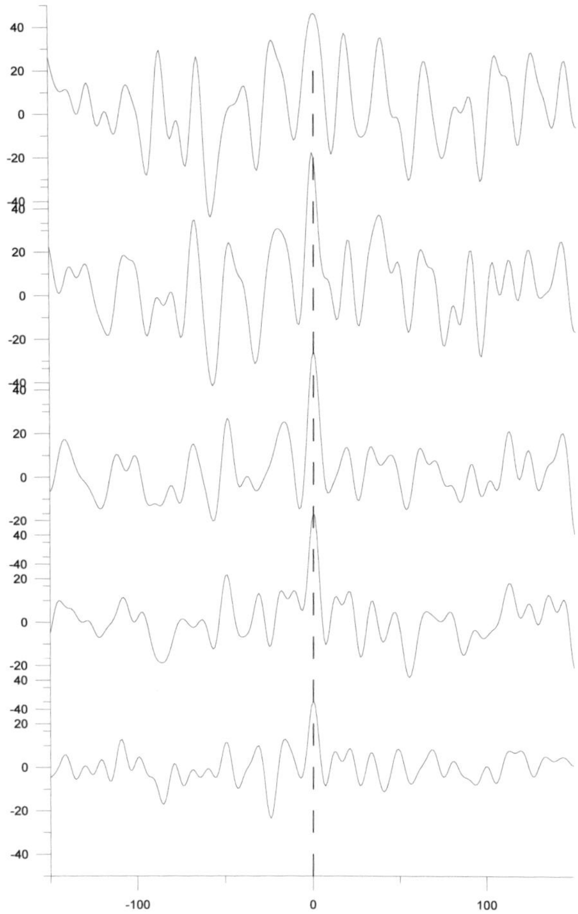

Fig. 19. Example of cross-correlation functions of jet boundaries for 5 rectangular areas in Fig. 18 in the same order - experiment with discharge rate 150 cm/s.

5. Conclusion

Investigation of hydrodynamic processes near submerged wastewater outfalls is an important scientific and engineering problem. Method of dye colouring for jet visualization has been extensively used both in laboratory and field conditions. It allowed observing and investigating evolution of jet integral parameters, in particular, jet spread and dilution laws, oscillations of buoyant jet fronts, both in homogeneous and stratified liquid. At the same time, precise and very specific experiments were needed to clarify the possibility of surface manifestations of the jets from submerged wastewater outfalls. This task demanded measurements of velocity fields of surface flows and in water column with high time and spatial resolution. Thus, optical PIV/PTV methods were optimal. This Chapter is devoted to application of these methods for investigation of dynamics of flows from submerged wastewater outfalls and their surface manifestations.

For this purpose a physical scale modeling of internal waves generation by the turbulent buoyant jets induced by submerged sewer in conditions of temperature stratification with shallow thermocline in LTST IAP RAS was performed. Velocity fields of the surface flows induced by internal waves were measured by the modified PTV-method. The modified method allowed to identify these very weak surface flows in the presence of large-scale background flows. The obtained experimental data are in good agreement with theoretical forecasts taking into account presence of a SAS film, which could not be eliminated completely. For the known parameters of the film and coefficients of scale modeling, we estimated the parameters of internal waves generated by a submerged sewer jet flows and the values of their surface manifestations for the nature conditions. The estimations of hydrodynamic contrasts (caused by manifestations) in the field of surface waves obtained in [Bondur 2004, 2006, 2011, Bondur & Grebenyuk 2001], show that such contrast could be detected with confidence by the modern remote sensing methods.

The experiments on measurements of submerged flows with PIV-methods were carried out for the purpose of studying mechanisms of internal waves generation by buoyant jets. At first preliminary test experiments were performed in small reservoir with saline stratification. It turned out that, when the jet approaches the pycnocline, a counterflow is generated at the edges. A stability analysis for the resulting profiles of flow velocities performed by the method of normal modes has revealed that, for the jet portions with counterflow, the condition of absolute instability criterion for axisymmetric jet oscillations is satisfied. The estimates for oscillation frequencies of the globally unstable mode are in good agreement with measured spectrum of the jet oscillations.

These experiments were continued as a laboratory scale modeling of submerged wastewater outfalls in LTST with application of PIV-technique for measurements of jets' parameters and contact methods for investigation of internal waves. It allowed to compare jet oscillations spectra, obtained by modified video images processing, with spectra of internal waves. The comparison confirmed that self-sustained jet oscillations serve as a source of internal waves. The investigation of mode structure of jet perturbations with application of modified PIV-method showed that the axisymmetric mode prevails and it effectively generates internal waves.

Within the framework of this study modified methods of using PIV/PTV technique were developed for the complicated investigations of turbulent submerged flows and its surface manifestations in the conditions of ambient stratification together with contact methods.

6. References

Adrian, R. J., (1991) Particle Imaging techniques for experimental fluid mechanics. *Annu. Rev. Fluid Mech.*, Vol. 23, p. 261–304.

Arabadzhi, V. V., Bogatyrev, S. D., Bakhanov, V. V, et al (1999) Laboratory Modeling of Hydrophysical processes in the Upper Ocean Layer (Large Thermostratified Tank

Institute of Applied Physics, Russian Academy of Sciences), in *Near-Surface Ocean Layer. Physical Processes of Remote Probing,* V. I. Talanov and E. N. Pelinovskii, Vol. 2, pp. 231–251, IAP RAS Press, Nizhnii Novgorod, [in Russian].

Beckers, M.R, Clercx, H., Heijst van G.J.F. & Verzicco, H.J.H. (2003) Evolution and instability of monopolar vortices in a stratified fluid. *Phys. Fluids,* Vol. 15, pp. 1033-1045.

Bondur V. G., (2004) Aerospace Methods in Modern Oceanology, in *New Ideas in Oceanology, Vol. 1: Physics. Chemistry. Biology,* pp. 55–117, Nauka Press, Moscow, [in Russian].

Bondur, V., (2006) Complex Satellite Monitoring of Coastal Water Areas *Proceedings of 31st International Symposium on Remote Sensing of Environment, ISRSE,* pp. 31 - 35.

Bondur, V.G. (2011) Satellite monitoring and mathematical modelling of deep runoff turbulent jets in coastal water areas In *Waste Water.* http://www.intechopen.com/articles/show/title/satellite-monitoring-and-mathematical-modelling-of-deep-runoff-turbulent-jets-in-coastal-water-areas, pp. 26-32, Intech.

Bondur, V. G. and Grebenyuk, Yu. V., (2001) Remote Indication of Anthropogenic Impacts on the Marine Environment Caused by Deep-Water Sewage Discharge *Issl. Zemli Kosmosa,* No. 6, pp. 1-12 [in Russian].

Bondur. V., Tsidilina M. (2006) Features of Formation of Remote Sensing and Sea truth Databases for The Monitoring of Anthropogenic Impact on Ecosystems of Coastal Water Areas. *Proceedings of 31st International Symposium on Remote Sensing of Environment, ISRSE,* pp. 192-195.

Bondur V., Keeler R., Gibson C. (2005) Optical satellite imagery detection of internal wave effects from a submerged turbulent outfall in the stratified ocean *GRL,* 32, L12610, doi:10.1029/2005GL022390.

Bondur, V. G., Zhurbas, V. M., and Grebenyuk, Yu. V. (2006), Mathematical Modeling of Turbulent Jets of Deep- Water Sewage Discharge into Coastal Basins, *Oceanology,* Vol. 46, No. 6, pp. 757–771.

Bondur, V.G., Zhurbas, V.M., Grebenuk Yu.V. (2009) Modeling and Experimental Research of Turbulent Jet Propagation in the Stratified Environment of Coastal Water Areas *Oceanology,* Vol. 49, No. 5, pp. 595–606.

Bondur V.G., Grebenyuk Yu.V., Ezhova E.V., Kazakov V.I., Sergeev D.A., Soustova I.A. and Troitskaya Yu.I. (2009) Surface Manifestations of Internal Waves Investigated by a Subsurface Buoyant Jet: 1. The Mechanism of Internal-Wave Generation *Izvestiya, Atmospheric and Oceanic Physics*: Vol. 45, No 6, pp. 779-790.

Bondur V.G., Grebenyuk Yu.V., Ezhova E.V., Kazakov V.I., Sergeev D.A., Soustova I.A. and Troitskaya Yu.I. (2010) Surface Manifestations of Internal Waves Investigated by a Subsurface Buoyant Jet: 2. Internal Wave Field *Izvestiya, Atmospheric and Oceanic Physics*: Vol. 46, pp. 768 - 779.

Bondur V.G., Grebenyuk Yu.V., Ezhova E.V., Kazakov V.I., Sergeev D.A., Soustova I.A. and Troitskaya Yu.I. (2010) Surface Manifestations of Internal Waves Investigated by a Subsurface Buoyant Jet: 3. Surface Manifestations of Internal Waves *Izvestiya, Atmospheric and Oceanic Physics*: Vol. 46, No. 4, pp. 482-491.

Briggs, R. J.(1964) *Electron-Stream Interaction with Plasmas* MIT Press, Cambridge.

Ermakov, S.A., Kijashko, S.V. (2006). Laboratory study of the damping of parametric ripples due to surfactant films In *Marine surface films.* pp.113-128. Springer. Germany.

Fan, L.N. (1968) Turbulent buoyant jet problems. In *Rep. № KH-R-18* Calif. Inst. Technol.USA.

Gibson C.H., Bondur V.G., Keeler R.N., Leung P.T. Energetics of the Beamed Zombie Turbulence Maser Action Mechanism for Remote Detection of Submerged Oceanic Turbulence. *Journal of Applied Fluid Mechanics*, Vol. 1, No. 1, pp. 11-42, 2006.

Gibson C.H., Bondur V.G., Keeler R.N., Leung P.T. Remote sensing of submerged oceanic turbulence and fossil turbulence. *International Journal of Dynamics of Fluids*, Vol.2, No.2 (2006), pp. 171-212.

Gibson C.H., Keeler R.N., Bondur V.G. (2007) Vertical stratified turbulent transport mechanism indicated by remote sensing. *Proceedings of SPIE, Coastal Remote Sensing. SPIE* Newsroom, Vol. 6680. 26-27 Aug.

Gibson, C.H., Bondur, V.G., Keeler, R.N., Leung, P.T., Prandke, H., Vithanage, D. (2007). Submerged turbulence detection with optical satellites, *Proceedings of SPIE, Coastal Remote Sensing. SPIE*, Vol. 6680, 6680X1-8. doi: 10.1117/12.732257 Aug. 26-27 2007

Heijst van G.J.F., Beckers, M., R. Verzicco, H.J.H. (2002) Dynamics of pancake-like vortices in a stratified fluid: experiments, model and numerical simulations. *Journal Fluid Mech*, Vol. 433, pp. 1-27.

Koh, C.Y. and Brooks, H.N. (1975) Fluid mechanics of waste-water disposal in the ocean *Annu .Rev. Fluid Mech*, Vol. 8, pp.187-211.

Monkewitz, P. A., Huerre P., Chomaz J.-M., (1993) Global Linear Stability Analysis of Weakly Non-Parallel Shear Flows, *J. Fluid Mech.* Vol. 251, pp. 1–20.

Ozmidov R. V., (1986) Diffusion of Admixture in the Ocean. Gidrometeoizdat, Leningrad, [in Russian].

Reul, N., Branger, H., Giovanangeli, J.P. (1999) Air flow separation over unsteady breaking waves *Phys. Fluids.* Vol. 11. pp. 1959–1961.

Sergeev, D.A., Troitskaya, Yu.I., (2011) APPLYING OF PIV/PTV-methods in Laboratory modeling of geophysical flows In *Modern optical methods of flows investigations* Ed. Rinkavichus B.S. pp. 330-347, Overley Press. Moscow.

Troitskaya, Yu. I. , Sergeev, D. A., Ezhova, E. V., Soustova, I. A., and Kazakov, V. I. (2008) Self-Induced Internal Waves Excited by Buoyant Plumes in a Stratified Tank, *Doklady Earth Sciences,* Vol. 419A, pp. 506 -510.

Turner, J. S. (1966) Jets and plumes with negative or reversing buoyancy *Journal Fluid. Mech.*, Vol. 26, p. 779-792.

Umeyama M., (2008) PIV Techniques for Velocity Fields of Internal Waves over a Slowly Varying Bottom Topography *Journal of Waterway Port Coastal and Ocean Engineering* Vol. 134, No. 5, pp. 286-298.

Veron, F., Saxena, G., Misra, S.K. (2007) Measurements of the viscous tangential stress in the airflow above wind waves *Geophys. Res. Lett.*, Vol. 34, L19603. doi: 10.1029/2007GL031242.

Zhang H. P., King B., and Harry L. Swinney Experimental study of internal gravity waves generated by supercritical Topography // *Physics of Fluids* (2007) 19, 096602

Yoda M., Hesselink L., Mungal M.D. The evolution and nature of large-scale structures in the turbulent jet // *Phys Fluids A*. 1992. V.4. No.4. P. 803-811.

Full Field Measurements in a River Mouth by Means of Particle Tracking Velocimetry

A. Ciarravano[1], E. Binotti[1], A. Bruschi[2], V. Pesarino[2], F. Lalli[2] and G.P. Romano[3]

[1]*Fluid Solutions – alternative Srl, Roma*
[2]*ISPRA – Italian Agency for Environment, Roma*
[3]*Department of Mechanical and Aerospace Engineering, University "La Sapienza", Roma*
Italy

1. Introduction

The measurement of marine streams is a difficult task even for well tested velocimetry methods based on advanced image analysis. The wide size of the measurement domain (from hundreds to thousands meters), the extended range of flow velocities (from mm/s to several m/s), the long time-scales (from minutes to weeks) and the related length and time scales of vortical structures involved in marine dynamics do not allow a straightforward application of Particle Image Velocimetry (PIV) and related methods. Due to the previous requirements, a natural choice is to consider Lagrangian "tracers" to be introduced in the fluid as candidates to sample the flow field in time. Therefore, the most useful measurement technique which enables to derive global velocity data from Lagrangian "tracers" is Particle Tracking Velocimetry, PTV (Tropea & Foss, 2007). As will be outlined in section 3, a large variety of Lagrangian buoys have been designed in the past to probe large-scale marine motion (larger than 50 km), whereas a lot of work is still necessary regarding the measurements at smaller scales. This aspect must be also considered within the debate on the effective role of mesoscales (around 10 km) and small-scales (100 m) in generating the "local" geophysical turbulence field (submesoscales) in the sea and how this dynamics is also dependent on the large-scales (Kanarska *et al.* 2007, Özgökmen 2011). Therefore, on-site measurements of flow at submesoscale level by using effective Lagrangian buoys are needed (Griffa *et al.* 2007, Haza *et al.* 2010, Wang 2010).

Moreover, there are also many specific questions to be solved at small-scale levels regarding for example the diffusion of pollutants into the sea. The general question to be solved is usually related to the spreading of a polluted river water into the marine water and how such spreading can be improved, from the point of view of water quality along the shoreline. The problem is complex, due to the interactions of the river flow with the sea stream and with marine structures. The major phenomena are the interactions between the flow from the river and the harbor structures. Thus, a classical river mouth configuration can be considered as the combination of two flow fields frequently encountered in fluid mechanics, known as the "jet in a cross-flow" and the "impact jet on a wall".

The relevance of the present work is that real field measurements are performed by using almost completely submerged floating buoys which have been designed for the following specific purpose: to measure the velocity of the water stream close to the free-surface by minimizing the sensitivity to wind contributions. Here, the interest is focused on barotropic features, *i.e.* stratification effects are not considered, even if the design of the floating buoys was conceived to consider submerged streams also. The buoys are designed, built and set-up to test the Pescara river mouth spreading into the Adriatic Sea. The river outlet and harbor have been recently modified and faced by a large breakwater which gave rise to several significant environmental effects, as for example concentrating fresh water along the near shore and changing the sea water quality along the shoreline. The aim of the present analysis is to test, by field experiments, the effective velocity field at the channel-harbor outlet to investigate some possible lay-out improvements of the situation.

2. Experimental studies on coastal dynamics

The study of the coastal dynamics is very complex and has been faced during the last decade to understand a number of phenomena of different nature (Shibayama, 2009). The focus, regarding the action of the sea on the coastline, is directed not only to the study of waves and sea currents, but also to aspects related to chemical and biological parameters, especially in recent years in which the human activity has led to effects both in the open sea and near the coastline. Some examples of these effects are coastal erosion, dispersion of pollutants, effects on marine biodiversity (Mc Nealy *et al.* 1995) and influence of marine works on the sedimentation (Wang *et al.* 2011).

In this framework, it is really important to understand the fluid dynamics of the coastal flows, firstly to evaluate in advance what would be the effects of human activity on this area, but also to derive solutions to existing problems. An example is that of harbors where it is necessary to modify and adapt the marine structures to solve problems of sedimentation and pollution of the surrounding area (Shibayama, 2011).

This type of study is performed in recent years by two complementary approaches: numerical models, *i.e.* by simulating with computer dedicated algorithms the coastal area (Apsley & Hu 2003, Hillman *et al.* 2007), or field experiments with sensors detecting kinematic, chemical and physical parameters directly in the sea to correlate the results with scale models of the area (Lacorata *et al.* 2001, Lalli *et al.* 2010).

Numerical simulations bring significant advantages in terms of costs and allow to reproduce different environmental conditions, such as the analysis of critical conditions. However, there is still a lot of work to be done to connect among them the different scales in the ocean dynamics and to correctly modeling the fully three-dimensional phenomena involved (Özgökmen 2011, Balas & Ozhan 2000, Lalli *et al.* 2001). The combination of numerical simulations with laboratory experiments is a very interesting and promising opportunity (Lalli *et al.* 2001, Miozzi *et al.* 2010). Nevertheless, the experimental approach on the field has the strong advantage to directly measure the values of interest. Considering the costs of the tests and the fact that the devices are sometimes very complex in terms of setup and data handling, in most cases the use of a work team made up of many people is required (as in the present activity).

Regarding such experiments, different techniques have been used by many years to measure the speed of marine streams and to detect physical and biological parameters. All of these are based on the two fluid dynamics frameworks, *i.e.* Eulerian or Lagrangian (Munson, 1994). In the Lagrangian methods, the variables are described in time following the trajectories of fluid elements, thus the observer is moving with the fluid particle motion. In the Eulerian approach, a control volume in a fixed reference system is defined within the fluid and its properties are measured as functions of space and time.

In the present study the Lagrangian framework was preferred because it is straightforward to assess for local fluid elements behavior. Otherwise, it would be necessary to install a large number of devices fixed in space, to be checked continuously in time.

3. A survey of Lagrangian buoys

In recent years, different types of Lagrangian buoys have been designed and manufactured with the aim of studying marine flows both in the open ocean and near to the coastline (Selsor, 1993). These devices were differentiated over the years depending on the purpose for which they were designed: some buoys were able to follow deep currents, whereas other types have been designed to follow the surface current. One of the most important parameter is the scale of the observed phenomenon, ocean large-scale, mesoscale, submesoscale and small-scale where sea bed effects must be also taken into account (Lalli *et al.* 2001). The design of a buoy must also take into account the residence time in water which can reach in some cases even more than a year for studies at oceanic scale (in general less in the Mediterranean sea and near the coastline). The Lagrangian buoys also differ depending on the type of on board sensor defining the range of use. Some are used to measure only the temperature, some only for the velocity field, while the most complex have onboard chemical and biological parameter sensors such as salinity, dissolved oxygen, pH *etc.* Series of Lagrangian buoys have been developed in different national and international research programs such as SVP (Surface Velocity Program) in the United States, for the investigation of ocean streams, or those within the European community concerning the observation of the Mediterranean sea. Hereafter, examples of developed Lagrangian buoy are reported.

SVP and "mini" SVP (Lumpkin & Pazos, 2006)

At present, there are two basic sizes of SVP drifters: the original, relatively heavy SVP drifter and the new "mini" version (Figure 1). The less expensive, easier-to-deploy mini design was proposed in 2002 and is currently produced alongside original SVP drifters by several manufacturers. The surface float ranges from 30.5 cm to 40 cm in diameter. It contains: batteries in 4-5 packs, each with 7-9 alkaline D-cell batteries; a transmitter; a thermistor to measure sea surface temperature; and possibly other instruments measuring barometric pressure, wind speed and direction, salinity, and/or ocean color. They also have a submergence sensor or a tether strain sensor to verify the presence of the drogue. The drogue is centered at 15 meters beneath the surface to measure mixed layer currents in the upper ocean. The outer surface of the drogue is made of nylon cloth. In the original design, it has seven sections, each 92 cm long and 92 cm in diameter, for a total length of 6.44 m. Mini drogues are not yet standardized among the manufacturers: they are 4 (Pacific Gyre) or 5 (Marlin-Yug) sections of original dimensions, or 4 (Clearwater) or 5 (Technocean) redesigned sections of diameter 61 cm, length 1.22 m per section. Throughout the drogue, rigid rings with spokes support the drogue's cylindrical shape. The drogue is a "holey-sock" and each drogue section contains two opposing holes, which are rotated 90 degrees from one section to the next. These holes act like the dimples of a golf ball by disrupting the formation of organized lee vortices. While the size of the surface float and drogue vary, the manufacturers all aim for a specific non-dimensional goal: a drag area ratio of 40. This ratio is the drag area (drag coefficient times cross-sectional area) of the drogue, divided by the drag area of all other components. At a drag area ratio of 40, the resulting downwind slip (defined later) is 0.7 cm/s in 10 m/s winds (Niiler and Paduan, 1995). Once deployed, a modern SVP drifter lives an average of around 400 days before ceasing transmission. Occasionally, drifters are picked up by fishermen or lose their drogue and run aground.

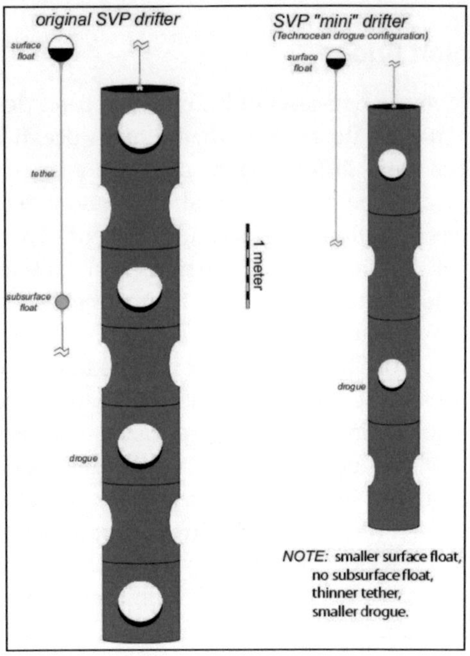

Fig. 1. SVP and Mini SVP Drifter Buoy.

Fig. 2. CMOD Buoy.

CMOD (Gerin et al. 2007)

The Compact Meteorological and Oceanographic Drifter (CMOD) (Figure 2) consists of a 60-cm-long aluminum cylindrical hull with a floatation collar (35-cm overall diameter). This is equipped with the sonobuoy case (62-cm-long and 12-cm-diameter) on a 100-m-long (4-m for a few of them) 0.5-in-diameter tether, resulting in a wet to dry area ratio of about 5 to 1. Air temperature is measured with a thermistor located in a radiation shield which houses the inlet for the barometer port and the Argos transmitting antenna and ground plane. This housing is on top of the mast about 50.8 cm above the surface. The sea surface temperature is measured at a subsurface depth of 44.5 cm.

MELBA (Dell'Erba, 2002)

The Lagrangian drifter used in the MELBA project is dedicated to the study in the Mediterranean Sea, in particular near the coastline (Figure 3). The body of the drifter consists of an aluminum tube, capable of withstanding up to 2000 meters deep, and containing the engine of the ascent and descent (single movement allowed to the buoy), the satellite communications system (active surface) and the mission control system. The accessory instrumentation sensors varies depending on the type of mission to perform: it may include measures of conductivity, salinity, temperature, chlorophyll, etc. The communication system, used to send the measurement data and to receive new subsequent missions is supported by two-way satellite constellation Orbcom. A GPS (Global Position System) is integrated for the geo-referencing of the measured data, once the drifter has emerged.

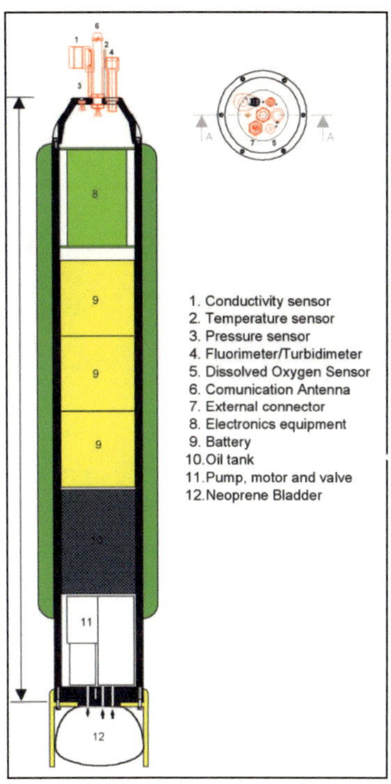

Fig. 3. MELBA Buoy.

"Code" Davis Drifter (Davis, 1985)

The "Code" Davis Drifter (conceived by Dr. Russ Davis of Scripps Institute of Oceanography) or "ARGODRIFTER" is a surface current monitoring Lagrangian drifter (Figure 4). It reports its position by several means, from a calculated position through the ARGOS/CLS system, or by transmitting a GPS location through either ARGOS/CLS or the IRIDIUM satellite system. The surrounding water temperature can also be monitored. Data transmissions can be varied through a wide spectrum of options at the selection of the user. The ARGODRIFTER consists of two orthogonal 1 meter cloth planes oriented vertically around a central instrument containing core. Four 10cm. diameter polyethylene foam floats provide positive buoyancy and are tethered at the end of the arms on Dacron lines 25cms long. These floats insure that the antennas, sometimes two are used when GPS is required, are sufficiently clear of seawater to insure adequate electronic transmissions. The antennas are mounted on 316 stainless steel springs to protect them in case of contact with foreign bodies, or they can "flex" in very high seas.

Fig. 4. "CODE" Davis Drifter.

In the present study, a Lagrangian buoy has been designed and tested by taking inspiration from the drifter "Code" Davis, since it was evaluated as a solution which meets the criteria of reliability and flexibility, especially in a measurement campaign carried out in a short period and close to the coastline. Starting from this configuration, some structural changes have been applied to allow a successful measurements of the sea surface stream (as detailed in section 5a).

4. Use and applications of experimental data

The experimental tests presented in this study represent a first approach to an important problem, *i.e.* the study of the coastal flows to highlight possible solutions to problems related to coastal hydrodynamics (Griffa *et al.*, 2007). Basically, the possible uses and implications related to the above issues can be classified into two approaches, diagnostic and prognostic which are also complementary.

The diagnostic approach is related to problems regarding monitoring of ocean streams, river plumes and discharges into the sea and analysis of biological and chemical parameters. The mark plume discharge into rivers and open waters is a typical phenomenon which can be investigated by a Lagrangian technique as Particle Tracking Velocimetry, PTV (Tropea & Foss, 2007). Simultaneously, with a prognostic approach the experimental tests are carried out also to validate the numerical models and codes used to simulate the phenomena. An example is given in this study in which the river freshwater flows into the sea having a different temperature and the determination of Lagrangian trajectories, provided by PTV, are also used to validate the numerical models.

Thus, the experimental investigation and the resulting data can be used in different ways and may be part of a process of analysis in which many aspects are involved, as numerical models validation, coastal erosion, sediments transport and deposition, marine works pre-planning, analysis for marine generators (renewable energy solutions).

5. Setup description

a. Drifter Buoy Description

The proximal sensing device used to derive the flow trajectories into the open field is a half-submerged buoy. The geometric and physical characterization have been defined to satisfy the main requirement, *i.e.* to follow the flow and to determine the trajectories close to the sea surface. The Lagrangian point of view allows to consider the buoys as macro-particles representing the flow behavior and transported by it while continuously acquiring the position with the on board electronics (GPS Data Logger). In order to verify the Lagrangian behavior, a system based on central fins has been designed with the objective of increasing as much as possible the wetted surface of the buoy, and consequently the drag produced by the flow.

A specific problem for the open sea measurements is to reduce the effects of the free surface where wind breeze effects are dominant. To minimize this disturb, the buoy have been weighted to reduce the area exposed to the wind, and the fins have been located about 20cm in depth, as displayed in Figure 5. The main dimensions of the buoy are reported in Table 1.

Height	41 cm
Width	50 cm
Cabinet Diameter	8 cm
Fins	25x25 cm
Optical Target	20x20 cm
Total Weight	1550 g
Cabinet Weight	1375 g
Appendages	175 g

Table 1. The designed buoy dimensions and weight.

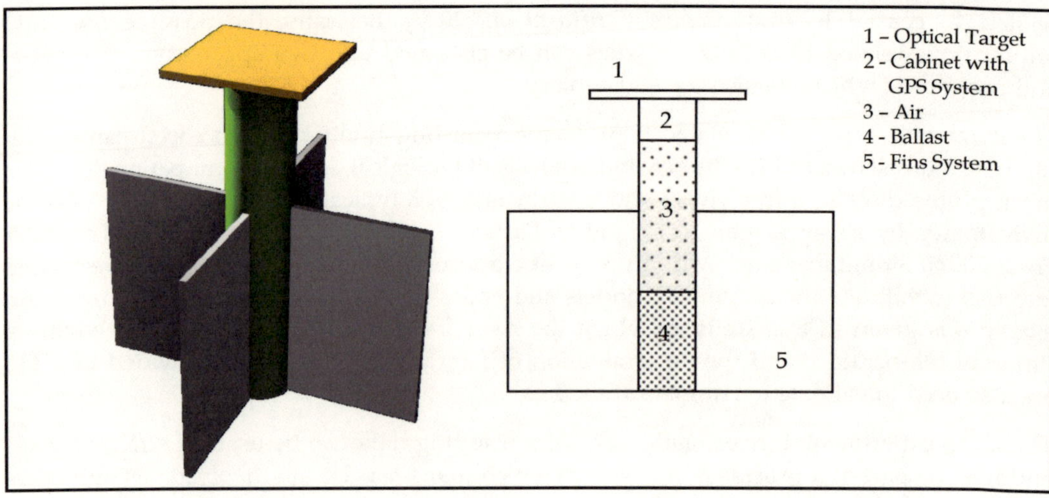

Fig. 5. Sketch of the Lagrangian buoy used in the present measurements.

BatteryLi-Ion, Charging time	3 h
Battery, Operation Time	32 h
Operating Conditions: temperature	-10°C to +60°C
Operating Conditions: humidity	5% to 90%
Accuracy on position	3.0 m
Accuracy on velocity	0.1 m/s
Velocity range	0 m/s to 515 m/s
Maximum acceleration	4 g
Protocols	NMEA-0183 (V3.01) - GGA, GSA, GSV, RMC (default)
Data bit	8
Size	46.5W x 72.2L x 20H mm
Weight	64 g
Data Log	up to 125,000 way points

Table 2. GPS Data Logger technical specifications

b. Electronic Transmission

The buoys are equipped with a GPS (Kaplan & Hegarty 2002) antenna Transystem i-Blue 747 which is an economic and functional satellite navigation system with storage of positions in time of the buoys on the on board data logger (maximum number of points equal to $1{,}25 \times 10^5$, one every second). This device is shown in Figure 6, while in table 2 the technical specifications are summarized. It uses a GPS Chip MTK with Frequency L1, 1575.42 MHz C/A Code 1.023MHz chip rate Channels 51 CH for tracking and Antenna Built-in patch antenna with LNA Datum WGS-84. The recording is performed switching the instrument to mode Nav-Log and storing the position data calculated by the GPS with an internal memory. The data are exported in ASCII format NMEA-0183 by connecting the device to the USB port of the PC.

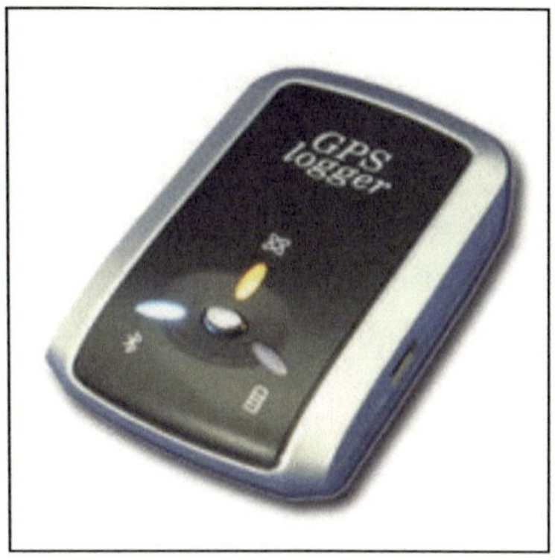

Fig. 6. GPS Data logger.

c. Preliminary Tests

The preliminary tests have been performed on two buoy models aiming primarily to verify the feasibility of the measurement campaign and to highlight possible critical points regarding the devices used and logistical activities related to the release and recovery of the Lagrangian buoys. The buoys have been released in marine water to verify their stability, floating level, flow traceability, data acquisition, visibility and handling. A picture of the floating buoys during the tests is given in Figure 7, while the measured trajectories are overlapped to those derived during the final tests and reported in Figure 10. As a result of the tests, the following specific aspects have been pointed out:

- the buoys are very stable in marine water (even in presence of waves) and seem to follow very well the local flow direction;
- the floating level is so that almost 90% of the buoy is submerged; the part coming out of water surface is required to recover the buoy after the end of the measurements;

Fig. 7. Two drifting buoys during the preliminary tests.

- the position where buoys are released is crucial and the best place is as close as possible to the center of the river, because close to the boundaries they move into the boundary layer, thus decreasing their speed and increasing the risk to stop somewhere under the docks;
- the time scale for the whole phenomenon to be completed, from the river mouth to the main breakwater, is at least 30 minutes and no more than 3 hours;
- to avoid unnecessary and difficult interpretation of data, it is useful to acquire the GPS signal only for the time required by the test, therefore to turn-on the GPS just before putting the buoy into the water and turn-off immediately after recovering;
- the management program bundled with the GPS (GPS Photo Tagger) cannot handle many tracks simultaneously, so an appropriate script is needed.

6. Measurement campaign

The measurement campaign was performed at the Pescara river mouth located in the shallow coastal environment of the Adriatic Sea. In Figure 8, a satellite image of the region considered in this campaign is presented. The main characteristics of this area are the presence of the harbor on the right side of the river mouth, the shoreline on the left and the big breakwater recently built, which influences the flow outcome and gives rise to significant environmental effects. Specifically, the dispersion of polluted water from the river, the possible increase of sediments and the loss of biodiversity should be addressed.

Fig. 8. Measurement region with the river mouth, the breakwater and the harbor.

The present activity aims to setup a method to measure the effective velocity field in order to investigate some possible change or modifications of the harbor and related structures. From the geometrical and dynamical point of views, such a problem is complex, due to the interactions of the river flow with the sea stream and with marine structure, as it is well visible by the satellite image. However, this activity has also a much general validity which can be adapted to similar conditions.

The final test was split into three different phases:

Test preparation. The first part of the test was dedicated to the preparation and mounting of the buoys and to the set up of the logistics with port authorities to stop vessels during the test period. At the end about 100 Lagrangian buoys were set-up, 30 of them mounting the GPS (plus fins and targets), while the other 70 are used for visual inspection of the phenomenon. All these buoys were led onto the supporting boats.

Buoy Release and data acquisition. After the preparation phase, the buoys were released at about 200 meters from the mouth of the river from the stern of the support boat as displayed in Figure 9 and data acquisition of their positions started. To measure the displacements of the buoys the devices have acquired samples at a data rate equal to 1Hz.

Buoy Recovery. After a period of about 3 hours, the buoys have been led back on board and all the devices were switched off. All the GPS were ready to be transported into the laboratory to start the processing phase.

Fig. 9. Buoys release along the river mouth at about 200 m from the outlet.

7. Data analysis and results

At this stage, data download and processing is performed in order to derive the buoys trajectories. In particular, the sequential steps of analysis consist of downloading GPS data, creating a comprehensive data set, data pre-processing, correcting GPS by reference position, data processing and filtering, deriving velocity fields and comparisons. To this end, GPS are connected to PC and the tracks downloaded in the format defined in the initial configuration. In our case, the format used was NMEA-0183 (National Marine Electronics Association). This format allows to obtain the position data including the altitude and speed. As mentioned above, scripts are used for data analysis as the software supplied with the GPS data logger was unable to handle a big amount of data. Once exported in ASCII format data are analyzed as follows:

Raw data acquisition. Once connected to the PC at USB port, the GPS devices allow to view the raw tracks stored in the data-logger and to store the data in the desired format. With the included software the downloaded raw tracks can be viewed as presented in Figure 10. The GPS data given in ASCII format (NMEA type), are easily exported into spreadsheet form, extracting only the fields of interest (sample number, latitude, longitude, altitude and speed). For each GPS a file containing the tracks of the stored trajectory is exported. *Differential correction of GPS data with reference position.* The GPS signal is affected both by random and bias errors. The main causes of failure are usually the variations of the orbits of satellites, the timing error of the satellites (different stages) and the atmospheric effects on the signal (weather, ionosphere disturbances). To reduce such bias errors, thus improving the precision, a differential GPS is used which provides a correction by referring to another reference GPS placed at a known position. In this way, it is possible to correct data to attain an accuracy up to 10 cm. The possible drawback is that this reference correction can be used only in conditions in which the position of the reference location is known very well.

Fig. 10. Raw trajectories of the Lagrangian buoys as measured by the GPS. The white and yellow line represent the two buoys trajectories during the preliminary tests.

In the present experiments, a fixed GPS, of the same type as those on board, has been placed at a point with known coordinates, and the position was acquired for the whole duration of the experiment. The chosen location has been identified on the outer edge of the dock of the river Pescara on the right side as displayed in Figure 11. It has the following WGS84 coordinates:

Reference position: coord_rif
Latitude: 42.469628 °
Longitude: 14.229803 °

Simultaneously to the GPS released with the buoys, also the acquisition of about 3500 samples of the fixed-point GPS position was performed and the results were analyzed in post-processing to make bias corrections. The measured positions by reference GPS were also corrected by knowing the exact point where the reference GPS is placed (differences around 1-10 m as displayed in Figure 11). Lastly, all the GPS position were computed as the difference between the true location and the median value of samples of fixed point reference. As a result, with this correction the effective accuracy of GPS data is under 1 m. It should be emphasized as the only possible correction on the data position is relative to the average of samples collected, thus reducing only the bias error. On the other hand, the use of the instantaneous data to correct even random errors (e.g. timing satellites) is not possible because it would require information on the phases of the satellites, currently not available with this low-cost technology.

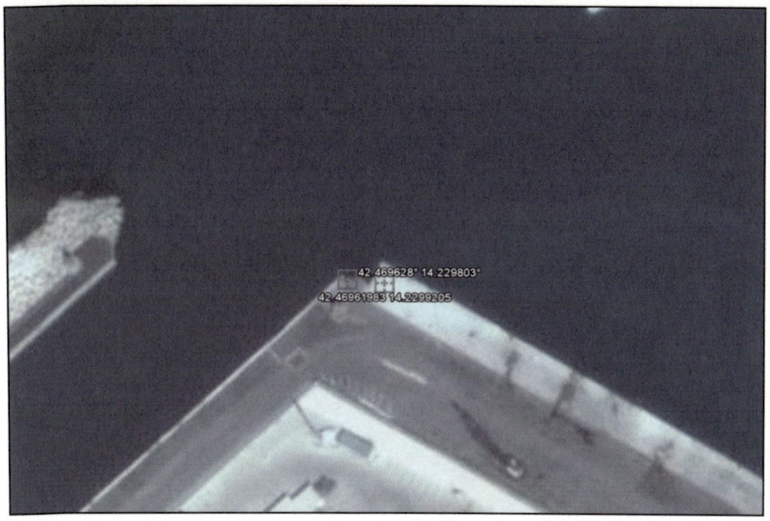

Fig. 11. Reference fixed positions along the harbor dock and coordinates.

Indeed, when analyzing the sample distribution of the position of the fixed point GPS in UTM coordinates, it is not possible to recover a well defined probability density distribution. Rather, data are largely scattered and the mean value does not represent the most probable or a meaningful value. In Figure 12, at the top, the coordinates on the (*latitude, longitude*) plane are presented showing that none of the acquired data coincides with the mean value, thus indicating the fact that different ensemble of data are grouped together. Considering that the GPS is linked to a given satellite configuration for time intervals not longer than a few minutes, it can be argued that the observed scattering id due to the change of such a configuration in time. This is confirmed by the probability density distributions of the latitude and longitude given in Figure 12 at the bottom, where the overlapping of two distinct almost Gaussian distributions is noticed for both. So far, the mean value is simply the one among the two distributions. It is possible to correct this additional bias error by selecting which of the distribution is related to each period of time, even if the correction is smaller than 0.3 m.

Processing and data filtering (smoothing spline). The data acquired by GPS once corrected by the geographical reference are still subject to random errors due to the factors previously mentioned. These errors give rise to some noise in the derived buoy trajectories. Graphically, this is represented in Figure 13 where all raw measured trajectories are shown. Although the general features of the flow field can be derived, these errors could give strong oscillations especially when differentiating the data to derive the buoy velocity. Therefore, a smoothing spline filter among the acquired n data has been applied by minimizing the quantity

$$\sum_{i=1}^{n}(Y_i - \hat{\mu}(x_i))^2 + \lambda\int_{x_1}^{x_n} \hat{\mu}''(x)^2 dx.$$

where Y_i are the measured data (at time x_i), μ is the smoothing spline estimator (to be determined), μ'' is its second-order derivative and λ is a positive coefficient taken equal to 2×10^3. The result of this spline is presented in Figure 14. The trajectories are now much smoother than before and the fluid-mechanics information can now be derived easily.

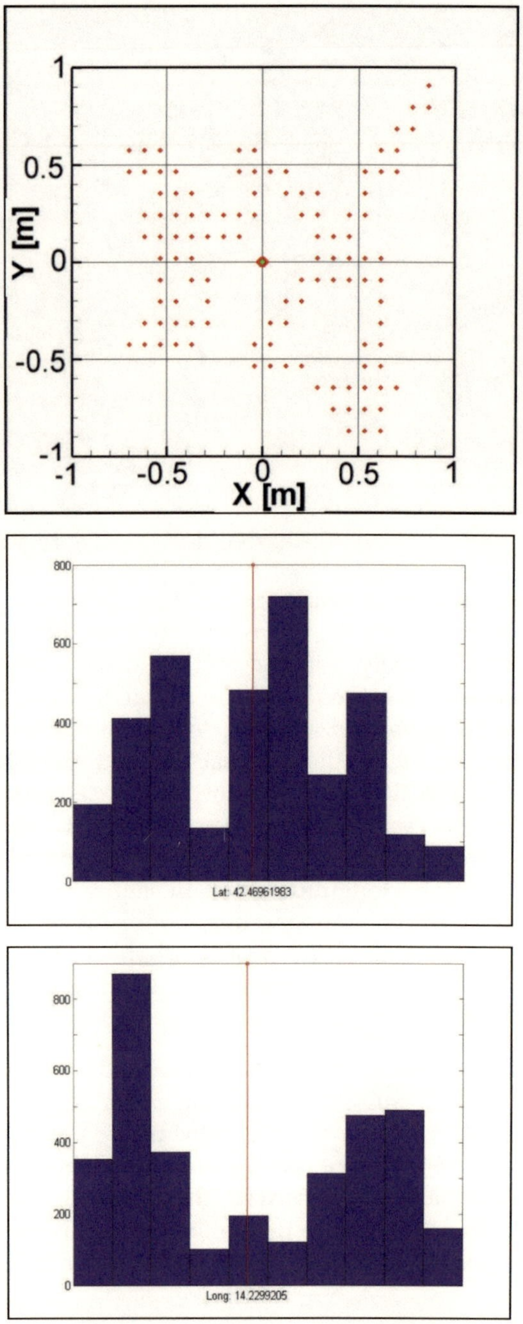

Fig. 12. Spatial distributions of the data acquired at the fixed reference GPS position (at the top) and statistical distributions of latitude and longitude of these data.

Coordinate transformation from WGS84 to UTM. After completing the previous differential correction, the coordinates of the GPS have been converted from WGS84 to UTM format (Universal Traverse Mercator).

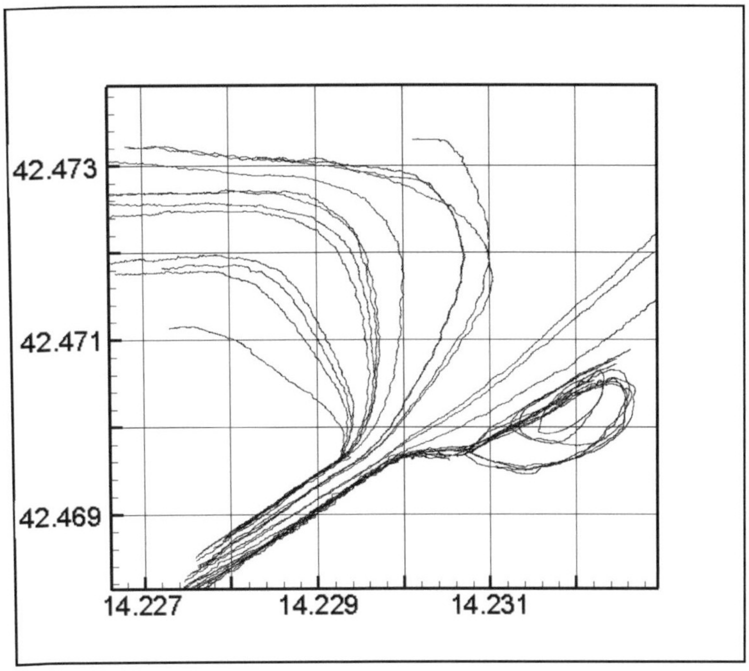

Fig. 13. Raw trajectories of the GPS buoys, before filtering.

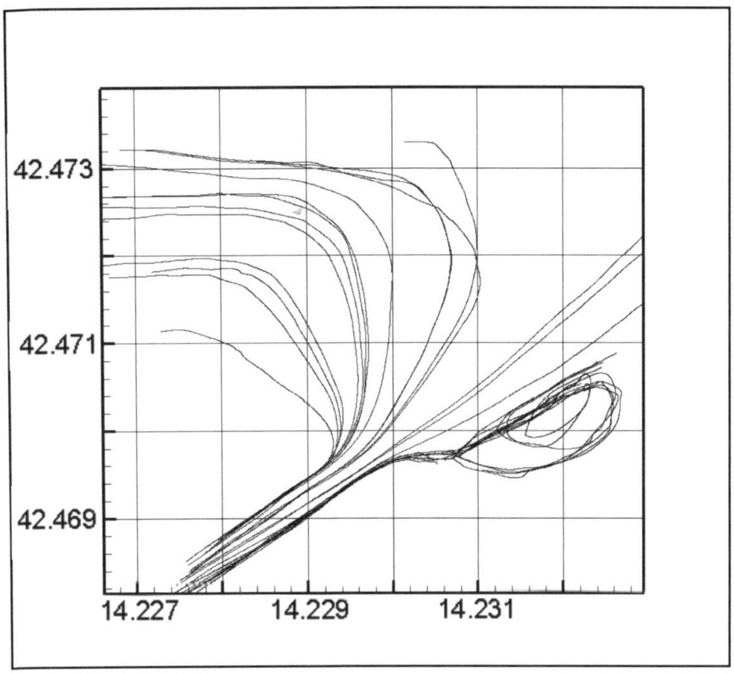

Fig. 14. Trajectories of the GPS buoys, after filtering.

Once obtained the final filtered coordinate data, these can be displayed in space (as in Figures 13 and 14) or in time as given in Figure 15 for one specific buoy. In this case the two coordinates are increasing in time with a similar behavior thus indicating that the buoy moved along a diagonal as the one colored in yellow in Figure 10. For increasing time, the horizontal coordinate has a maximum and then decreases, thus indicating that the buoy is turning on the left side of the field, whereas the vertical coordinate continues to increase departing from the river mouth. A displacement equal to several hundred meters is attained after about 1500s, thus indicating an average velocity equal to 0.3 m/s for both velocity components. Similar information can be derived for the other buoys. From plots like that given in Figure 15, it is possible to derive the velocity of the buoy. The two velocity components in the measurement plane can be computed by centered differences on the positions dividing by the time interval among data acquisition (1 s). An example is given in Figures 16 for the two velocity components separately and in Figure 17 for the absolute value of the velocity (the trajectory is the one displayed in Figure 15). At a first sight, the velocity behavior appears very noise regardless of the filtering applied to the data in position. However, it is still possible to derive meaningful behaviors if interpolation are performed as reported in the figure (consider that the large oscillations are on a time scale of 1 s, whereas the useful information are on time scales one or two order of magnitude larger, so that averaging over 10-100 samples is reasonable). As expected from the previous raw computation, the velocity components have values ranging in the interval (0 - 0.9) m/s and the maximum value is observed just after 500 s as in Figure 15. After about 2000 s the horizontal velocity is almost vanished, whereas the vertical one attains a constant value around 0.5 m/s. Similar information can be derived from the plot given in Figure 17.

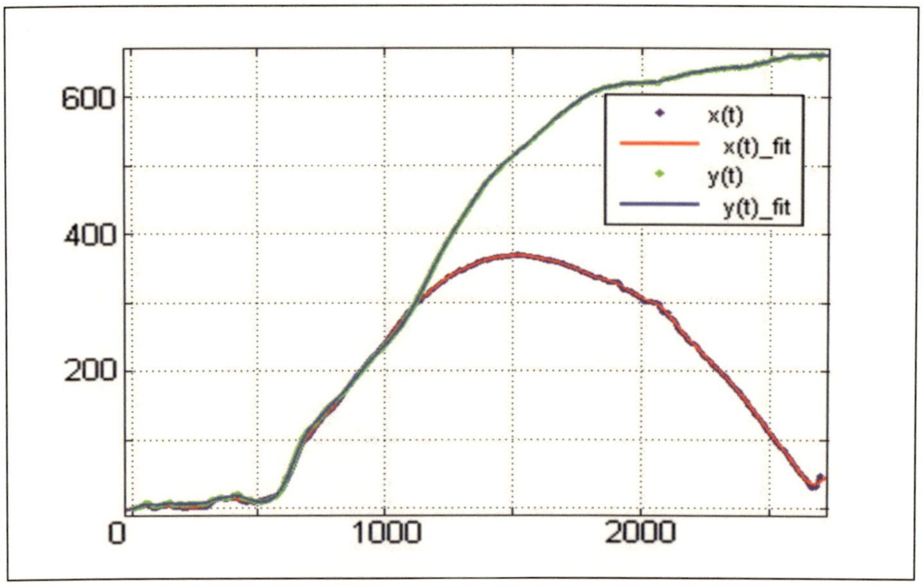

Fig. 15. Values of x and y position in time (in meters) of GPS buoys *vs* time (in seconds) with interpolating fits.

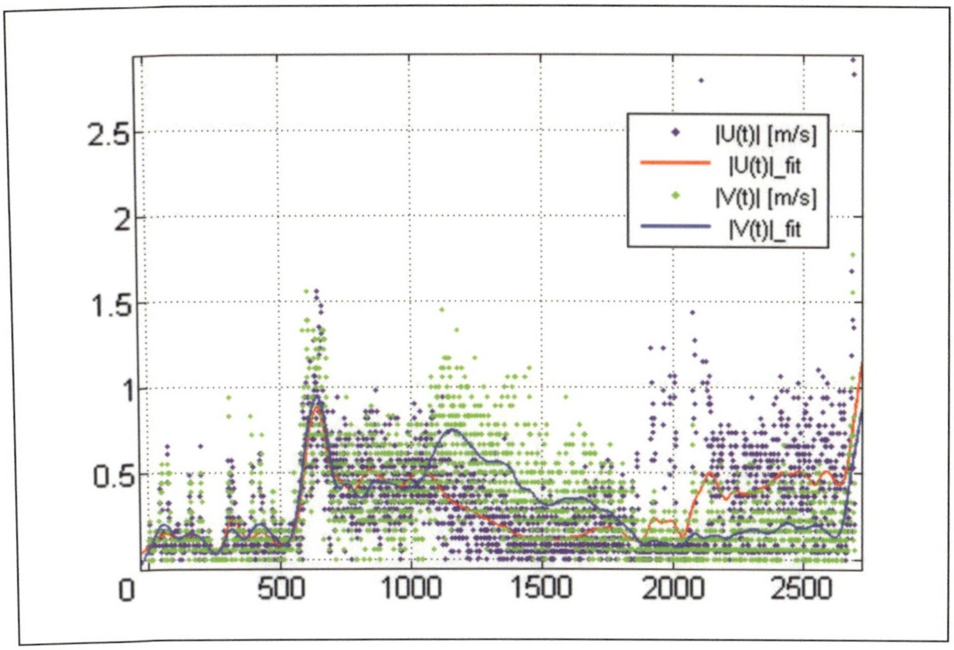

Fig. 16. Modules of the buoy velocity for x and y components (in m/s) *vs* time (in seconds), GPS raw data and interpolating lines.

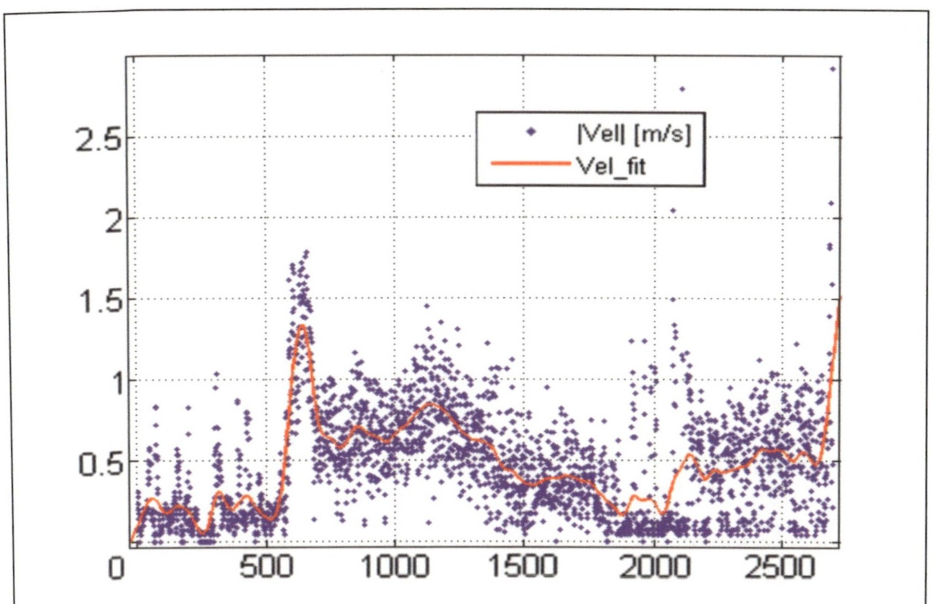

Fig. 17. Absolute value of the buoy velocity (in m/s) over time (in seconds), GPS raw data and interpolating line.

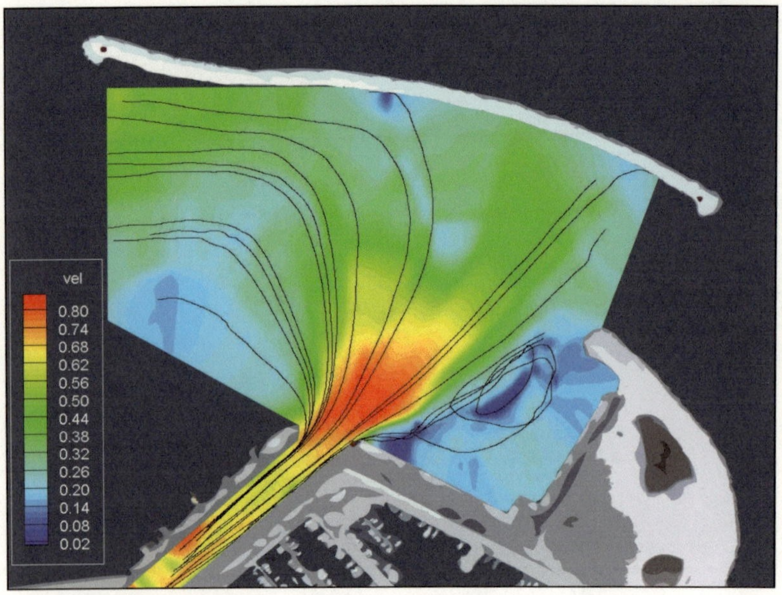

Fig. 18. Trajectories of buoys and geographical distribution of the absolute value of the velocity (in m/s) calculated on filtered data.

Therefore, it is possible to establish that after about 500 s, this specific buoy felt an increase of the velocity which is then decreasing slowly to a constant value after about 2000 s.

To connect this behavior in time to the spatial position and to a general view of the flow field, the buoy trajectories have been reported on the geographical reference of the river mouth. This has been performed in Figure 18, where the entire set of buoy trajectories is displaced. The color code represents the velocity absolute value. It is possible to notice several aspects:

- first of all, the spatial distribution of the buoys is rather homogeneous thanks to the care adopted for the initial release into the river, nonetheless the small number of employed buoys;
- secondarily, the buoys are dispersed all over the measurement field, either in the recirculation region on the right side of the river mouth, or towards the breakwater and the left part of the field;
- as already noticed, the velocity, which is almost constant within the river, is increased just at the river mouth (and this is a possible effect of the specific bathymetry and sedimentation), then decreasing especially when moving orthogonally to the river axis as in jet flows;
- the largest part of the buoys are forced on the left side of the field (and a possible effect of the sea stream and waves on this phenomenon cannot be ruled completely out) or trapped into the recirculation region close to the harbor dock.

The quality of these results is high and the data contains a large quantity of information, so that the measurement of Lagrangian trajectories by means of floating boys equipped with GPS fulfills all the initial requirements. Therefore, from such an analysis it is clear how large are the potential applications of the tested technique in similar or different conditions.

8. Conclusions and future developments

The hydrodynamic field of the mouth of the river Pescara has been investigated, covering an area of 2 km x 1 km and performing measurement on floating buoys in order to get information on flow diffusion from the river onto the sea.

The Particle Tracking Velocimetry (PTV) technique has been successfully applied to this phenomenology, by using a set of drifting buoy equipped with GPS system measuring consecutive positions in time. The determination of the best buoy to be used for this specific purpose, the consequent set-up and tests have been performed step by step.

Once tested, the buoys have been released into the sea for a measurement campaign, obtaining a characterization of the hydrodynamic field in terms of trajectories and velocity components in the plane.

After data pre and post-processing, the reconstructed field allows to derive information on the behavior close or far from the river mouth and so on the possible consequences on the environment.

Specifically, the velocity is clearly larger at the river mouth, decreasing when the river water spreads into the sea. Other two important features to be pointed out, are that the tracer are mostly directed towards the left part of the field and this could depend on the sea stream and waves during the experiments, as well on the curved breakwater placed in front of the river. Secondly, probably independently on the sea conditions, there is a considerable number of tracers which are trapped on the right part of the field and start recirculating with also negative velocities as an effect of the recently added harbor docks and structures.

The experimental setup together with the methodology have demonstrated the possibility to derive the fluid dynamics features of the mouth of the river using very simple and low cost devices (drifting buoys) that have been studied and designed for the specific survey. In this context important conclusions have been achieved especially observing the interactions between the flow and the marine structures, thus representing a high value result to improve the harbor impact on the coastal zone.

The increase of the number of buoys (thus increasing the spatial resolution of the measurement), the possible additional simultaneous measurement of temperature and other atmospheric and sea parameters, the possibility of using a variable height of the fins to measure also the flow stream at different depths are some of the possible improvements that the authors are going to test in the near future.

9. References

Apsley, D. and Hu, W., 2003, CFD simulation of two- and three-dimensional free-surface flow. International *Journal for Numerical Methods in Fluids*, vol. 42, pp. 465–491.

Balas L., Ozhan E., 2000: "An Implicit three-Dimensional Numerical Model to Simulate Transport Processes in Coastal Water Bodies", *Journal of Computational Physics*, vol. 34, pp. 307-339.

Davis, R. E., 1985, Drifter observations of coastal currents during CODE. The method and descriptive view, *Journal of Geophysical Research*, vol. 90, pp. 4741-4755.

Dell'Erba R, 2002, MELBA: una boa lagrangiana per lo studio del mare mediterraneo. Equazione del moto e sua soluzione numerica. *Internal Report ENEA Centro Ricerche Casaccia.*

Gerin, R.; Poulain, P., M.; Taupier-Letage, I.; Millot, C.; Ben Ismail, S.; Sammari, C., 2007, Surface circulation in the Eastern Mediterranean using Lagrangian drifters. *Internal Report Istituto Nazionale di Oceanografia e di Geofisica,* vol. 38

Griffa, A., A.D. Kirwan, A.J. Mariano, T.M. Özgökmen and T. Rossby, 2007: *Lagrangian Analysis and Prediction of Coastal and Ocean Dynamics.* Cambridge University Press

Haza, A.C., T.M. Özgökmen, A. Griffa, A. Molcard, P.M. Poulain, and G. Peggion, 2010: Transport properties in small scale coastal flows: relative dispersion from VHF radar measurements in the Gulf of La Spezia. *Ocean Dynamics,* vol. 60, pp. 861-882.

Hillman G, Rodriguez A, Pagot M, Tyrrell D, Corral M, Oroná C, and Möller O, 2007, 2D Numerical Simulation of Mangueira Bay Hydrodynamics. Journal of Coastal Research: *Special Issue 47 - ECOSUD: Estuaries & Coastal Areas: Basis and Tools for a More Sustainable Development,* pp. 108 – 115.

Lumpkin, R. and M. Pazos, 2006, Measuring surface currents with Surface Velocity Program drifters: the instrument, its data, and some recent results. *Lagrangian Analysis and Prediction of Coastal and Ocean Dynamics* (LAPCOD), Chapter 2, ed. A. Griffa, A. D. Kirwan, A. J. Mariano, T. Ozgokmen, and T. Rossby.

Kanarska Y, Shchepetkin A, & McWilliams JC, 2007: Algorithm for non-hydrostatic dynamics in the Regional Oceanic Modeling System, *Ocean Modelling,* vol. 18, pp. 143-174.

Kaplan E and Hegarty C, 2002, *Understanding GPS: Principles and Applications,* Second Edition, Artech House.

Lacorata, G., Aurell, E., Vulpiani, A (2001) Drifter dispersion in the Adriatic Sea: Lagrangian data and chaotic model. *Annales Geophysicae* vol. 19(1), pp. 121-9.

Lalli F., Berti D., Miozzi M., Miscione F., Porfidia B., Serva L., Vittori E., Romano G.P. (2001), Analysis of Breakwater-Induced Environmental Effects at Pescara (Adriatic Sea, Italy) Channel-Harbor, *11th International Offshore and Polar Engineering Conference,* Stavanger, Norway.

Lalli F, Bruschi A, Lama R, Liberti L, Mandrone S, Pesarino V, 2010, Coanda effect in coastal flows, *Coastal Engineering,* vol. 57, pp. 278-289.

McNeely J. A., Gadgil M., Leveque C., Padock C., Redford K., 1995, Human influence on biodiversity. In *Global biodiversity assessment.* (V. H. Heywood and R. T. Watson, ed.), pp. 771–821. Cambridge University Press, Cambridge.

Miozzi M, Lalli F, Romano GP, 2010, Experimental investigation of a free-surface turbulent jet with Coanda effect. *Experiments in Fluids,* vol. 49(1) pp. 341-343.

Munson BR, Young DF, Okiishi TH, 1994, *Fundamentals of Fluid Mechanics,* Wiley and Sons.

Niiler, P. P. , J. D. Paduan, 1995, Wind-driven motions in the northeast Pacific as measured by Lagrangian drifters. *J. Phys. Oceanogr.* Vol. 25, pp. 2819-2830.

Özgökmen T, 2011, *Large Eddy Simulations of Submesoscale Flows,* La Londe Coastal Oceanography Summer School, September 2011.

Selsor, H. D., 1993. Data from the sea: Navy drift buoy program. *Sea Technology,* vol. 34(12), pp. 53-58.

Shibayama T, (2009), *Coastal Processes – Concepts in Coastal Engineering and their Applications to Multifarious Environments,* World Scientific.

Tropea C and Foss JF, 2007, *Handbook of Experimental Fluid Mechanics,* Springer- Verlag.

Wang Y, 2009, *Remote Sensing of Coastal Environment,* CRC Press Taylor & Francis Group.

Wang Y, Yuan D, Nie H, 2011, Numerical simulation of sediment transport in Bohai Bay, *Water Resource and Environmental Protection* (ISWREP), 2011